INTERPOLATION

AND

APPROXIMATION

INTERPOLATION
AND
APPROXIMATION

PHILIP J. DAVIS

Brown University

DOVER PUBLICATIONS, INC.
NEW YORK

Published in Canada by General Publishing Company, Ltd., 30 Lesmill Road, Don Mills, Toronto,
Ontario.
Published in the United Kingdom by Constable
and Company, Ltd., 10 Orange Street, London WC 2.

This Dover edition, first published in 1975, is an
unabridged republication, with minor corrections,
of the work originally issued by the Blaisdell Publishing Company, a Division of Ginn and Company,
Waltham, Massachusetts, in 1963. A new Preface
(with bibliography) has been prepared by the
author especially for the Dover edition.

International Standard Book Number: 0-486-62495-1
Library of Congress Catalog Card Number: 75-2568

Manufactured in the United States of America
Dover Publications, Inc.
180 Varick Street
New York, N.Y. 10014

To the Memory of my Father and Mother

FRANK DAVIS *and* ANNIE DAVIS

In the play between content and form,

they sought a middle road.

Preface to the Dover Edition

This edition, apart from minor corrections, is identical with the fourth printing of the book originally published by Blaisdell Publishing Company (later: Xerox College).

In the fifteen years since the preparation of the original manuscript the field of interpolation and approximation has flourished greatly. Perhaps several dozens of subtopics hardly mentioned in this book have been developed to the point where each is now a separate discipline in its own right.

Granted a copious reservoir of animal energy, it might indeed have been possible to update the present work so as to take notice of some of the significant recent achievements. But I think that the present work, limited as it is, has a scope and a point of view which remain of importance.

For the reader who wishes to pursue some of the recent developments, I append here a list of books which cover a wide variety of topics.

Ahlberg, J. H., Nilson, E. N., and Walsh, J. L., The Theory of Splines and Their Applications, Academic Press, N.Y., 1967.

Bézier, P., Numerical Control: Math. and Applications, Wiley & Sons, N. Y., 1972.

Butzer, P. L., and Nessel, R. J., Fourier Analysis and Approximation, Birkhaüser Verlag, Basel, 1971.

Cheney, E. W., Introduction to Approximation Theory, McGraw-Hill, N.Y., 1966.

Davis, P. J., and Rabinowitz, P., Numerical Integration, Academic Press, N.Y., 1975.

Forest, A. R., "On Coons' and Other Methods for the Representation of Curved Surfaces," Computer Graphics and Image Processing, Vol. I, 1972, pp. 341–359.

Freud, G., Orthogonal Polynomials, Pergamon Press, Oxford, 1971.

Goldstein, A. A., Constructive Real Analysis, Harper & Row, N.Y., 1967.

Karlin, S., and Studden, W., Tschebyscheff Systems with Applications in Analysis and Statistics, Interscience, 1966.

Laurent, P. J., Approximation et Optimisation, Hermann, Paris, 1972.

Lorentz, G. G., Approximation of Functions, Holt, Rinehart and Winston, N.Y., 1966.

Luke, Y. L., The Special Functions and Their Approximation, Academic Press, N.Y., 1969.

Meinardus, G., Approximation of Functions: Theory and Numerical Methods, Springer, N.Y., 1967.

Nachbin, L., Elements of Approximation Theory, Van Nostrand, N.Y., 1967.

Olver, F. W. J., Asymptotics and Special Functions, Academic Press, N.Y., 1974.

Rice, J. R., The Approximation of Functions, Addison-Wesley, Reading, Mass., Vol. I : 1964, Vol. II : 1969.

Rivlin, T. J., An Introduction to the Approximation of Functions, Blaisdell, Waltham, Mass., 1969.

Sard, A., Linear Approximation, Am. Math. Soc., Providence, R.I., Math. Surveys, No. 9, 1963.

Schoenberg, I. J., Cardinal Spline Interpolation, Society for Industrial and Applied Mathematics, Philadelphia, 1973.

Shapiro, H. S., Smoothing and Approximation, Van Nostrand, N.Y., 1969.

_____, Topics in Approximation Theory, Springer, N.Y., 1971.

Singer, I., Best Approximation in Normed Linear Spaces by Elements of Linear Subspaces, Grund. Math. Wissen, Springer, N.Y., 1970.

Talbot, A. (ed.), Approximation Theory, Academic Press, N.Y., 1970.

Todd, J., Introduction to the Constructive Theory of Functions, Birkhaüser Verlag, Basel, 1963.

Wermer, John, Banach Algebras and Several Complex Variables, Markham, Chicago, 1971.

Werner, H., Vorlesung über Approximationstheorie, Lecture Notes in Mathematics, Vol. 14, Springer, N.Y. 1966.

P. J. D.

Providence, R.I.

Fall, 1974

Foreword

During the past few decades, the subject of interpolation and approximation has not been overly popular in American universities. Neglected in favor of more abstract theories, it has been taught only where some staff member has actively engaged in research in the field. This has resulted in a scarcity of English language books at the intermediate level. Since the development of high speed computing machinery, the flame of interest in interpolation and approximation has burned brighter, and the realization that portions of the theory are best presented through functional analysis has added additional fuel to the flame. It has been my intention, therefore, to prepare a book which would be at the level of students who have had some real variable, some complex variable, some linear algebra, and, perhaps, a bit of integration theory. The book would merge, insofar as possible, the real and the complex, the concrete and the abstract, and would provide a place for general results of these previous courses to find application and to pass in review.

A one semester course can be based on Chapters 2, 3, 4, 6, 7, and 8. The problems, on the whole, are simple, and are intended to secure the material presented rather than to extend the coverage of the book. The illustrative examples are an integral part of the text, but a number of them lack complete details and can be used as additional problems.

The fields of interpolation and approximation have been cultivated for centuries. The amount of information available is truly staggering. Take, for instance, the subject of orthogonal polynomials. A bibliography prepared in the late 1930's contains several hundred pages of references, and the topic continues to grow. I have had to anthologize. I have sought breadth rather than depth, and have tried to display a variety of analytical techniques. To some extent, I have been guided by what I consider "useful." Accordingly, I have developed neither the calculus of finite differences, nor L^p spaces ($p \neq 2$, ∞), nor approximation on infinite sets, for in my work with computation, I have rarely dealt with these things. On the other hand, I am aware that utility cannot be made into a principle of selection for a mathematics book. It comes down to this: I have included the topics that have caught my imagination. I hope that the selection will introduce the student to some of the best and encourage the scholar to seek the rest.

A word is in order on the portions of the book that are devoted to functional analysis. This subject is generally presented with spectral theory as its culmination. Here, the elementary geometric portions are developed for ultimate application to approximation theory. This is by no means new. In 1930, Dunham Jackson ended his book on the theory of approximation with a chapter on the geometry of function space, and over the years, more and more emphasis has been given to functional analysis. At its best, functional analysis unifies many seemingly diverse situations in a wonderful way and is a genuine principle of research. At its worst, it is a scintillating wrapper that provides attractive packages, and camouflages with glamorous language the fact that their content may be small or may have been originally obtained in the drab workshops of hard analysis. Perhaps functional analysis is analogous to Cartesian geometry which put an end to the synthetic drudgery of Apollonius though the major theorems on conics are Greek. Though the welding of functional analysis to conventional analysis has produced an imperfect seam that is visible in practically every chapter of this book, I believe that functional analysis is a good way to present many of the topics, and that it can and should be introduced at an early stage of a student's career.

This book derives from several teaching experiences. In the Spring of 1957 and again in the Spring of 1959, I presented a series of lectures on approximation theory to the members of training programs in Numerical Analysis that were held at the National Bureau of Standards in Washington, D.C. under the sponsorship of the National Science Foundation. In 1959, I gave a course in interpolation and approximation at the Harvard Summer School, and have given shorter courses in special computer programs at Wayne University and the University of Pennsylvania. Out of these varied experiences and out of a day to day exposure to live computation in the Applied Mathematics Division of the National Bureau of Standards, the plan of the present book emerged. The liberal policy of the National Bureau of Standards, encouraging study, research, and writing has enabled me to carry this plan to fruition. The following chapters were prepared at the Bureau: 5, and 9–14.

My debt to various works is clear, but I must single out Gontcharoff, Natanson, Szegö, and Walsh for special mention. My debt to my teachers should be made more explicit. My concern with these matters extends back to dissertation days when Ralph P. Boas interested me in interpolatory function theory. The lectures of Stefan Bergman opened my eyes to the beauty of orthogonal functions and the kernel function. The lectures and subsequent work with J. L. Walsh deepened my interest in problems in the complex plane. I had the pleasure of knowing and working with Michael Fekete in the last years of his life. A wonderful man of that wonderful school of Hungarian mathematicians, his single-minded insistence on simplicity and elegance made an immediate impression on me.

To discharge my debt to my colleagues, I must thank Dr. Oved Shisha for his kind attentions to this book. I have profited greatly from many discussions with him and from his detailed criticism of the manuscript.

From students (may their number increase) I have learned that though mathematics proceeds from false starts and bungling, it is presented backwards, as a fait accompli. This may provide clean copy and heighten the dramatic effect, but it taxes the understanding. There are many places in this book where the masters of analysis work their magic. How did they happen to get such and such an idea? This question has been asked me over and over by students just beyond the advanced calculus stage. It can rarely be answered, but one should say: tackle the problem yourself and you may learn. The road to understanding is rough; to smooth it too much denies the reality of creative genius.

Thanks go to Ellen Rigby who helped me with the figures and to Richard Strafella for various tasks with the manuscript. These were done as part of their Antioch College plan of quarters devoted to work.

PHILIP J. DAVIS
Washington, D.C.
Fall, 1961

Contents

INTERPOLATION
AND
APPROXIMATION

CHAPTER I

Introduction

This chapter contains material from algebra and analysis that will be of use in the later portions of the book. It is presented here for ready reference and for review. The reader is assumed to be familiar with some of the theorems. Other theorems may be less familiar and their proofs have been given. Though L^p spaces are mentioned in Theorem 1.4.0, they do not reappear until Chapter VII, and only the elementary portions of measure and integration theory are used.

1.1 Determinants. Let v_i designate the n-tuple of numbers $(a_{i1}, a_{i2}, \ldots, a_{in})$. For a constant α, we shall mean by αv_i the n-tuple $(\alpha a_{i1}, \alpha a_{i2}, \ldots, \alpha a_{in})$, while by $v_i + v_j$, we shall mean the n-tuple $(a_{i1} + a_{j1}, a_{i2} + a_{j2}, \ldots, a_{in} + a_{jn})$. The letters $e_1, \ldots, e_n$ will designate the unit n-tuples $(1, 0, 0, \ldots, 0)$, $(0, 1, 0, \ldots, 0)$, $\ldots$, $(0, 0, 0, \ldots, 1)$. The function of the n^2 variables a_{ij} $(i, j = 1, 2, \ldots, n)$, known as the determinant of those quantities, is generally written

$$D = \begin{vmatrix} a_{11} & a_{12} & \cdots & a_{1n} \\ a_{21} & a_{22} & \cdots & a_{2n} \\ \cdot & & & \cdot \\ \cdot & & & \cdot \\ \cdot & & & \cdot \\ a_{n1} & a_{n2} & \cdots & a_{nn} \end{vmatrix} = |a_{ij}| = D(v_1, v_2, \ldots, v_n). \qquad (1.1.1)$$

The determinant is completely characterized by the following three properties

(a) $D(v_1, v_2, \ldots, v_i, \ldots, v_n) = D(v_1, v_2, \ldots, v_i + v_j, \ldots, v_n)$

  $(i \neq j)$ (Invariance).

(b) $D(v_1, v_2, \ldots, \alpha v_i, \ldots, v_n) = \alpha D(v_1, v_2, \ldots, v_i, \ldots, v_n)$ $\qquad (1.1.2)$

  (Homogeneity).

(c) $D(e_1, e_2, \ldots, e_n) = 1$ (Normalization).

The whole of determinant theory can be built up from this starting point and is related to the theory of the volume of an n-dimensional parallelotope.

Given an n by n matrix $A = (a_{ij})$, the determinant associated with this matrix is designated by $|A|$ or det A. If from the array A we delete the ith row and the jth column, a certain $(n-1)$ by $(n-1)$ submatrix A_{ij} will remain. The determinant associated with this submatrix is known as the *minor* of the element a_{ij}. The quantity $(-1)^{i+j} |A_{ij}|$ is the *cofactor* of a_{ij}. For the cofactor we write $A_{ij}*$. The following rules of computation for determinants are fundamental.

(a) $|A| = |A'|$, $A' = $ transpose of $A = (a_{ji})$.

(b)   If two rows (or columns) of A are interchanged, producing a matrix A_1, then $|A| = -|A_1|$.

(c)   If two rows (or columns) of A are identical, then $|A| = 0$.

(d)   If a row (or column), v, of A is replaced by kv producing a matrix A_1, then $|A_1| = k\,|A|$.

(e)   If a scalar multiple kv_i of the ith row (or column) is added to the jth row (or column) v_j, $(i \neq j)$ and   (1.1.3) the matrix A_1 results, then $|A| = |A_1|$.

(f)   A determinant may be evaluated in terms of cofactors:

$$|A| = \sum_{i=1}^{n} a_{ij}A_{ij}* \qquad 1 \leq j \leq n$$

$$= \sum_{j=1}^{n} a_{ij}A_{ij}* \qquad 1 \leq i \leq n.$$

The expansion (f) is of considerable utility for it reduces an $n \times n$ determinant to a sum of n determinants of order $n - 1$. Coupled with the elementary equation $|a_{11}| = a_{11}$, it contains within it a recursive definition of a determinant. The complete expansion of a determinant in terms of the matrix elements is less useful theoretically and hardly at all numerically.

1.2 Solution of Linear Systems of Equations.   Consider the system of n linear equations in n unknowns $x_1, x_2, \ldots, x_n$

$$\sum_{j=1}^{n} a_{ij}x_j = b_i \qquad (i = 1, 2, \ldots, n). \tag{1.2.1}$$

THEOREM 1.2.1 (Cramer's Rule).   *If* $|A| = |a_{ij}| \neq 0$, *then* (1.2.1) *possesses a unique solution given by*

$$x_r = \frac{\sum\limits_{i=1}^{n} A_{ir}* b_i}{|A|} \qquad r = 1, 2, \ldots, n. \tag{1.2.2}$$

THEOREM 1.2.2 (The Alternative Theorem).   *The homogeneous system*

$$\sum_{j=1}^{n} a_{ij}x_j = 0 \qquad (i = 1, 2, \ldots, n) \tag{1.2.3}$$

possesses a non-trivial solution (i.e., a solution other than $x_1 = x_2 = \cdots = x_n = 0$) *if and only if* $|A| = 0$. *If for a fixed* $A = (a_{ij})$ *there are solutions to the non-homogeneous system* (1.2.1) *for every selection of the quantities* b_i, *then* $|A| \neq 0$ *and the homogeneous system has only the trivial solution.*

1.3 Linear Vector Spaces. It will be useful to formulate many questions of interpolation and approximation theory within an abstract framework. The notion of a linear vector space over a field F is therefore a basic one.

DEFINITION 1.3.1.   A *linear vector space* (or a linear space) X is a set of elements (or vectors) $x, y, \ldots$, for which two types of operation are possible. Any two elements $x, y \in X$ determine a unique element $x + y \in X$ as their sum. Each element $x \in X$ and each scalar α of a given field F determine a unique element $\alpha x \in X$ as a scalar product. Vector sums and scalar products are required to obey the following laws.

(a) $x + y = y + x$.
(b) $x + (y + z) = (x + y) + z$.
(c)   There exists a unique element $0 \in X$ such that
       $x + 0 = x$ for all $x \in X$.
(d)   To each $x \in X$ there exists a unique inverse $-x$ such
       that $x + (-x) = 0$.   (1.3.1)
(e) $\alpha(\beta x) = (\alpha\beta)x$ for all $\alpha, \beta \in F$, $x \in X$.
(f) $\alpha(x + y) = \alpha x + \alpha y$.
(g) $(\alpha + \beta)x = \alpha x + \beta x$.
(h) $1(x) = x$.

Conditions (a)–(d) are frequently summed up by saying that the elements of X form an *Abelian group* under addition. The element 0 is called the *zero vector*.

In this book, the underlying field F of scalars will be either (1) the field of real numbers, or (2) the field of complex numbers. We can, therefore, speak either of a real or a complex linear vector space.

DEFINITION 1.3.2.   An expression of the form

$$\alpha_1 x_1 + \alpha_2 x_2 + \cdots + \alpha_n x_n; \quad \alpha_i \in F, \, x_i \in X$$

is called a *linear combination* of the x's.

DEFINITION 1.3.3.   A finite set of vectors $x_1, \ldots, x_n$ is *linearly dependent* if we can find constants (i.e., scalars) $\alpha_1, \alpha_2, \ldots, \alpha_n$, not all zero such that $\alpha_1 x_1 + \alpha_2 x_2 + \cdots + \alpha_n x_n = 0$. If such is not the case, the vectors are called *independent*.

DEFINITION 1.3.4.   Let n be a positive integer. Suppose that we can find n vectors $x_1, x_2, \ldots, x_n \in X$ which are independent while every $n + 1$ vectors are dependent. Then X is said to be a linear space of dimension n. If no such n exists, then X is called an *infinite dimensional* space.

DEFINITION 1.3.5.   A set of elements $x_1, x_2, \ldots$, is said to be a *basis* for X if the x_i are independent and if every $x \in X$ can be expressed (uniquely) as a linear combination of the x_i.

THEOREM 1.3.1.   *X has finite dimension n if and only if it has a basis of n elements. If X has dimension n any n independent elements constitute a basis.*

Ex. 1.   *The Real, n-dimensional, Cartesian Space R_n.* This consists of vectors x which are n-tuples of real numbers: $x = (x_1, x_2, \ldots, x_n)$. Let $y = (y_1, y_2, \ldots, y_n)$ be a second vector. (x and y are considered equal if and only if $x_i = y_i$, $i = 1, 2, \ldots, n$.) Vector addition is defined by

$$x + y = (x_1 + y_1, x_2 + y_2, \ldots, x_n + y_n).$$

Scalar multiplication is defined by $\alpha x = (\alpha x_1, \alpha x_2, \ldots, \alpha x_n)$. We set $0 = (0, 0, \ldots, 0)$ and $-x = (-x_1, -x_2, \ldots, -x_n)$. The n vectors, $e_1 = (1, 0, \ldots, 0)$, $e_2 = (0, 1, \ldots, 0), \ldots, e_n = (0, 0, \ldots, 1)$, known as *unit vectors*, are independent.

Ex. 2.   *The Complex, n-dimensional, Cartesian Space C_n.* This consists of n-tuples of complex numbers: $(z_1, z_2, \ldots, z_n)$. The laws of combination are as in Ex. 1.

Ex. 3.   *Linear Spaces of Functions.* In this example, a function, considered as a whole, is thought of as constituting an element of a space. Let S designate a point set lying on the real axis. Consider the totality of real-valued functions with domain S. Call this totality T. For $f, g \in T$, define their sum $f + g$ by means of

$$(f + g)(x) = f(x) + g(x), x \in S. \tag{1.3.2}$$

Define a scalar product by means of

$$(\alpha f)(x) = \alpha f(x), x \in S. \tag{1.3.3}$$

Let the zero vector be the function of T that vanishes identically. Let $-f$ designate the function defined by

$$(-f)(x) = -f(x), x \in S. \tag{1.3.4}$$

With these definitions, T is a linear vector space. If S contains more than a finite number of points, T is of infinite dimension.

1.4 The Hierarchy of Functions. Our dealings will be almost exclusively with functions of a single real or complex variable. We shall work with finite intervals in the real case and bounded sets in the complex case. The deeper analytical properties of interpolation and approximation depend to a great extent on what may be called "the degree of smoothness" of the function approximated. In order of increasing smoothness, we shall deal with: L^p functions, bounded functions, continuous functions, functions satisfying a Lipschitz condition, differentiable functions, n-times differentiable functions, infinitely differentiable functions, analytic functions, entire functions, polynomials of restricted degree, constants. It will become apparent in subsequent chapters that the processes of interpolation and approximation become stronger when applied to functions further down this list. We shall now define these classes of functions and recall some basic facts about them.

DEFINITION 1.4.0.   Let $p > 0$. The class of functions $f(x)$ which are measurable and for which $|f(x)|^p$ is integrable over $[a, b]$ is known as $L^p[a, b]$. If $p = 1$, the class is designated by $L[a, b]$.

THEOREM 1.4.0.   (a) $L^p[a, b]$ is a linear space. (b) If $f \in L[a, b]$, then $f \geq 0$ and $\int_a^b f(x)\,dx = 0$ imply $f = 0$ almost everywhere. (c) If $f \in L[a, b]$, then $g(x) = \int_a^x f(x)\,dx$ is continuous. (d) If $-\infty < a < b < \infty$, then $f \in L^p[a, b]$, $p' < p$, implies $f \in L^{p'}[a, b]$. (e) If $f \in L^p[a, b]$ with $p \geq 1$, we can find an absolutely continuous function $g(x)$ such that $\int_a^b |f(x) - g(x)|^p\,dx \leq \varepsilon$ for arbitrary $\varepsilon > 0$.

For these results, the reader is referred to standard texts on integration theory.

Let S denote a point set in R_n or in the complex plane and P a point in that set. Though Definitions 1.4.1–1.9.1 are meaningful for complex valued functions of a real variable, we shall generally deal with real valued functions whenever S is in R_n.

DEFINITION 1.4.1.   A function f is *bounded* on S if there exists a constant M such that

$$|f(P)| \leq M \text{ for all } P \in S. \tag{1.4.1}$$

If no such constant exists, the function is said to be *unbounded on S*. The class of functions which are bounded on S will be designated by $B(S)$. $B(S)$ is a linear space.

Ex. 1.   The function $y = \sin x^2$ is bounded on $-\infty < x < \infty$.

Ex. 2.   The Gamma function $y = \Gamma(x)$ is unbounded on the interval $0 < x \leq 1$, and on the interval $1 \leq x < \infty$.

Ex. 3.   If $[a, b]$ is a finite interval, $f \in B[a, b]$ and measurable there implies $f \in L^p[a, b]$ for all $p > 0$.

DEFINITION 1.4.2.   Let the function f be defined on the set S. It is continuous at a point P_0 of S if

$$\lim_{n \to \infty} f(P_n) = f(P_0) \tag{1.4.2}$$

whenever $P_n \to P_0$, $P_n \in S$. If f is defined on an interval $[a, b]$ and is *continuous* at $x_0 \in [a, b]$, then given an $\varepsilon > 0$, we can find a δ such that

$$|f(x) - f(x_0)| \leq \varepsilon \tag{1.4.3}$$

for all $|x - x_0| \leq \delta$ in $[a, b]$. The δ will depend upon f, x_0, and ε. The class of functions continuous on $I : [a, b]$ will be designated by $C[a, b]$. It is a linear space.

It may occur that for a given f and ε we can find a δ for which (1.4.3) holds independently of x_0. This leads to the notion of uniform continuity.

DEFINITION 1.4.3.   A function f is *uniformly continuous* over a set S if given an $\varepsilon > 0$, we can find a δ such that

$$|f(x_1) - f(x_2)| \leq \varepsilon \tag{1.4.4}$$

for all $|x_1 - x_2| \leq \delta$; $x_1, x_2 \in S$.

In one important case, the notions of continuity and uniform continuity coincide:

THEOREM 1.4.1.   *A function which is continuous on a compact (i.e., closed and bounded) point set is uniformly continuous there.*

Ex. 1.   The function $A(x) = |x|$ is continuous on $-\infty < x < \infty$.

Ex. 2.   The function $f(x) = (1 + e^{1/x})^{-1}$ is discontinuous at $x = 0$ however $f(0)$ may be defined, for $\lim_{x \to 0^+} f(x) = 0$ while $\lim_{x \to 0^-} f(x) = 1$. It is continuous elsewhere.

Ex. 3.   The function $f(x) = \dfrac{1}{x}$ is continuous on the open interval $(0, \frac{1}{2})$ but is not uniformly continuous there.

Ex. 4.   The function $f(x) = \dfrac{1}{1 + x^2}$ is uniformly continuous over the whole line $-\infty < x < \infty$, for we have

$$\left| \frac{1}{1 + x_1^2} - \frac{1}{1 + x_2^2} \right| = \frac{|x_2^2 - x_1^2|}{(1 + x_1^2)(1 + x_2^2)} \leq |x_2 - x_1| \frac{(|x_1| + |x_2|)}{(1 + x_1^2)(1 + x_2^2)}.$$

Inasmuch as $\qquad \dfrac{|x_1|}{1 + x_1^2} \leq \dfrac{1}{2}, \dfrac{|x_1| + |x_2|}{(1 + x_1^2)(1 + x_2^2)} \leq 1.$

Thus, $|f(x_1) - f(x_2)| \leq \varepsilon$ whenever $|x_2 - x_1| \leq \varepsilon$.

THEOREM 1.4.2 (First Mean Value Theorem for Integrals).

Let f, $g \in C\,[a, b]$. Suppose moreover that $g \geq 0$ there. Then,

$$\int_a^b f(x)g(x)\, dx = f(\xi) \int_a^b g(x)\, dx \tag{1.4.5}$$

for some ξ with $a \leq \xi \leq b$. The theorem is also true if $g \in L[a, b]$, $g \geq 0$ a.e.
It is occasionally useful to have information about the best ε which goes with a given δ in the definition of uniform continuity.

DEFINITION 1.4.4.   Let $f(x)$ be defined on an interval I. Set

$$w(\delta;f) = w(\delta) = \sup |f(x_1) - f(x_2)| \tag{1.4.6}$$

where the sup is taken over all pairs x_1, $x_2 \in I$ for which $|x_1 - x_2| \leq \delta$. The function $w(\delta)$ (which depends on f) is called the *modulus of continuity of f on I.*

Ex. 5. $f(x) = x^2$, $I = (0, 1)$, $w(\delta) = 2\delta - \delta^2$.

Ex. 6. $f(x) = \dfrac{1}{x}$, $I = (0, 1)$, $w(\delta) = +\infty$.

Ex. 7. $f(x) = \sin\dfrac{1}{x}$, $I = (0, 1)$, $w(\delta) \equiv 2$.

THEOREM 1.4.3.   *Let $f(x) \in C[a, b]$. The modulus of continuity has the following properties*

$$w(0) = 0 \tag{1.4.7}$$

If $0 < \delta_1 < \delta_2$ then $w(\delta_1) \leq w(\delta_2)$ (Monotonicity) $\tag{1.4.8}$

$$w(\delta_1 + \delta_2) \leq w(\delta_1) + w(\delta_2) \ (Subadditivity) \tag{1.4.9}$$

$$w(n\delta) \leq nw(\delta). \tag{1.4.10}$$

Moreover, $w(\delta) \in C[0, b - a]$.

Proof: (1.4.7) is obvious. Since $|x_1 - x_2| \leq \delta_1$ implies $|x_1 - x_2| \leq \delta_2$, the corresponding sup cannot decrease, and (1.4.8) follows. To prove (1.4.9), observe that if $0 \leq x_2 - x_1 \leq \delta_1$, then $|f(x_1) - f(x_2)| \leq w(\delta_1) \leq w(\delta_1) + w(\delta_2)$. On the other hand, if $\delta_1 < x_2 - x_1 \leq \delta_1 + \delta_2$, then $x_1 + \delta_1 < x_2$ and $x_2 - (x_1 + \delta_1) < \delta_2$. But,

$$|f(x_1) - f(x_2)| \leq |f(x_1) - f(x_1 + \delta_1)| + |f(x_1 + \delta_1) - f(x_2)|$$
$$\leq w(\delta_1) + w(x_2 - (x_1 + \delta_1)) \leq w(\delta_1) + w(\delta_2).$$

Therefore, $w(\delta_1 + \delta_2) = \sup\limits_{0 \le x_2 - x_1 \le \delta_1 + \delta_2} |f(x_1) - f(x_2)| \le w(\delta_1) + w(\delta_2)$.
(1.4.10) follows immediately from (1.4.9) by induction. From (1.4.8 and 9),
$0 \le w(\delta + \delta_1) - w(\delta) \le w(\delta_1)$. Now, by Theorem 1.4.1, $\lim\limits_{\delta_1 \to 0} w(\delta_1) = 0$
and hence w is continuous at δ.

1.5 Functions Satisfying a Lipschitz Condition

DEFINITION 1.5.1. Let $f(x)$ be defined on an interval I and suppose we
can find two positive constants M and α such that

$$|f(x_1) - f(x_2)| \le M |x_1 - x_2|^\alpha \text{ for all } x_1, x_2 \in I. \tag{1.5.1}$$

Then f is said to satisfy a *Lipschitz Condition of order* α. The class of such
functions will be designated by Lip α. When it is useful to put the constant
M in evidence, one writes $\text{Lip}_M \alpha$.

THEOREM 1.5.1. Lip α *is a linear space. If* $f \in$ Lip α *on* I, *then* f *is continuous; indeed, uniformly continuous on* I. *If* $f \in$ Lip α *with* $\alpha > 1$ *then*
$f = $ *constant. If* $f \in$ Lip α, *it may fail to be differentiable, but if it possesses a
derivative satisfying* $|f'(x)| \le M$ *then* $f \in \text{Lip}_M 1$. *If* $\alpha < \beta$ *then* Lip $\alpha \supset$ Lip β.
The conditions $f \in \text{Lip}_M \alpha$ *and* $w(\delta) \le M \delta^\alpha$ *are equivalent.*

Ex. 1. Let $0 < \alpha < 1$. Let $x > 0, h > 0$. Then $\dfrac{d}{dx}[(x + h)^\alpha - x^\alpha] =$
$\alpha[(x + h)^{\alpha-1} - x^{\alpha-1}] < 0$. Therefore $(x + h)^\alpha - x^\alpha$ is decreasing for all $x \ge 0$
and hence $(x + h)^\alpha - x^\alpha \le h^\alpha$. This means that $x^\alpha \in$ Lip α on any positive
interval.

1.6 Differentiable Functions

DEFINITION 1.6.1. Let $f(x)$ be defined on an interval I. It is said to be
differentiable at a point $x_0 \in I$ if the following limit exists

$$\lim\limits_{x \to x_0} \frac{f(x) - f(x_0)}{x - x_0} = f'(x_0). \tag{1.6.1}$$

If x_0 is an end point of I then the limit in (1.6.1) is replaced by an appropriate one-sided limit. The function $f(x)$ is *differentiable on* I if it is differentiable at each point of I.

Ex. 1. $A(x) = |x|$ is differentiable at all $x \ne 0$. At $x = 0$ it possesses right
and left hand derivatives $\lim\limits_{x \to 0^+}, \lim\limits_{x \to 0^-} \dfrac{A(x) - A(0)}{x - 0}$.

Ex. 2. $S(x) = \begin{cases} 0 & x \le 0 \\ 1 & x > 0 \end{cases}$ is discontinuous at $x = 0$ and is not differentiable
there. Elsewhere it is differentiable.

Ex. 3. $f(x) = x^{\frac{1}{3}}$. Though continuous at $x = 0, f(x)$ fails to be differentiable there. It is sometimes convenient to write $f'(0) = +\infty$.

For differentiable functions, we have Rolle's Theorem and the Mean Value Theorem:

THEOREM 1.6.1 (Rolle). *Let $f(x) \in C[a, b]$ and be differentiable at each point of (a, b). If $f(a) = f(b)$ then there is a point $x = \xi$ with $a < \xi < b$ for which $f'(\xi) = 0$.*

THEOREM 1.6.2. *Let $f(x) \in C[a, b]$ be differentiable at each point of (a, b). Then we can find a ξ with $a < \xi < b$ such that*

$$f(b) = f(a) + (b - a)f'(\xi). \qquad (1.6.2)$$

If f is differentiable at each point of I, its derivative $f'(x)$ may exhibit a wide variety of smoothness properties. A particularly noteworthy case is where $f'(x)$ is itself continuous on I. The class of functions that have a continuous derivative on $[a, b]$ is designated by $C^1[a, b]$. More generally,

DEFINITION 1.6.2. *If $f(x)$ is n times differentiable on $[a, b]$ and if $f^{(n)}(x)$ is itself continuous on $[a, b]$, we shall write $f(x) \in C^n[a, b]$.*
$C^n[a, b]$ is a linear space of functions.

Ex. 4. Let
$$f(x) = \begin{cases} x^k & x \geq 0 \\ 0 & x < 0. \end{cases}$$

Then, $f \in C^{k-1}$ on $-\infty < x < \infty$. But $f \notin C^k$ on any interval containing the origin.

Ex. 5. Let $f(x) = |x|^{\frac{5}{2}}$. Then $f \in C^2$, but $f \notin C^3$ on any interval containing the origin.

For functions having higher derivatives we have the following generalized Rolle's Theorem.

THEOREM 1.6.3. *Let $n \geq 2$. Suppose that $f \in C[a, b]$ and let $f^{(n-1)}(x)$ exist at each point of (a, b). Suppose that $f(x_1) = f(x_2) = \cdots = f(x_n) = 0$ for $a \leq x_1 < x_2 < \cdots < x_n \leq b$. Then there is a point $\xi, x_1 < \xi < x_n$ such that $f^{(n-1)}(\xi) = 0$.*

Proof: We give the proof for $n = 3$. The general case is similar. Let $f(x_1) = f(x_2) = f(x_3) = 0$. Since f is differentiable in $x_1 < x < x_3$, we can find ξ_1 and ξ_2 such that $x_1 < \xi_1 < x_2 < \xi_2 < x_3$ and $f'(\xi_1) = 0, f'(\xi_2) = 0$.

Since f'' also exists, a second application of Rolle's Theorem yields a ξ, $\xi_1 < \xi < \xi_2$ with $f^{(2)}(\xi) = 0$.

Taylor's Theorem with the exact remainder and the various expressions for the remainder involving higher derivatives constitute generalizations of the Mean Value Theorem.

THEOREM 1.6.4.   *Let $f(x) \in C^{n+1}[a, b]$ and let $x_0 \in [a, b]$. Then for all*

$$a \leq x \leq b, f(x) = f(x_0) + f'(x_0)(x - x_0) + \frac{f''(x_0)}{2!}(x - x_0)^2 + \cdots$$

$$+ \frac{f^{(n)}(x_0)}{n!}(x - x_0)^n + \frac{1}{n!}\int_{x_0}^{x} f^{(n+1)}(t)(x - t)^n \, dt. \quad (1.6.3)$$

THEOREM 1.6.5.   *Let $f(x) \in C^n[a, b]$ and let $f^{(n+1)}(x)$ exist in (a, b). Then there is a ξ with $a < \xi < b$ such that*

$$f(b) = f(a) + f'(a)(b - a) + \frac{f''(a)}{2!}(b - a)^2 + \cdots$$

$$+ \frac{f^{(n)}(a)}{n!}(b - a)^n + \frac{f^{(n+1)}(\xi)}{(n + 1)!}(b - a)^{n+1}. \quad (1.6.4)$$

A form of the remainder theorem (sometimes referred to as Young's form) is useful on occasion.

THEOREM 1.6.6.   *Let $f(x)$ be $n + 1$ times differentiable at $x = x_0$. Then,*

$$f(x) = f(x_0) + f'(x_0)(x - x_0) + \cdots + \frac{f^{(n)}(x_0)}{n!}(x - x_0)^n$$

$$+ \frac{(x - x_0)^{n+1}}{(n + 1)!}[f^{(n+1)}(x_0) + \varepsilon(x)] \quad (1.6.5)$$

where $$\lim_{x \to x_0} \varepsilon(x) = 0.$$

Proof: Set

$$R(x) = f(x) - f(x_0) - f'(x_0)(x - x_0) - \cdots - \frac{(x - x_0)^{n+1}}{(n + 1)!} f^{(n+1)}(x_0).$$

Then (1.6.5) is equivalent to showing that $\lim_{x \to x_0} \dfrac{R(x)}{(x - x_0)^{n+1}} = 0$. By differentiating, we find that

$$R(x_0) = R'(x_0) = \cdots = R^{(n+1)}(x_0) = 0. \quad (1.6.6)$$

Let $\varepsilon > 0$. The functions

$$P(x) = R(x) + \varepsilon(x - x_0)^{n+1}, \quad Q(x) = R(x) - \varepsilon(x - x_0)^{n+1} \quad (1.6.7)$$

are $n + 1$ times differentiable at $x = x_0$. Moreover, $P^{(k)}(x_0) = 0, Q^{(k)}(x_0) = 0,$ $k = 0, 1, \ldots, n$, while $P^{(n+1)}(x_0) = \varepsilon(n + 1)! > 0,$

$$Q^{(n+1)}(x_0) = -(n + 1)! \, \varepsilon < 0.$$

This implies that $P(x)$ increases monotonically in some interval $(x_0, x_0 + \delta)$ while $Q(x)$ decreases monotonically in $(x_0, \ x_0 + \delta)$. Therefore, for x in $(x_0, x_0 + \delta)$,

$$R(x) + \varepsilon(x - x_0)^{n+1} > 0$$
$$R(x) - \varepsilon(x - x_0)^{n+1} < 0.$$

Therefore,

$$-\varepsilon < \frac{R(x)}{(x - x_0)^{n+1}} < \varepsilon. \tag{1.6.8}$$

Since ε is arbitrary, (1.6.8) implies that $\displaystyle\lim_{x \to x_0^+} \frac{R(x)}{(x - x_0)^{n+1}} = 0$. A similar argument shows that $\displaystyle\lim_{x \to x_0^-} \frac{R(x)}{(x - x_0)^{n+1}} = 0$, and the proof is complete.

1.7   Infinitely Differentiable Functions

DEFINITION 1.7.1.   If $f(x) \in C^n[a, b]$ for $n = 0, 1, 2, \ldots$, then f is called *infinitely differentiable in* $[a, b]$. We shall write $C^\infty[a, b]$ for the class of such functions.

Ex. 1. $f(x) = x^2$ is infinitely differentiable on $-\infty < x < \infty$.

Ex. 2. $f(x) = \dfrac{1}{1 + x^2}$ is infinitely differentiable on $-\infty < x < \infty$.

Ex. 3. $f(x) = \displaystyle\sum_{n=1}^{\infty} \frac{\cos nx}{n^{\log n}}$ is infinitely differentiable on $-\infty < x < \infty$. For since $|\cos nx| \leq 1$ and $\displaystyle\sum_{n=1}^{\infty} \frac{1}{n^{\log n}} < \infty$, the original series converges absolutely and uniformly. Since moreover, for any integer $p \geq 0$ $\displaystyle\sum_{n=1}^{\infty} \frac{n^p}{n^{\log n}} < \infty$, the differentiated series of all orders converge uniformly and hence represent the respective derivatives of $f(x)$.

The functions of class $C^\infty[a, b]$ form a linear space.
If $f \in C^\infty[a, b]$ and $x_0 \in [a, b]$ we may form the Taylor expansion

$$f(x) \sim \sum_{k=0}^{\infty} \frac{f^{(n)}(x_0)}{n!} (x - x_0)^n. \tag{1.7.1}$$

For a given x this series may or may not converge. If it converges, it may or may not converge to $f(x)$. The famous function that displays this behavior is

$$f(x) = e^{-x^{-2}}, \quad x \neq 0; \quad f(0) = 0. \tag{1.7.2}$$

This function is in $C^{\infty}(-\infty, \infty)$ and

$$f^{(n)}(0) = 0 \qquad n = 0, 1, 2, \ldots. \tag{1.7.3}$$

With $x_0 = 0$, (1.7.1) converges to 0 for every x. There are an infinity of functions of class C^{∞} for which (1.7.3) holds. If (1.7.1) converges to $f(x)$ over an interval, we are led to the notion of an analytic function.

1.8   Functions Analytic on the Line

DEFINITION 1.8.1.   Let $f(x)$ be defined on $[a, b]$ and assume that at each point $x_0 \in [a, b]$ there is a power series expression of $f(x)$ valid in some interval:

$$f(x) = a_0 + a_1(x - x_0) + a_2(x - x_0)^2 + \cdots, |x - x_0| < p(x_0). \tag{1.8.1}$$

Then $f(x)$ is said to be *analytic* on the interval. We write $f(x) \in A[a, b]$.

Ex. 1. $f(x) = [(x)(x - 1)]^{-\frac{1}{2}} \in A[\varepsilon, 1 - \varepsilon], 0 < \varepsilon < 1 - \varepsilon.$

Ex. 2. $f(x) = \displaystyle\int_0^x e^{t^2} \, dt$ is analytic over the entire line $-\infty < x < \infty.$

Ex. 3. $A(x) = |x|$ is not analytic over an interval containing $x = 0$ in its interior. But it is "piece-wise" analytic.

THEOREM 1.8.1. $A[a, b]$ *is a linear space. If* $f(x) \in A[a, b]$ *then*

$$f(x) \in C^{\infty}[a, b].$$

The constants a_n of (1.8.1) are

$$a_n = \frac{1}{n!} f^{(n)}(x_0) \qquad n = 0, 1, \ldots. \tag{1.8.2}$$

It does not follow conversely that if $f \in C^{\infty}[a, b]$ then $f \in A[a, b]$. This is demonstrated by the example (1.7.2). Another example is $f(x) = \displaystyle\sum_{k=1}^{\infty} \frac{\cos nx}{n^{\log n}}$ which, as we have seen, is infinitely differentiable on $-\infty < x < \infty$ and of period 2π. The ideas of Theorem 12.3.2 will show that $f(x) \notin A[-\pi, \pi]$.

1.9   Functions Analytic in a Region

DEFINITION 1.9.1.   Let R be a region of the complex plane and let $f(z)$ be a single valued function of the complex variable z defined in R. If $z_0 \in R, f(z)$ is said to be *analytic* at z_0 (or *regular* at z_0) if it has a representation of the form

$$f(z) = \sum_{n=0}^{\infty} a_n(z - z_0)^n \qquad (1.9.1)$$

valid in some neighborhood of z_0: $|z - z_0| < p(z_0)$. If $z_0 = \infty$, we require an expansion of the form

$$f(z) = \sum_{n=0}^{\infty} a_n z^{-n}, \ |z| > p. \qquad (1.9.2)$$

A function is *analytic* (or *regular*) *in* R if it is analytic at each point of R. We shall write $A(R)$ for the class of such functions. $A(R)$ is a linear space.

Ex. 1. The function $f(z) = \dfrac{1}{1 + z^2}$ is analytic in any region not containing the points $z = \pm i$.

Ex. 2. The function $f(z) = \displaystyle\int_0^z e^{t^2}\, dt$ is analytic in any region not containing $z = \infty$.

Ex. 3. A branch of the function $f(z) = (z(z - 1))^{\frac{1}{2}}$ may be selected that is regular in any rectangle $0 < \varepsilon \le x \le 1 - \varepsilon$, $-R \le y \le R$.

The relationship between functions analytic on a line and functions analytic in a region is given by the following theorem.

Theorem 1.9.1. *Let $f(x) \in A[a, b]$. Then we can find a region R containing $[a, b]$ into which $f(x)$ can be continued analytically such that $f(z) \in A(R)$.*

Proof: For each point $x_0 \in [a, b]$ there is a quantity $p(x_0)$ and an expansion

$$f(x) = \sum_{n=0}^{\infty} a_n(x - x_0)^n \text{ valid in } |x - x_0| < p(x_0). \qquad (1.9.3)$$

When x is replaced by $z = x + iy$, (1.9.3) defines an analytic continuation of $f(x)$ into the circle $|z - x_0| < p(x_0)$. Let x_0 run through the interval $[a, b]$. The circles $|z - x_0| < p(x_0)$ cover $[a, b]$. Let R be the union of these circles. R is an open set and is arcwise connected. For if p, $q \in R$, join p to x_1 and q to x_2, the centers of their respective circles. Then the arc px_1x_2q lies in R. R is therefore a region and $f(x)$ can be continued analytically into it.

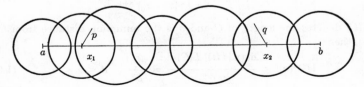

Figure 1.9.1.

Cauchy's Theorem is a basic tool in complex analysis.

THEOREM 1.9.2. *Let R be a simply connected region and let $f(z) \in A(R)$. Let z_0 lie in R and suppose that C is a simple, closed, rectifiable curve which lies in R and which goes around z_0 in the positive sense. Then*

$$f^{(n)}(z_0) = \frac{n!}{2\pi i} \int_C \frac{f(z)}{(z - z_0)^{n+1}} \, dz. \tag{1.9.4}$$

Whenever the Cauchy integral formula is employed, it will be understood that C satisfies the above conditions.

Analytic functions may be completely characterized by the growth of their derivatives, and this provides a second approach independent of power series.

THEOREM 1.9.3 (Pringsheim). *Let $f(x) \in C^\infty[a, b]$. A necessary and sufficient condition that $f \in A[a, b]$ is that there exist a constant $r > 0$ such that*

$$|f^{(n)}(x)| \leq r^n n! \quad a \leq x \leq b, \quad n = 0, 1, \dots . \tag{1.9.5}$$

Proof: Sufficiency. Let x_0 be a fixed point in $[a, b]$ and suppose that (1.9.5) holds. By Theorem 1.6.5 we have for $x \in [a, b]$,

$$f(x) = \sum_{k=0}^{n-1} \frac{f^{(k)}(x_0)}{k!} (x - x_0)^k + \frac{f^{(n)}(\xi)}{n!} (x - x_0)^n. \tag{1.9.6}$$

This holds for all n, and $\xi = \xi(n, x)$ is between x and x_0. In view of (1.9.5),

$$\left| \frac{f^{(n)}(\xi)}{n!} (x - x_0)^n \right| \leq r^n |x - x_0|^n,$$

so that if $|x - x_0| < \dfrac{1}{r}$, the remainder in (1.9.6) will converge to 0. The function f possesses a power series expansion valid in a neighborhood of x_0. This means that $f \in A[a, b]$. Necessity. If $f \in A[a, b]$, then by Theorem 1.9.1, we can find a simply connected region R containing $[a, b]$ in which f is analytic. Let C be a curve surrounding $[a, b]$ and lying in R. Then, for $x_0 \in [a, b]$, we have from (1.9.4),

$$|f^{(n)}(x_0)| \leq \frac{n!}{2\pi} \int_C \frac{|f(z)|}{|z - x_0|^{n+1}} \, ds. \tag{1.9.7}$$

If $L(C)$ denotes the length of C and δ is the minimum distance from C to $[a, b]$, then

$$|f^{(n)}(x)| \leq \frac{\max_{z \in C} |f(z)| \, L(C) n!}{2\pi \, \delta^{n+1}} = \frac{Mn!}{\delta^n}, \quad a \leq x \leq b \tag{1.9.8}$$

where M is a constant independent of n. It is now clear that we can find an r that makes (1.9.5) true.

Ex. 4.   Suppose $f \in A[-1, 1]$. It is impossible to have $f^{(n)}(0) = (n!)^2$. For then, $(n!)^2 \leq r^n n!$, $n = 0, 1, \ldots$. But, by Stirling's Theorem, $\sqrt[n]{n!} \to \infty$ so that we cannot find such an r.

On the other hand, if f is analytic only in the semi-open interval $0 < x \leq 1$ but is in $C^\infty[0, 1]$, we may very well have $f^{(n)}(0) = (n!)^2$. A theorem to this effect is developed in Chapter V.

Of great importance is the class of functions that are analytic in a circle C_r: $|z| < r$. Here we have the fundamental theorem of Cauchy-Hadamard.

THEOREM 1.9.4.   *Let $f(z) \in A(C_r)$ but $\notin A(C_{r'})$ if $r' > r$. This holds if and only if*

$$r^{-1} = \lim_{n \to \infty} \sup \left| \frac{f^{(n)}(0)}{n!} \right|^{1/n} \tag{1.9.9}$$

THEOREM 1.9.5 (Maximum Principle).   *Let $f(z)$ be analytic in a region R and not be constant there. Let z_0 lie in R. Then in any neighborhood of z_0 there exists a point z_1 where $|f(z_1)| > |f(z_0)|$. If $f(z_0) \neq 0$, then in any neighborhood of z_0 there is a point z_2 where $|f(z_2)| < |f(z_0)|$.*

1.10   Entire Functions

DEFINITION 1.10.1.   A function $f(x)$ is called *entire* if it has a representation of the form

$$f(z) = \sum_{k=0}^{\infty} a_k z^k \quad \text{valid for } |z| < \infty. \tag{1.10.1}$$

We shall designate this class of functions by E. E is a linear space.

Ex. 1.   Some examples of entire functions are

$$\frac{\sin z}{z}, \ 2^z, \int_0^z e^{t^2} \, dt, \ \frac{1}{\Gamma(z)}, \text{ the Bessel function } J_n(z).$$

THEOREM 1.10.1.   *The function $f(z) = \sum_{k=0}^{\infty} a_k z^k$ is entire if and only if*

$$\lim_{n \to \infty} |a_n|^{1/n} = 0. \tag{1.10.2}$$

Proof: This follows from (1.9.9).

1.11   Polynomials

DEFINITION 1.11.1.   By a polynomial of degree n is meant a function of the form

$$p_n(z) = a_0 z^n + a_1 z^{n-1} + \cdots + a_n, \quad a_0 \neq 0. \tag{1.11.1}$$

The class of polynomials of degree $\leq n$ will be designated by $\mathscr{P}_n$.

One might distinguish between the classes of polynomials with real co-efficients and with complex coefficients. It will usually be clear from the context which class we are dealing with and separate notations will not be introduced.

$\mathscr{P}_n$ is a linear space.

The following basic facts about polynomials should be recalled.

THEOREM 1.11.1 (Fundamental Theorem of Algebra). *If $n \geq 1$, a polynomial of degree n possesses a complex root.*

THEOREM 1.11.2 (Factorization Theorem). *If $p_n(z)$ is a polynomial of degree n then we may find n complex numbers $z_1, z_2, \ldots, z_n$ such that*

$$p_n(z) = a_0 z^n + a_1 z^{n-1} + \cdots + a_n$$

$$\equiv a_0(z - z_1)(z - z_2) \cdots (z - z_n) \quad (a_0 \neq 0).$$

The quantities z_i need not be distinct. If there are $r \leq n$ distinct roots $z_1, z_2, \ldots, z_r$, then for appropriate positive integers $\alpha_1, \alpha_2, \ldots, \alpha_r$, satisfying $\alpha_1 + \alpha_2 + \cdots + \alpha_r = n$, we have

$$p_n(z) = a_0(z - z_1)^{\alpha_1}(z - z_2)^{\alpha_2} \cdots (z - z_r)^{\alpha_r}. \tag{1.11.2}$$

The α_i are uniquely determined and the zero z_i is known as an α_i-fold zero. We have

$$p_n(z_i) = p_n'(z_i) = \cdots = p_n^{(\alpha_i - 1)}(z_i) = 0, \quad p_n^{(\alpha_i)}(z_i) \neq 0. \tag{1.11.3}$$

Conversely, these derivative conditions imply the above factorization.

THEOREM 1.11.3 (Uniqueness). *If $f(z) \in \mathscr{P}_n$ and f vanishes at more than n distinct points then it vanishes identically.*

Proof: Let the degree of f be $k \leq n$. By Theorem 1.11.2,

$$f(z) = a_0(z - z_1) \cdots (z - z_k).$$

By hypothesis, we can find a point $z^* \neq z_1, z_2, \ldots, z_k$ such that $f(z^*) = 0$. Then, $0 = a_0(z^* - z_1) \cdots (z^* - z_k)$ so that $a_0 = 0$. This implies that

$$f(z) \equiv 0.$$

1.12 Linear Functionals and the Algebraic Conjugate Space. In many problems, we must associate a number with a function extracted from a given class of functions. For instance, to each function $f(x)$ that has a continuous derivative on $[a, b]$, we may want to associate the number $\int_a^b (1 + [f'(x)]^2)^{\frac{1}{2}} \, dx$. To each function $f(x, y)$ that is twice continuously differentiable over a closed bounded region B, we may have to form the number $\displaystyle\iint_B \left(\frac{\partial f}{\partial x}\right)^2 + \left(\frac{\partial f}{\partial y}\right)^2 dx\, dy$ or even more simply, $f(x_0, y_0)$ where $(x_0, y_0) \in B$. Such an association is known as a *functional*. An important restriction is that the association behave linearly, and this leads to the following definition.

DEFINITION 1.12.1. Let X be a linear vector space and to each x let there be associated a unique real (or complex) number designated by $L(x)$. If for $x, y \in X$ and for all real (or complex) α, β we have

$$L(\alpha x + \beta y) = \alpha L(x) + \beta L(y), \tag{1.12.1}$$

then L is called a *linear functional* over X.

Ex. 1. $X = C[a, b]$. The elements of X are functions $f(x)$.

$$L(f) = \int_a^b f(x)\, dx \quad \text{or} \quad L(f) = \int_a^b x^2 f(x)\, dx.$$

Ex. 2. $X = C^2[a, b]$. $L(f) = f''(a) + f'(b) - f\left(\dfrac{a+b}{2}\right)$.

Ex. 3. $X = A[a, b]$. $L(f) = \displaystyle\int_a^b f(x)\, dx - \sum_{i=1}^n a_i f(x_i),\ a \le x_i \le b$.

Ex. 4. $X = A(R)$ where R is a region of the complex plane. Let C be a rectifiable curve lying in R.

$$L(f) = \int_C f(z)\, dz.$$

Ex. 5. $X = R_n$. $x = (x_1, x_2, \ldots, x_n)$. Let $a_1, \ldots, a_n$ be fixed constants and set

$$L(x) = \sum_{i=1}^n a_i x_i.$$

Interpolation theory is concerned with reconstructing functions on the basis of certain functional information assumed known. In many cases, the functionals are linear.

Functionals can be added to one another and scalar products can be formed. If, for instance, $f \in C^1[a, b]$ and

$$L_1(f) = \int_a^b f(x)\, dx \quad \text{and} \quad L_2(f) = f'\left(\frac{a+b}{2}\right),$$

we can identify the functional

$$L(f) = \alpha \int_a^b f(x)\, dx + \beta f'\left(\frac{a+b}{2}\right)$$

with the expression $\alpha L_1 + \beta L_2$. L is itself a linear functional. These observations form the basis for the following definition.

DEFINITION 1.12.2. Let X be a given linear space and let L_1 and L_2 be two linear functionals defined on X. The sum of L_1 and L_2 and the scalar product of α and L_1 are defined by

$$\text{(a)} \quad (L_1 + L_2)(x) = L_1(x) + L_2(x), \ x \in X$$

$$\text{(b)} \quad (\alpha L_1)(x) = \alpha L_1(x). \tag{1.12.2}$$

It is a simple matter to show that the set of all linear functionals defined on X combined by the above rules constitute a second linear space.

DEFINITION 1.12.3. Let X be a given linear space. The set of linear functionals defined on X and combined by (1.12.2) forms a linear space called the *algebraic conjugate space* of X and denoted by X^*.

X^*, then, has elements that are linear functionals. We can speak of linear combinations, linear independence, dimension, bases, etc., for linear functionals.

Ex. 6. $X = C[a, b]$. Let $x_1, x_2, \ldots, x_n$ be n distinct points lying in $[a, b]$. Let $L_k(f) = f(x_k)$ for $f \in X$. Then $L_1, L_2, \ldots, L_n$ are independent in X^*. For otherwise, for constants $a_1, \ldots, a_n$ not all zero, $a_1 L_1 + a_2 L_2 + \cdots + a_n L_n = 0$ (the 0 functional). Thus, for all $f \in C[a, b]$, $a_1 f(x_1) + a_2 f(x_2) + \cdots + a_n f(x_n) = 0$. This is impossible. For if $a_k \neq 0$, we may find a continuous function for which $f(x_k) = 1, f(x_i) = 0, \ i \neq k$. This leads to the contradiction $a_k = 0$.

Ex. 7. $X = \mathscr{P}_{n-2}[a, b]$. The above n functionals are linearly dependent. This is a consequence of the Lagrange interpolation formula in Chapter II.

THEOREM 1.12.1. *If X has dimension n then X^* has dimension n also.*

Proof: Let $x_1, x_2, \ldots, x_n$ be a basis (n independent elements). Then for any $x \in X$, $x = a_1 x_1 + a_2 x_2 + \cdots + a_n x_n$ in a unique way. Therefore, $L(x) = a_1 L(x_1) + \cdots + a_n L(x_n)$. For any $x \in X$ set

$$L_1(x) = a_1$$
$$L_2(x) = a_2$$
$$\cdot$$
$$\cdot \tag{1.12.3}$$
$$\cdot$$
$$L_n(x) = a_n.$$

L_i are linear functionals defined on X. They are independent, for, if not, we would have $\beta_1 L_1 + \beta_2 L_2 + \cdots + \beta_n L_n = 0$ with some $\beta_j \neq 0$. Then,

$$\beta_1 L_1(x_j) + \beta_2 L_2(x_j) + \cdots + \beta_j L_j(x_j) + \cdots + \beta_n L_n(x_j) = 0(x_j) = 0.$$

But $L_i(x_j) = \delta_{ij} = \begin{cases} 1 \text{ if } i = j \\ 0 \text{ if } i \neq j \end{cases}$ so that we obtain $\beta_j = 0$, a contradiction.

This shows that the dimension of X^* is at least n. We next show it is at most n.

Suppose we have $n + 1$ functionals, $L_1, L_2, \ldots, L_{n+1}$. Consider the $n + 1$ n-tuples

$$[L_i(x_1), L_i(x_2), \ldots, L_i(x_n)], \qquad i = 1, 2, \ldots, n + 1.$$

Since R_n (or C_n) is of dimension n, these n-tuples cannot be independent. Hence we can find numbers $\alpha_1, \ldots, \alpha_{n+1}$ not all zero such that

$$\alpha_1[L_1(x_1), \ldots, L_1(x_n)] + \cdots + \alpha_{n+1}[L_{n+1}(x_1), \ldots, L_{n+1}(x_n)]$$
$$= 0 = [0, 0, \ldots, 0].$$

Therefore

$$(\alpha_1 L_1 + \cdots + \alpha_{n+1} L_{n+1})(x_i) = 0, \quad \text{for} \quad i = 1, 2, \ldots, n.$$

By taking linear combinations,

$$(\alpha_1 L_1 + \cdots + \alpha_{n+1} L_{n+1})(x) = 0 \quad \text{for} \quad x \in X.$$

Therefore $L_1, \ldots, L_{n+1}$ must be dependent and the dimension of X^* is at most, and hence, precisely n.

This theorem tells us that over a space X of dimension n any linear functional can be expressed as a linear combination of n fixed independent linear functionals.

1.13   Some Assorted Facts.   Two special conformal maps.

A.

$$w = \tfrac{1}{2}(z + z^{-1}). \tag{1.13.1}$$

Set $w = u + iv$ and $z = \rho e^{i\theta}$. The exterior of the unit circle, $|z| > 1$, is mapped conformally onto the w-plane with the interval $-1 \leq u \leq 1$ deleted. The image of the point $(\rho \cos \theta, \rho \sin \theta)$ is the point

$$(\tfrac{1}{2}(\rho + \rho^{-1}) \cos \theta, \tfrac{1}{2}(\rho - \rho^{-1}) \sin \theta).$$

The circle $|z| = \rho > 1$ maps onto the ellipse

$$u = \tfrac{1}{2}(\rho + \rho^{-1}) \cos \theta, v = \tfrac{1}{2}(\rho - \rho^{-1}) \sin \theta, 0 \leq \theta \leq 2\pi. \tag{1.13.2}$$

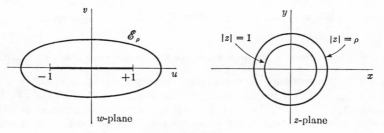

Figure 1.13.1.

DEFINITION 1.13.1.   The ellipse (1.13.2) will be designated by $\mathscr{E}_\rho$, $(\rho > 1)$. The semi-axes of $\mathscr{E}_\rho$ are, respectively

$$a = \tfrac{1}{2}(\rho + \rho^{-1})$$
$$b = \tfrac{1}{2}(\rho - \rho^{-1})$$

(1.13.3)

and hence

$$\rho = a + b = \text{sum of semi-axes of } \mathscr{E}_\rho.$$

(1.13.4)

The foci of $\mathscr{E}_\rho$ are at $u = \pm 1$ so that $\mathscr{E}_\rho$, $\rho > 1$, forms a confocal family of ellipses. The image of the unit circle under (1.13.1) is the interval $-1 \le u \le 1$ traced from 1 to -1, thence back to 1.

When z is solved for w, we obtain

$$z = w + \sqrt{w^2 - 1}.$$

(1.13.5)

For values of z outside the unit circle, that branch of the root must be taken which leads to $z(\infty) = \infty$.

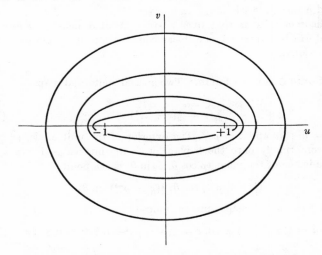

Figure 1.13.2   The Family $\mathscr{E}_\rho$ of Confocal Ellipses.

B.

$$z = \cos w = \cos(u + iv) = \cos u \cosh v - i \sin u \sinh v. \quad (1.13.6)$$

Let R be the rectangle in the w plane with vertices at $w = \sigma i$, $\sigma i + \pi$, $-\sigma i + \pi$, $-\sigma i$. R is mapped onto the ellipse $\mathscr{E}_\rho$, $\rho = e^\sigma$, with the two intervals $[1, a]$, $[-a, -1]$, $a = \cosh \sigma$, deleted. As a point w traces out the vertical sides of R, the image point z traces each of these two intervals twice.

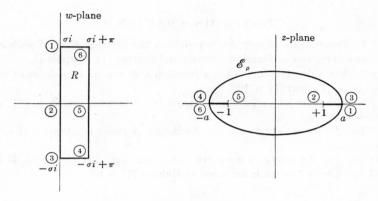

Figure 1.13.3.

Zorn's Lemma. Today, a mathematics book without this lemma would be like an 18th century gentleman without his sword.

DEFINITION 1.13.2. A *partial ordering* of a set X is a binary relation between elements designated by '$\leq$' and such that

$$x \leq y, \quad y \leq z \quad \text{implies} \quad x \leq z \tag{1.13.7}$$

$$x \leq x \tag{1.13.8}$$

$$x \leq y, \quad y \leq x \quad \text{implies} \quad x = y.$$

If in addition for any $x, y \in X$ it is true that either

$$x \leq y \quad \text{or} \quad y \leq x, \tag{1.13.9}$$

the set is called *totally ordered* (or, simply ordered).

If A is a subset of a partially ordered set, and if an element z satisfies

$$x \leq z \quad \text{for all} \quad x \in A, \tag{1.13.10}$$

then z is called an *upper bound* for A.

If z is an element of a partially ordered set X such that no element $x \in X$, $x \neq z$ satisfies

$$z \leq x \tag{1.13.11}$$

then z is called a *maximal* element of X.

THEOREM 1.13.1 (ZORN'S LEMMA). *Let X be a partially ordered set and suppose that every totally ordered subset of X has an upper bound in X. Then X has a maximal element.*

Zorn's Lemma is known to be equivalent to the Axiom of Choice.

NOTES ON CHAPTER I

1.1   Determinant theory developed from the point of view of n-dimensional volume can be found in Schreier and Sperner [1], Chapter II.

1.4   For a discussion of when a function $w(\delta)$ can be a modulus of continuity see Tieman [1], p. 109.

1.7   For more on infinitely differentiable functions, see Boas [4], pp. 150–156.

1.10   An up-to-date account of the theory of entire functions is given in Boas [3].

1.12   For the algebraic conjugate space, see, e.g., Taylor [3], pp. 34–35.

1.13   Zorn's Lemma is discussed in Halmos [2], p. 62.

PROBLEMS

1. For what values of a and b are the curves $y = ax^b$ bounded on $[0, 1]$?

2. For what values of a and b are the curves $y = \dfrac{1}{x^2 + ax + b}$ bounded on $[-1, 1]$?

3. Show that $\dfrac{1}{x} \sin x$ and $x \sin \dfrac{1}{x}$ (properly defined at $x = 0$) are continuous over any finite interval.

4. Show that $y = e^{-x}$ is uniformly continuous over the infinite interval

$$0 \le x < \infty.$$

5. Let $f \in C[a, b]$. Use the first mean value theorem to show that

$$\lim_{n \to \infty} \int_a^b f(x) \,|\sin nx|\, dx = \lim_{n \to \infty} \int_a^b f(x) \,|\cos nx|\, dx = \frac{2}{\pi} \int_a^b f(x)\, dx. \quad \text{(Fejér)}.$$

6. Compute $w(f; \delta)$ for $f(x) = \sin x$ on $-\infty < x < \infty$.

7. Compute $w(f; \delta)$ for $f(x) = x^2 - 3x + 1$ on $-1 \le x \le 1$.

8. Let $f(x) \in C^1[a, b]$ and let $f'(x)$ be increasing and positive. Find $w(\delta)$.

9. Let $f(z)$ be analytic in $|z| \le 1$. Show that $w(f; \delta) \le M\delta$ for some M. Generalize.

10. Let $f(x)$ be periodic and integrable. Define *the moving average of f* by means of

$$f_h(x) = \frac{1}{2h} \int_{x-h}^{x+h} f(t)\, dt.$$

Prove: 1. $f_h(x)$ is periodic.
    2. If $f(x) \in C^n$ then $f_h(x) \in C^{n+1}$.
    3. $w(f_h; \delta) \le w(f; \delta)$ and hence f_h is "smoother" than f.
    4. If f is sufficiently smooth, $(f_h(x))' = (f')_h$.

11. Let $f(x), g(x) \in \text{Lip } \alpha$ on $[a, b]$. Then the same is true of $f(x)\, g(x)$.

12. Does $x^\alpha \log x$, $\alpha > 0$, satisfy a Lipschitz condition on $[0, 1]$?

13. If
$$f(x) \in C^2[a, b],$$
then
$$(a - b)f'(x) = f(a) - f(b) + \tfrac{1}{2}(b - x)^2 f''(\xi_1) - \tfrac{1}{2}(a - x)^2 f''(\xi_2)$$
for $x \in (a, b)$, $a < \xi_2 < x$, $x < \xi_1 < b$.

14. Use the last result to show that
$$M_1 \leq \frac{2M_0}{h} + \tfrac{1}{2}M_2 h$$
where $M_i = \max\limits_{a \leq x \leq b} |f^{(i)}(x)|$ and $h = b - a$. (Hadamard)

15. If $\lim\limits_{h \to 0} \dfrac{f(x + h) - 2f(x) + f(x - h)}{h^2}$ exists, $f(x)$ is said to have a *second Riemann derivative* at x. Use Theorem 1.6.6 to show that if $f''(x)$ exists then the above limit exists and equals it. Show, however, that there are many functions that do not have a second derivative at x but have a second Riemann derivative.

16. Let $f \in C^1[a, b]$ and let $f''(x)$ exist at each point of (a, b). Suppose $f(a) = f'(a) = 0$ and $f(b) = 0$. Then there is a point ξ, $a < \xi < b$ with $f''(\xi) = 0$. Generalize to functions having higher order zeros at k points.

17. If $\lim\limits_{x \to \infty} f(x) = a$ and $\lim\limits_{x \to \infty} f'''(x) = 0$ prove that $\lim\limits_{x \to \infty} f'(x) = 0$ and $\lim\limits_{x \to \infty} f''(x) = 0$.

18. Let $f(x), g(x) \in C^\infty[a, b]$ and $a \leq x_0 \leq b$. If $f^{(n)}(x_0) = 0$, $n = 0, 1, 2, \ldots$, then $\dfrac{d^n}{dx^n}(f(x)g(x))\big|_{x_0} = 0$, $n = 0, 1, \ldots$.

19. If $\sum\limits_{n=0}^{\infty} a_n z^n$, $\sum\limits_{n=0}^{\infty} b_n z^n$ are analytic in $|z| < 1$, so is $\sum\limits_{n=0}^{\infty} a_n b_n z^n$.

20. Use Theorem 1.9.3 to show that $x^{\frac{1}{2}}$ is in $A[a, b]$ for any $0 < a < b < \infty$.

21. Show directly that $e e^x$ satisfies the conditions of Theorem 1.9.3 on $[0, 1]$.

22. Make use of Theorem 1.10.1 to show that $f(z) = \displaystyle\int_0^z e^{t^2}\, dt$ is entire.

23. If $f(z)$ is entire and satisfies $|f(z)| \geq m |z|^n$ for all $|z| > r$, then f is a polynomial and its degree is at least n.

24. Let $f \in \mathscr{P}_2$ and suppose that $f(a) = f'(a) = 0$, $f(b) = 0$ $b \neq a$. Then $f \equiv 0$. In general, if $f \in \mathscr{P}_n$ and has roots of total multiplicity $> n$, then $f \equiv 0$.

25. Prove that 2^x can coincide with a polynomial at only a finite number of points. Is this true when x is replaced by z?

26. Let $f \in A(-\infty, \infty)$ and $f^{(k)}(x) > 0$ $k = 0, 1, \ldots$. Then $f(x)$ cannot coincide with a polynomial infinitely often. Generalize.

27. If $f(x)$ is a polynomial, then $\lim\limits_{n \to \infty} f^{(n)}(x) = 0$ for all x. Is the converse true?

28. The spaces $C^n[a, b]$, $C^\infty[a, b]$, $A[a, b]$, $A(R)$, E are all infinite dimensional.

29. $\mathscr{P}_n$ defined on $[a, b]$ has dimension $n + 1$. What about $\mathscr{P}_n$ defined on a set S consisting of k points?

30. Let A_r designate the set of functions that are analytic in $|z| < r$ but in no disc $|z| < r'$ with $r' > r$. Is A_r a linear space?

31. Let N be the space of all functions that are analytic in $|z| < R$ and have $|z| = R$ as a natural boundary. Is N a linear space?

CHAPTER II

Interpolation

2.1 Polynomial Interpolation. This whole book can be regarded as a theme and variation on two theorems: an interpolation theorem of great antiquity and Weierstrass' approximation theorem of 1885. The simple theorem of polynomial interpolation upon which much practical numerical analysis rests says, in effect, that a straight line can be passed through two points, a parabola through three, a cubic through four, and so on.

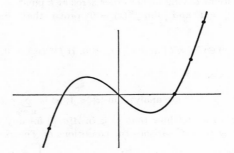

Figure 2.1.1.
Polynomial Interpolation.

THEOREM 2.1.1. *Given $n + 1$ distinct (real or complex) points $z_0, z_1, \ldots, z_n$ and $n + 1$ (real or complex) values $w_0, w_1, \ldots, w_n$. There exists a unique polynomial $p_n(z) \in \mathscr{P}_n$ for which*

$$p_n(z_i) = w_i \qquad i = 0, 1, \ldots, n. \tag{2.1.1}$$

Proof: Set up a polynomial $a_0 + a_1 z + \cdots + a_n z^n$ with $n + 1$ undetermined coefficients a_i. The conditions (2.1.1) lead to the system of $n + 1$ linear equations in the a_i:

$$a_0 + a_1 z_i + \cdots + a_n z_i^n = w_i \qquad i = 0, \ldots, n. \tag{2.1.2}$$

The determinant of the system is the Vandermonde determinant formed from $z_0, \ldots, z_n$:

$$V(z_0, z_1, \ldots, z_n) = \begin{vmatrix} 1 & z_0 & z_0^2 \cdots z_0^n \\ 1 & z_1 & z_1^2 \cdots z_1^n \\ \cdot & \cdot & \cdot & \cdot \\ \cdot & \cdot & \cdot & \cdot \\ \cdot & \cdot & \cdot & \cdot \\ 1 & z_n & z_n^2 \cdots z_n^n \end{vmatrix}. \tag{2.1.3}$$

To evaluate V, we may proceed as follows. Consider the function

$$V(z) = V(z_0, z_1, \ldots, z_{n-1}, z) = \begin{vmatrix} 1 & z_0 & \cdots & z_0^n \\ \cdot & & & \cdot \\ \cdot & & & \cdot \\ \cdot & & & \cdot \\ 1 & z_{n-1} & \cdots & z_{n-1}^n \\ 1 & z & \cdots & z^n \end{vmatrix}. \tag{2.1.4}$$

$V(z)$ is obviously in $\mathscr{P}_n$. Furthermore it vanishes at $z_0, z_1, \ldots, z_{n-1}$, for inserting these values in place of z yields two identical rows in the determinant. Thus,

$$V(z_0, z_1, \ldots, z_{n-1}, z) = A(z - z_0)(z - z_1) \cdots (z - z_{n-1}) \tag{2.1.5}$$

where A depends only on $z_0, z_1, \ldots, z_{n-1}$. To evaluate A, expand the determinant in (2.1.4) by minors of its last row. We then see that the coefficient of z^n is $V(z_0, \ldots, z_{n-1})$. Thus, we have

$$V(z_0, z_1, \ldots, z_{n-1}, z) = V(z_0, \ldots, z_{n-1})(z - z_0)(z - z_1) \cdots (z - z_{n-1}) \tag{2.1.6}$$

and hence we have the recursion formula

$$V(z_0, z_1, \ldots, z_{n-1}, z_n) = V(z_0, \ldots, z_{n-1})(z_n - z_0)(z_n - z_1) \cdots (z_n - z_{n-1}). \tag{2.1.7}$$

Since $V(z_0, z_1) = z_1 - z_0$, we have from (2.1.7),

$$V(z_0, z_1, z_2) = (z_1 - z_0)(z_2 - z_0)(z_2 - z_1)$$

and by multiple applications of (2.1.7),

$$V(z_0, z_1, \ldots, z_n) = \prod_{i > j}^n (z_i - z_j). \tag{2.1.8}$$

By assumption, the points $z_0, z_1, \ldots, z_n$ are distinct. Therefore $V \neq 0$. There is consequently a unique solution to the system (2.1.2).

Here is a second proof that contains a useful line of reasoning. Consider the system (2.1.2). If, when the right-hand side is 0 ($w_i = 0$), the system possesses only the trivial zero solution, Theorem 1.2.2 tells us that its determinant does not vanish. Hence for an arbitrary right-hand side there is one and only one solution. Now a zero right-hand side to (2.1.2) means that

$p_n(z)$ vanishes at $n + 1$ distinct points. By Theorem 1.11.3, $a_k = 0$, $k = 0, 1, \ldots, n$. The homogeneous equation possesses only the trivial solution and the rest follows.

2.2 The General Problem of Finite Interpolation.

In Theorem 2.1.1 we have reconstructed a polynomial $\in \mathscr{P}_n$ on the basis of $n + 1$ values. Can we do it on the basis of $n + 1$ arbitrary pieces of linear information? Can we do it for functions other than polynomials? These questions lead to the following general problem.

Let X be a linear space of dimension n and let $L_1, L_2, \ldots, L_n$ be n given linear functionals defined on X. For a given set of values $w_1, w_2, \ldots, w_n$, can we find an element of X, say x, such that

$$L_i(x) = w_i \qquad i = 1, 2, \ldots, n? \tag{2.2.1}$$

The answer is yes if the L_i are independent in X^*.

LEMMA 2.2.1. *Let X have dimension n. If $x_1, \ldots, x_n$ are independent in X and $L_1, \ldots, L_n$ are independent in X^* then*

$$|L_i(x_j)| \neq 0. \tag{2.2.2}$$

Conversely, if either $x_1, \ldots, x_n$ or $L_1, \ldots, L_n$ are independent and (2.2.2) holds then the other set is also independent.

Proof: Suppose that $|L_i(x_j)| = 0$. Then also $|L_j(x_i)| = 0$. The system

$$a_1 L_1(x_1) + a_2 L_2(x_1) + \cdots + a_n L_n(x_1) = 0$$
$$\cdot \qquad\qquad\qquad\qquad \cdot$$
$$\cdot \qquad\qquad\qquad\qquad \cdot$$
$$\cdot \qquad\qquad\qquad\qquad \cdot$$
$$a_1 L_1(x_n) + a_2 L_2(x_n) + \cdots + a_n L_n(x_n) = 0$$

would have a nontrivial solution $a_1, \ldots, a_n$.
Then,

$$(a_1 L_1 + a_2 L_2 + \cdots + a_n L_n)(x_i) = 0 \qquad i = 1, 2, \ldots, n.$$

Since $x_1, \ldots, x_n$ form a basis for X,

$$(a_1 L_1 + a_2 L_2 + \cdots + a_n L_n)(x) = 0 \qquad x \in X$$

and hence $a_1 L_1 + \cdots + a_n L_n = 0$.
Therefore, $L_1, \ldots, L_n$ are dependent contrary to our assumption.
To show the converse, we may trace the argument backwards.

THEOREM 2.2.2. *Let a linear space X have dimension n and let $L_1, L_2, \ldots, L_n$ be n elements of X^*. The interpolation problem (2.2.1) possesses a solution for arbitrary values $w_1, w_2, \ldots, w_n$ if and only if the L_i are independent in X^*. The solution will be unique.*

Proof: In this generality, the theorem is nothing but a rewording of Theorem 1.2.2. If the L_i are independent and if $x_1, \ldots, x_n$ are independent, then $|L_i(x_j)| \neq 0$ by Lemma 2.2.1. Hence the system

$$L_i(a_1 x_1 + a_2 x_2 + \cdots + a_n x_n) = w_i \qquad i = 1, 2, \ldots, n$$

or

$$a_1 L_i(x_1) + a_2 L_i(x_2) + \cdots + a_n L_i(x_n) = w_i \qquad (2.2.3)$$

possesses a solution $a_1, \ldots, a_n$ and the element $\sum\limits_{i=1}^{n} a_i x_i$ solves the interpolation problem. Conversely, if the problem (2.2.1) has a solution for arbitrary w_i, then (2.2.3) has a solution for arbitrary w_i. By Theorem 1.2.2, this implies that $|L_i(x_j)| \neq 0$ and hence by Lemma 2.2.1, the L_i are independent.

The determinant $|L_i(x_j)|$ is a *generalized Gram determinant* (cf. Chapter 8.7) and its nonvanishing is synonomous with the possibility of solution of the interpolation problem. We may speak of independent systems of functionals as having the "interpolation property." In the next section, we shall study some spaces and functionals for which the interpolation problem can be solved. But before passing to it, we should rid ourselves of the naive hope that an interpolation problem can always be solved providing the number of parameters equals the number of conditions.

Ex. 1.   Let X designate the set of functions of the form $a_0 + a_2 x^2$ defined on $[-1, 1]$. X has dimension 2. If $L_1(f) = f(x_1)$ and $L_2(f) = f(x_2)$, $-1 \leq x_1, x_2 \leq 1$, then the generalized Gram determinant for the independent elements 1, x^2 is

$$\begin{vmatrix} 1 & x_1{}^2 \\ 1 & x_2{}^2 \end{vmatrix} = (x_2 - x_1)(x_2 + x_1).$$

This vanishes if $x_1 = x_2$ or $x_1 = -x_2$. In these cases L_1 and L_2 are not independent. The first case would be excluded trivially, but the second tells us that we cannot force the even functions $a_0 + a_2 x^2$ to take on arbitrary values at distinct points.

Ex. 2.   The strength of Theorem 2.1.1 is brought out by noting that it cannot be extended as it stands to polynomial interpolation in several variables. Let the powers in two real variables be listed as follows: $p_0(x, y) = 1$, $p_1(x, y) = x$, $p_2(x, y) = y$, $p_3(x, y) = x^2$, $p_4(x, y) = xy$, $p_5(x, y) = y^2$, $p_6(x, y) = x^3$, $\ldots$. It is not always possible, having been given n arbitrary distinct points (x_i, y_i), to find a linear combination of $p_0, \ldots, p_{n-1}$ that takes on preassigned values at these points.

2.3 Systems Possessing the Interpolation Property.

Many spaces of functions and related systems of independent functionals are known and have been studied in detail. We shall list some of the more common ones.

Ex. 1.   (Interpolation at discrete points)
$$X = \mathscr{P}_n. \quad L_0(f) = f(z_0), \, L_1(f) = f(z_1), \, \ldots, \, L_n(f) = f(z_n).$$
We assume that $z_i \neq z_j, \, i \neq j$.

Ex. 2.   (Taylor interpolation)
$$X = \mathscr{P}_n. \quad L_0(f) = f(z_0), \, L_1(f) = f'(z_0), \, \ldots, \, L_n(f) = f^{(n)}(z_0).$$

Ex. 3.   (Abel-Gontscharoff Interpolation)
$$X = \mathscr{P}_n. \quad L_0(f) = f(z_0), \, L_1(f) = f'(z_1), \, L_2(f) = f''(z_2), \, \ldots, \, L_n(f) = f^{(n)}(z_n).$$

Ex. 4.   (Lidstone Interpolation)
$$X = \mathscr{P}_{2n+1}. \quad L_1(f) = f(z_0), \, L_2(f) = f(z_1)$$
$$L_3(f) = f''(z_0), \, L_4(f) = f''(z_1)$$

$$\cdot \qquad \qquad \cdot$$
$$\cdot \qquad \qquad \cdot$$

$$L_{2n+1}(f) = f^{(2n)}(z_0), \, L_{2n+2}(f) = f^{(2n)}(z_1), \, (z_0 \neq z_1).$$

Ex. 5.   (Simple Hermite or Osculatory Interpolation)
$$X = \mathscr{P}_{2n-1}. \quad L_1(f) = f(z_1), \, L_2(f) = f'(z_1)$$
$$L_3(f) = f(z_2), \, L_4(f) = f'(z_2)$$

$$\cdot \qquad \qquad \cdot$$
$$\cdot \qquad \qquad \cdot$$

$$L_{2n-1}(f) = f(z_n), \, L_{2n}(f) = f'(z_n), \, (z_i \neq z_j, \, i \neq j).$$

Ex. 6.   (Full Hermite Interpolation)
$X = \mathscr{P}_N$. To avoid indexing difficulties, we list the functional information employed without using the symbol L.
$$f(z_0), f'(z_0), \, \ldots, f^{(m_0)}(z_0)$$
$$f(z_1), f'(z_1), \, \ldots, f^{(m_1)}(z_1)$$

$$\cdot$$
$$\cdot$$

$$f(z_n), f'(z_n), \, \ldots, f^{(m_n)}(z_n)$$
$$(z_i \neq z_j, \, N = m_0 + m_1 + \cdots + m_n + n).$$

Ex. 7.   (Generalized Taylor Interpolation)
X consists of the linear combinations of the $n + 1$ linearly independent functions $\varphi_0(z), \varphi_1(z), \ldots, \varphi_n(z)$ that are analytic at z_0.

$$L_0(f) = f(z_0), \; L_1(f) = f'(z_0), \ldots,$$
$$L_n(f) = f^{(n)}(z_0).$$

$|\varphi_i^{(j)}(z_0)| \neq 0.$

Ex. 8. (Trigonometric Interpolation)

A linear combination of $1, \cos x, \ldots, \cos nx, \sin x, \sin 2x, \ldots, \sin nx$ is known as a *trigonometric polynomial of degree* $\leq n$. The corresponding linear space will be designated by $\mathscr{T}_n$. It has dimension $2n + 1$.

$$X = \mathscr{T}_n. \quad L_0(f) = f(x_0), \; L_1(f) = f(x_1), \ldots, L_{2n}(f) = f(x_{2n}),$$
$$-\pi \leq x_0 < x_1 < \cdots < x_{2n} < \pi.$$

Ex. 9. (Fourier Series)

$$X = \mathscr{T}_n. \quad L_{2k}(f) = \int_{-\pi}^{\pi} f(x) \cos kx \, dx, \, k = 0, 1, \ldots, n.$$

$$L_{2k-1}(f) = \int_{-\pi}^{\pi} f(x) \sin kx \, dx, \, k = 1, 2, \ldots, n.$$

Before demonstrating that these functionals are independent over the respective spaces, a few remarks are in order. Ex. 1 is, of course, Theorem 2.1.1. Exs. 1, 2, 5 are special cases of Ex. 6. Ex. 2 is a special case of Ex. 7 if we select $\varphi_k(z) = z^k$. Ex. 9 is not generally thought of as an interpolation process since the usual interpolatory processes make use of point data. But it—and indeed all orthogonal expansions—fit into the present pattern, and so we have listed it here.

The most direct way to show that the interpolation problem formed from these examples has a solution is to exhibit the solution explicitly. For some of the examples, we shall do this in subsequent sections. But it suffices to show that the generalized Gram determinant does not vanish, (2.2.2), or to apply the Alternative Theorem 1.2.2 directly.

Ex. 6. We shall show that if $p \, \epsilon \, \mathscr{P}_N$ and satisfies

$$p(z_0) = 0, \, p'(z_0) = 0, \ldots, p^{m_0}(z_0) = 0$$
$$p(z_1) = 0, \, p'(z_1) = 0, \ldots, p^{m_1}(z_1) = 0$$
$$\cdot \qquad\qquad\qquad\qquad \cdot$$
$$\cdot \qquad\qquad\qquad\qquad \cdot \qquad\qquad (2.3.1)$$
$$\cdot \qquad\qquad\qquad\qquad \cdot$$
$$p(z_n) = 0, \, p'(z_n) = 0, \ldots, p^{m_n}(z_n) = 0$$

where $N = m_0 + m_1 + \cdots + m_n + n$, then p must vanish identically. By the Factorization Theorem, if p satisfies all conditions of (2.3.1) with the exception of the last, i.e., $p^{m_n}(z_n) = 0$, then we must have

$$p(z) = A(z)(z - z_0)^{m_0+1}(z - z_1)^{m_1+1} \cdots (z - z_{n-1})^{m_{n-1}+1}(z - z_n)^{m_n},$$

$$A(z) = \text{polynomial}.$$

By examining the degree of this product, it appears that $A = $ constant. Since, moreover,

$$p^{(m_n)}(z_n) = A(m_n)! \, (z_n - z_0)^{m_0+1} \cdots (z_n - z_{n-1})^{m_{n-1}+1} = 0$$

and $z_i \neq z_j$, $i \neq j$, we have $A = 0$ and therefore $p \equiv 0$. The homogeneous interpolation problem has the zero solution only and hence the nonhomogeneous problem possesses a unique solution.

Ex. 3.   The generalized Gram determinant is

$$\begin{vmatrix} 1 & z_0 & z_0^2 \cdots z_0^n \\ 0 & 1 & 2z_1 \cdots nz_1^{n-1} \\ 0 & 0 & 2 \cdots n(n-1)z_2^{n-2} \\ \cdot & & \cdot \\ \cdot & & \cdot \\ \cdot & & \cdot \\ 0 & 0 & 0 \cdots n! \end{vmatrix} = 1! \, 2! \cdots n! \neq 0.$$

Ex. 4.   Let $p \in \mathscr{P}_{2n+1}$. If $p^{(2j)}(z_0) = 0$ for $j = 0, 1, \ldots, n$, then by Theorem 1.6.4, $p(z) = a_1(z - z_0) + a_3(z - z_0)^3 + \cdots + a_{2n+1}(z - z_0)^{2n+1}$.   If now, $p^{(2n)}(z_1) = 0$ then $a_{2n+1} = 0$ and $p^{(2j)}(z_1) = 0$, $j = n - 1, n - 2, \ldots, 0$ implies, by recurrence, that the remaining coefficients are 0. The homogeneous problem possesses the 0 solution only, and so the nonhomogeneous problem has a solution and it is unique.

As far as Ex. 7 is concerned, no proof is required, for condition (2.2.2) has been built into the hypothesis. In this example, the crucial determinant reduces to the *Wronskian* of the functions $\phi_0, \ldots, \phi_n$ and we postulate that it does not vanish at z_0.

Ex. 8.   The crucial determinant here is

$$G = \begin{vmatrix} 1 & \cos x_0 & \sin x_0 & \cos 2x_0 & \sin 2x_0 & \cdots & \cos nx_0 & \sin nx_0 \\ 1 & \cos x_1 & \sin x_1 & \cos 2x_1 & \sin 2x_1 & \cdots & \cos nx_1 & \sin nx_1 \\ \cdot & & & & & & & \cdot \\ \cdot & & & & & & & \cdot \\ \cdot & & & & & & & \cdot \\ 1 & \cos x_{2n} & \sin x_{2n} & \cos 2x_{2n} & \sin 2x_{2n} & \cdots & \cos nx_{2n} & \sin nx_{2n} \end{vmatrix}. \qquad (2.3.2)$$

To evaluate G we reduce its elements to complex form. Multiply the 3rd, 5th, . . . columns by i and add them respectively to the 2nd, 4th, . . . columns. We obtain

$$G = |1 \quad e^{ix_j} \quad \sin x_j \quad e^{2ix_j} \quad \sin 2x_j \quad \ldots e^{nix_j} \quad \sin nx_j|.$$

Multiply the 3rd, 5th, . . . columns by $-2i$ and to them add the 2nd, 4th, . . .

columns respectively:

$$(-2i)^n G = |1 \quad e^{ix_j} \quad e^{-ix_j} \quad e^{2ix_j} \quad e^{-2ix_j} \cdots e^{nix_j} \quad e^{-nix_j}|.$$

Interchange the columns:

$$(-1)^{n(n+1)}(-2i)^n G = |e^{-nix_j} e^{-(n-1)ix_j} \cdots 1 \cdots e^{(n-1)ix_j} e^{nix_j}|.$$

Multiply the jth row by e^{nix_j}, $j = 0, \ldots, 2n$:

$$e^{ni(x_0+x_1+\cdots+x_{2n})}(-1)^{n(n+1)}(-2i)^n G = |1 \; e^{ix_j} e^{2ix_j} \cdots e^{2nix_j}|.$$

The determinant in the last line is a Vandermonde. Hence from (2.1.8),

$$e^{ni(x_0+x_1+\cdots+x_{2n})}(-1)^{n(n+1)}(-2i)^n G = \prod_{j>k}^{2n} (e^{ix_j} - e^{ix_k}).$$

In view of the conditions on the x_j, $e^{ix_j} \neq e^{ix_k}$, $j \neq k$, and so $G \neq 0$.

Ex. 9.   In view of the orthogonality of the sines and cosines (Chap. 8.3, Ex. 3), the crucial determinant has positive quantities on the main diagonal and 0's elsewhere and hence does not vanish.

2.4   Unisolvence.   Let the functions $f_1(x)$, $f_2(x)$, $\ldots, f_n(x)$ be defined on an interval I. Given n distinct points $x_1, \ldots, x_n \in I$ and n values $w_1, \ldots, w_n$, we will be able to solve uniquely the interpolation problem

$$\sum_{i=0}^{n} a_i f_i(x_j) = w_j \qquad j = 1, 2, \ldots, n \tag{2.4.1}$$

if and only if

$$|f_i(x_j)| \neq 0. \tag{2.4.2}$$

DEFINITION 2.4.1.   A system of n functions $f_1, \ldots, f_n$ defined on a point set S is called *unisolvent* on S if (2.4.2) holds for every selection of n distinct points lying in S.

Pointwise interpolation can always be carried out uniquely with a unisolvent system.

It follows that $f_1, \ldots, f_n$ is unisolvent on S if and only if the only linear combination of the f's that vanishes on n distinct points of S vanishes identically.

Ex. 1.   The system $1, x^2$ is unisolvent on $[0, 1]$ but not on $[-1, 1]$.

Ex. 2.   The system $1, x, x^2, \ldots, x^n$ is unisolvent over any interval $[a, b]$.

Ex. 3.   Suppose that $w(x)$ does not vanish on $[a, b]$. Then

$$w(x), \; xw(x), \; x^2w(x), \; \ldots, \; x^nw(x)$$

is unisolvent on $[a, b]$.

Ex. 4.   The system of complex powers $1, z, z^2, \ldots, z^n$ is unisolvent over any region.

Ex. 5.   The trigonometric system

$$1, \cos x, \cos 2x, \ldots, \cos nx, \sin x, \sin 2x, \ldots, \sin nx$$

is unisolvent on $-\pi \leq x < \pi$.

Ex. 6.   Let a_i be distinct values not in $[a, b]$. Then the system

$$\frac{1}{x + a_1}, \frac{1}{x + a_2}, \ldots, \frac{1}{x + a_n}$$

is unisolvent in $[a, b]$. For we shall prove in Chap. 11.3 that

$$\left| \frac{1}{x_i + a_j} \right| = \prod_{i>j}^{n} (x_i - x_j)(a_i - a_j) \bigg/ \prod_{i,j=1}^{n} (x_i + a_j).$$

As far as functions of one variable are concerned, unisolvent systems are reasonably plentiful. In several dimensions, the situation is vastly different. We have already had a hint of this in 2.2, Ex. 2 where we noticed that the fundamental theorem of polynomial interpolation does not go over directly to several variables.

THEOREM 2.4.1 (Haar).   *Let S be a point set in a Euclidean space of n-dimension, R_n, $n \geq 2$. Suppose that S contains an interior point p. Let $f_1, f_2, \ldots, f_n$ $(n > 1)$ be defined on S and continuous in a neighborhood of p. Then this set of functions cannot be unisolvent on S.*

Proof: Let U be a ball with center at p and contained in S and sufficiently small so that the f_i are continuous in U. Select n distinct points $p_1, p_2, \ldots, p_n \in U$. We may assume that $|f_i(p_j)| \neq 0$, for otherwise the system is surely not unisolvent. Hold the points $p_3, p_4, \ldots, p_n$ fixed. Now move the points p_1 and p_2 continuously through U in such a manner that the positions of p_1 and p_2 are interchanged. Since U has dimension ≥ 2, it is clear that this can be carried out in such a manner that p_1 and p_2 coincide neither with one another nor with the remaining points. In this way we induce an interchange of two columns of the determinant $|f_i(p_j)|$. Its sign therefore changes. Since the functions are continuous, there must be some intermediate position of p_1 and p_2 for which the value of the determinant is zero.

In order to carry out this argument, it is not necessary to have an interior point. It suffices if the set S contains a "ramification point;" that is to say, a point p at which three arcs meet. Then by a process of "train switching" we may carry out the same argument. It is surprising that unisolvence has this topological aspect.

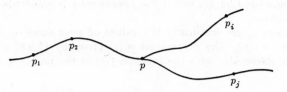

Figure 2.4.1.

2.5 Representation Theorems: The Lagrange Formula. Let z_0, $z_1, \ldots, z_n$ be distinct and introduce the following polynomials of degree n:

$$l_k(z) = \frac{(z - z_0)(z - z_1) \cdots (z - z_{k-1})(z - z_{k+1}) \cdots (z - z_n)}{(z_k - z_0)(z_k - z_1) \cdots (z_k - z_{k-1})(z_k - z_{k+1}) \cdots (z_k - z_n)},$$

$$k = 0, 1, \ldots, n. \quad (2.5.1)$$

It is clear that

$$l_k(z_j) = \delta_{kj} = \begin{cases} 0 \text{ if } k \neq j \\ 1 \text{ if } k = j \end{cases}. \quad (2.5.2)$$

For given values $w_0, w_1, \ldots, w_n$, the polynomial

$$p_n(z) = \sum_{k=0}^{n} w_k l_k(z) \quad (2.5.3)$$

is in $\mathscr{P}_n$ and takes on these values at the points z_i:

$$p_n(z_k) = w_k \qquad k = 0, 1, \ldots, n. \quad (2.5.4)$$

Formula (2.5.3) is the *Lagrange Interpolation Formula*. Since the interpolation problem (2.5.4) has a unique solution, all other representations of the solution must, upon rearrangement of terms, coincide with the Lagrange polynomial.

An alternate form is useful. Introduce

$$w(z) = (z - z_0)(z - z_1) \cdots (z - z_n). \quad (2.5.5)$$

Then,

$$w'(z_k) = (z_k - z_0)(z_k - z_1) \cdots (z_k - z_{k-1})(z_k - z_{k+1}) \cdots (z_k - z_n) \quad (2.5.6)$$

and hence from (2.5.1),

$$l_k(z) = \frac{w(z)}{(z - z_k)w'(z_k)}. \quad (2.5.7)$$

The formula (2.5.3) becomes

$$p_n(z) = \sum_{k=0}^{n} w_k \frac{w(z)}{(z - z_k)w'(z_k)}. \quad (2.5.8)$$

The polynomials $l_k(z)$ are called the *fundamental* polynomials for pointwise interpolation.

The numbers w_i are frequently the values of some function $f(z)$ at the points z_i: $w_i = f(z_i)$. The polynomial $p_n(z)$ given by (2.5.8) and formed with these w's coincides with the function $f(z)$ at the points $z_0, z_1, \ldots, z_n$. That is, if

$$p_n(z) = \sum_{k=0}^{n} f(z_k)l_k(z) = \sum_{k=0}^{n} f(z_k) \frac{w(z)}{(z - z_k)w'(z_k)}, \qquad (2.5.9)$$

then

$$p_n(z_k) = f(z_k) \qquad k = 0, 1, \ldots, n. \qquad (2.5.10)$$

DEFINITION 2.5.1.   We shall designate the unique polynomial of class $\mathscr{P}_n$ that coincides with f at $z_0, \ldots, z_n$ by $p_n(f; z)$.

Suppose that $q(z) \in \mathscr{P}_n$. Then q is uniquely determined by the $n + 1$ values $q(z_i)$, $i = 0, \ldots, n$. Hence we must have

$$p_n(q; z) \equiv q(z). \qquad (2.5.11)$$

Now take $q(z) = (z - u)^j$, $j = 0, 1, \ldots, n$ and regard u as an independent variable. From (2.5.11) and (2.5.9),

$$(z - u)^j = \sum_{k=0}^{n} (z_k - u)^j l_k(z) \qquad j = 0, 1, \ldots, n \qquad (2.5.12)$$

holding identically in z and u.

By selecting $u = z$ we obtain

$$\sum_{k=0}^{n} l_k(z) \equiv 1 \qquad (2.5.13)$$

$$\sum_{k=0}^{n} (z_k - z)^j l_k(z) \equiv 0, \qquad j = 1, 2, \ldots, n.$$

The $n + 1$ identities (2.5.13) are the *Cauchy relations* for the fundamental polynomials $l_k(z)$.

The importance of the fundamental polynomials lies in the identity (2.5.2) and the resulting simple explicit solution (2.5.9) of the interpolation problem. If we set

$$L_0(f) = f(z_0), \ L_1(f) = f(z_1), \ldots, L_n(f) = f(z_n),$$

then (2.5.2) can be written as

$$L_i(l_j) = \delta_{ij}. \qquad (2.5.14)$$

In anticipation of certain geometric developments in Chapter VIII, we will say that the polynomials $l_i(z)$ and the functionals L_i are *biorthonormal*. For a given set of independent functionals, we can always find a related biorthonormal set of polynomials. Indeed, we have the following generalization of Lagrange's formula.

THEOREM 2.5.1.   *Let* X *be a linear space of dimension* n. *Let* $L_1, L_2, \ldots,$ L_n *be* n *independent functionals in* X^*. *Then, there are determined uniquely* n *independent elements of* X, $x_1^*, x_2^*, \ldots, x_n^*$, *such that*

$$L_i(x_j^*) = \delta_{ij}. \tag{2.5.15}$$

For any $x \in X$ *we have*

$$x = \sum_{i=1}^{n} L_i(x) x_i^*. \tag{2.5.16}$$

For every choice of $w_1, \ldots, w_n$, *the element*

$$x = \sum_{i=1}^{n} w_i x_i^* \tag{2.5.17}$$

is the unique solution of the interpolation problem

$$L_i(x) = w_i, \qquad i = 1, 2, \ldots, n. \tag{2.5.18}$$

Proof: Let $x_1, \ldots, x_n$ be a basis for X. By Lemma 2.2.1, $|L_i(x_j)| \neq 0$. If we set $x_j^* = a_{j1}x_1 + \cdots + a_{jn}x_n$, then this determinant condition guarantees that the system (2.5.15) can be solved for a_{ji} to produce a set of elements $x_1^*, \ldots, x_n^*$. By Theorem 2.2.2, the solution to the interpolation problem (2.5.15) is unique, for each j, and by Lemma 2.2.1, the x_i^* are independent.

Denote $y = \sum_{i=1}^{n} L_i(x) x_i^*$. Then $L_j(y) = \sum_{i=1}^{n} L_i(x) L_j(x_i^*)$. Hence, by (2.5.15), $L_j(y) = L_j(x)$, $j = 1, 2, \ldots, n$. Again, since interpolation with the n conditions L_i is unique, $y = x$ and this establishes (2.5.16). Equation (2.5.18) is established similarly.

In this theorem and throughout the remainder of the book an asterisk (*) will be applied to the symbol of an element whenever the element is one of a *biorthonormal* or an *orthonormal* set. (Cf. Def. 8.3.1.) An asterisk on the symbol of a space will be used to denote the *conjugate space*. (Cf. Def. 1.12.3.)

The solution to the interpolation problem (2.5.18) can be given in determinantal form.

THEOREM 2.5.2.   *Let the hypotheses of Theorem 2.5.1 hold and let* $x_1, \ldots,$ x_n *be a basis for* X. *If* $w_1, \ldots, w_n$ *are arbitrary numbers then the element*

$$x = -\frac{1}{G} \begin{vmatrix} 0 & x_1 & x_2 & \cdots & x_n \\ w_1 & L_1(x_1) & L_1(x_2) & \cdots & L_1(x_n) \\ \cdot & \cdot & \cdot & & \cdot \\ \cdot & \cdot & \cdot & & \cdot \\ \cdot & \cdot & \cdot & & \cdot \\ w_n & L_n(x_1) & L_n(x_2) & \ldots & L_n(x_n) \end{vmatrix} \tag{2.5.19}$$

satisfies $L_i(x) = w_i$, $i = 1, 2, \ldots, n$.

Proof: It is clear that x is a linear combination of $x_1, \ldots, x_n$ and hence is in X. Furthermore, we have

$$L_j(x) = -\frac{1}{G}\begin{vmatrix} 0 & L_j(x_1) & L_j(x_2) & \cdots & L_j(x_n) \\ w_1 & L_1(x_1) & L_1(x_2) & \cdots & L_1(x_n) \\ \cdot & \cdot & \cdot & & \cdot \\ \cdot & \cdot & \cdot & & \cdot \\ \cdot & \cdot & \cdot & & \cdot \\ w_j & L_j(x_1) & L_j(x_2) & \cdots & L_j(x_n) \\ \cdot & \cdot & \cdot & & \cdot \\ \cdot & \cdot & \cdot & & \cdot \\ \cdot & \cdot & \cdot & & \cdot \\ w_n & L_n(x_1) & L_n(x_2) & \cdots & L_n(x_n) \end{vmatrix}.$$

Expand this determinant by minors of the 1st column. The minor of each nonzero element, with the exception of w_j, is 0, for it contains two identical rows. The cofactor of w_j is $-G$. Hence, $L_j(x) = w_j$, $j = 1, 2, \ldots, n$.

Ex. 1.   (Taylor Interpolation)

The polynomials $\dfrac{z^n}{n!}$, $n = 0, 1, \ldots$, and the functionals $L_n(f) = f^{(n)}(0)$, $n = 0, 1, \ldots$, are biorthonormal.

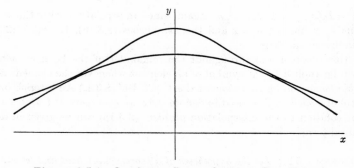

Figure 2.5.1   Osculatory Interpolation at Two Points

$$p(x) = \tfrac{3}{4} - \tfrac{1}{4}x^2, \qquad f(x) = \frac{1}{1 + x^2}$$

$$p(+1) = f(+1), \qquad p'(+1) = f'(+1)$$

$$p(-1) = f(-1), \qquad p'(-1) = f'(-1).$$

Ex. 2.   (Osculatory Interpolation)

Set $w(z) = (z - z_1)(z - z_2) \cdots (z - z_n)$, $l_k(z) = \dfrac{w(z)}{(z - z_k)w'(z_k)}$.

The polynomials $\left[1 - \dfrac{w''(z_k)}{w'(z_k)}(z - z_k)\right] l_k{}^2(z)$, $(z - z_k)l_k{}^2(z)$ of degree $2n - 1$ and the functionals

$$L_k(f) = f(z_k), \quad M_k(f) = f'(z_k), \qquad k = 1, 2, \ldots, n$$

are biorthonormal.

The resulting expansion of type (2.5.17) is, therefore,

$$p_{2n-1}(z) = \sum_{k=1}^{n} w_k \left[1 - \frac{w''(z_k)}{w'(z_k)}(z - z_k)\right] l_k{}^2(z) + \sum_{k=1}^{n} w_k'(z - z_k)l_k{}^2(z), \qquad (2.5.20)$$

and produces the unique element of $\mathscr{P}_{2n-1}$ which solves the "osculatory" interpolation problem

$$\begin{aligned} p(z_k) &= w_k \\ p'(z_k) &= w_k' \end{aligned} \qquad k = 1, 2, \ldots, n. \qquad (2.5.21)$$

Ex. 3. (Two Point Taylor Interpolation)

Let a and b be distinct points. The polynomial

$$p_{2n-1}(z) = (z - a)^n \sum_{k=0}^{n-1} \frac{B_k(z - b)^k}{k!} + (z - b)^n \sum_{k=0}^{n-1} \frac{A_k(z - a)^k}{k!} \qquad (2.5.22)$$

$$\begin{aligned} A_k &= \frac{d^k}{dz^k}\left[\frac{f(z)}{(z - b)^n}\right]_{z=a} \\ B_k &= \frac{d^k}{dz^k}\left[\frac{f(z)}{(z - a)^n}\right]_{z=b} \end{aligned} \qquad (2.5.23)$$

is the unique solution in $\mathscr{P}_{2n-1}$ of the interpolation problem

$$\begin{aligned} p_{2n-1}(a) &= f(a),\, p'_{2n-1}(a) = f'(a),\, \ldots,\, p_{2n-1}^{(n-1)}(a) = f^{(n-1)}(a) \\ p_{2n-1}(b) &= f(b),\, p'_{2n-1}(b) = f'(b),\, \ldots,\, p_{2n-1}^{(n-1)}(b) = f^{(n-1)}(b). \end{aligned} \qquad (2.5.24)$$

Ex. 4. Exs. 1, 2, 3 are, of course, special cases of the general Hermite interpolation problem. (Cf. Ex. 6, 2.3.) Let $z_1, z_2, \ldots, z_n$ be n distinct points, $\alpha_1, \ldots, \alpha_n$ be n integers ≥ 1 and $N = \alpha_1 + \alpha_2 + \cdots + \alpha_n - 1$. Set $w(z) = \prod\limits_{i=1}^{n}(z - z_i)^{\alpha_i}$ and

$$l_{ik}(z) = w(z)\frac{(z - z_i)^{k-\alpha_i}}{k!}\frac{d^{(\alpha_i-k-1)}}{dz^{(\alpha_i-k-1)}}\left[\frac{(z - z_i)^{\alpha_i}}{w(z)}\right]_{z=z_i} \qquad (2.5.25)$$

$$p_N(z) = \sum_{i=1}^{n} r_i l_{i0}(z) + \sum_{i=1}^{n} r_i' l_{i1}(z) + \cdots + \sum_{i=1}^{n} r_i^{(\alpha_i-1)}l_{i\alpha_i-1}(z) \qquad (2.5.26)$$

is the unique member of $\mathscr{P}_N$ for which

$$p_N(z_1) = r_1,\, p'_N(z_1) = r_1',\, \ldots,\, p_N^{(\alpha_1-1)}(z_1) = r_1^{(\alpha_1-1)}$$

$$\begin{matrix} \cdot & & \cdot \\ \cdot & & \cdot \\ \cdot & & \cdot \end{matrix} \qquad (2.5.27)$$

$$p_N(z_n) = r_n,\, p'_N(z_n) = r_n',\, \ldots,\, p_N^{(\alpha_n-1)}(z_n) = r_n^{(\alpha_n-1)}.$$

Ex. 5.   Given the $2n + 1$ points

$$-\pi \leq x_0 < x_1 < \cdots < x_{2n} < \pi.$$

Construct the functions $t_j(x) = \prod_{\substack{k=0 \\ k \neq j}}^{2n} \sin \tfrac{1}{2}(x - x_k) \Big/ \prod_{\substack{k=0 \\ k \neq j}}^{2n} \sin \tfrac{1}{2}(x_j - x_k)$, $j = 0$, $1, \ldots, 2n$. If $L_j(f) = f(x_j)$, then t_j and L_j are biorthonormal.

Each function $t_j(x)$ is a linear combination of $1, \cos x, \ldots, \cos nx, \sin x, \ldots,$ $\sin nx$ and hence is an element of $\mathcal{T}_n$.

To show this, observe that the numerator of t_j is the product of $2n$ factors of the form $\sin \tfrac{1}{2}(x - x_k) = \alpha e^{ix/2} + \beta e^{-ix/2}$ for appropriate constants α and β. The product is therefore of the form $\sum_{k=-n}^{n} c_k e^{ikx}$, and is a combination of the required form. The function

$$T(x) = \sum_{k=0}^{2n} w_k t_k(x) \tag{2.5.28}$$

is therefore an element of $\mathcal{T}_n$ and is the unique solution of the interpolation problem $T(x_k) = w_k$, $k = 0, 1, \ldots, 2n$. Formula (2.5.28) is known as the *Gauss formula of trigonometric interpolation*.

Ex. 6.   Given $n + 1$ distinct points

$$0 \leq x_0 < x_1 < \cdots < x_n < \pi. \text{ Set}$$

$$C_j(x) = \prod_{\substack{k=0 \\ k \neq j}}^{n} (\cos x - \cos x_k) \Big/ \prod_{\substack{k=0 \\ k \neq j}}^{n} (\cos x_j - \cos x_k). \tag{2.5.29}$$

Then C_j is a cosine polynomial of order $\leq n$ (i.e., a function of the form $\sum_{k=0}^{n} a_k \cos kx$) for which $C_j(x_k) = \delta_{jk}$. Given $n + 1$ distinct values $w_0, w_1, \ldots, w_n$ there is a unique cosine polynomial of order $\leq n$, $C(x)$, for which $C(x_k) = w_k$, $k = 0$, $1, \ldots, n$. It is

$$C(x) = \sum_{k=0}^{n} w_k C_k(x). \tag{2.5.30}$$

Ex. 7.   Given n distinct points $0 < x_1 < \cdots < x_n < \pi$. Set

$$S_j(x) = \sin x \prod_{\substack{k=1 \\ k \neq j}}^{n} (\cos x - \cos x_k) \Big/ \prod_{\substack{k=1 \\ k \neq j}}^{n} (\cos x_j - \cos x_k). \tag{2.5.31}$$

Then $S_j(x)$ is a sine polynomial of order $\leq n$ for which $S_j(x_k) = \delta_{jk}$. Given n distinct values $w_1, w_2, \ldots, w_n$, there is a unique sine polynomial of order $\leq n$, $S(x)$, for which $S(x_k) = w_k$ and it is

$$S(x) = \sum_{k=1}^{n} w_k S_k(x). \tag{2.5.32}$$

Ex. 8.   Let $z_0, z_1, \ldots, z_n$ be $n + 1$ distinct real (or complex) points. Let $w_0, w_1, \ldots, w_m$ be a second such set of $m + 1$ points. Set

$$P(z) = (z - z_0) \cdots (z - z_n),$$
$$Q(w) = (w - w_0) \cdots (w - w_m),$$
$$P_j(z) = P(z)/(z - z_j),$$
$$Q_k(w) = Q(w)/(w - w_k).$$

The $(m + 1)(n + 1)$ polynomials

$$l_{jk}(z, w) = \frac{P_j(z)Q_k(w)}{P_j(z_j)Q_k(w_k)} \qquad (2.5.33)$$

satisfy

$$l_{jk}(z_r, w_s) = \delta_{jr}\delta_{ks}. \qquad (2.5.34)$$

Hence

$$p(z, w) = \sum_{j=0}^{n} \sum_{k=0}^{m} \mu_{jk} l_{jk}(z, w) \qquad (2.5.35)$$

is a polynomial of degree $\leq mn$ which satisfies the $(m + 1)(n + 1)$ interpolation conditions

$$p(z_j, w_k) = \mu_{jk} \qquad \begin{array}{l} j = 0, 1, \ldots, n \\ k = 0, 1, \ldots, m. \end{array} \qquad (2.5.36)$$

Formula (2.5.35) may be regarded as the generalization of the Lagrange formula to two dimensions. Extensions to any number of variables will follow in a similar fashion. It shows that by taking a sufficiently large number of powers of several variables polynomial interpolation can be achieved.

2.6   Representation Theorems: The Newton Formula.   The La-
grange formula (2.5.3) or (2.5.17) has one drawback. If we desire to pass from a space of dimension n to a space of one higher dimension, we must determine an entirely new set of elements $y_1^*, y_2^*, \ldots, y_{n+1}^*$ that are not related in a simple fashion to the old set $x_1^*, x_2^*, \ldots, x_n^*$. A representation of Newton gets around this difficulty by taking linear combinations of both the basis elements $x_1, x_2, \ldots,$ and the prescribed functionals $L_1, L_2, \ldots$. We shall first study this representation in the case of polynomial interpolation.

Let $z_0, \ldots, z_n$ be $n + 1$ distinct points and form the $n + 1$ independent Newton polynomials

$$1, \quad z - z_0, \quad (z - z_0)(z - z_1), \ldots, (z - z_0)(z - z_1) \cdots (z - z_{n-1}).$$

For given values $w_0, w_1, \ldots, w_n$ there is a unique member of $\mathscr{P}_n$ for which $p(z_i) = w_i$, $i = 0, 1, \ldots, n$. Let us see if we can represent it in the form

$$p(z) = a_0 + a_1(z - z_0) + a_2(z - z_0)(z - z_1) + \cdots$$
$$+ a_n(z - z_0)(z - z_1) \cdots (z - z_{n-1}). \qquad (2.6.1)$$

To determine the constants a_i, set $z = z_0$, $z = z_1, \ldots$, successively, and solve the resulting linear equations:

$$a_0 = w_0$$

$$a_1 = \frac{w_1 - w_0}{z_1 - z_0}$$

$$a_2 = \frac{1}{z_2 - z_1}\left(\frac{w_2 - w_0}{z_2 - z_0} - \frac{w_1 - w_0}{z_1 - z_0}\right). \tag{2.6.2}$$

$$\vdots$$

Note that for a fixed set of points $z_0, \ldots, z_n$, each a_i is a certain linear combination of the w_i, and that, furthermore, a_0 involves only w_0 and z_0; a_1 involves only w_0, w_1, z_0, z_1; a_2 involves only $w_0, w_1, w_2, z_0, z_1, z_2$, etc.

DEFINITION 2.6.1.   The constant a_j is called the *divided difference of the jth order of* $w_0, w_1, \ldots, w_j$ *with respect to* $z_0, z_1, \ldots, z_j$. *It is designated by*

$$a_j = [w_0, w_1, \ldots, w_j]. \tag{2.6.3}$$

A compact formula for a_j can be found by comparing (2.6.1) with the Lagrange formula (2.5.8) with which it must coincide. The coefficient of z^n in (2.6.1) is a_n. The coefficient of z^n in (2.5.8) is seen to be $\sum_{k=0}^{n} \frac{w_k}{w'(z_k)}$. Therefore,

$$a_n = [w_0, w_1, \ldots, w_n] = \sum_{k=0}^{n} \frac{w_k}{w'(z_k)} \tag{2.6.4}$$

where $w(z) = w_n(z) = (z - z_0)(z - z_1) \cdots (z - z_n)$. Thus, again, from (2.6.4),

$$a_0 = w_0$$

$$a_1 = \frac{w_0}{z_0 - z_1} + \frac{w_1}{z_1 - z_0}$$

$$a_2 = \frac{w_0}{(z_0 - z_1)(z_0 - z_2)} + \frac{w_1}{(z_1 - z_0)(z_1 - z_2)} + \frac{w_2}{(z_2 - z_0)(z_2 - z_1)} \tag{2.6.5}$$

$$\vdots$$

If the w_i are taken as the value of a function f at z_i: $f(z_i) = w_i$, then we may combine (2.6.1) and (2.6.3) to obtain

$$p_n(f; z) = \sum_{k=0}^{n} [f(z_0), f(z_1), \ldots, f(z_k)](z - z_0)(z - z_1) \cdots (z - z_{k-1}). \tag{2.6.6}$$

$$w_{-1}(z) \equiv 1.$$

This form of the interpolating polynomial is the *finite Newton series* for a function $f(z)$.

With $z_0, z_1, \ldots, z_n$ fixed, introduce the linear functionals

$$L_0(f) = f(z_0)$$

$$L_1(f) = \frac{f(z_0)}{z_0 - z_1} + \frac{f(z_1)}{z_1 - z_0} \qquad (2.6.7)$$

.

.

.

according to the scheme in (2.6.5). Then (2.6.6) becomes

$$p_n(f; z) = \sum_{k=0}^{n} L_k(f) w_{k-1}(z). \qquad (2.6.8)$$

Since $w_j(z) \in \mathscr{P}_n$ if $0 \le j \le n - 1$ it follows that $w_j(z) = p_n(w_j(z); z)$ and hence

$$w_j(z) \equiv \sum_{k=0}^{n} L_k(w_j(z)) w_{k-1}(z) \qquad (2.6.9)$$

By setting $j = 0, 1, 2, \ldots$, in (2.6.9) successively we obtain

$$L_k(w_{j-1}(z)) = \delta_{kj}. \qquad (2.6.10)$$

Comparing the biorthogonality relationships (2.6.10) and (2.5.14) our introductory remarks become clear. Whereas $w_k(z)$ depends only on the points $z_0, \ldots, z_k$, $l_k(z) = l_{k,n}(z)$ depends upon all the points $z_0, \ldots, z_k, \ldots, z_n$. In the Lagrange representation, we add an additional point and increase the degree of the interpolating polynomial at the cost of changing all the fundamental polynomials. In the Newton representation, this can be accomplished by adding one more term. The Newton representation has a *permanence property*, and this is characteristic of Fourier series and other orthogonal and biorthogonal expansions. (See 8.5.) The price that is paid for the convenience of the permanence property is that the multipliers of the individual polynomials are no longer simple values at a point, but certain linear combinations of these values.

This type of biorthonormality and permanence can be obtained in a general setting.

THEOREM 2.6.1 (Biorthonormality Theorem or Generalized Newton Representation). *Let X be a linear space of infinite dimension. Let $x_1, x_2, \ldots$ be a sequence of elements of X such that for each n, $x_1, \ldots, x_n$ are independent, Suppose, further, that $L_1, L_2, \ldots$, is a sequence of linear functionals in X^* such that for each n, the $n \times n$ determinant*

$$|L_i(x_j)|_{i,j=1}^{n} \ne 0. \qquad (2.6.11)$$

Then there is determined uniquely two triangular systems of constants a_{ij}, b_{ij} with $a_{ii} \neq 0$ such that if

$$
\begin{aligned}
L_1{}^* &= a_{11}L_1 & x_1{}^* &= x_1 \\
L_2{}^* &= a_{21}L_1 + a_{22}L_2 & x_2{}^* &= b_{21}x_1 + x_2 \\
L_3{}^* &= a_{31}L_1 + a_{32}L_2 + a_{33}L_3 & x_3{}^* &= b_{31}x_1 + b_{32}x_2 + x_3
\end{aligned}
\tag{2.6.12}
$$

$$
\begin{array}{cc}
\cdot & \cdot \\
\cdot & \cdot \\
\cdot & \cdot \\
\cdot & \cdot
\end{array}
$$

we have

$$
L_i{}^*(x_j{}^*) = \delta_{ij}, \quad i,j = 1, 2, \cdots . \tag{2.6.13}
$$

Proof: We want $L_1{}^*(x_1{}^*) = 1$. Therefore, $a_{11}L_1(x_1) = 1$ or

$$
a_{11} = (L_1(x_1))^{-1} \neq 0.
$$

The denominator does not vanish by (2.6.11). We shall now carry out an inductive proof. Assume that we have already determined

$$
\begin{array}{cccc}
a_{11} & & & \\
a_{21} & a_{22} & & \\
\cdot & \cdot & & \\
\cdot & \cdot & & \\
\cdot & \cdot & & \\
a_{n1} & a_{n2} & \cdots & a_{nn}
\end{array}
\qquad
\begin{array}{ccccc}
1 & & & & \\
b_{21} & 1 & & & \\
\cdot & \cdot & & & \\
\cdot & \cdot & & & \\
\cdot & \cdot & & & \\
b_{n1} & b_{n2} & \cdots & b_{n,n-1} & 1
\end{array}
$$

with $a_{11}a_{22} \cdots a_{nn} \neq 0$ and such that

$$
L_i{}^*(x_j{}^*) = \delta_{ij}, \quad i,j = 1, 2, \ldots, n. \tag{2.6.14}
$$

We will show that we can obtain first $b_{n+1,1}, b_{n+1,2}, \ldots, b_{n+1,n}, 1$, and from a knowledge of these values can then obtain $a_{n+1,1}, a_{n+1,2}, \ldots, a_{n+1,n+1}$ with $a_{n+1,n+1} \neq 0$ and such that

$$
L_i{}^*(x_j{}^*) = \delta_{ij}, \qquad i,j = 1, 2, \ldots, n+1. \tag{2.6.15}
$$

The conditions included in (2.6.15) that are not already contained in (2.6.14) are

$$
\begin{aligned}
L_i{}^*(x_{n+1}^*) &= 0 & i &= 1, 2, \ldots, n \quad \text{and} \\
L_{n+1}^*(x_i{}^*) &= 0 & i &= 1, 2, \ldots, n, \\
L_{n+1}^*(x_{n+1}^*) &= 1. &&
\end{aligned}
\tag{2.6.16}
$$

The first n equations in (2.6.16) yield the system

$$b_{n+1,1}L_1^*(x_1) + b_{n+1,2}L_1^*(x_2) + \cdots + b_{n+1,n}L_1^*(x_n) = -L_1^*(x_{n+1})$$

$$\vdots$$

$$b_{n+1,1}L_n^*(x_1) + b_{n+1,2}L_n^*(x_2) + \cdots + b_{n+1,n}L_n^*(x_n) = -L_n^*(x_{n+1}).$$

This system has a unique solution providing $|L_i^*(x_j)|_{i,j=1}^n \neq 0$. But from Problem 19, Chapter II,

$$|L_i^*(x_j)| = \begin{vmatrix} a_{11} & 0 & \cdots & 0 \\ a_{21} & a_{22} & \cdots & 0 \\ \cdot & & & \cdot \\ \cdot & & & \cdot \\ \cdot & & & \cdot \\ a_{n1} & a_{n2} & \cdots & a_{nn} \end{vmatrix} \cdot |L_i(x_j)|_{i,j=1}^n$$

$$= a_{11}a_{22}\cdots a_{nn}\, |L_i(x_j)|_{i,j=1}^n \neq 0.$$

Having determined the b's, or equivalently x_{n+1}^*, consider the second group of $n+1$ equations in (2.6.16). This yields

$$a_{n+1,1}L_1(x_1^*) + \cdots + a_{n+1,n+1}L_{n+1}(x_1^*) = 0$$

$$\vdots$$

$$a_{n+1,1}L_1(x_n^*) + \cdots + a_{n+1,n+1}L_{n+1}(x_n^*) = 0$$
$$a_{n+1,1}L_1(x_{n+1}^*) + \cdots + a_{n+1,n+1}L_{n+1}(x_{n+1}^*) = 1$$

This system has a unique solution providing that

$$|L_i(x_j^*)|_{i,j=1}^{n+1} \neq 0.$$

But, again by Problem 19,

$$|L_i(x_j^*)|_{i,j=1}^{n+1} = \begin{vmatrix} 1 & 0 & \cdots & 0 \\ b_{21} & 1 & \cdots & 0 \\ \cdot & & & \cdot \\ \cdot & & & \cdot \\ \cdot & & & \cdot \\ b_{n+1,1} & b_{n+2,1} & \cdots & 1 \end{vmatrix} \cdot |L_i(x_j)|_{i,j=1}^{n+1}$$

$$= |L_i(x_j)|_{i,j=1}^{n+1} \neq 0.$$

Furthermore, $a_{n+1,n+1} = a_{11}a_{22} \cdots a_{nn} |L_i(x_j)|_{i,j=1}^n / |L_i(x_j)|_{i,j=1}^{n+1} \neq 0$. We observe finally that at no stage is there any arbitrariness in our determination of the constants and hence the solution is unique.

COROLLARY 2.6.2. *Let X_n designate the subspace of X spanned by $x_1, \ldots,$* x_n (*i.e., the set of all linear combinations $a_1x_1 + \cdots + a_nx_n$). If $x \in X_n$ then*

$$x = \sum_{k=1}^n L_k{}^*(x)x_k{}^*.$$

Proof: If $y = \sum_{k=1}^n L_k{}^*(x)x_k{}^*$ then $L_j{}^*(y) = \sum_{k=1}^n L_k{}^*(x)L_j{}^*(x_k{}^*) = L_j{}^*(x)$ by (2.6.13). Since $x_1, \ldots, x_n$ are independent, it follows from (2.6.12) that $x_1{}^*, \ldots, x_n{}^*$ are independent. Hence from (2.6.13) and Lemma 2.2.1 it follows that $L_1, \ldots, L_n$ and consequently $L_1{}^*, \ldots, L_n{}^*$ are independent. In view of this, $L_j{}^*(y - x) = 0$, $j = 1, 2, \ldots, n$ implies $y = x$.

For a given $x \in X$, the formal series

$$x \sim \sum_{k=1}^\infty L_k{}^*(x)x_k{}^* \tag{2.6.17}$$

is a *biorthogonal expansion* of the element x. In particular cases, the relation of the series to x has been the object of vast investigations.

It will help to grasp the difference between the biorthonormality of Lagrange type (2.5.15) and that of Newton type (2.6.15) if each is expressed in the language of matrices. Let G designate the matrix $(L_i(x_j))$. Let A designate the matrix (a_{ij}) where the a_{ij} are the quantities appearing in the proof of Theorem 2.5.1. I is the unit matrix. Then, (2.5.15) may be expressed as

$$GA' = I, \quad A' = \text{transpose of } A. \tag{2.6.18}$$

On the other hand, if A and B designate the lower triangular matrices taken from the coefficient scheme of (2.6.12) then

$$AGB' = I. \tag{2.6.19}$$

Note that (2.6.19) is equivalent to

$$G = A^{-1}(B')^{-1}. \tag{2.6.20}$$

Now, A^{-1} is a lower triangular matrix with non-zero elements on its principal diagonal and $(B')^{-1}$ is an upper triangular matrix with 1's on its principal diagonal. (2.6.20) has a matrix formulation. A matrix $G = (g_{ij})$ is said to be

regularly arranged if none of its principal minors $\left(\text{i.e., } g_{11}, \begin{vmatrix} g_{11} & g_{12} \\ g_{21} & g_{22} \end{vmatrix}, \ldots \right)$

vanishes. If G is regularly arranged, then it can be expressed as the product of a lower triangular matrix by an upper triangular matrix with 1's on its principal diagonal. This is known as an *LU-decomposition* of G.

The result of biorthogonalization can be expressed by means of determinants.

THEOREM 2.6.3. *With the notation of the previous theorem, let*

$$G_r = |L_i(x_j)|_{i,j=1}^r \; .$$

Then,

$$x_j{}^* = \frac{1}{G_{j-1}} \begin{vmatrix} L_1(x_1) & L_1(x_2) & \cdots & L_1(x_j) \\ \cdot & \cdot & & \cdot \\ \cdot & \cdot & & \cdot \\ \cdot & \cdot & & \cdot \\ L_{j-1}(x_1) & L_{j-1}(x_2) & \cdots & L_{j-1}(x_j) \\ x_1 & x_2 & \cdots & x_j \end{vmatrix}$$

$$\tag{2.6.21}$$

$$L_j{}^* = \frac{1}{G_j} \begin{vmatrix} L_1(x_1) & L_2(x_1) & \cdots & L_j(x_1) \\ \cdot & \cdot & & \cdot \\ \cdot & \cdot & & \cdot \\ L_1(x_{j-1}) & L_2(x_{j-1}) & \cdots & L_j(x_{j-1}) \\ L_1 & L_2 & \cdots & L_j \end{vmatrix} .$$

Proof: Expand these determinants according to the minors of the last row. We see that $x_j{}^*$ is a linear combination of $x_1, x_2, \ldots, x_j$ and $L_j{}^*$ is a linear combination of $L_1, L_2, \ldots, L_j$. Moreover the coefficient of x_j in $x_j{}^*$ is 1. Fix a $j > 1$. We shall show that $L_i{}^*(x_j{}^*) = 0$ for $i < j$. It suffices to show that $L_i(x_j{}^*) = 0$ for $i < j$. But

$$L_i(x_j{}^*) = \frac{1}{G_{j-1}} \begin{vmatrix} L_1(x_1) & L_1(x_2) & \cdots & L_1(x_j) \\ \cdot & \cdot & & \cdot \\ \cdot & \cdot & & \cdot \\ \cdot & \cdot & & \cdot \\ L_{j-1}(x_1) & L_{j-1}(x_2) & \cdots & L_{j-1}(x_j) \\ L_i(x_1) & L_i(x_2) & \cdots & L_i(x_j) \end{vmatrix} = 0$$

inasmuch as two rows are identical. Similarly, we can show that for fixed $i > 1$, $L_i{}^*(x_j{}^*) = 0$ for $j < i$. It remains to show that $L_i{}^*(x_i{}^*) = 1$. Since $x_i{}^* = b_{i1}x_1 + b_{i2}x_2 + \cdots + b_{i,i-1}x_{i-1} + x_i$, it suffices to show that $L_i{}^*(x_i) = 1$. Now, from the second equation of (2.6.21), $L_i{}^*(x_i) = G_i/G_i = 1$. The above biorthogonal representation is unique and the theorem follows.

We now give some examples of biorthogonal systems of the Newton type.

Ex. 1.   (Newton Polynomials). X is the space of all polynomials in z,

$$x_1 = 1, x_2 = z, x_3 = z^2, \ldots, L_1(f) = f(z_0), L_2(f) = f(z_1), \ldots,$$

where z_i are distinct points. Then,

$$G_r = \begin{vmatrix} 1 & z_0 & z_0^2 & \cdots & z_0^{r-1} \\ 1 & z_1 & z_1^2 & \cdots & z_1^{r-1} \\ \cdot & \cdot & \cdot & & \cdot \\ \cdot & \cdot & \cdot & & \cdot \\ \cdot & \cdot & \cdot & & \cdot \\ 1 & z_{r-1} & z_{r-1}^2 & \cdots & z_{r-1}^{r-1} \end{vmatrix} = \prod_{i>j}^{r-1} (z_i - z_j) \qquad (2.6.22)$$

$$x_j^* = \frac{1}{G_{j-1}} \begin{vmatrix} 1 & z_0 & z_0^2 & \cdots & z_0^{j-1} \\ \cdot & \cdot & \cdot & & \cdot \\ \cdot & \cdot & \cdot & & \cdot \\ \cdot & \cdot & \cdot & & \cdot \\ 1 & z_{j-2} & z_{j-2}^2 & \cdots & z_{j-2}^{j-1} \\ 1 & z & z^2 & \cdots & z^{j-1} \end{vmatrix}$$

This is a polynomial $q(z)$ of degree $j - 1$ in z. Now $q(z_0) = q(z_1) = \cdots = q(z_{j-2}) = 0$ inasmuch as two rows are identical. Hence $x_j^* = (z - z_0)(z - z_1) \cdots (z - z_{j-1})$. Corresponding to L_j^* we have the divided differences of f at the points $z_0, z_1, \ldots$. Formula (2.6.21) yields the representation

$$a_n = [f(z_0), f(z_1), \ldots, f(z_n)]$$

$$= \begin{vmatrix} 1 & 1 & \cdots & 1 \\ z_0 & z_1 & \cdots & z_n \\ \cdot & \cdot & & \cdot \\ \cdot & \cdot & & \cdot \\ \cdot & \cdot & & \cdot \\ z_0^{n-1} & z_1^{n-1} & \cdots & z_n^{n-1} \\ f(z_0) & f(z_1) & \cdots & f(z_n) \end{vmatrix} : \begin{vmatrix} 1 & 1 & \cdots & 1 \\ z_0 & z_1 & \cdots & z_n \\ \cdot & \cdot & & \cdot \\ \cdot & \cdot & & \cdot \\ \cdot & \cdot & & \cdot \\ z_0^{n-1} & z_1^{n-1} & \cdots & z_n^{n-1} \\ z_0^n & z_1^n & \cdots & z_n^n \end{vmatrix} \qquad (2.6.23)$$

for divided differences.

Ex. 2 (Abel-Gontscharoff Polynomials).   These polynomials, $Q_n(z)$, arise from biorthogonalizing the powers $1, z, z^2, \ldots$, against the functionals $L_0(f) = f(z_0), L_1(f) = f'(z_1), L_2(f) = f''(z_2), \ldots$. We have

$$G_r = \begin{vmatrix} 1 & z_0 & z_0^2 & \cdots & & z_0^{r-1} \\ 0 & 1 & 2z_1 & \cdots & & (r-1)z_1^{r-2} \\ 0 & 0 & 2 & \cdots & & (r-1)(r-2)z_2^{r-3} \\ \cdot & \cdot & & & & \cdot \\ \cdot & \cdot & & & & \cdot \\ \cdot & & & & & \\ 0 & 0 & & \cdots & & (r-1)! \end{vmatrix} = 1! \, 2! \, 3! \cdots (r-1)! \qquad (2.6.24)$$

$$L_n{}^*(f) = \frac{1}{G_{n+1}} \begin{vmatrix} 1 & 0 & 0 & 0 \\ z_0 & 1 & 0 & 0 \\ z_0^2 & 2z_1 & 2 & 0 \\ \cdot & \cdot & \cdot & \cdot \\ \cdot & \cdot & \cdot & \cdot \\ \cdot & \cdot & \cdot & \cdot \\ z_0^{n-1} & (n-1)z_1^{n-2} & (n-1)! & 0 \\ f(z_0) & f'(z_1) & f^{(n-1)}(z_{n-1}) & f^{(n)}(z_n) \end{vmatrix} = \frac{f^{(n)}(z_n)}{n!}$$

(2.6.25)

$$Q_n(z) = \frac{1}{G_n} \begin{vmatrix} 1 & z_0 & z_0^2 & \cdots & z_0^n \\ 0 & 1 & 2z_1 & \cdots & nz_1^{n-1} \\ 0 & 0 & 2 & \cdots & n(n-1)z_2^{n-2} \\ \cdot & \cdot & \cdot & & \cdot \\ \cdot & \cdot & \cdot & & \cdot \\ \cdot & \cdot & \cdot & & \cdot \\ 1 & z & z^2 & \cdots & z^n \end{vmatrix}$$

(2.6.26)

Thus, in particular, we have

$$Q_0(z) = 1$$
$$Q_1(z) = z - z_0$$
$$Q_2(z) = z^2 - 2z_1 z + z_0(2z_1 - z_0).$$

(2.6.27)

.
.
.

The polynomials $Q_n(z)$ are the *Abel-Gontscharoff polynomials*. A more tractable representation can be found for $Q_n(z)$ in terms of iterated integration. Consider the n-fold iterated integral

$$T_n(z) = n! \int_{z_0}^z dz' \int_{z_1}^{z'} dz'' \int_{z_2}^{z''} dz''' \cdots \int_{z_{n-1}}^{z^{(n-1)}} dz^{(n)}, \, n \geq 1.$$

(2.6.28)

It is clear that T_n is a polynomial of degree n with leading coefficient 1. Furthermore, $T_n(z_0) = 0$, and by successive differentiation,

$$T_n'(z_1) = 0, \ldots, \frac{T_n^{(n)}(z_n)}{n!} = 1,$$

$$\frac{T_n^{(r)}(z_r)}{r!} = 0 \quad \text{for} \quad r > n.$$

Thus, the biorthogonality conditions hold for T_n, and hence $T_n \equiv Q_n$.

Ex. 3 (Bernoulli Polynomials). Let $f(z)$ be analytic at $z = 0$ and assume that $f(0) \neq 0$. By Leibnitz' rule

$$\frac{d^n}{dz^n}(f(z)e^{sz}) = \sum_{k=0}^n \binom{n}{k} s^k e^{sz} f^{(n-k)}(z).$$

(2.6.29)

Hence,

$$\frac{d^n}{dz^n}\left(f(z)e^{sz}\right)\bigg|_{z=0} = \sum_{k=0}^{n}\binom{n}{k}s^k f^{(n-k)}(0) \tag{2.6.30}$$

is a polynomial of degree n in s. In particular, select $f(z) = \dfrac{z}{e^z-1}$, a function that is analytic in $|z| < 2\pi$ and $f(0) \neq 0$. Hence, we may write

$$\frac{ze^{sz}}{e^z-1} = \sum_{n=0}^{\infty}\frac{1}{n!}\left(\frac{ze^{sz}}{e^z-1}\right)_0^{(n)}z^n = \sum_{n=0}^{\infty}\frac{1}{n!}B_n(s)z^n \tag{2.6.31}$$

for certain polynomials $B_n(s)$ of degree n. (2.6.31) is valid for $|z| < 2\pi$ and can be shown to hold uniformly in s and z provided s is confined to a closed bounded region and z to a closed subset of $|z| < 2\pi$. The polynomials $B_n(z)$ are the *Bernoulli Polynomials*. The *generating function* (2.6.31) provides a convenient way to define them. For a general $f(z)$, the resulting polynomials $p_n(s)$ defined by

$$f(z)e^{sz} = \sum_{n=0}^{\infty}p_n(s)z^n \tag{2.6.32}$$

are known as *Appel Polynomials*.

Differentiating (2.6.31) j times with respect to s we obtain

$$\frac{z^{j+1}e^{sz}}{e^z-1} = \sum_{n=0}^{\infty}\frac{B_n^{(j)}(s)z^n}{n!}, \quad |z| < 2\pi. \tag{2.6.33}$$

Set $s = 0, 1$ in (2.6.33), and subtract,

$$z^{j+1}\left(\frac{e^z-1}{e^z-1}\right) = \sum_{n=0}^{\infty}\frac{B_n^{(j)}(1)-B_n^{(j)}(0)}{n!}z^n, \quad |z| < 2\pi, j = 0, 1, \ldots$$

By the uniqueness theorem for power series we must have

$$\frac{B_{j+1}^{(j)}(1)-B_{j+1}^{(j)}(0)}{(j+1)!} = 1 \quad \text{while} \quad B_r^{(j)}(1)-B_r^{(j)}(0) = 0, \quad r \neq j+1. \tag{2.6.34}$$

We see now that the polynomials $B_0(x)$, $B_1(x)$, . . . , and the functionals

$$L_0(f) = f(0), L_1(f) = f(1) - f(0),$$

$$L_2(f) = \frac{f'(1)-f'(0)}{2!}, L_3(f) = \frac{f''(1)-f''(0)}{3!}, \ldots,$$

form a biorthogonal set.

Ex. 4 (Orthogonal Polynomials).   Though these polynomials will be treated in detail in Chapter X, it is interesting to note how they fit in with Theorem 2.6.1 and Corollary 2.6.2. Let $X = C[a, b]$. Let $w(x)$ be a fixed positive weight function for which the integrals $\int_a^b w(x)x^n\,dx$, $n = 0, 1, 2, \ldots$, all exist. Introduce the functionals $L_n(f) = \int_a^b w(x)x^n f(x)\,dx$, $n = 0, 1, 2, \ldots$. These are the *weighted moments* of f. It will then be possible to biorthonormalize the powers $1, x, x^2, \ldots$, against these functionals. (In Chapter VIII it will be shown that the determinant condition for this is fulfilled.) We then obtain a's and b's such that

$$L_n^*(f) = \int_a^b w(x)(a_{n0} + a_{n1}x + \cdots + a_{nn}x^n)f(x)\,dx \qquad (2.6.35)$$

$$p_n(x) = b_{n0} + b_{n1}x + \cdots + b_{nn-1}x^{n-1} + x^n$$
$$L_n^*(p_j(x)) = \delta_{nj}.$$

A glance at the determinants (2.6.21) shows that the a's and b's in (2.6.35) are proportional. Indeed, since $L_i(x^j) = \int_a^b w(x)x^i x^j\,dx = L_j(x^i)$, the minors corresponding to elements x_i and L_i are equal. After accounting for the factors G_{n-1} and G_n in front of these determinants, we find

$$L_n^*(f) = \int_a^b w(x)\frac{G_{n-1}}{G_n}p_n(x)f(x)\,dx. \qquad (2.6.36)$$

If we now set

$$p_n^*(x) = \sqrt{\frac{G_{n-1}}{G_n}}\,p_n(x) \qquad (2.6.37)$$

we shall have

$$\sqrt{\frac{G_{j-1}}{G_j}}\sqrt{\frac{G_n}{G_{n-1}}}\,L_n^*(p_j) = \delta_{nj}$$

$$= \int_a^b w(x)\sqrt{\frac{G_{n-1}}{G_n}}\,p_n(x)\sqrt{\frac{G_{j-1}}{G_j}}\,p_j(x)\,dx$$

$$= \int_a^b w(x)p_n^*(x)p_j^*(x)\,dx. \qquad (2.6.38)$$

The polynomials p_n^* are called *orthonormal over $[a, b]$ with respect to the weight $w(x)$.* They are determined up to a factor of ± 1. (2.6.21) and (2.6.37) now give us the following determinant representation for the orthonormal polynomials

$$p_n^*(x) = C_n \begin{vmatrix} (1, 1) & (1, x) & \cdots & (1, x^n) \\ (x, 1) & (x, x) & \cdots & (x, x^n) \\ \cdot & & & \cdot \\ \cdot & & & \cdot \\ (x^{n-1}, 1) & (x^{n-1}, x) & \cdots & (x^{n-1}, x^n) \\ 1 & x & \cdots & x^n \end{vmatrix} \qquad (2.6.39)$$

where

$$(x^i, x^j) = \int_a^b w(x)x^{i+j}\,dx$$

and

$$C_n = (G_n G_{n-1})^{-\frac{1}{2}}$$

with

$$G_n = \big|(x^i, x^j)\big|_{i,j=0}^n.$$

2.7   Successive Differences

DEFINITION 2.7.1.   Let there be given a sequence of values $y_0, y_1, \ldots$.
The *differences* of adjacent values are designated by

$$\Delta y_k = y_{k+1} - y_k, \quad k = 0, 1, \cdots. \tag{2.7.1}$$

Higher differences are defined similarly

$$\Delta^2 y_k = \Delta(\Delta y_k) = \Delta y_{k+1} - \Delta y_k = (y_{k+2} - y_{k+1}) - (y_{k+1} - y_k). \tag{2.7.2}$$

In general,

$$\Delta^{n+1} y_k = \Delta(\Delta^n y_k) = \Delta^n y_{k+1} - \Delta^n y_k.$$

We define $\Delta^0 y_k = y_k$.

THEOREM 2.7.1.   *We have*

$$\Delta^0 y_k = y_k$$
$$\Delta^1 y_k = y_{k+1} - y_k$$
$$\Delta^2 y_k = y_{k+2} - 2y_{k+1} + y_k$$
$$\Delta^3 y_k = y_{k+3} - 3y_{k+2} + 3y_{k+1} - y_k$$

In general,

$$\Delta^n y_k = \sum_{r=0}^{n} (-1)^{n-r} \binom{n}{r} y_{k+r}, \quad \binom{n}{r} = \frac{n!}{r!\,(n-r)!}. \tag{2.7.4}$$

Proof: Formula (2.7.4) holds trivially for $n = 0$, and this begins an inductive proof. Assume (2.7.4) true for n. Then,

$$\Delta^{n+1} y_k = \Delta(\Delta^n y_k) = \Delta\left(\sum_{r=0}^{n} (-1)^{n-r} \binom{n}{r} y_{k+r} \right)$$

$$= \sum_{r=0}^{n} (-1)^{n-r} \binom{n}{r} \Delta y_{k+r} = \sum_{r=0}^{n} (-1)^{n-r} \binom{n}{r} (y_{k+1+r} - y_{k+r})$$

$$= \sum_{r=1}^{n+1} (-1)^{n+1-r} \binom{n}{r-1} y_{k+r} - \sum_{r=0}^{n} (-1)^{n-r} \binom{n}{r} y_{k+r}$$

$$= \sum_{r=1}^{n} (-1)^{n+1-r} y_{k+r} \left[\binom{n}{r-1} + \binom{n}{r} \right] + \binom{n}{n} y_{k+n+1} - (-1)^n \binom{n}{0} y_k$$

$$= \sum_{r=0}^{n+1} (-1)^{n+1-r} y_{k+r} \binom{n+1}{r}.$$

Thus, if the formula is true for n, it must be true for $n + 1$, and the induction is complete.

COROLLARY 2.7.2.   *For* $k = 0$ *we have*

$$\Delta^n y_0 = \sum_{r=0}^{n} (-1)^{n-r} \binom{n}{r} y_r. \tag{2.7.5}$$

In the case of interpolation at abscissas $z_0, z_1, \ldots,$ that are spaced evenly:

$$z_0 = a, z_1 = a + h, z_2 = a + 2h, \ldots, z_n = a + nh, \tag{2.7.6}$$

the divided differences may be given an elegant expression in terms of successive differences. If $w(z) = (z - z_0)(z - z_1) \cdots (z - z_n)$, then

$$w'(z_k) = (z_k - z_0)(z_k - z_1) \cdots (z_k - z_{k-1}) (z_k - z_{k+1}) \cdots (z_k - z_n).$$

Since $z_i - z_j = (i - j)h$,

$$w'(z_k) = (kh)(k - 1)h \cdots (h)(-h)(-2h) \cdots (-(n - k)h)$$
$$= h^n k! \, (n - k)! \, (-1)^{n-k}. \tag{2.7.7}$$

Therefore from (2.6.4),

$$a_n = [y_0, y_1, \ldots, y_n] = \sum_{k=0}^{n} \frac{y_k}{w'(z_k)}$$
$$= \sum_{k=0}^{n} \frac{(-1)^{n-k} y_k}{h^n k! \, (n - k)!} = \frac{1}{n! \, h^n} \sum_{k=0}^{n} (-1)^{n-k} \binom{n}{k} y_k$$
$$= \frac{\Delta^n y_0}{n! \, h^n}. \tag{2.7.8}$$

We can therefore prove the following theorem.

THEOREM 2.7.3.   *Let* $p(z)$ *be the unique polynomial of* $\mathscr{P}_n$ *that takes on the values* $y_0, y_1, \ldots, y_n$ *at the* $n + 1$ *points* $a, a + h, \ldots, a + nh$. *Then*

$$p_n(z) = y_0 + \frac{\Delta y_0}{h} (z - a) + \frac{\Delta^2 y_0}{2! \, h^2} (z - a)(z - a - h) + \cdots$$
$$+ \frac{1}{n! \, h^n} \Delta^n y_0 (z - a)(z - a - h) \cdots (z - a - (n - 1)h). \tag{2.7.9}$$

If $p_n(f; z)$ interpolates to f at $a, a + h, \ldots, a + nh$ then

$$p_n(f; z) = f(a) + \frac{\Delta f(a)}{h} (z - a) + \frac{1}{2! \, h^2} \Delta^2 f(a)(z - a)(z - a - h) + \cdots$$
$$+ \frac{1}{n! \, h^n} \Delta^n f(a)(z - a)(z - a - h) \cdots (z - a - (n - 1)h). \tag{2.7.10}$$

We have written

$$\Delta f(a) = f(a + h) - f(a)$$
$$\Delta^2 f(a) = f(a + 2h) - 2f(a + h) + f(a), \text{ etc.} \tag{2.7.11}$$

Formulas (2.7.9) and (2.7.10) are known as *Newton's forward difference formulas.*

If $f(x)$ is defined at $a, a + h, a + 2h, \ldots$, the formal series

$$f(x) \sim \sum_{k=0}^{\infty} \frac{\Delta^k f(a)}{k!\, h^k} (z - a)(z - a - h) \cdots (z - a - (k - 1)h) \quad (2.7.12)$$

is called a *Newton series for f.*

Ex. 1. If $f(x) = x^n$, then $\Delta f(x) = nhx^{n-1} + \cdots$. The first difference is therefore a polynomial of degree $n - 1$. Similarly, we find $\Delta^n x^n = n!\, h^n$ and $\Delta^p x^n = 0$ for $p > n$.

Ex. 2. If $f(x) = e^{\sigma x}$ then $\Delta f(x) = (e^{\sigma h} - 1)e^{\sigma x}$. Iterating this, $\Delta^n f(x) = (e^{\sigma h} - 1)^n e^{\sigma x}$.

Ex. 3. If $f(x) \in \mathscr{P}_n$ then the series (2.7.12) converges to $f(x)$. From Ex. 1, the series reduces to a sum of $n + 1$ terms and, by Theorem 2.7.3, is that member of $\mathscr{P}_n$ which interpolates to f at $a, a + h, \ldots, a + nh$. By uniqueness, it must coincide with f.

Ex. 4. On the other hand, the function $f(x) = \sin \pi x$ has zeros at $0, \pm 1, \pm 2, \ldots$, so that with $a = 0, h = 1, \Delta^k f(0) \equiv 0$. The series (2.7.12) is identically zero and does not represent $f(x)$ over any interval. An entire function may still not be sufficiently restricted in its behavior to be represented by its Newton series.

NOTES ON CHAPTER II

2.1 The discussion of polynomial interpolation in Chapters II and III can be amplified by related material in any text on numerical analysis. Mention should be made also of the numerous practical articles of H. E. Salzer related to interpolation.

2.3 Abel-Gontscharoff Interpolation: J. M. Whittaker [1], p. 38; V. L. Gontscharoff [1], pp. 84–86. Lidstone Interpolation: D. V. Widder [3], R. P. Boas, Jr. [2]. Hermite's Interpolation: A. A. Markoff [1]; Gontscharoff [1], p. 64. Hermite's formulas are rediscovered and republished every few years. Generalized Taylor Interpolation: D. V. Widder [1], [2], I. M. Sheffer [1]. Trigonometric Interpolation, A. Zygmund [1], Vol. II.

2.4 For additional examples of unisolvent systems, see Pólya and Szegö [1], vol. II, pp. 45–52. Further theory is presented in Achieser [1], p. 67 et seq. and in Motzkin [1]. References to recent work related to Haar's Theorem can be found in Buck [2].

2.5–2.6 General formulae of Lagrange and Newton type have been given implicitly and explicitly by many authors. For instance, see the articles

by Widder and Sheffer cited in 2.3. Also: W. E. Milne [1]. H. B. Curry [1] develops these notions and contains some further references.
Bernoulli Polynomials: N. E. Nörlund [1].
Appel Polynomials: Boas and Buck [1], E. D. Rainville [1].
2.7 For the algebraic side of differences, consult books on difference calculus such as Fort [1]. There are extensive studies of the convergence of interpolation series some of which are found in books: Nörlund [1], Whittaker [1]. A. O. Gelfand [1] has a noteworthy treatment of Newton series and allied questions. See also Buck [1].

PROBLEMS

1. If $V(x_1, x_2, \ldots, x_n)$ designates the Vandermonde determinant, show that $V(1, 2, 3, \ldots, n) = 1!\, 2!\, 3! \cdots (n - 1)!$

2. Can a parabola p be found for which $p(0), p''(0), p'''(0)$ have preassigned values? For which $\int_{-1}^{1} p(x)\, dx, p(0), p'(0)$ have preassigned values?

3. Construct a polynomial in $\mathscr{P}_3$ for which $p(0) = 1, p(1) = 3, p'(-1) = 4, p''(0) = 0$. Is the answer unique?

4. Three points lie on a nonvertical line. What happens when you try to fit a parabola to them? A cubic? Formulate a general statement.

5. Show that we can not always find a function of the form $f(x) = \dfrac{A + Bx}{1 + Cx}$ that passes through three points with distinct abscissas.

6. Is it possible to fit a curve of the form $f(x) = A + Be^{Cx}$ to the data $f(0) = 0, f(1) = 1, f(2) = \frac{1}{2}$?

7. X consists of all functions of the form $a_0 + a_1 x + a_2 y + a_3 x^2 + a_4 xy + a_5 y^2$ defined on $-1 \leq x \leq 1, -1 \leq y \leq 1$. Find a basis for X^*.

8. Let $X = \mathscr{P}_n$ considered on $0 \leq x \leq 1$. Let $0 < x_0 < \cdots < x_n \leq 1$. Prove that $L_j(f) = \int_0^{x_j} f(t)\, dt, j = 0, 1, \ldots, n$, are independent over X^*.

9. Select the constants $A, \ldots, E$ so that $\dfrac{A + Bx + Cx^2}{1 + Dx + Ex^2}$ agrees with the Maclaurin series expansion of e^x as far as x^4. How close is the resulting rational function to e^x over the interval $|x| \leq \frac{1}{10}$?

10. If $R(x) = \dfrac{A + Bx}{1 + Cx}$, can the interpolation problem $R(0) = f(0), R'(0) = f'(0), R''(0) = f''(0)$ always be solved? What about a similar problem for rational functions of higher degree? The resulting rational functions are called the *Padé Approximants to* $f(x)$.

11. Let $x_0, x_1, \ldots, x_n$ be fixed. Let $p(x) = a_0 + a_1 x + \cdots + a_n x^n$ and $p(x_i) = y_i$. Given an $\varepsilon > 0$, we can find a δ such that $|y_i| \leq \delta$ implies $|a_i| \leq \varepsilon$.

12. Discuss the possibility of trigonometric interpolation with Taylor conditions.

13. Discuss the possibility of osculatory trigonometric interpolation.

14. Let $T(x) = a_0 + a_1 \cos x + b_1 \sin x + \cdots + a_n \cos nx + b_n \sin nx$. Consider $e^{inx}T(x) = P_{2n}(e^{ix})$ and show that the number of real roots modulo 2π of $T(x)$, each root counted with its multiplicity, is $\leq 2n$.

15. If $a_0 + a_1 \cos x + \cdots + a_n \cos nx$ vanishes at $n + 1$ points,

$$0 \le x_0 < x_1 \cdots < x_n < \pi,$$

it vanishes identically.

16. If $b_1 \sin x + \cdots + b_n \sin nx$ vanishes at n points, $0 < x_1 < \cdots < x_n < \pi$, it vanishes identically.

17. Let $s_1, s_2, \ldots, s_n$ be distinct. Then the set $e^{s_1 x}, \ldots, e^{s_n x}$ is unisolvent over any interval.

18. Show that $p_n(x^{n+1}; x) = x^{n+1} - (x - x_0)(x - x_1) \cdots (x - x_n)$.

19. Let $G(x_1, \ldots, x_n) = (L_i(x_j))$. Set $y_k = \sum_{i=1}^{n} a_{ki} x_i$ and $T = (a_{ij})$. Then, $G(y_1, \ldots, y_n) = G(x_1, \ldots, x_n)T'$, $T' =$ transpose of T. Prove a corresponding result for a linear transformation of the L's.

20. In Theorem 2.5.1, determine the $x_i{}^*$ explicitly in terms of determinants.

21. Prove the following "dual" of Theorem 2.5.1. Let X be a linear space of dimension n and let $x_1, x_2, \ldots, x_n$ be n independent elements. Then, there are determined uniquely n independent elements of $X^*, L_1{}^*, \ldots, L_n{}^*$, such that $L_i{}^*(x_j) = \delta_{ij}$.

22. Show that if the abscissas $x_1 \ne x_2$, then $L_1{}^*(f) = \dfrac{x_2 f(x_1) - x_1 f(x_2)}{x_2 - x_1}$ and $L_2{}^*(f) = \dfrac{f(x_2) - f(x_1)}{x_2 - x_1}$ are biorthonormal to the functions $1, x$. Interpret in $X = \mathscr{P}_1$.

23. The functions $\dfrac{(-1)^n \sin z}{z - n\pi}$ and the functionals

$$L_n(f) = f(n\pi) \quad n = 0, \pm 1, \pm 2, \ldots,$$

are biorthonormal. The infinite expansion of form (2.6.17) is

$$f(z) \sim \sin z \sum_{n=-\infty}^{\infty} \frac{(-1)^n f(n\pi)}{z - n\pi}.$$

It is called the *Cardinal Series* for f.

24. Biorthogonalize $1, x, x^2, \ldots$, against

$$L_0(f) = \int_0^1 f(x)\, dx, \quad L_1(f) = \int_0^2 f(x)\, dx, \ldots.$$

Compute the first three polynomials.

25. Let $\left| L_i(x_j) \right|_{i,j=1}^{n} \ne 0$. Then there is a permutation of the elements

$$x_1, x_2, \ldots, x_n \colon x_1', x_2', \ldots, x_n'$$

such that $\left| L_i(x_j') \right|_{i,j=1}^{k} \ne 0$, $k = 1, 2, \ldots, n$. (Cf. the hypotheses of Theorem 2.6.1.)

26. Compute the first four Bernoulli polynomials from (2.6.30) or (2.6.31).

27. Calculate the nth divided difference for $f(x) = \dfrac{1}{x}$.

28. Prove that $f(n) = \sum_{k=0}^{n} \binom{n}{k} \Delta^k f(0)$.

29. Express x^2, x^3, x^4 as linear combinations of 1, x, $x(x-1)$, $x(x-1)(x-2)$, $x(x-1)(x-2)(x-3)$.

30. Verify the formal Newton Series

$$e^{\sigma z} \sim 1 + (e^\sigma - 1)z + \frac{(e^\sigma - 1)^2}{2!} z(z-1) + \cdots,$$

$$\frac{1}{t-z} \sim \frac{1}{t} + \frac{z}{t(t-1)} + \frac{z(z-1)}{t(t-1)(t-2)} + \cdots.$$

31. Verify the formal Newton series

$$\log \Gamma(1+z) \sim \frac{\log 2}{2!} z(z-1) + \frac{(\log 3 - 2\log 2)}{3!} z(z-1)(z-2) + \cdots.$$

(Hermite).

CHAPTER III

Remainder Theory

The results of the previous chapter are purely algebraic. They relate to the possibility of carrying out interpolatory processes. But once these processes have been carried out, how good are the approximations that result? Remainder theory deals with this question and is consequently of great importance to numerical analysis as well as to various parts of pure analysis.

3.1 The Cauchy Remainder for Polynomial Interpolation

THEOREM 3.1.1. *Let $f(x) \in C^n[a, b]$ and suppose that $f^{(n+1)}(x)$ exists at each point of (a, b).*

If, $a \leq x_0 < x_1 < \cdots < x_n \leq b$, then

$$f(x) - p_n(f; x) = \frac{(x - x_0)(x - x_1) \cdots (x - x_n)}{(n + 1)!} f^{(n+1)}(\xi) \qquad (3.1.1)$$

where $\min (x, x_0, x_1, \ldots, x_n) < \xi < \max (x, x_0, x_1, \ldots, x_n)$. *The point ξ depends upon $x, x_0, x_1, \ldots, x_n$ and f.*

Proof: Since $p_n(f; x_k) = f(x_k)$, the function $f(x) - p_n(f; x)$ vanishes at $x = x_0, x = x_1, \ldots, x = x_n$. Let x be fixed and $\neq x_0, x_1, \ldots, x_n$. Set

$$K(x) = \frac{f(x) - p_n(f; x)}{(x - x_0)(x - x_1) \cdots (x - x_n)} \qquad (3.1.2)$$

and consider the following function of t:

$$W(t) = f(t) - p_n(f; t) - (t - x_0)(t - x_1) \cdots (t - x_n)K(x). \qquad (3.1.3)$$

The function $W(t)$ vanishes at $t = x_0, t = x_1, \ldots, t = x_n$. In addition, in virtue of (3.1.2), it vanishes at the additional point $t = x$. By the generalized Rolle's Theorem 1.6.3, the function $W^{(n+1)}(t)$ must vanish at a point ξ with $\min (x, x_0, \ldots, x_n) < \xi < \max (x, x_0, \ldots, x_n)$. But from (3.1.3)

$$W^{(n+1)}(t) = f^{(n+1)}(t) - (n + 1)! \, K(x)$$

so that

$$0 = W^{(n+1)}(\xi) = f^{(n+1)}(\xi) - (n + 1)! \, K(x) \qquad (3.1.4)$$

and therefore

$$K(x) = \frac{1}{(n + 1)!} f^{(n+1)}(\xi). \qquad (3.1.5)$$

Inserting this in (3.1.2) we obtain (3.1.1). If $x = x_k$, (3.1.1) holds trivially with any ξ.

COROLLARY 3.1.2. (Error in Linear Interpolation). *Let* $f(x) \in C'[a, b]$ *and suppose that* $f''(x)$ *exists at each point of* (a, b). *Then, for* $a \le x \le b$,

$$f(x) - \left(\frac{b - x}{b - a}f(a) + \frac{x - a}{b - a}f(b)\right) = \frac{(x - a)(x - b)}{2}f''(\xi), \quad a < \xi < b.$$

$$(3.1.6)$$

In most instances, the value of ξ is not known exactly, and the following estimate becomes of importance.

COROLLARY 3.1.3.   *Let*

$$R_n(f; x) = f(x) - p_n(f; x). \qquad (3.1.7)$$

Then if $f(x) \in C^{n+1}[a, b]$,

$$\left|R_n(f; x)\right| \le \left\{\max_{a \le t \le b}\left|f^{(n+1)}(t)\right|\right\}\frac{|x - x_0|\,|x - x_1|\cdots|x - x_n|}{(n + 1)!}. \qquad (3.1.8)$$

EX. 1.   A value for arcsin (.5335) is obtained by interpolating linearly between the values for $x = .5330$ and $x = .5340$. Estimate the error committed. We have $(\arcsin x)'' = x(1 - x^2)^{-\frac{3}{2}}$ and $(\arcsin x)''' = (1 + 2x^2)(1 - x^2)^{-\frac{5}{2}}$. Since the 3rd derivative is positive over $.533 \le x \le .534$, the maximum value of the 2nd derivative occurs at $x = .534$. From (3.1.8),

$$|R_1| \le \frac{.534}{(1 - (.534)^2)^{\frac{3}{2}}}\frac{(.0005)^2}{2} \le 1.2 \times 10^{-7}.$$

A direct computation shows that the true error is 1.101×10^{-7}.

This example points out the following facts. In order to use the estimate (3.1.8) in practical work, it is necessary to have an expression for a higher derivative of the function interpolated, and it is necessary to obtain an upper bound for the value of this high derivative over a certain interval. This might be a formidable task even for quite elementary functions. Think of obtaining the 8th derivative of arcsin x or, worse still, of

$$(1 + (x + 2)^{\frac{1}{2}} + (x + 3)^{\frac{1}{2}})^{\frac{1}{3}}!$$

There are several ways in which this difficulty can be overcome. This applies not only to the error estimate for interpolation, but to all error estimates of mean value type, i.e., those involving higher derivatives. If we are working with a tabulated function, we can estimate derivatives by means of differences. The justification for this procedure is found in Corollary 3.4.4. Secondly, if we are working with analytic functions and if we are in a position to obtain an upper bound for the values of the function in the complex plane, then we can use (1.9.8) to estimate the derivative. This process is summed up by the following result.

COROLLARY 3.1.4.   *Let* $f(x) \in A(R)$ *where R is a region that contains* $[a, b]$. *Let C be a closed curve that contains* $[a, b]$ *in its interior and set* $L(C) =$ *length of C,* $M_C = \max_{z \in C} |f(z)|$, $\delta = $ *minimum distance from C to* $[a, b]$. *Then,*

$$|R_n(f; x)| = |f(x) - p_n(f; x)| \leq \frac{L(C)M_C}{2\pi\delta^{n+2}} |x - x_0| |x - x_1| \cdots |x - x_n|.$$

$$(3.1.9)$$

Ex. 2.   Let $f(x) = [e^{x^2-4} - 1]^{\frac{1}{3}}$. $[a, b] = [-1, 1]$, $n = 4$, $x_0 = -1$, $x_1 = -\frac{1}{2}$, $x_2 = 0$, $x_3 = \frac{1}{2}$, $x_4 = 1$. Estimate the error committed at $x = \frac{1}{4}$ by interpolation at these points. Now $f(z)$ is analytic in $|z| < 2$, and

$$\begin{aligned}|f(z)| &= |e^{z^2-4} - 1|^{\frac{1}{3}} \\ &\leq (|e^{z^2-4}| + 1)^{\frac{1}{3}} = (e^{Re(z^2-4)} + 1)^{\frac{1}{3}} \\ &= (e^{x^2-y^2-4} + 1)^{\frac{1}{3}}.\end{aligned}$$

If $C: |z| = \rho$, $1 < \rho < 2$,

$$M_C = \max_{z \in C} |f(z)| \leq (e^{\rho^2-4} + 1)^{\frac{1}{3}} \leq 2^{\frac{1}{3}}. \quad L(C) = 2\pi\rho,$$

and $\delta = \rho - 1$. Write $\rho = 2 - \varepsilon$. Then

$$\frac{L(C)M_C}{2\pi\delta^{n+2}} \leq \frac{(2\pi)(2 - \varepsilon)2^{\frac{1}{3}}}{2\pi(1 - \varepsilon)^6}.$$

Since (3.1.9) is valid for any $0 < \varepsilon < 2$, we may select $\varepsilon = 0$, leading to

$$|R_4(f; \tfrac{1}{4})| \leq 2^{\frac{4}{3}}(\tfrac{5}{4})(\tfrac{3}{4})(\tfrac{1}{4})(\tfrac{1}{4})(\tfrac{3}{4}) \approx .11.$$

3.2   Convex Functions.   Here we make a different sort of application of the remainder theorem.

DEFINITION 3.2.1.   Let $f(x)$ be defined on $[a, b]$. Then f is said to be *convex* on this interval if an arbitrary chord joining two points of the curve is never below the curve.

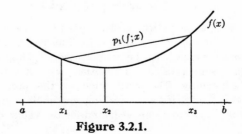

Figure 3.2.1.

Ex. 1.   The parabola $y = x^2$ is convex on any interval $[a, b]$.

This definition can be recast in the language of interpolation. Let $a \le x_1 < x_2 < x_3 \le b$ and let $p_1(f; x)$ be that element of $\mathscr{P}_1$ which coincides with f at x_1 and x_3; then

$$f(x_2) - p_1(f; x_2) \le 0, \quad x_1 < x_2 < x_3. \tag{3.2.1}$$

THEOREM 3.2.1.   *Let $f''(x)$ exist on (a, b). Then $f(x)$ is convex in every closed subinterval if and only if $f''(x) \ge 0$ on (a, b).*

Proof: By (3.1.6),

$$f(x_2) - p_1(f; x_2) = \tfrac{1}{2}(x_2 - x_1)(x_2 - x_3)f''(\xi), \quad x_1 < \xi < x_3,$$

Since $(x_2 - x_1)(x_2 - x_3) < 0$, $f''(x) \ge 0$ implies that the left hand side is nonpositive and hence f is convex.

Suppose, conversely that f is convex but that $f''(x) = k < 0$ for some $a < x < b$. Then, by definition of the second derivative,

$$\lim_{h \to 0^+} \frac{f'(x + h) - f'(x)}{h} = k$$

$$\lim_{h \to 0^+} \frac{f'(x - h) - f'(x)}{-h} = k.$$

Hence, $$\lim_{h \to 0^+} \frac{f'(x + h) - f'(x - h)}{h} = 2k$$

Since $k < 0$, for sufficiently small h, say for $0 < h \le h_1$, we must have $f'(x + h) - f'(x - h) < k_1 h, k_1 < 0$.

Hence, $$\int_0^{h_1} [f'(x + h) - f'(x - h)]\, dh < \int_0^{h_1} k_1 h\, dh = \frac{k_1 h_1{}^2}{2}.$$

Therefore $$f(x + h_1) - 2f(x) + f(x - h_1) < 0.$$

This tells us that the chord extended from $x - h_1$ to $x + h_1$ lies below the curve at x and this contradicts the assumption of convexity.

If f lacks a second derivative, we can at least say that second differences are nonnegative.

THEOREM 3.2.2.   *Let $f(x)$ be convex in $[a, b]$.*

$$\text{If } a \le x_0 < x_0 + h < x_0 + 2h \le b,$$

then

$$\Delta^2 f(x_0) = f(x_0 + 2h) - 2f(x_0 + h) + f(x_0) \ge 0. \tag{3.2.2}$$

Proof: This inequality asserts that the midpoint of any chord lies above or on the curve.

3.3   Best Real Error Estimates; The Tschebyscheff Polynomials.

The error estimate (3.1.8) for polynomial interpolation splits into two parts. The first part, $\max_{a \leq x \leq b} |f^{(n+1)}(x)|$, depends upon the function interpolated, but is independent of the manner in which the interpolation is carried out. The second part, $\frac{1}{(n+1)!} |x - x_0| \, |x - x_1| \cdots |x - x_n|$, is independent of the function, but depends upon the points. The estimate (3.1.8) was obtained by replacing $|f^{(n+1)}(\xi)|$ by $\max_{a \leq x \leq b} |f^{(n+1)}(x)|$. This was a pure expedient, and in many cases, of course, the error predicted by (3.1.8) will be far greater than the actual error.

But since the first part is, so to speak, beyond our control, let us look at the second part. A small error estimate will also result from a small second part. Consider the quantity $\max_{a \leq x \leq b} |(x - x_0)(x - x_1) \cdots (x - x_n)|$. It depends upon $x_0, x_1, \ldots, x_n$, and this leads us to the following important and interesting question: how can we select points $x_0, x_1, \ldots, x_n$ in $[a, b]$ so that the maximum is as small as possible? As far as the estimate (3.1.8) is concerned, this selection of points will be the best possible selection. Indeed, it turns out that this choice of points is a happy choice as far as a number of questions in interpolation theory are concerned. The answer to the problem just posed is given by the zeros of the Tschebyscheff Polynomials, and we now turn to their theory.

In deMoivre's formula $(\cos \theta + i \sin \theta)^n = \cos n\theta + i \sin n\theta$, set $\cos \theta = x$. If $0 \leq \theta \leq \pi$, $\sin \theta = \sqrt{1 - x^2} \geq 0$. Then,

$$\cos n\theta + i \sin n\theta = (x + i\sqrt{1 - x^2})^n.$$

If we expand this expression by the Binomial Theorem, and take the real parts of the resulting equation, we obtain

$$\cos n(\arccos x) = \cos n\theta = x^n + \binom{n}{2} x^{n-2}(x^2 - 1)$$

$$+ \binom{n}{4} x^{n-4}(x^2 - 1)^2 + \ldots. \quad (3.3.1)$$

Thus, $\cos n\theta$ is a certain polynomial of degree n in $\cos \theta$.

DEFINITION 3.3.1.   The Tschebyscheff polynomial of degree n is defined by

$$T_n(x) = \cos (n \arccos x) = x^n + \binom{n}{2} x^{n-2}(x^2 - 1) + \cdots (n = 0, 1, \ldots).$$

$$(3.3.2)$$

There are a number of distinct families (in fact, infinitely many) of polynomials that go by the name of "Tschebyscheff Polynomials." The polynomials defined by (3.3.2) are the Tschebyscheff Polynomials, *par excellence.*

When we have occasion to deal with other types of Tschebyscheff Polynomials, we shall include some qualifying expression.

It is easy to compute the first few Tschebyscheff polynomials explicitly using (3.3.2). We find

$$
\begin{aligned}
T_0(x) &= 1 \\
T_1(x) &= x \\
T_2(x) &= 2x^2 - 1 \\
T_3(x) &= 4x^3 - 3x \\
T_4(x) &= 8x^4 - 8x^2 + 1 \\
T_5(x) &= 16x^5 - 20x^3 + 5x \\
T_6(x) &= 32x^6 - 48x^4 + 18x^2 - 1
\end{aligned} \tag{3.3.3}
$$

The Tschebyscheff polynomials satisfy a three term recurrence relation.

THEOREM 3.3.1.

$$
T_{n+1}(x) = 2xT_n(x) - T_{n-1}(x) \quad n = 1, 2, \ldots . \tag{3.3.4}
$$

Proof:

$$
\cos (n + 1)\theta = \cos n\theta \cos \theta - \sin n\theta \sin \theta
$$
$$
\cos (n - 1)\theta = \cos n\theta \cos \theta + \sin n\theta \sin \theta
$$

Adding and rearranging,

$$
\cos (n + 1)\theta = 2 \cos n\theta \cos \theta - \cos (n - 1)\theta
$$

Now set $\cos \theta = x$, $\cos n\theta = T_n(x)$, and (3.3.4) is obtained.

COROLLARY 3.3.2.

$$
T_n(x) = 2^{n-1}x^n + \textit{terms of lower degree.} \tag{3.3.5}
$$

THEOREM 3.3.3. $T_n(x)$ *has simple zeros at the n points*

$$
x_k = \cos \frac{2k - 1}{2n} \pi \quad k = 1, 2, \ldots, n. \tag{3.3.6}
$$

On the closed interval $-1 \leq x \leq 1$, $T_n(x)$ *has extreme values at the* $n + 1$ *points*

$$
x_k' = \cos \frac{2k}{2n} \pi \quad k = 0, 1, \ldots, n \tag{3.3.7}
$$

where it assumes the alternating values $(-1)^k$.

Proof: $T_n(x_k) = \cos \left(n \arccos \left(\cos \frac{2k - 1}{2n} \pi \right) \right) = \cos \left(\frac{2k - 1}{2} \pi \right) = 0$,

$k = 1, 2, \ldots, n$. Now

$$
T_n'(x) = \frac{n}{\sqrt{1 - x^2}} \sin (n \arccos x). \tag{3.3.8}
$$

Hence, $T_n{}'(x_k) = \dfrac{n}{\sqrt{1 - x_k{}^2}} \sin\left(\dfrac{2k - 1}{2}\,\pi\right) \neq 0$ and the zeros must be simple. Moreover,

$$T_n{}'(x_k{}') = n\left(1 - \cos^2 \dfrac{k\pi}{n}\right)^{-\frac{1}{2}} \sin\,(k\pi) = 0$$

for $k = 1, 2, \ldots, n - 1$. Now,

$$T_n(x_k{}') = \cos\left(n \text{ arc cos }\left(\cos \dfrac{k\pi}{n}\right)\right) = \cos\,(k\pi) = (-1)^k.$$

This is valid for $k = 0, 1, \ldots, n$. But for $-1 \leq x \leq 1$,

$$T_n(x) = \cos\,(n \text{ arc cos } x)$$

and hence $|T_n(x)| \leq 1$. This shows that the points $x_k{}'$ are extreme points. It is easily shown that $x_k{}'$ are the only extreme points in $-1 \leq x \leq 1$.

DEFINITION 3.3.2. $\tilde{T}_n(x) = \dfrac{1}{2^{n-1}}\, T_n(x).$

Note that $\tilde{T}_n(x) = x^n +$ terms of lower degree.

THEOREM 3.3.4 (Tschebyscheff).   *Let $\widetilde{\mathscr{P}}_n$ designate the class of all polynomials of degree n with leading coefficient 1. Then, for any $p \in \widetilde{\mathscr{P}}_n$,*

$$\max_{-1 \leq x \leq 1} |\tilde{T}_n(x)| \leq \max_{-1 \leq x \leq 1} |p(x)|.$$

Proof: On $-1 \leq x \leq 1$, $|\tilde{T}_n|$ assumes its maximum value, $\dfrac{1}{2^{n-1}}$, $n + 1$ times at the points $x_k{}' = \cos\left(\dfrac{k\pi}{n}\right)$ $k = 0, 1, \ldots, n$.

Suppose there were a $p \in \widetilde{\mathscr{P}}_n$, with $\max\limits_{-1 \leq x \leq 1} |p(x)| < \dfrac{1}{2^{n-1}}$. Form the difference $Q(x) = \tilde{T}_n(x) - p(x)$. Clearly $Q(x) \in \mathscr{P}_{n-1}$. Now

$$Q(x_k{}') = \tilde{T}_n(x_k{}') - p(x_k{}') = \dfrac{(-1)^k}{2^{n-1}} - p(x_k{}'), \quad k = 0, 1, \ldots, n.$$

These quantities are alternatively $+$ and $-$ inasmuch as $|p(x_k{}')| < \dfrac{1}{2^{n-1}}$. Therefore, there are $n + 1$ points where $Q(x)$ takes values with alternating signs. $Q(x)$ therefore has n zeros. Since $Q \in \mathscr{P}_{n-1}$, it must vanish identically. Thus, $p(x) \equiv \tilde{T}_n(x)$. This yields

$$\dfrac{1}{2^{n-1}} = \max_{-1 \leq x \leq 1} |\tilde{T}_n(x)| = \max_{-1 \leq x \leq 1} |p(x)| < \dfrac{1}{2^{n-1}}.$$

This is a contradiction.

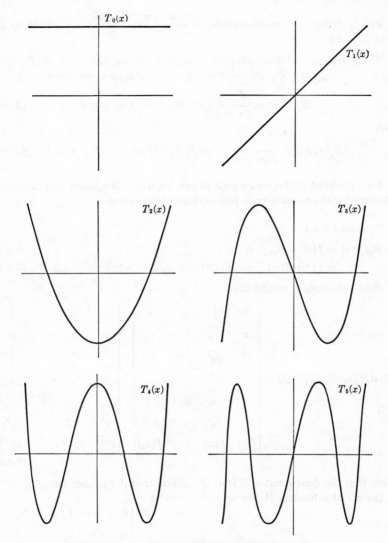

Figure 3.3.1 The Tschebyscheff Polynomials.

COROLLARY 3.3.5. $\displaystyle\max_{-1 \leq x \leq 1} |x^n + a_1 x^{n-1} + \cdots + a_n| \geq \frac{1}{2^{n-1}}$.

COROLLARY 3.3.6. $\displaystyle\max_{a \leq x \leq b} |a_0 x^n + a_1 x^{n-1} + \cdots + a_n| \geq |a_0| \frac{(b-a)^n}{2^{2n-1}}$.

Proof: Make the transformation $x = \dfrac{b + a}{2} + \dfrac{b - a}{2} t$ converting $[a, b]$ into $[-1, 1]$.

When polynomial interpolation is carried out on the zeros of $T_{n+1}(x)$: $x_0, x_1, \ldots, x_n$, then $\tilde{T}_{n+1}(x) = (x - x_0) \cdots (x - x_n)$ and we have

$$R_n(f; x) = \frac{\tilde{T}_{n+1}(x)}{(n + 1)!} f^{(n+1)}(\xi), \quad -1 < \xi < 1, \tag{3.3.9}$$

and

$$|R_n(f; x)| \leq \frac{1}{2^n(n + 1)!} \max_{-1 \leq x \leq 1} |f^{(n+1)}(x)|, \quad -1 \leq x \leq 1. \tag{3.3.10}$$

3.4 Divided Differences and Mean Values.   We begin with a formal identity for the remainder in polynomial interpolation.

THEOREM 3.4.1.

$$R_n(f; z) = f(z) - p_n(f; z)$$
$$= [f(z), f(z_0), \ldots, f(z_n)] (z - z_0)(z - z_1) \cdots (z - z_n). \tag{3.4.1}$$

Proof: According to (2.6.23),

$$[f(z), f(z_0), \ldots, f(z_n)] = \begin{vmatrix} 1 & 1 & \cdots & 1 \\ z & z_0 & \cdots & z_n \\ z^2 & z_0^2 & \cdots & z_n^2 \\ \cdot & \cdot & & \cdot \\ \cdot & \cdot & & \cdot \\ \cdot & \cdot & & \cdot \\ z^n & z_0^n & \cdots & z_n^n \\ f(z) & f(z_0) & \cdots & f(z_n) \end{vmatrix} \div \begin{vmatrix} 1 & 1 & \cdots & 1 \\ z & z_0 & \cdots & z_n \\ \cdot & \cdot & & \cdot \\ \cdot & \cdot & & \cdot \\ \cdot & \cdot & & \cdot \\ z^n & z_0^n & \cdots & z_n^n \\ z^{n+1} & z_0^{n+1} & \cdots & z_n^{n+1} \end{vmatrix} \tag{3.4.2}$$

Note that the denominator D is in $\mathscr{P}_{n+1}$, and that it vanishes for $z_0, z_1, \ldots, z_n$ (identical columns). Hence we have

$$D = (-1)^{n+1}(z - z_0)(z - z_1) \cdots (z - z_n) \begin{vmatrix} 1 & \cdots & 1 \\ z_0 & \cdots & z_n \\ \cdot & & \cdot \\ \cdot & & \cdot \\ \cdot & & \cdot \\ z_0^n & \cdots & z_n^n \end{vmatrix}.$$

It follows from this and by expanding the numerator of (3.4.2) in minors of the first column that the function

$$\Phi(z) = [f(z), f(z_0), \ldots, f(z_n)](z - z_0)(z - z_1) \cdots (z - z_n)$$

equals $f(z) + q(z)$, where $q(z) \in \mathscr{P}_n$. Moreover, $\Phi(z_i) = 0$ $i = 0, 1, \ldots, n$. Hence $-q(z_i) = f(z_i)$. By the uniqueness of interpolation, $q(z) = -p_n(f; z)$ and (3.4.1) follows.

COROLLARY 3.4.2.   *Let* $f(x) \in C[a, b]$ *and suppose that* $f^{(n+1)}(x)$ *exists at each point of* (a, b). *If* $a \leq x_0 < x_1 < \cdots < x_n \leq b$, *and* $x \in [a, b]$, *then*

$$[f(x), f(x_0), \ldots, f(x_n)] = \frac{f^{(n+1)}(\xi)}{(n + 1)!} \tag{3.4.3}$$

where $\min (x, x_0, \ldots, x_n) < \xi < \max (x, x_0, \ldots, x_n)$.

Proof: Combine Theorem 3.4.1 with Theorem 3.1.1.

Divided differences may be regarded as generalizations of derivatives. More precisely,

COROLLARY 3.4.3.   *Let* $f(x) \in C^{n+1}[a, b]$. *Then if* $x \in [a, b]$,

$$\lim_{\substack{x_i \to x \\ i = 0, 1, 2, \ldots, n}} [f(x), f(x_0), \ldots, f(x_n)] = \frac{f^{(n+1)}(x)}{(n + 1)!} \tag{3.4.4}$$

In the case of equally spaced points, we obtain a mean value theorem for successive differences. Let $x_0 = a$, $x_1 = a + h$, $\ldots$, $x_n = a + nh$. Then, from (2.7.8), $[f(x_0), f(x_1), \ldots, f(x_n)] = \dfrac{\Delta^n f(x_0)}{n!\, h^n}$. Combining this with Corollary 3.4.2 leads to

COROLLARY 3.4.4.   *Let* $f(x) \in C[a, b]$ *and suppose that* $f^{(n)}(x)$ *exists at each point of* (a, b); *then*

$$\Delta^n f(x_0) = h^n f^{(n)}(\xi) \qquad x_0 < \xi < x_n. \quad \text{For } n = 1, \tag{3.4.5}$$

this is the simple mean value theorem.

Ex. 1.   Tables of functions frequently list the first few differences. Suppose that f has been tabulated at an interval of h and suppose that we obtain the value of f at a point x between successive abscissas a and $a + h$ by linear interpolation. By Cor. 3.1.2, the error committed, R_1, is

$$\frac{(x - a)(x - a - h)}{2} f''(\xi), \quad a < \xi < a + h.$$

By (3.4.5) $\Delta^2 f(a) = h^2 f''(\xi_1)$ so that if "h is sufficiently small" $f''(\xi) \approx \dfrac{1}{h^2} \Delta^2 f(a)$.

Since $\displaystyle\max_{a \leq x \leq a+h} |(x-a)(x-a-h)| = \frac{h^2}{4}$, it follows that

$$|R_1| \leq \tfrac{1}{8}|\Delta^2 f(a)|.$$

This leads to a rule of thumb long employed by computers: *the error in linear interpolation does not exceed $\tfrac{1}{8}$ of the 2nd difference.*

3.5 Interpolation at Coincident Points.

In formulating the fundamental problem of polynomial interpolation, we have assumed that the interpolating points are distinct. With a proper convention as to what interpolation at coincident points means, this restriction can be overcome. The convention arises from considerations of the following sort. Suppose we interpolate to $f(x)$ at the distinct points $x_0, x_1, \ldots, x_n$. Then,

$$p_n(f; x) = \sum_{k=0}^{n} [f(x_0), f(x_1), \ldots, f(x_k)](x - x_0) \cdots (x - x_{k-1}). \quad (3.5.1)$$

If we make $x_1, \ldots, x_n$ "coincide" at x_0 by allowing $x_1, x_2, \ldots, x_n \to x_0$, then, by (3.4.4), the limiting expression on the right hand side will be

$$p_n(f; x) = \sum_{k=0}^{n} \frac{f^{(k)}(x_0)}{k!} (x - x_0)^k. \quad (3.5.2).$$

The interpolation polynomial approaches the truncated Taylor expansion for $f(x)$ at x_0. This is an interpolation in which the values $f(x_0), \ldots, f^{(n)}(x_0)$ have been prescribed. This provides an interpretation for interpolation at $n + 1$ coincident points.

In an analogous way, the following convention is introduced. Suppose that among the $n + 1$ points $x_0, x_1, \ldots, x_n$ only $j + 1$ of them, $x_0, x_1, \ldots, x_j$, are distinct. Suppose that in the list of points x_0 occurs n_0 times, $\ldots, x_j$ occurs n_j times so that $n_0 + n_1 + \cdots + n_j = n + 1$. Then, by interpolation to $f(x)$ at $x_0, x_1, \ldots, x_n$ we shall understand the determination of the unique polynomial of degree $\leq n$, $p_n(x)$, for which

$$p_n(x_0) = f(x_0), \; p_n'(x_0) = f'(x_0), \ldots, p_n^{(n_0-1)}(x_0) = f^{(n_0-1)}(x_0)$$

$$\cdot$$
$$\cdot \qquad\qquad\qquad\qquad\qquad\qquad (3.5.3)$$
$$\cdot$$

$$p_n(x_j) = f(x_j), \; p_n'(x_j) = f'(x_j), \ldots, p_n^{(n_j-1)}(x_j) = f^{(n_j-1)}(x_j).$$

This is a problem of Hermite interpolation and the existence and uniqueness of this polynomial is guaranteed by Ex. 4, Ch. 2.5.

To justify this convention from the point of view of a limiting process, in the way in which (3.5.2) was derived from (3.5.1), we should have to study the limiting expressions of the divided differences when each of several groups of arguments approach distinct limits. This would lead to the notion of *generalized divided differences* and is a topic that will not be pursued in this book.

Having identified "coincident point" interpolation with Hermite interpolation, we point out that a remainder formula analogous to (3.1.1) is easily obtained.

THEOREM 3.5.1. *Let $x_0, x_1, \ldots, x_n$ be $n + 1$ distinct points in $[a, b]$. Let $n_0, n_1, \ldots, n_n$ be $n + 1$ integers ≥ 0. Let $N = (n_0 + n_1 + \cdots + n_n) + n$. Designate by $H_N(f; x)$ the unique element of $\mathscr{P}_N$ for which*

$$H_N^{(k)}(f; x_i) = f^{(k)}(x_i) \quad k = 0, 1, \ldots, n_i, \quad i = 0, 1, \ldots, n. \quad (3.5.4)$$

Let $f(x) \in C[a, b]$ and suppose that $f^{(N+1)}(x)$ exists in (a, b); then

$$f(x) - H_N(f; x) = \frac{f^{(N+1)}(\xi)}{(N+1)!} (x - x_0)^{n_0+1}(x - x_1)^{n_1+1} \cdots (x - x_n)^{n_n+1}$$

$$(3.5.5)$$

where $\min(x, x_0, \ldots, x_n) < \xi < \max(x, x_0, \ldots, x_n).$

Proof: Consider $(f(x) - H_N(f; x))/(x - x_0)^{n_0+1} \cdots (x - x_n)^{n_n+1}$ and proceed as in the proof of Theorem 3.1.1.

3.6 Analytic Functions: Remainder for Polynomial Interpolation.

Let $f(z)$ be analytic in a closed simply connected region R. Let C be a simple, closed, rectifiable curve that lies in R and contains the distinct points $z_0, z_1, \ldots, z_n$ in its interior. Consider the integral

$$I = \frac{1}{2\pi i} \int_C \frac{f(z)}{(z - z_0)(z - z_1) \cdots (z - z_n)} \, dz. \quad (3.6.1)$$

The integrand is analytic or has simple poles at $z_0, z_1, \ldots, z_n$. Hence, by the residue theorem,

$$I = \sum_{k=0}^{n} \frac{f(z_k)}{(z_k - z_0) \cdots (z_k - z_{k-1})(z_k - z_{k+1}) \cdots (z_k - z_n)} \quad (3.6.2)$$

Compare this with (2.6.4), (2.5.6) and obtain

$$[f(z_0), f(z_1), \ldots, f(z_n)] = I. \quad (3.6.3)$$

By the same token,

$$[f(z), f(z_0), \ldots, f(z_n)] = \frac{1}{2\pi i} \int_C \frac{f(t)}{(t - z)(t - z_0), \ldots, (t - z_n)} \, dt, \quad z \in R. \quad (3.6.4)$$

From (3.4.1),

$$\begin{aligned}
R_n(f; z) &= f(z) - p_n(f; z) \\
&= [f(z), f(z_0), \ldots, f(z_n)](z - z_0)(z - z_1) \cdots (z - z_n) \\
&= \frac{(z - z_0)(z - z_1) \cdots (z - z_n)}{2\pi i} \int_C \frac{f(t) \, dt}{(t - z)(t - z_0) \cdots (t - z_n)}.
\end{aligned}$$

We have therefore proved

THEOREM 3.6.1 (Hermite). *Under the above regularity conditions,*

$$f(z) - p_n(f; z) = R_n(f; z)$$

$$= \frac{1}{2\pi i} \int_C \frac{(z - z_0)(z - z_1) \cdots (z - z_n) f(t) \, dt}{(t - z_0)(t - z_1) \cdots (t - z_n)(t - z)} \tag{3.6.5}$$

COROLLARY 3.6.2.

$$p_n(f; z) = \frac{1}{2\pi i} \int_C \frac{w(t) - w(z)}{w(t)(t - z)} f(t) \, dt \tag{3.6.6}$$

where $w(z) = (z - z_0)(z - z_1) \cdots (z - z_n)$.

Proof: $f(z) = \dfrac{1}{2\pi i} \displaystyle\int_C \dfrac{f(t)}{t - z} \, dt$. Subtract (3.6.5) from this.

COROLLARY 3.6.3. *Formula (3.6.6) has meaning if the points* $z_0, z_1, \ldots, z_n$ *are not distinct and yields the polynomial that interpolates to f in the generalized sense explained in 3.5.*

Proof: We shall give only a brief indication of how this goes. For simplicity, suppose that z_0 and z_1 are coincident and the other z's are distinct. Then, $w(z) = (z - z_0)^2(z - z_2) \cdots (z - z_n)$ and so, $w(z_0) = 0$, $w'(z_0) = 0$. From (3.6.6),

$$p_n(f; z_0) = \frac{1}{2\pi i} \int_C \frac{w(t) f(t)}{w(t)(t - z_0)} \, dt = f(z_0)$$

and

$$p_n'(f; z_0) = \frac{1}{2\pi i} \int_C \frac{\partial}{\partial z} \left(\frac{w(t) - w(z)}{(t - z)} \right)_{z = z_0} \frac{f(t) \, dt}{w(t)}$$

$$= \frac{1}{2\pi i} \int_C \frac{f(t)}{(t - z_0)^2} \, dt = f'(z_0).$$

Now $p_n(f; z_i)$ $i = 2, \ldots, n$ is easily computed to be $f(z_i)$ and therefore $p_n(f; z)$ takes on interpolatory values as required.

COROLLARY 3.6.4. *If f is analytic at* z_0 *then*

$$\lim_{z_1, z_2, \ldots, z_n \to z_0} [f(z_0), f(z_1), \ldots, f(z_n)] = \frac{1}{n!} f^{(n)}(z_0) \tag{3.6.7}$$

Proof: In the limit, I becomes $\dfrac{1}{2\pi i} \displaystyle\int_C \dfrac{f(z)}{(z - z_0)^{n+1}} = \dfrac{1}{n!} f^{(n)}(z_0)$. This is the complex analog of (3.4.4).

Ex. 1. The limiting form of (3.6.5) as $z_1, \ldots, z_n \to z_0$ is the Taylor Series with the exact complex remainder

$$f(z) - \sum_{k=0}^{n} \frac{f^{(k)}(z_0)}{k!} (z - z_0)^k = \frac{1}{2\pi i} \int_C \left(\frac{z - z_0}{t - z_0}\right)^{n+1} \frac{f(t)}{t - z} \, dt.$$

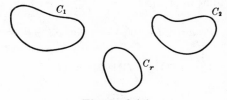

Figure 3.6.1.

A moment's consideration of the residue theorem should convince the reader of the validity of the following generalization of formulas (3.6.5) and (3.6.6). Let C consist of a finite number of mutually exterior curves $C_1, \ldots, C_r$. Let $f_i(z)$ be a single valued analytic function in and on C_i $i = 1, 2, \ldots, r$. The functions f_i need not be analytic continuations of one another. The total configuration of functions $f_i(z)$ will be designated by $f(z)$, and f will be thought of as a certain *analytic configuration*. If, now, each of the points $z_0, \ldots, z_n$ is contained in the interior of some C_i, and if interpolation to $f(z)$ is carried out at these points, then formulas (3.6.3), (3.6.5), and (3.6.6) are still valid.

Ex. 2. Formula (3.6.5) provides a complex analog of the estimate (3.1.8). If $\delta_i = $ minimum distance from z_i to C and $\delta = $ minimum distance from z to C then,

$$|R_n(f; z)| \leq \frac{L(C)M_C|z - z_0| \, |z - z_1| \cdots |z - z_n|}{2\pi \, \delta \, \delta_0 \, \delta_1 \cdots \delta_n}.$$

3.7 Peano's Theorem and Its Consequences. If we examine, once again, the Cauchy remainder for polynomial interpolation (3.1.1), we may note the prominent role played by the portion $f^{(n+1)}(\xi)$. If, for instance, $f \in \mathscr{P}_n$, then $f^{(n+1)} \equiv 0$, and the remainder vanishes identically as it should. For a fixed x, we may consider the remainder $R_n(f; x) = f(x) - p_n(f; x)$ as a linear functional which operates on f and which annihilates all elements of $\mathscr{P}_n$. Peano observed that if a linear functional has this property, then it must also have a simple representation in terms of $f^{(n+1)}$.

Without striving for full generality, consider functions of class $C^{n+1}[a, b]$, and let linear functionals of the following type be defined over this class.

$$L(f) = \int_a^b [a_0(x)f(x) + a_1(x)f'(x) + \cdots + a_n(x)f^{(n)}(x)]\, dx$$

$$+ \sum_{i=1}^{j_0} b_{i0}f(x_{i0}) + \sum_{i=1}^{j_1} b_{i1}f'(x_{i1}) + \cdots + \sum_{i=1}^{j_n} b_{in}f^{(n)}(x_{in}). \quad (3.7.1)$$

The functions $a_i(x)$ are assumed to be piecewise continuous over $[a, b]$ and the points x_{ij} to lie in $[a, b]$.

THEOREM 3.7.1 (Peano). *Let* $L(p) = 0$ *for all* $p \in \mathscr{P}_n$. *Then, for all* $f \in C^{n+1}[a, b]$,

$$L(f) = \int_a^b f^{(n+1)}(t)K(t)\, dt \quad (3.7.2)$$

where

$$K(t) = \frac{1}{n!} L_x[(x - t)_+^n] \quad (3.7.3)$$

and

$$(x - t)_+^n = (x - t)^n \qquad x \geq t$$
$$(x - t)_+^n = 0 \qquad\qquad x < t. \qquad (3.7.4)$$

The notation $L_x[(x - t)_+^n]$ *means that the functional* L *is applied to* $(x - t)_+^n$ *considered as a function of* x.

Proof: Taylor's Theorem with the exact remainder tells us that

$$f(x) = f(a) + f'(a)(x - a) + \cdots$$

$$+ \frac{f^{(n)}(a)(x - a)^n}{n!} + \frac{1}{n!} \int_a^x f^{(n+1)}(t)(x - t)^n\, dt.$$

We may evidently write the last term as $\dfrac{1}{n!} \displaystyle\int_a^b f^{(n+1)}(t)(x - t)_+^n\, dt$. Now apply L to both sides of this expansion and recall that L vanishes for all elements of $\mathscr{P}_n$. This yields

$$L(f) = \frac{1}{n!} L \int_a^b f^{(n+1)}(t)(x - t)_+^n\, dt. \quad (3.7.5)$$

Since we have assumed a form (3.7.1) for L, we are working under hypotheses which allow an interchange of the functional L with the integral in (3.7.5). Hence,

$$L(f) = \frac{1}{n!} \int_a^b f^{(n+1)}(t)L_x[(x - t)_+^n]\, dt. \quad (3.7.6)$$

The function $K(t)$ is called the *Peano Kernel* associated with the functional L.

COROLLARY 3.7.2. *If, in addition to the above hypotheses, the kernel $K(t)$ does not change its sign on $[a, b]$ then for all $f \in C^{n+1}[a, b]$,*

$$L(f) = \frac{f^{(n+1)}(\xi)}{(n + 1)!} L(x^{n+1}), \qquad a \leq \xi \leq b \tag{3.7.7}$$

Proof: From (3.7.2) and (1.4.5),

$$L(f) = f^{(n+1)}(\xi) \int_a^b K(t)\, dt \qquad a \leq \xi \leq b \tag{3.7.8}$$

Insert $f = x^{n+1}$ in (3.7.8) and obtain

$$L(x^{n+1}) = (n + 1)! \int_a^b K(t)\, dt \tag{3.7.9}$$

Combining these yields (3.7.7).

A functional that satisfies the conditions $K(t) \geq 0$ (or $K(t) \leq 0$) on $[a, b]$ is known as a *positive* (or *negative*) *functional of order n.* Many of the error functionals that occur in numerical analysis are of this type.

Ex. 1. *Kowalewski's Exact Remainder for Polynomial Interpolation.* Let $x, x_0, \ldots, x_n$ be fixed in $[a, b]$. Let $L(f) = R_n(f; x) = f(x) - \sum_{k=0}^{n} f(x_k)\ell_k(x)$. (See (2.5.9)).
Then,

$$n!\, K(t) = L_x(x - t)_+^n = (x - t)_+^n - \sum_{k=0}^{n} (x_k - t)_+^n \ell_k(x)$$

$$= \sum_{k=0}^{n} [(x - t)_+^n - (x_k - t)_+^n]\ell_k(x).$$

The last equality follows from (2.5.13). We now put this in a more convenient form. For fixed k we have by (3.7.4),

$$\int_a^b [(x - t)_+^n - (x_k - t)_+^n]f^{(n+1)}(t)\, dt$$

$$= \int_a^x [(x - t)^n - (x_k - t)^n]f^{(n+1)}(t)\, dt + \int_{x_k}^x (x_k - t)^n f^{(n+1)}(t)\, dt.$$

Hence,

$$n! \int_a^b K(t) f^{(n+1)}(t)\, dt$$

$$= \int_a^x f^{(n+1)}(t) \sum_{k=0}^{n} [(x - t)^n - (x_k - t)^n]\ell_k(x)\, dt + \sum_{k=0}^{n} \ell_k(x) \int_{x_k}^x (x_k - t)^n f^{(n+1)}(t)\, dt.$$

The inner sum in the second integral may be transformed by (2.5.13):

$$\sum_{k=0}^{n} [(x - t)^n - (x_k - t)^n]\ell_k(x) = (x - t)^n - \sum_{k=0}^{n} (x_k - t)^n \ell_k(x).$$

Since $\sum_{k=0}^{n} (x_k - t)^n \ell_k(x) = p_n((x-t)^n; x) = (x-t)^n$, the inner sum vanishes identically. Thus, finally,

$$L(f) = f(x) - p_n(f; x) = \frac{1}{n!} \sum_{k=0}^{n} \ell_k(x) \int_{x_k}^{x} (x_k - t)^n f^{(n+1)}(t)\, dt, \quad f \in C^{n+1}[a, b]. \tag{3.7.10}$$

Ex. 2.   (Integral remainder for linear interpolation.) The case $n = 1$, $x_0 = a$, $x_1 = b$ is particularly noteworthy. Then $\ell_0(x) = \dfrac{x-b}{a-b}$, $\ell_1(x) = \dfrac{x-a}{b-a}$. From (3.7.10),

$$f(x) - \frac{x-b}{a-b} f(a) - \frac{x-a}{b-a} f(b)$$

$$= \frac{x-b}{b-a} \int_a^x (t-a) f''(t)\, dt + \frac{x-a}{b-a} \int_x^b (t-b) f''(t)\, dt. \tag{3.7.11}$$

Introduce the following function defined over the square $a \le x \le b, a \le t \le b$

$$G(x, t) = \begin{cases} \dfrac{(t-a)(x-b)}{b-a} & t \le x \\[2mm] \dfrac{(x-a)(t-b)}{b-a} & x \le t. \end{cases} \tag{3.7.12}$$

Then we may write (3.7.11) in the form

$$R_1(f; x) = \int_a^b G(x, t) f''(t)\, dt \tag{3.7.13}$$

The function $G(x, t)$ is, for fixed x, the Peano kernel for $R_1(f)$.

Let $h(x) \in C[a, b]$ and $H''(x) = h(x)$. Set

$$\phi(x) = \int_a^b G(x, t) h(t)\, dt. \tag{3.7.14}$$

Then, by (3.7.13),

$$\phi(x) = H(x) - p_1(H; x) \quad \text{so that} \quad \phi''(x) = H''(x) = h(x).$$

Furthermore, $\phi(a) = R_1(H; a) = 0$, $\phi(b) = R_1(H; b) = 0$.

Therefore the integral (3.7.14) solves the differential problem

$$\begin{aligned} \phi''(x) &= h \\ \phi(a) &= \phi(b) = 0. \end{aligned} \tag{3.7.15}$$

The function $G(x, t)$ is known as *the Green's function* for the differential system (3.7.15). These remarks indicate the close relationship between Peano's kernels and Green's functions, and hence between interpolation theory and the theory of linear differential equations. Unfortunately, we shall not be able to pursue this relationship.

Ex. 3.   Let $x_1 = x_0 + h$, $x_2 = x_0 + 2h$, $x_3 = x_0 + 3h$ and set

$$L(f) = -f(x_0) + 3f(x_1) - 3f(x_2) + f(x_3) = \Delta^3 f(x_0).$$

L annihilates all elements of $\mathscr{P}_2$. Hence, $n = 2$ and

$$K(t) = \frac{1}{2!} L(x - t)^2_+.$$

If we write this out explicitly we find

$$\begin{aligned}
2K(t) &= (x_3 - t)^2 - 3(x_2 - t)^2 + 3(x_1 - t)^2 = (t - x_0)^2, & x_0 \le t \le x_1 \\
&= (x_3 - t)^2 - 3(x_2 - t)^2 & x_1 \le t \le x_2 \quad (3.7.16) \\
&= (x_3 - t)^2 & x_2 \le t \le x_3
\end{aligned}$$

The kernel $K(t)$ consists of 3 parabolic arches and is of class $C^1[x_0, x_3]$. Thus, for $f \in C^3[x_0, x_3]$,

$$L(f) = \Delta^3 f(x_0) = \int_{x_0}^{x_3} K(t) f'''(t)\, dt. \qquad (3.7.17)$$

Note that $K(t) \ge 0$. We may apply (3.7.7) yielding

$$\Delta^3 f(x_0) = \frac{f^{(3)}(\xi)}{3!} \Delta^3(x^3) = h^3 f^{(3)}(\xi). \quad \text{Cf. (3.4.5).}$$

Similar formulas hold for differences of all orders.

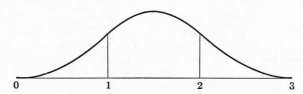

Figure 3.7.1   Peano Kernel for 3rd Difference, $x_0 = 0$, $h = 1$.

Ex. 4.   The Trapezoidal Rule and the Euler-MacLaurin Summation Formula. Let

$$L(f) = \int_a^b f(x)\, dx - \frac{b - a}{2} [f(a) + f(b)] \qquad (3.7.18)$$

be the error in estimating the definite integral $\int_a^b f(x)\, dx$ by the trapezoidal rule $\frac{1}{2}(b - a)[f(a) + f(b)]$. The rule is exact for linear functions, and, in particular, for constants. If we select $n = 0$, we have

$$(x - t)^0_+ = S(x, t) = \begin{cases} 1 & \text{for } x \ge t \\ 0 & \text{for } x < t \end{cases}$$

Then

$$\begin{aligned}
L_x(S(x, t)) &= \int_a^b S(x, t)\, dx - \frac{b - a}{2} [S(a, t) + S(b, t)] \\
&= \int_t^b dx - \frac{b - a}{2} [0 + 1] = \tfrac{1}{2}(a + b) - t, \quad t > a.
\end{aligned}$$

Therefore

$$L(f) = -\int_a^b (t - \tfrac{1}{2}(a + b))f'(t)\, dt. \tag{3.7.19}$$

Consider, next, the extended trapezoidal rule,

$$L(f) = \int_a^b f(x)\, dx - \frac{b - a}{n}\left[\frac{f(a)}{2} + f(a + \sigma) + f(a + 2\sigma) + \cdots \right.$$

$$\left. + f(a + (n - 1)\sigma) + \frac{f(b)}{2}\right], \quad \sigma = \frac{1}{n}(b - a). \tag{3.7.20}$$

An expression analogous to (3.7.19) is most conveniently obtained by adding expressions of this form for each subinterval.

$$L(f) = -\sum_{k=0}^{n-1}\int_{a+k\sigma}^{a+(k+1)\sigma} (t - (a + (k + \tfrac{1}{2})\sigma))f'(t)\, dt. \tag{3.7.21}$$

In particular, if we select $a = 0$, $b = n$, $\sigma = 1$, then over $k \le t \le k + 1$, $t - k - \tfrac{1}{2}$ becomes $t - [t] - \tfrac{1}{2}$, where $[t]$ is the largest integer contained in t, and we rewrite (3.7.21) as

$$\int_0^n f(x)\, dx + \frac{f(0) + f(n)}{2} - [f(0) + f(1) + \cdots + f(n)]$$

$$= \int_0^n ([t] - t + \tfrac{1}{2})f'(t)\, dt. \tag{3.7.22}$$

This is the simplest version of the Euler-MacLaurin summation formula.

Ex. 5.   Remainder in Simpson's Rule. Let

$$L(f) = \int_{-1}^{+1} f(x)\, dx - \tfrac{1}{3}f(-1) - \tfrac{4}{3}f(0) - \tfrac{1}{3}f(1). \quad L(p) = 0 \quad \text{if} \quad p \in \mathscr{P}_3.$$

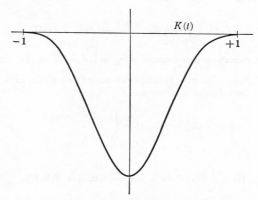

Figure 3.7.2 Kernel for Simpson's Rule.

Applying (3.7.3) we find $6K(t) = L_x[(x - t)_+^3]$ or,

$$K(t) = -\tfrac{1}{72}(1 - t)^3(3t + 1) \qquad 0 \le t \le 1$$
$$K(t) = K(-t) \qquad -1 \le t \le 0$$
$$(3.7.23)$$

Note that $K(t) \le 0$ so that Corollary 3.7.2 is applicable:

$$L(f) = \frac{1}{4!} f^{(4)}(\xi) L(x^4) = \frac{f^{(4)}(\xi)}{4!} \left(-\frac{4}{15} \right) = -\frac{f^{(4)}(\xi)}{90} .$$

This leads to the following error for Simpson's rule:

$$\int_{-1}^{1} f(x)\, dx = \tfrac{1}{3} f(-1) + \tfrac{4}{3} f(0) + \tfrac{1}{3} f(1) - \frac{f^{(4)}(\xi)}{90}, \quad -1 \le \xi \le 1. \quad (3.7.24)$$

3.8 Interpolation in Linear Spaces; General Remainder Theorem.

We cannot say too much in the general case, for the underlying structure is too meagre. But it is instructive to derive what result we can, and this will round off the formal algebraic work of 2.5, 2.6.

Given an element x in a linear space X, we interpolate to x by an appropriate linear combination of $x_1, \ldots, x_n$, $a_1 x_1 + \cdots + a_n x_n$, such that $L_i(a_1 x_1 + \cdots + a_n x_n) = L_i(x)$ $i = 1, 2, \ldots, n$. Let

$$x_R = x - (a_1 x_1 + \cdots + a_n x_n).$$

Then, $L_i(x_R) = 0$ $i = 1, 2, \ldots, n$.

THEOREM 3.8.1. *Under the assumption that* $|L_i(x_j)| \neq 0$, *we have*

$$x_R = \begin{vmatrix} x & x_1 & \cdots & x_n \\ L_1(x) & L_1(x_1) & \cdots & L_1(x_n) \\ \cdot & & & \cdot \\ \cdot & & & \cdot \\ \cdot & & & \cdot \\ L_n(x) & L_n(x_1) & \cdots & L_n(x_n) \end{vmatrix} \div \begin{vmatrix} L_1(x_1) & L_1(x_2) & \cdots & L_1(x_n) \\ \cdot & \cdot & & \cdot \\ \cdot & \cdot & & \cdot \\ \cdot & \cdot & & \cdot \\ L_n(x_1) & L_n(x_2) & \cdots & L_n(x_n) \end{vmatrix} \quad (3.8.1)$$

Proof: It is clear by expanding the numerator of (3.8.1) by the minors of its first row that the right hand side of (3.8.1) is a linear combination of $x, x_1, \ldots, x_n$, and that the coefficient of x is precisely 1. Applying L_i to the right hand side (which we may do by letting it operate on each element of the first row), we see that this row is identical with the $(i + 1)$st row and hence $L_i(x_R) = 0$, $i = 1, 2, \ldots, n$. Thus, the expression (3.8.1) has all the properties the remainder element x_R should have.

NOTES ON CHAPTER III

3.2 Hardy, Littlewood, and Pólya [1], pp. 70–75. R. P. Boas, Jr. [4] pp. 142–150.

3.3 The Tschebyscheff polynomials are everywhere dense in the literature of numerical analysis, approximation theory, and special function theory. The National Bureau of Standards, Table of Chebyshev Polynomials [1] has an introduction by C. Lanczos that summarizes the properties of these famous polynomials and indicates their use in numerical analysis.

3.6 J. L. Walsh [2] p. 50. For coincident points of interpolation, p. 53.

3.7 G. Peano [1], [2]. In recent years, Arthur Sard has called attention to the utility of Peano's Theorem. A. Sard [1], [2]. Kowalewski's Remainder: G. Kowalewski [1] pp. 21–24. For some of the kernels see Sard [1], Milne [1], Kuntzmann [1] pp. 44–49, 152–157.

3.8 See references to 2.5, 2.6.

PROBLEMS

1. Log_{10} 12.7 has been computed by looking up $\log_{10}$ 12 and $\log_{10}$ 13 and interpolating linearly between these values. Show that the error incurred is $\leq .004$.

2. Formulate rules of thumb for the accuracy of quadratic, cubic, and quartic interpolation on equidistant points.

3. A polynomial of degree n, $p_n(x)$, coincides with e^x at the points $\dfrac{0}{n}, \dfrac{1}{n}, \ldots,$ $\dfrac{n-1}{n}, \dfrac{n}{n}$. How large shall n be taken so as to insure that $\left| e^x - p_n(x) \right| \leq 10^{-6}$ over $0 \leq x \leq 1$?

4. Let $f(z) = e^{z^2}, z_i = \dfrac{i}{10}, i = 0, 1, \ldots, 10$. Estimate $R_{10}(f; \frac{1}{20})$.

5. Same problem with $f(z) = \sqrt{9 + \sqrt{z}}$.

6. Write explicitly the remainder for simple osculatory interpolation at the $n + 1$ points $x_0, x_1, \ldots, x_n$. If $f^{(2n+2)}(x) \geq 0$, the interpolant never exceeds the function over the range of interest.

7. If $p \geq 1, y = x^p$ is convex on $[0, a]$ for any $a > 0$.

8. A monotone increasing function of a convex function is itself convex. True or false?

9. Find necessary and sufficient conditions on the a's in order that $\sum\limits_{n=0}^{4} a_n x^n$ be convex on $-1 \leq x \leq 1$.

10. $$T_m(T_n(x)) = T_n(T_m(x)) = T_{mn}(x).$$
In particular,
$$T_n(2x^2 - 1) = 2T_n{}^2(x) - 1.$$

11. $$T_m(x)T_n(x) = \tfrac{1}{2}[T_{m+n}(x) + T_{m-n}(x)].$$

12. Prove the identities $\displaystyle\int T_0(x)\, dx = T_1(x), \quad \int T_1(x)\, dx = \tfrac{1}{4}T_2(x),$

$$\int T_n(x)\, dx = \frac{1}{2}\left(\frac{T_{n+1}(x)}{n+1} - \frac{T_{n-1}(x)}{n-1} \right) \quad n > 1.$$

13. $f'(0) - \dfrac{1}{2h}[f(h) - f(-h)] = \displaystyle\int_{-1}^{1} f^{(3)}(x)K(x)\, dx \quad 0 < h < 1$

$$K(x) = \begin{cases} -\dfrac{(x+h)^2}{4h} & -h \le x \le 0 \\[2mm] -\dfrac{(x-h)^2}{4h} & 0 \le x \le h \\[2mm] 0 & \text{otherwise} \end{cases}$$

14. Derive this formula directly by integration by parts.

15. Show that $f(-1) = f(0) - f'(0) + \tfrac{1}{2}f''(0) + \displaystyle\int_{-1}^{1} f'''(t)K(t)\, dt$

where
$$K(t) = \begin{cases} -\tfrac{1}{4}(1-t)^2 & 0 < t \le 1 \\[2mm] -\tfrac{1}{4}(1+t)^2 & -1 \le t \le 0 \end{cases}$$

16. Show that $-\tfrac{1}{4}f(-1) - \tfrac{1}{2}f(0) + \tfrac{3}{4}f(1) - f'(\tfrac{1}{4}) = \displaystyle\int_{-1}^{1} f^{(3)}(x)K(x)\, dx$

where
$$K(x) = \begin{cases} \tfrac{1}{8}(1+x)^2 & -1 \le x < 0 \\[2mm] \tfrac{1}{8}(1+2x+3x^2) & 0 \le x < \tfrac{1}{4} \\[2mm] \tfrac{3}{8}(1-x)^2 & \tfrac{1}{4} \le x \le 1 \end{cases}$$

17. Find the Peano kernel for $\Delta^4 f(x_0)$, $h = 1$.

18. Derive the following formula of Euler-MacLaurin type

$$\int_a^b f(x)\, dx = \tfrac{1}{2}(b-a)(f(a) + f(b)) + \tfrac{1}{12}(b-a)^2(f'(a) - f'(b))$$

$$+ \frac{1}{4!}\int_a^b (x-a)^2(x-b)^2 f^{(4)}(x)\, dx.$$

19. Study the continuity class of $K(t)$ for various functionals.

CHAPTER IV

Convergence Theorems for Interpolatory Processes

4.1 Approximation by Means of Interpolation. Suppose that we have been given certain information about a function. Perhaps we know its values, or the values of its derivatives at certain points. Perhaps we know its moments. How can we use this information to construct other functions that will approximate it? This is the practical problem of numerical analysis. Theoretical analysis can go beyond. Having been given an infinite number of facts about our function, how can we reconstruct the function completely? Borrowing some terms from harmonic analysis, we may say that the process of extracting functional information constitutes an *analysis* of the function, while the process of reconstructing the function from given functional information is a *synthesis* of the function.

One of the most surprising facts in the theory of interpolation and approximation is that the simplest and most natural approach to synthesis leads to failure, or rather, to an impossibility. Given a function of class $C[a, b]$, what is more natural than to think that if a sequence of polynomials $p_n(f; x)$ is set up that duplicates the function at $n + 1$ points of the interval, then as $n \to \infty$, $p_n(f; x)$ will converge to $f(x)$? Yet, this may not be the case. One of the first indications of this came around the turn of the century when Méray and later Runge, investigated interpolation to certain meromorphic functions. Runge looked at the function $f(x) = \dfrac{1}{1 + x^2}$ and found the following to be true: if $p_n(f; x)$ interpolates to f at $n + 1$ equidistant points of the interval $|x| \leq 5$, p_n converges to f only in the interval $|x| \leq 3.63 \cdots$ and diverges outside this interval. Although f is analytic on the whole real axis, its singularities at $\pm i$ induce, so to speak, this divergence. In 1912, Bernstein proved that equidistant interpolation over $|x| \leq 1$ to the function $y = |x|$ diverges for $0 < |x| < 1$.

These results relate to equidistant points of interpolation. The possibility was still open that a more felicitous choice of points would give rise to a convergent interpolation process. Some indications of this are contained in (3.3.10) where interpolation on the zeros of the Tschebyscheff polynomials minimizes certain error estimates. The hopes for this idea vanished when Bernstein and Faber simultaneously discovered in 1914 that if any triangular system of interpolation points is prescribed in advance, we can construct a continuous function for which the interpolation process carried out on these points cannot converge uniformly to this function. Even the Tschebyscheff zeros as interpolation points fare badly, for in 1937, Marcinkiewicz gave an example of a continuous function for which interpolation at these zeros diverges at every point of $(-1, 1)$.

Yet, the damage is not as great as one might think and can be repaired in several ways. The first is to change the way the interpolation is carried out by not insisting that for $n + 1$ points a polynomial of class $\mathscr{P}_n$ be employed. Fejér proved a remarkable theorem showing how an interpolation process with controlled derivative values can converge properly for all continuous functions. The second way is not to insist on working with the class of continuous functions, but to assume some smoothness properties. The Tschebyscheff zeros (and the zeros of other orthogonal polynomials) actually are a remarkable system of interpolating points. Bernstein showed in 1916 that if f is a continuous function for which $\lim_{\delta \to 0} w(\delta; f) \log \delta = 0$, interpolation at these points produces a properly convergent sequence of polynomials. If interpolation is carried out on bounded sets, it suffices to assume that our functions are analytic in certain regions. If interpolation is carried out on unbounded sets, say the integers, then for convergence, we shall have to assume that our functions are entire and of severely restricted growth. Thus, if we are to have convergence, there must be a subtle interplay between the distribution of points of interpolation and the smoothness or growth properties of the interpolated function. Though much is known about this interplay, we shall be able to develop in this chapter only a few of its broader features.

4.2 Triangular Interpolation Schemes. We first describe an interpolation scheme of great generality. Let there be given a triangular sequence of real or complex points

$$T: \begin{matrix} z_{00} \\ z_{10} & z_{11} \\ z_{20} & z_{21} & z_{22} \\ \cdot \\ \cdot \\ \cdot \end{matrix} \tag{4.2.1}$$

Suppose that a function $f(z)$ has been defined on a region containing the points of T, and let $p_n(f; z)$ be that element of $\mathscr{P}_n$ for which

$$p_n(f; z_{ni}) = f(z_{ni}) \qquad \begin{matrix} i = 0, 1, \ldots, n \\ n = 0, 1, \ldots. \end{matrix} \tag{4.2.2}$$

In other words, $p_n(f; z)$ interpolates to f at the points of the $(n + 1)$st row of T. The numbers in the rows of T may or may not be distinct. If they are not distinct, then the interpolation polynomial is to be formed in accordance with the convention explained in 3.5.

We now ask the question, does

$$\lim_{n \to \infty} p_n(f; z) = f(z)?$$ (4.2.3)

In such generality, the answer is a resounding no, but the problem is to delineate those circumstances under which the answer is yes.

In many cases of interest, the matrix T degenerates by having its elements not depend upon the row, but only upon the column. In such a case, we can drop the double indexing, and write the scheme as follows

$$
S: \begin{array}{ccc}
z_0 & & \\
z_0 & z_1 & \\
z_0 & z_1 & z_2 \\
\cdot & & \\
\cdot & & \\
\cdot & &
\end{array}
$$ (4.2.4)

For a scheme of this type, we have

$$p_n(f; z) = \sum_{k=0}^{n} [f(z_0), f(z_1), \ldots, f(z_k)](z - z_0), \ldots, (z - z_{k-1})$$ (4.2.5)

and so the existence of $\lim_{n \to \infty} p_n(f; z)$ is identical with the convergence of the *interpolation series*

$$f(z) \sim \sum_{k=0}^{\infty} [f(z_0), f(z_1), \ldots, f(z_k)](z - z_0) \cdots (z - z_{k-1}).$$ (4.2.6)

In order to appreciate the kinds of things that may occur with triangular interpolation schemes, we shall consider a few examples.

Ex. 1.   A scheme S is used with $z_0 = 1, z_1 = \frac{1}{2}, \ldots, z_n = \dfrac{1}{n + 1} \cdots$. The interpolated function is $f(x) = x \sin \dfrac{\pi}{x}$ which is continuous in $-\infty < x < \infty$. Since $f(z_k) = 0$, the interpolation polynomials $p_n(f; x)$ are all identically zero. The sequence $p_n(f; x)$ converges, but not to f.

Ex. 2.   On the other hand if $f \in \mathscr{P}_n$, then no matter how the matrix T is constituted, we shall always have $p_m(f; z) \equiv f(z)$ for $m \geq n$. Hence convergence takes place to the proper value. In other words, if the class of interpolated functions is sufficiently small (the class of all polynomials) a triangular scheme is always convergent.

Ex. 3.   A degenerate case of S is where all the points have a common value z_0. By our convention, the interpolating polynomials are the partial sums of the Taylor expansion of $f: f(z_0) + f'(z_0)(z - z_0) + \cdots + \dfrac{f^{(n)}(z_0)}{n!}(z - z_0)^n$. We have convergence to $f(z)$ if and only if f is analytic at z_0 and the convergence holds throughout the largest circle $|z - z_0| < \rho$ in which $f(z)$ is analytic. On the other hand, if f is merely of class C^∞, we may have divergence or convergence to a wrong value.

Ex. 4. For $z_{n0}, \ldots, z_{nn}$ select the $(n + 1)$st roots of 1. Call them w_j. Choose $f(z) = \dfrac{1}{z}$. Then, $p_n(f; z) \equiv z^n$ inasmuch as $w_j{}^n = \dfrac{1}{w_j} = f(z_{nj})$. Notice that $p_n(f; x)$ converges to 0 in $|z| < 1$, diverges for $|z| > 1$. The sequence converges to f only at $z = 1$.

Ex. 5. A scheme S is used with $z_0 = 0, z_1 = 1, z_2 = 2, \ldots$. Select $f(z) = (1 + \sigma)^z$, for a fixed σ, with $\sigma \neq -1$. It is easily verified through (2.7.10) that

$$p_n(f; z) = 1 + \sigma z + \sigma^2 \frac{z(z - 1)}{2!} + \cdots + \sigma^n \frac{z(z - 1) \cdots (z - n + 1)}{n!}.$$

The convergence of the scheme is equivalent to the convergence of the series $\sum_{n=0}^{\infty} \sigma^n \dfrac{z(z - 1) \cdots (z - n + 1)}{n!}$. For fixed z, this is the power series expansion of $(1 + \sigma)^z$ about $\sigma = 0$. Now if $z \neq 0, 1, 2, \ldots, (1 + \sigma)^z$ is analytic in $|\sigma| < 1$ and has a singularity at $\sigma = -1$. The series, and therefore $p_n(f; z)$, is convergent for $|\sigma| < 1$ and divergent for $|\sigma| > 1$.

4.3 A Convergence Theorem for Bounded Triangular Schemes.

If the points of interpolation are all confined to a bounded region of the plane and if the function we are interpolating is analytic in a sufficiently large region, then we shall have uniform convergence in a sub-region. This theorem is of interest in itself and also because it illustrates the use of the complex remainder (3.6.5) in estimating errors. This is a technique that can be put to practical use in numerical analysis.

Theorem 4.3.1. *Let R, S, and T be bounded simply connected regions, $R \subset S \subset T$, whose boundaries are C_R, C_S, and C_T, respectively. C_T is a simple, closed, rectifiable curve, and C_S and C_T are assumed to be disjoint.*

Let $\delta = minimum$ distance from C_T to C_R, $\Delta = maximum$ distance from C_S to C_R and assume that $\dfrac{\Delta}{\delta} < 1$.

Let the points of a triangular system lie in R and let $f(z)$ be analytic in and on C_T. Then $p_n(f; z)$ converges to f uniformly in S.

Proof: From (3.6.5),

$$f(z) - p_n(f; z) = R_n(f; z)$$

$$= \frac{1}{2\pi i} \int_{C_T} \frac{(z - z_{n0})(z - z_{n1}) \cdots (z - z_{nn}) f(t)\, dt}{(t - z_{n0})(t - z_{n1}) \cdots (t - z_{nn})(t - z)}. \tag{4.3.1}$$

Hence,

$$|R_n(f;z)| \leq \frac{1}{2\pi} \int_{C_T} \frac{|z - z_{n0}| \cdots |z - z_{nn}| \, |f(t)| \, ds}{|t - z_{n0}| \cdots |t - z_{nn}| \, |t - z|} . \tag{4.3.2}$$

For $z_{ik} \in R$ and $z \in C_S$, $|z - z_{ik}| < \Delta$. For $z_{ik} \in R$ and $t \in C_T$, $|t - z_{ik}| > \delta$. If we set $M = \max\limits_{t \in C_T} |f(t)|$, $d = \min\limits_{\substack{t \in C_T \\ z \in S}} |t - z|$, then,

$$|R_n(f;z)| \leq \frac{1}{2\pi} \int_{C_T} \frac{\Delta^{n+1}}{\delta^{n+1}} \frac{M}{d} \, ds = \frac{ML(C_T)}{2\pi d} \left(\frac{\Delta}{\delta}\right)^{n+1}, \tag{4.3.3}$$

where $L(C_T)$ = length of C_T. This estimate holds uniformly for $z \in S$. Since $\dfrac{\Delta}{\delta} < 1$, $\lim\limits_{n \to \infty} R_n(f;z) = 0$ uniformly in S.

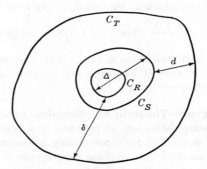

Figure 4.3.1.

Ex. 1. Let the points of the interpolation scheme all lie on the real segment $I = [-a, a]$. Select a $\delta > 2a$ and let T be the set of points whose distance from I is $\leq \delta$. If $f(z) \in A(T)$, the interpolatory scheme converges uniformly to f on I. This is independent of the distribution of the interpolating points.

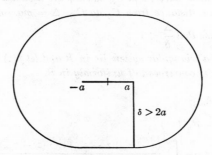

Figure 4.3.2.

4.4 Lemniscates and Interpolation. More penetrating theorems can be obtained by postulating a more regular distribution of the points in the triangular scheme. An examination of equations (4.3.2) and (4.3.3) reveals that the asymptotic behavior of the expression

$$|(z - z_{n0})(z - z_{n1}) \cdots (z - z_{nn})|^{1/(n+1)}$$

contains the key to convergence. We will therefore investigate convergence under the hypothesis that the following limit exists on certain sets of the complex plane:

$$\lim_{n \to \infty} |(z - z_{n0})(z - z_{n1}) \cdots (z - z_{nn})|^{1/(n+1)} = \sigma(z). \qquad (4.4.1)$$

Here are some examples of triangular distributions of points with this property.

Ex. 1. Let all the points z_{ij} consist of a single point z_0. Then $\sigma(z) = |z - z_0|$.

Ex. 2. Let $z_{n0}, z_{n1}, \ldots, z_{nn}$ be the $n + 1$ zeros of the $(n + 1)$st Tschebyscheff polynomial $T_{n+1}(z)$. Then, $(z - z_{n0})(z - z_{n1}) \cdots (z - z_{nn}) = \tilde{T}_{n+1}(z)$ (Def. 3.3.2). Let $z = \rho e^{i\theta}$, $w = \frac{1}{2}(z + z^{-1}) = \frac{1}{2}(\rho e^{i\theta} + \rho^{-1}e^{-i\theta})$. Then,

$$T_n(w) = \frac{1}{2}(\rho^n e^{in\theta} + \rho^{-n}e^{-in\theta}). \qquad (4.4.2)$$

This may be proved by induction. It is true for $n = 0$ by (3.3.3). Assume it is true for $0, 1, 2, \ldots, n$. By (3.3.4),

$$\begin{aligned}
T_{n+1}(w) &= (\rho e^{i\theta} + \rho^{-1}e^{-i\theta})\tfrac{1}{2}(\rho^n e^{in\theta} + \rho^{-n}e^{-in\theta}) \\
&\quad - \tfrac{1}{2}(\rho^{n-1}e^{i(n-1)\theta} + \rho^{-n+1}e^{-i(n-1)\theta}) \\
&= \tfrac{1}{2}(\rho^{n+1}e^{(n+1)i\theta} + \rho^{-(n+1)}e^{-(n+1)i\theta}).
\end{aligned}$$

This proves the identity for $n + 1$.

From (4.4.2)

$$\begin{aligned}
\operatorname{Re} T_n(w) &= \tfrac{1}{2}(\rho^n + \rho^{-n}) \cos n\theta \\
\operatorname{Im} T_n(w) &= \tfrac{1}{2}(\rho^n - \rho^{-n}) \sin n\theta
\end{aligned} \qquad (4.4.3)$$

and so

$$|T_n(w)| = \tfrac{1}{2}(\rho^{2n} + 2 \cos 2n\theta + \rho^{-2n})^{\frac{1}{2}}. \qquad (4.4.4)$$

Therefore, for w on $\mathscr{E}_\rho$ (cf. Def. (1.13.1)) we have

$$\lim_{n \to \infty} |T_n(w)|^{1/n} = \rho, \text{ uniformly.} \qquad (4.4.5)$$

Since $\tilde{T}_n = \dfrac{1}{2^{n-1}} T_n$,

$$\sigma(z) = \lim_{n \to \infty} |\tilde{T}_n(z)|^{1/n} = \rho/2, \, z \in \mathscr{E}_\rho. \qquad (4.4.6)$$

Ex. 3. A similar limit holds on $\mathscr{E}_\rho$ for the Legendre polynomials (cf. Theorem 12.4.5).

Ex. 4. The points $z_{n0}, \ldots, z_{nn}$ are evenly distributed on $[-1, 1]$: $z_{nk} = -1 + \dfrac{2k}{n}$. Here we have

$$\sigma_n(z) = \left|(z - z_{n0}) \cdots (z - z_{nn})\right|^{1/(n+1)} = \left|(z + 1)\left(z + 1 - \frac{2}{n}\right) \cdots (z + 1 - 2)\right|^{1/(n+1)}$$

$$= 2\left|\left(\frac{z+1}{2} - \frac{0}{n}\right)\left(\frac{z+1}{2} - \frac{1}{n}\right) \cdots \left(\frac{z+1}{2} - \frac{n}{n}\right)\right|^{1/(n+1)}$$

Therefore $\log \frac{1}{2}\sigma_n(z) = \dfrac{1}{n + 1} \displaystyle\sum_{k=0}^{n} \log \left|\dfrac{z+1}{2} - \dfrac{k}{n}\right|$. By the definition of the inte-

gral, $\displaystyle\lim_{n\to\infty} \log \frac{1}{2}\sigma_n(z) = \int_0^1 \log \left|\dfrac{z+1}{2} - t\right| dt = \frac{1}{2}\int_{-1}^{+1} \log |z - u| \, du - \log 2$.

Hence, for all z outside $[-1, 1]$, $\sigma(z) = \exp \frac{1}{2}\displaystyle\int_{-1}^{+1} \log |z - u| \, du$.

Ex. 5. A wide generalization of Ex. 1 is contained in the next theorem.

THEOREM 4.4.1. *Let $z_1, z_2, \ldots$, be a sequence that has k limit points ζ_1, $\zeta_2, \ldots, \zeta_k$ approached cyclically. That is,*

$$\lim_{n\to\infty} z_{nk+1} = \zeta_1$$

$$\lim_{n\to\infty} z_{nk+2} = \zeta_2$$

$$.$$
$$.$$
$$.$$

$$\lim_{n\to\infty} z_{nk+k} = \zeta_k. \tag{4.4.7}$$

Then

$$\sigma(z) = \lim_{n\to\infty} |(z - z_1) \cdots (z - z_n)|^{1/n} = |(z - \zeta_1) \cdots (z - \zeta_k)|^{1/k} \tag{4.4.8}$$

uniformly on any set S that is bounded and such that

$$\inf_{t\in S} |z_i - t| > \delta > 0 \qquad i = 1, 2, \ldots.$$

Proof: We begin with the following observation which is really a slight modification of the "consistency of Cesáro summability."

Let $\phi_k(z)$ be a sequence of functions defined on a point set S and which converges uniformly on S to a function $\phi(z)$. Assume that $\phi_k(z)$ and $\phi(z)$ are all bounded on S. Then, the sequence of arithmetic means

$$(1/n)[\phi_1(z) + \cdots + \phi_n(z)]$$

also converges uniformly to $\phi(z)$ on S. For, given an $\varepsilon > 0$, we can find an

$m = m(\varepsilon)$ such that $|\phi(z) - \phi_n(z)| \leq \varepsilon$ for all $z \in S$ and all $n \geq m$. Now, for $n \geq m$,

$$D_n(z) = (1/n)[\phi_1(z) + \cdots + \phi_n(z)] - \phi(z)$$

$$= (1/n)[\phi_1(z) + \cdots + \phi_m(z)] + \frac{n - m}{n} \cdot \frac{1}{n - m}[(\phi_{m+1}(z) - \phi(z))$$

$$+ \cdots + (\phi_n(z) - \phi(z))] - (m/n)\phi(z).$$

Let $M_m = \max_{z \in S}(|\phi_1(z)| + \cdots + |\phi_m(z)|)$, $M = \max_{z \in S}|\phi(z)|$. Hence,

$$|D_n(z)| \leq \frac{M_m}{n} + \frac{n - m}{n} \cdot \frac{1}{n - m}(n - m)\varepsilon + \frac{mM}{n}, \quad z \in S.$$

Keep m fixed and let $n \to \infty$. We obtain, for n sufficiently large, i.e., for

$$\left(n \geq \max\left(\frac{M_m}{\varepsilon}, \frac{mM}{\varepsilon}\right)\right),$$

$$|D_n(z)| \leq \varepsilon + \varepsilon + \varepsilon = 3\varepsilon, \quad z \in S.$$

This inequality implies the uniform convergence stated.

We now turn to the proof of the theorem.

Let $\phi_i(z) = \log|z - z_i|$. For $z \in S$, $\log|z - z_i| \geq \log \delta, i = 1, 2, \ldots$. In view of the cyclic limit conditions, the z_i are bounded. Since S is bounded, $\log|z - z_i| \leq B$ for $z \in S$, $i = 1, 2, \ldots$. Thus, $\phi_i(z)$ are uniformly bounded in S; that is, we can find an M such that $|\phi_i(z)| \leq M$, $z \in S$, $i = 1, 2, \ldots$.

The function $\log|z - \zeta_i|$ is also bounded in S since the ζ_i are limit points of the z_k and the latter are bounded away from S.

Let $N = nk + p$, $0 \leq p < k$ and consider n and p as functions of N. Then

$$\log|(z - z_1) \cdots (z - z_N)|^{1/N} = \frac{1}{N} \sum_{i=1}^{N} \phi_i(z)$$

$$= \frac{n}{N} \frac{1}{n} \sum_{j=0}^{n-1} \phi_{jk+1} + \cdots + \frac{n}{N} \frac{1}{n} \sum_{j=0}^{n-1} \phi_{jk+k} + \frac{n}{N} \frac{1}{n} \sum_{j=1}^{p} \phi_{nk+j}.$$

(If $p = 0$, the last sum is taken to be 0.) In view of (4.4.7), $\lim_{n \to \infty} \phi_{nk+i} = \log|z - \zeta_i|$ $i = 1, 2, \ldots, k$, uniformly on S. Hence, their mean values $\frac{1}{n} \sum_{j=0}^{n-1} \phi_{jk+i}$ also approach this limit uniformly. Furthermore,

$$\left|\frac{1}{n} \sum_{j=1}^{p} \phi_{nk+j}\right| \leq \frac{1}{n} \sum_{j=1}^{p} |\phi_{nk+j}| \leq \frac{pM}{n} \leq \frac{kM}{n}.$$

As $n \to \infty$, $\dfrac{1}{n} \sum\limits_{j=1}^{p} \phi_{nk+j}$ approaches 0 uniformly in S. Since, finally, $\lim\limits_{N \to \infty} \dfrac{n}{N} = \dfrac{1}{k}$ we have $\lim\limits_{N \to \infty} \log |(z - z_1) \cdots (z - z_N)|^{1/N} = \dfrac{1}{k} \sum\limits_{i=1}^{k} \log |z - \zeta_i|$ uniformly in S. The proof is completed by exponentiation.

Actually, (4.4.8) holds uniformly on any set S that is bounded and such that $\inf\limits_{t \in S} |\zeta_i - t| > \delta > 0$, $i = 1, \ldots, k$.

Motivated by (4.4.8), we shall next study the loci given by

$$|(z - z_1)(z - z_2) \cdots (z - z_n)| = \text{constant}.$$

DEFINITION 4.4.1. Let $z_1, \ldots, z_n$ be n complex numbers not necessarily distinct. For $r > 0$, the set of points satisfying

$$|(z - z_1)(z - z_2) \cdots (z - z_n)| = r^n \tag{4.4.9}$$

is called a *lemniscate* and will be designated by Γ_r. The points z_i are called the *foci* of the lemniscate and r its *radius*. The set of points satisfying the inequality

$$|(z - z_1)(z - z_2) \cdots (z - z_n)| < r^n \tag{4.4.10}$$

will be designated by $\mathscr{L}_r$.

With z_i fixed and r varying, we may speak of a *family of confocal lemniscates*. Note that if $r_1 < r_2$ then $\mathscr{L}_{r_1} \subset \mathscr{L}_{r_2}$.

Ex. 6. $k = 1$. $|z - z_1| = r$ is a family of concentric circles centered at z_1.

Ex. 7.   Let $k = 2$ and z_1, z_2 be distinct. Then, $|z - z_1| |z - z_2| = r^2$ is the locus of points which move in such a way that the product of their distances to z_1 and to z_2 is constant. If $0 < r < \frac{1}{2} |z_2 - z_1|$, then the locus consists of two mutually exterior ovals, one surrounding z_1 and the other z_2. These are the *Ovals of Cassini*. When $r = \frac{1}{2} |z_2 - z_1|$, we obtain the *Lemniscate of Bernoulli*, a figure 8 with a double point at $\frac{1}{2}(z_1 + z_2)$. When $r > \frac{1}{2} |z_2 - z_1|$, the locus consists of one closed contour containing z_1 and z_2 in its interior.

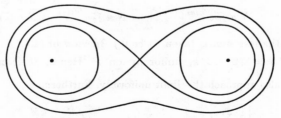

Figure 4.4.1   Confocal Lemniscates with Two Foci.

The behavior of confocal families of lemniscates follows the pattern of Example 7. If $z_1, \ldots, z_n$ are n distinct points and if r is sufficiently small, the locus Γ_r consists of n closed contours surrounding precisely one of the foci. As r increases, the contours increase in size until two or more of them touch and then coalesce, reducing the number of contours. This coalescence continues until for sufficiently large r there is but one contour surrounding all the foci. As $r \to \infty$, the single contour becomes more and more circular in its shape. If the z_i are not all distinct, this picture can be modified in an appropriate way.

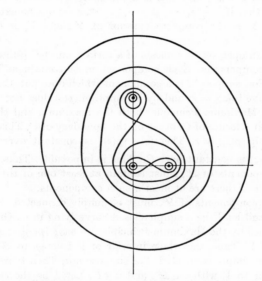

Figure 4.4.2   Confocal Lemniscates with 3 Foci
$$|z(z-1)(z-2i)| = r^3.$$

We shall now sketch the proofs of these facts. Consider the set $\mathscr{L}_r$. It is nonempty, for it contains the foci $z_1, \ldots, z_n$. It is open, for suppose $z \in \mathscr{L}_r$. That is, $|z - z_1| \cdots |z - z_n| < r^n$. By the continuity of the absolute value, this inequality must hold in a neighborhood of z. $\mathscr{L}_r$ is bounded. For let a circle C_1 contain the foci. Draw a second circle C_2, concentric with C_1, and such that the difference of the radii of C_2 and C_1 is greater than r. For any point z exterior to C_2, $|z - z_i| > r$. Hence $|z - z_1| \cdots |z - z_n| > r^n$. Therefore no point exterior to C_2 can be in $\mathscr{L}_r$.

Let Δ_r designate the set of points for which $|z - z_1| \cdots |z - z_n| > r^n$. Again, by continuity, if $z \in \Delta_r$, there is a whole neighborhood of z in Δ_r.

Hence every point of Δ_r is an exterior point of $\mathscr{L}_r$. The exterior of $\mathscr{L}_r$ is precisely Δ_r. For suppose z is in the exterior of $\mathscr{L}_r$ and that

$$|(z - z_1) \cdots (z - z_n)| \equiv |p(z)| = r^n.$$

Since $p(z) \neq 0$, there is a whole neighborhood N of z whose closure lies in the exterior of $\mathscr{L}_r$ and in which $p(z)$ has no zero. By the Maximum Principle (Theorem 1.9.5) $p(z)$ cannot have a maximum or a minimum in the interior of N. Therefore there is a point z' on the boundary of N at which

$$|p(z')| < |p(z)| = r^n.$$

This means that $z' \in \mathscr{L}_r$ and this is a contradiction since z' is exterior to $\mathscr{L}_r$. It follows that if $|z - z_1| \cdots |z - z_n| = r^n$, z cannot be exterior to $\mathscr{L}_r$. Thus, finally, Δ_r is the complete exterior of $\mathscr{L}_r$ and Γ_r is its complete boundary.

Since $\mathscr{L}_r$ is an open set, it consists of a certain number (finite or infinite) of connected components. Each component must contain at least one of the points z_i. For, suppose C is a component which does not. On the boundary of C we have $|p(z)| = r$, and by assumption, $p(z)$ does not vanish in C. Hence, by the Maximum Principle, both the maximum and the minimum of $|p(z)|$ in $\bar{C}$, the closure of C, occur on the boundary of C. This means that $\max_{z \in \bar{C}} |p(z)| = \min_{z \in \bar{C}} |p(z)| = r$. Therefore $|p(z)|$ is constant over C and this implies that $p(z)$ is constant over C. This is impossible. Thus, each of the connected components of $\mathscr{L}_r$ must contain at least one of the points z_i in its interior, and so there are at most n such components.

Each of the components of $\mathscr{L}_r$ must be simply connected. For, let C be a component and let Γ be a simple closed curve lying in C. On Γ we have $|p(z)| < r$. Hence by the Maximum Principle, we have $|p(z)| < r$ throughout the interior of Γ. Thus, the whole interior of Γ belongs to $\mathscr{L}_r$. It follows that C must be simply connected. For suppose not. Then there would be a point a interior to Γ with $a \in \mathscr{L}_r$ and $a \notin C$. Let I be the closure of the interior of Γ. Then $A = I \cup C$ is a connected set that is in $\mathscr{L}_r$ and contains C properly. This is impossible since C is a maximal connected subset. $\mathscr{L}_r$ therefore consists of a number (not exceeding n) of simply connected regions and the boundary of $\mathscr{L}_r$ consists of a number of contours each of which is the complete boundary of a simply connected region.

At a point of Γ_r where the derivative of $\prod_{i=1}^{n} (z - z_i)$ does not vanish, one can show, using the implicit function theorem for analytic functions, that the lemniscate is an analytic curve.

What is the situation when r is sufficiently small? Assume that the points $z_1, \ldots, z_n$ are distinct. For r sufficiently small, the n circles $|z - z_i| < r$ have no common points. Then if $\rho < r$, $\mathscr{L}_\rho$ must be contained in the union of these circles. For otherwise, $|z - z_i| \geq r$ and hence $\prod_{i=1}^{n} |z - z_i| \geq r^n > \rho^n$.

Moreover, since the points z_i are interior to the components of $\mathscr{L}_\rho$, it follows that $\mathscr{L}_\rho$ consists of exactly n components with precisely one component lying in each one of these circles. The lemniscate Γ_ρ consists of precisely n ovals, one in each circle.

What happens when r is sufficiently large? Let a circle C of diameter D contain all the points z_i. If z is in C then $|z - z_i| < D$ so that

$$\prod_{i=1}^{n} |z - z_i| < D^n.$$

Therefore C is contained in any $\mathscr{L}_\rho$ with $\rho > D$. Since we know that each component of $\mathscr{L}_\rho$ must contain at least one of the z_i in it, it follows that $\mathscr{L}_\rho$ must consist of a single region containing this circle. The corresponding lemniscate Γ_ρ consists of a single contour.

As r becomes very large, so also must $|z|$. With z far removed from z_i, the points z_i may be regarded as a single iterated point and the lemniscate is "almost" a circle.

The multiple points of lemniscates occur at the solutions of $p'(z) = 0$, $p(z) \equiv \prod_{i=1}^{n} (z - z_i)$. But it would take us out of our way to discuss the interesting geometrical facts related to them.

LEMMA 4.4.2. *Let $\phi_n(z)$, $\phi(z)$ be functions of a complex variable and suppose that*

$$\lim_{n \to \infty} |\phi_n(z)|^{1/n} = |\phi(z)| \qquad (4.4.11)$$

on a set S and uniformly on a subset $S' \subseteq S$. Let $\{a_n\}$ be a sequence of complex numbers for which

$$\lim_{n \to \infty} \sup |a_n|^{1/n} = 1/r, \qquad 0 \leq r < \infty. \qquad (4.4.12)$$

Then the series $\sum\limits_{n=0}^{\infty} a_n \phi_n(z)$ converges at all points of S where $|\phi(z)| < r$ and diverges at all points of S where $|\phi(z)| > r$. It converges uniformly at all points of S' where $|\phi(z)| \leq s < r$.

Proof: Given a $z \in S \cap \{z : |\phi(z)| < r\}$. For all n sufficiently large, we have from (4.4.11), $|\phi_n(z)|^{1/n} \leq r'$ for some $0 < r' < r$. Select an r'' with $r' < r'' < r$. Then for all n sufficiently large, we have from (4.4.12),

$$|a_n|^{1/n} \leq \frac{1}{r''}.$$

Thus,

$$|a_n \phi_n(z)| \leq \left(\frac{r'}{r''}\right)^n.$$

Our infinite series is majorized by a convergent geometric series and must itself be convergent.

If

$$z \in S' \cap \{z : |\phi(z)| \leq s\}, \qquad s < r,$$

then the estimate $|\phi_n(z)|^{1/n} \leq r'$, $s < r' < r$ holds uniformly in this set and the same reasoning allows us to conclude uniform convergence there.

Let $z \in S$ be such that $|\phi(z)| > r$. Then for all n sufficiently large and for some $r' > r$, $|\phi_n(z)|^{1/n} > r'$. Select an r'' with $r' > r'' > r$. Since

$$\lim_{n \to \infty} \sup |a_n|^{1/n} = \frac{1}{r},$$

we can find a sequence of integers $n_1, n_2, \ldots \to \infty$, with $|a_{n_k}|^{1/n_k} \geq \frac{1}{r''}$. Hence $|a_{n_k} \phi_{n_k}(z)| \geq \left(\frac{r'}{r''}\right)^{n_k}$ for all k sufficiently large. Since $r'/r'' > 1$, the general term of the series does not approach 0 and the series is divergent.

Ex. 8.   Let $\phi_n(z) = z^n$. Then $\lim_{n \to \infty} |\phi_n(z)|^{1/n} = |z|$ for all $|z| < \infty$ and the limit is uniform on any set. If $\lim_{n \to \infty} \sup |a_n|^{1/n} = \frac{1}{r}$, we have convergence of $\sum_{n=0}^{\infty} a_n z^n$ for $|z| < r$ and divergence for $|z| > r$. Our lemma is therefore a simple modification of the Cauchy-Hadamard Theorem for the radius of convergence of a power series (Theorem 1.9.4).

Ex. 9. $\phi_n(z) = T_n(z) =$ the nth Tschebyscheff polynomial. From (4.4.5), $|\phi(z)| = \lim_{n \to \infty} |T_n(z)|^{1/n} = \rho$ for z on $\mathscr{E}_\rho$. If $\lim_{n \to \infty} \sup |a_n|^{1/n} = \frac{1}{r}$, $1 < r \leq \infty$, we conclude that the series $\sum_{n=0}^{\infty} a_n T_n(z)$ converges in the interior of $\mathscr{E}_r$, diverges in its exterior and converges uniformly if any $\mathscr{E}_{r'}$, $r' < r$. In the case of expansions in Tschebyscheff polynomials, we have *confocal ellipses of convergence* as the analog of circles of convergence for power series.

Ex. 10.   Expansions in Legendre polynomials have ellipses of convergence. Cf. Theorem 12.4.7.

THEOREM 4.4.3.   *Let* $\mathscr{L}_\rho$ *designate the lemniscate interior*

$$|(z - \zeta_1)(z - \zeta_2) \cdots (z - \zeta_k)| < \rho^k.$$

Let $z_0, z_1, \ldots,$ *lie in* $\mathscr{L}_\rho$ *and approach* $\zeta_1, \ldots, \zeta_k$ *cyclically; i.e.,* (4.4.7) *holds. Let* $f(z)$ *be analytic in* $\mathscr{L}_\rho$, *but not in any* $\mathscr{L}_{\rho_1}$ *with* $\rho_1 > \rho$. *If* $\mathscr{L}_\rho$

consists of several mutually exterior regions, then $f(z)$ is assumed to be a general analytic configuration (cf. 3.6). *If*

$$p_n(f; z) = a_0 + a_1(z - z_0) + \cdots + a_n(z - z_0)(z - z_1) \cdots (z - z_{n-1})$$

(4.4.13)

interpolates to f at $z_0, \ldots, z_n$, we have

$$\lim_{n \to \infty} p_n(f; z) = f(z) \qquad z \in \mathscr{L}_\rho$$

(4.4.14)

and uniformly in any closed set lying in $\mathscr{L}_\rho$. Exterior to $\mathscr{L}_\rho$, the limit does not exist. More precisely, we have

$$|f(z) - p_n(f; z)| \le M(\rho'/\rho)^n \quad \text{for} \quad z \in \mathscr{L}_{\rho'}, \ \rho' < \rho.$$

(4.4.15)

Furthermore

$$\lim_{n \to \infty} \sup |a_n|^{1/n} = \frac{1}{\rho}.$$

(4.4.16)

Proof: A closed set S lying in $\mathscr{L}_\rho$ must also lie in some $\mathscr{L}_{\rho'}$ with $\rho' < \rho$. Hence (4.4.15) implies (4.4.14), uniformly in S. We therefore prove (4.4.15). Select a ρ' sufficiently close to ρ so that all the points $z_0, z_1, \ldots,$ lie in $\mathscr{L}_{\rho'}$. Select ρ'' with $\rho' < \rho'' < \rho$. $\Gamma_{\rho''}$ consists of one or at most a finite number of mutually exterior curves which may, possibly, meet at a finite number of points. From the remarks at the end of 3.6, we have

$$f(z) - p_n(f; z) = \frac{1}{2\pi i} \int_{\Gamma_{\rho''}} \frac{(z - z_0)(z - z_1) \cdots (z - z_n) f(t) \, dt}{(t - z_0)(t - z_1) \cdots (t - z_n)(t - z)}.$$

(4.4.17)

For $z \in \Gamma_{\rho'}$ we have by (4.4.8),

$$\lim_{n \to \infty} |(z - z_0) \cdots (z - z_n)|^{1/(n+1)} = |(z - \zeta_1) \cdots (z - \zeta_k)|^{1/k} = \rho'$$

uniformly. Hence, for some ρ''' with $\rho' < \rho''' < \rho''$, and for n sufficiently large, we have

$$|(z - z_0)(z - z_1) \cdots (z - z_n)| \le (\rho''')^{n+1}, \qquad z \in \Gamma_{\rho'}.$$

(4.4.18)

By the Maximum Principle, this inequality holds throughout $\mathscr{L}_{\rho'}$.

Similarly we have

$$\lim_{n \to \infty} |(t - z_0) \cdots (t - z_n)|^{1/(n+1)} = \rho'', \qquad t \in \Gamma_{\rho''}$$

uniformly. Hence, for some ρ^{iv} with $\rho''' < \rho^{iv} < \rho''$, and n sufficiently large,

$$|(t - z_0) \cdots (t - z_n)| \ge (\rho^{iv})^{n+1}, \qquad t \in \Gamma_{\rho''}.$$

(4.4.19)

We now use (4.4.18) and (4.4.19) to estimate (4.4.17):

$$|f(z) - p_n(f; z)| \le \frac{1}{2\pi} \int_{\Gamma_{\rho''}} \left(\frac{\rho'''}{\rho^{iv}}\right)^{n+1} \frac{|f(t)| \, ds}{|t - z|}, \qquad z \in \mathscr{L}_{\rho'}.$$

(4.4.20)

Let $L(\Gamma_{\rho''}) = $ length of $\Gamma_{\rho''}$, $m = \max\limits_{t \in \Gamma_{\rho''}} |f(t)|$, $\delta = $ minimum distance from $\Gamma_{\rho'}$ to $\Gamma_{\rho''}$ and we have

$$|f(z) - p_n(f; z)| \leq \frac{mL(\Gamma_{\rho''})}{2\pi\,\delta} \left(\frac{\rho'''}{\rho^{iv}}\right)^{n+1} \leq M\left(\frac{\rho'}{\rho}\right)^n \qquad (4.4.21)$$

for an appropriate constant M independent of n.

Consider next the number μ defined by

$$\limsup_{n \to \infty} |a_n|^{1/n} = \frac{1}{\mu}. \qquad (4.4.22)$$

We shall prove $\mu = \rho$ and establish (4.4.16). Suppose, first, that $\mu < \rho$. By the above work, $\lim\limits_{n \to \infty} p_n(f; z)$ exists in $\mathscr{L}_\rho$. But by Lemma 4.4.2 and Theorem 4.4.1, this limit does not exist at a point z exterior to $\mathscr{L}_\mu$ and distinct from $\zeta_1, \ldots, \zeta_k$. Hence, at a point z ($\neq \zeta_1, \ldots, \zeta_k$) belonging to $\mathscr{L}_\rho$ but exterior to $\mathscr{L}_\mu$, the limit both exists and does not exist. This is impossible. Suppose, secondly, that $\mu > \rho$. By Lemma 4.4.2, $p_n(f; z)$ converges uniformly in closed subregions of $\mathscr{L}_\mu$ to a function that is analytic. This is impossible, since by hypothesis $\mathscr{L}_\mu$ is the largest lemniscate of analyticity. Hence $\mu = \rho$. This establishes (4.4.16) and the fact that we have divergence exterior to $\mathscr{L}_\rho$.

This theorem has a long history, going back, in various amounts of generality, about a century. In the form stated, it is due to Walsh. More general assumptions as to the "strength" of the limit points can and have been made.

Interpolation processes may be found that yield expansions of analytic functions in quite general regions. We shall, however, approach such approximations by other methods.

Ex. 11. Let $z_{2n} = 0, z_{2n+1} = 1$ $n = 0, 1, \ldots$. Then $k = 2$ and $\zeta_1 = 0$, $\zeta_2 = 1$. Define $f(z)$ as 0 in a neighborhood of ζ_1 and 1 in a neighborhood of ζ_2. We then have

$$p_{2n-1}(f; z) = a_0 + a_1 z + a_2 z(z - 1) + a_3 z^2(z - 1)$$
$$+ a_4 z^2(z - 1)^2 + \cdots + a_{2n-1} z^n(z - 1)^{n-1}$$
$$p_{2n}(f; z) = p_{2n-1}(f; z) + a_{2n} z^n (z - 1)^n.$$

In view of (2.6.6), (3.6.1), (3.6.3), we have

$$a_{2n-1} = \frac{1}{2\pi i} \int_C \frac{dz}{z^n(z - 1)^n}, \; a_{2n} = \frac{1}{2\pi i} \int_C \frac{dz}{z^{n+1}(z - 1)^n}$$

where C is any contour containing ζ_2 in its interior and ζ_1 in its exterior. Hence,

$$a_{2n-1} = \frac{1}{(n-1)!} \frac{d^{n-1}}{dz^{n-1}}\left(\frac{1}{z^n}\right)\Big|_{z=1} = (-1)^{n-1}\binom{2n-2}{n-1}$$

$$a_{2n} = \frac{1}{(n-1)!} \frac{d^{n-1}}{dz^{n-1}}\left(\frac{1}{z^{n+1}}\right)\Big|_{z=1} = (-1)^{n-1}\binom{2n-1}{n} \qquad n = 1, 2, \cdots$$

$$a_0 = 0.$$

The interpolation series is therefore

$$0 + z + z(z - 1) - 2z^2(z - 1) - 3z^2(z - 1)^2 + 6z^3(z - 1)^2 + \cdots.$$

In view of the fact that $\lim\limits_{n \to \infty} \sup |a_n|^{1/n} = 2 = 1/\tfrac{1}{2}$, the series converges in the interior of the leminiscate $|z(z - 1)| = \tfrac{1}{4}$ and diverges in its exterior. The sum of the series is 0 in the left lobe and 1 in the right lobe.

The fairly "arbitrary" shape of lemniscates displayed in Figure 4.4.2 leads one to suspect that quite general curves can be approximated by them. This is indeed the case.

DEFINITION 4.4.2.   Let E be a closed, bounded, and nonempty set in the complex plane. By the *closed ρ neighborhood of E*, $E(\rho)$, is meant the set of all points whose distance from E is $\leq \rho$. That is,

$$E(\rho) = \bigcup_{w \in E} \{z \colon |z - w| \leq \rho\}.$$

THEOREM 4.4.4.   *Let E be a closed and bounded set in the complex plane. For any $\rho > 0$, we can find a lemniscate Γ such that $\Gamma \subseteq E(\rho)$ — E and each component of E is separated from the exterior component of the complement of $E(\rho)$ by a component of Γ.*

This theorem (which we shall not use in the sequel and whose proof we shall not give) goes back to Hilbert, and early proofs make use of potential theory. In the generality stated, the theorem is due to Fekete. His methods are somewhat more elementary. This theorem can be made the basis of interpolation processes that yield expansions in general regions.

NOTES ON CHAPTER IV

4.2, 4.4. Walsh [2], Chap. III, IV. Gontscharoff [1], Chap. V. For the "strength" of limit points, see Korovkin [1]. Fekete's proof will be found in Fekete [1].

PROBLEMS

1. Cosh z is approximated by a cubic polynomial found by interpolating at the points $\pm\tfrac{1}{2}$, $\pm\tfrac{1}{2}i$. Estimate the error over the unit circle.

2. Let z_{nk} be the $n + 1$st roots of unity. Compute $\sigma(z)$. (Cf. (4.4.1))

3. Sketch the family Γ_r whose foci are at ± 1, $\pm i$.

4. Expand $(1 + u)^{\frac{1}{2}}$ and set $u = z^2 - 1$ to obtain

$$(z^2)^{\frac{1}{2}} = 1 + \frac{z^2 - 1}{2} + \cdots + (-1)^{n-1} \frac{1 \cdot 3 \cdot 5 \cdots (2n - 3)}{2 \cdot 4 \cdot 6 \cdots (2n)} (z^2 - 1)^n + \cdots.$$

Show that the series converges to z in the right open lobe of $|z^2 - 1| = 1$ and to $-z$ in the left open lobe. What are the interpolation properties of the series?

5. Following Ex. 11, let $z_{3n} = 0$, $z_{3n+1} = 1$, $z_{3n+2} = -1$. Define $f(z)$ as $-1, 0, 1$ in the neighborhood of $-1, 0, 1$ respectively. Discuss.

6. Interpret the mode of interpolation and the statement of Theorem 4.4.3 when $z_0, z_1, \ldots$, consists of the k points $\zeta_1, \zeta_2, \ldots, \zeta_k$ repeated cyclically.

7. Discuss the convergence of series of the form

$$a_0 + a_1(z - 1) + a_2(z - 1)(2z - 1) + a_3(z - 1)(2z - 1)(3z - 1) + \cdots.$$

8. Let $m(z)$ map a simply connected region B conformally onto the unit circle. Discuss the convergence of series of the form $\sum\limits_{n=0}^{\infty} a_n (m(z))^n$.

CHAPTER V

Some Problems of Infinite Interpolation

5.1 Characteristics of Such Problems. In Chapters 2 and 3, we considered interpolation problems with a finite number of conditions. In Chapter 4, we allowed the number of conditions to grow and, under certain favorable circumstances, we obtained solutions in the form of infinite series of polynomials. Not all problems involving an infinity of interpolating conditions can be treated in this manner, and the present chapter explores several alternate approaches.

In passing from a problem with a finite number of conditions to one with an infinity of conditions, analytic as well as algebraic difficulties arise to complicate the situation. If we look for a solution within a given class of functions, we may be unsuccessful, or we may be too successful, for the solution may not be unique. The following examples illustrate these possibilities.

Ex. 1. Find a function analytic in $|z| < r$, for which $f^{(n)}(0) = (n!)^2$ $n = 0, 1, \ldots$. We must have

$$f(z) = \sum_{n=0}^{\infty} \frac{1}{n!}(n!)^2 z^n = \sum_{n=0}^{\infty} n! \, z^n.$$

This series has a zero radius of convergence and so the problem has no solution. From (1.9.9), the interpolation problem $f^{(n)}(0) = a_n$ has a solution analytic in $|z| < r$ if and only if $\limsup_{n \to \infty} \frac{1}{n}|a_n|^{1/n} \leq \frac{1}{re}$. If it has a solution, the solution is unique.

Ex. 2. Given a set of points $0 \leq x_1 < x_2 < \cdots < 1$ with $\lim_{n \to \infty} x_n = 1$ and a set of values $y_1, y_2, \ldots$, find a function of class $C[0, 1]$ for which $f(x_i) = y_i$. It should be clear that a necessary and sufficient condition that this interpolation problem have a solution is that $\lim_{n \to \infty} y_n$ exists. Assuming this, we may then solve the problem in an infinity of ways.

Ex. 3. Find a function of class $C[0, 1]$ for which

$$\int_0^1 x^n f(x) \, dx = n, \; n = 0, 1, \ldots.$$

By Theorem 1.4.2,

$$\left| \int_0^1 x^n f(x) \, dx \right| = \left| f(\xi) \int_0^1 x^n \, dx \right| = \frac{|f(\xi)|}{n+1} \leq \frac{\max\limits_{0 \leq x \leq 1} |f(x)|}{n+1} .$$

The sequence of moments $\int_0^1 x^n f(x) \, dx$ therefore approaches 0 as a limit and the interpolation problem has no solution.

5.2 Guichard's Theorem. A natural generalization of the fundamental theorem for polynomial interpolation (Theorem 2.1.1) is the following theorem of Guichard.

THEOREM 5.2.1. *Let* $z_0, z_1, \ldots,$ *be a sequence of distinct complex numbers such that* $\lim\limits_{n \to \infty} z_n = \infty$. *Let* $w_0, w_1, \ldots,$ *be a completely arbitrary sequence of complex values. Then there exists an entire function* $f(z)$ *such that*

$$f(z_n) = w_n \qquad n = 0, 1, \ldots. \tag{5.2.1}$$

We shall give two proofs of this theorem. The first is function-theoretic in nature and is based upon theorems of Weierstrass and Mittag-Leffler that have an interpolatory character. We shall state these theorems, but refer to standard texts on complex variable for their proofs.

THEOREM 5.2.2 (The Weierstrass Product Theorem). *Let* $z_0 = 0, z_1, \ldots,$ *be a sequence of distinct complex numbers for which* $\lim\limits_{n \to \infty} z_n = \infty$. *Let* $n_0, n_1, \ldots,$ *be a sequence of integers* ≥ 1. *Then, for an appropriate selection of integers* p_k, *the product*

$$f(z) = z^{n_0} \prod_{k=1}^{\infty} \left(1 - \frac{z}{z_k} \right)^{n_k} \left[\exp \left(\frac{z}{z_k} + \frac{1}{2} \left(\frac{z}{z_k} \right)^2 + \cdots + \frac{1}{p_k} \left(\frac{z}{z_k} \right)^{p_k} \right) \right]^{n_k} \tag{5.2.2}$$

converges for $|z| < \infty$ *to an entire function that has a zero of order* n_k *at* z_k, $k = 0, 1, \ldots.$

THEOREM 5.2.3 (Mittag-Leffler's Partial Fraction Theorem). *Let* $z_0, z_1, \ldots,$ *be a sequence of distinct points, and let* $\lim\limits_{n \to \infty} z_n = \infty$. *For* $i = 0, 1, \ldots,$ *let* $a_{i1}, a_{i2}, \ldots, a_{in_i}$ *be given complex numbers. Then there exists a meromorphic function having at each* z_i *a principal part*

$$\frac{a_{i,1}}{(z - z_i)} + \frac{a_{i,2}}{(z - z_i)^2} + \cdots + \frac{a_{i,n_i}}{(z - z_i)^{n_i}} \tag{5.2.3}$$

and analytic everywhere else.

Proof of Theorem 5.2.1: By Theorem 5.2.2, construct a function $g(z)$ that is entire and has simple zeros at z_i: $g(z_i) = 0$ and $g'(z_i) \neq 0$, $i = 0, 1, 2 \cdots$. By Theorem 5.2.3, construct a meromorphic function $h(z)$ whose principal part at z_i is $\dfrac{w_i}{g'(z_i)(z - z_i)}$ and which is analytic everywhere else. Then the function $f(z) = g(z)h(z)$ solves the interpolation problem, for the zeros of g cancel the poles of h so that f is an entire function. Moreover, in a neighborhood of $z = z_i$ we have

$$f(z) = g(z)h(z) = [g'(z_i)(z - z_i) + \text{higher powers of } (z - z_i)]$$

$$\times \left[\frac{w_i}{g'(z_i)(z - z_i)} + r(z) \right]$$

where $r(z)$ is analytic in a neighborhood of z_i. Hence $f(z_i) = w_i$ as required.

5.3 A Second Approach: Infinite Systems of Linear Equations in Infinitely Many Unknowns.

A simple-minded approach to the problem in Theorem 5.2.1 is the following. We are looking for an entire function

$$f(z) = a_0 + a_1 z + a_2 z^2 + \cdots \tag{5.3.1}$$

for which

$$f(z_i) = w_i, \qquad i = 1, 2, \ldots. \tag{5.3.2}$$

Regard the a's of (5.3.1) as unknowns and determine them so that the conditions (5.3.2) hold. These conditions lead to

$$
\begin{aligned}
a_0 + a_1 z_1 + a_2 z_1{}^2 + \cdots &= w_1 \\
a_0 + a_1 z_2 + a_2 z_2{}^2 + \cdots &= w_2 \\
a_0 + a_1 z_3 + a_2 z_3{}^3 + \cdots &= w_3 \\
\cdot \qquad\qquad\quad \cdot \\
\cdot \qquad\qquad\quad \cdot \\
\cdot \qquad\qquad\quad \cdot
\end{aligned}
\tag{5.3.3}
$$

an infinite system of linear equations in the unknowns $a_0, a_1, \ldots$. Assuming, for the moment, that we have succeeded in producing numbers a_0, $a_1, \ldots$, which make the left hand of (5.3.3) converge to the right hand, it follows from the properties of power series that the series (5.3.1) converges absolutely for $|z| < |z_1|$, for $|z| < |z_2|$, etc. Since $\lim z_n = \infty$, $f(z)$ will be entire and $f(z_i) = w_i$. The matter therefore hinges upon our ability to solve the system (5.3.3). It should be clear that infinite problems of *linear* interpolation theory can always be reduced to such systems.

Questions relating to the existence and uniqueness of solutions of finite systems of linear equations have been completely resolved. Not so for infinite systems. While much is known, much remains unknown. There is no all-encompassing theory, but rather many different theories that have their origins in the variety of assumptions that can be made about the growth properties of the coefficients and of the solution. We shall prove a theorem of Pólya which gives a sufficient condition for the existence of a solution of an infinite system. Pólya's Theorem is interesting because it has numerous applications to interpolation problems, and also because it is one of the few theorems about infinite systems in which nothing is assumed about the right-hand side.

THEOREM 5.3.1 (Pólya).   *Let there be given an infinite set of linear equations in infinitely many unknowns* $x_1, x_2, \ldots$:

$$a_{11}x_1 + a_{12}x_2 + \cdots = b_1$$
$$a_{21}x_1 + a_{22}x_2 + \cdots = b_2$$

$$\cdot \qquad\qquad \cdot$$
$$\cdot \qquad\qquad \cdot \tag{5.3.4}$$
$$\cdot \qquad\qquad \cdot$$

No assumptions are made about the b's, *but as far as the* a's *are concerned we assume*

(A)   *Let* $q \geq 0$ *and* $n \geq 1$ *be arbitrary integers. From the infinite array of coefficients*

$$a_{1,q+1}\, a_{1,q+2} \ldots$$
$$\cdot$$
$$\cdot \tag{5.3.5}$$
$$\cdot$$
$$a_{n,q+1}\, a_{n,q+2} \ldots$$

we may select n *columns such that the determinant formed by these columns does not vanish.*

(B)   *For* $j = 2, 3, \ldots,$ *we have*

$$\lim_{k \to \infty} \frac{a_{j-1,k}}{a_{jk}} = 0 \tag{5.3.6}$$

Under assumptions (A) *and* (B), *we may find a solution* x_i *to* (5.3.4) *with all the infinite series absolutely convergent.*

A solution will be constructed in a "blockwise" fashion. This will require a preliminary description and a lemma. The first block of unknowns will be

$$x_1, x_2, \ldots, x_{q_1}. \tag{5.3.7}$$

The second block will be

$$x_{q_1+1}, x_{q_1+2}, \ldots, x_{q_2}. \tag{5.3.8}$$

In general, the nth block of unknowns will be

$$x_{q_{n-1}+1}, x_{q_{n-1}+2}, \ldots, x_{q_n}. \tag{5.3.9}$$

The numbers q_i are certain integers satisfying $1 \leq q_1 < q_2 < \cdots$, and will be specified precisely in the proof below.

The first block of unknowns will be assumed to satisfy

$$a_{11}x_1 + a_{12}x_2 + \cdots + a_{1,q_1}x_{q_1} = b_1 \tag{5.3.10}$$

The second block will be assumed to satisfy

$$a_{1,q_1+1}x_{q_1+1} + a_{1,q_1+2}x_{q_1+2} + \cdots + a_{1,q_2}x_{q_2} = 0$$

$$a_{2,q_1+1}x_{q_1+1} + a_{2,q_1+2}x_{q_1+2} + \cdots + a_{2,q_2}x_{q_2}$$
$$= b_2 - (a_{21}x_1 + a_{22}x_2 + \cdots + a_{2,q_1}x_{q_1}) \tag{5.3.11}$$

By adding (5.3.10) to (5.3.11), we see that the first two blocks satisfy

$$a_{11}x_1 + \cdots + a_{1,q_2}x_{q_2} = b_1$$
$$a_{21}x_1 + \cdots + a_{2,q_2}x_{q_2} = b_2 \tag{5.3.12}$$

In general, the nth block will be assumed to satisfy

$$a_{1,q_{n-1}+1}x_{q_{n-1}+1} + \cdots + a_{1,q_n}x_{q_n} = 0$$
$$\vdots \qquad\qquad\qquad \vdots \tag{5.3.13}$$
$$a_{n-1,q_{n-1}+1}x_{q_{n-1}+1} + \cdots + a_{n-1,q_n}x_{q_n} = 0$$
$$a_{n,q_{n-1}+1}x_{q_{n-1}+1} + \cdots + a_{n,q_n}x_{q_n}$$
$$= b_n - (a_{n,1}x_1 + a_{n,2}x_2 + \cdots + a_{n,q_{n-1}}x_{q_{n-1}}) = b_n{}'$$

By addition, the first n blocks of unknowns will satisfy the n conditions

$$a_{11}x_1 + a_{12}x_2 + \cdots + a_{1,q_n}x_{q_n} = b_1$$
$$\vdots \qquad\qquad\qquad \vdots \tag{5.3.14}$$
$$a_{n1}x_1 + a_{n2}x_2 + \cdots + a_{n,q_n}x_{q_n} = b_n$$

But, in order to provide absolute convergence, we need more. We shall require that (5.3.13) be solved subject to the condition that the terms of the first $n-1$ of its left-hand members be uniformly small in absolute value.

LEMMA 5.3.2. *Let n and q be positive integers. Let b be arbitrary and* $\varepsilon > 0$. *Under conditions* (A) *and* (B) *of Theorem* 5.3.1, *we can find an integer* $q' > q$ *and values* $x_{q+1}, x_{q+2}, \ldots, x_{q'}$ *such that*

$$a_{1,q+1}x_{q+1} \; + a_{1,q+2}x_{q+2} \; + \cdots + a_{1,q'}x_{q'} \; = 0$$
$$\vdots \qquad\qquad\qquad\qquad\qquad\qquad \vdots \qquad\qquad\qquad (5.3.15)$$
$$a_{n-1,q+1}x_{q+1} + a_{n-1,q+2}x_{q+2} + \cdots + a_{n-1,q'}x_{q'} = 0$$
$$a_{n,q+1}x_{q+1} \; + a_{n,q+2}x_{q+2} \; + \cdots + a_{n,q'}x_{q'} \; = b$$

and such that

$$|a_{1,q+1}x_{q+1}| \; + \cdots + |a_{1,q'}x_{q'}| \; < \varepsilon$$
$$\vdots \qquad\qquad\qquad\qquad \vdots \qquad\qquad\qquad (5.3.16)$$
$$|a_{n-1,q+1}x_{q+1}| + \cdots + |a_{n-1,q'}x_{q'}| < \varepsilon.$$

Proof: Let $t_1, \ldots, t_n$ be n independent variables. In view of condition (A), we can select n integers $k_1, k_2, \ldots, k_n$ with $q < k_1 < k_2 < \cdots < k_n$ in such a fashion that the determinant of the system

$$a_{1k_1}x_{k_1} + a_{1k_2}x_{k_2} + \cdots + a_{1k_n}x_{k_n} = t_1$$
$$\vdots \qquad\qquad\qquad\qquad\qquad \vdots \qquad\qquad\qquad (5.3.17)$$
$$a_{nk_1}x_{k_1} + a_{nk_2}x_{k_2} + \cdots + a_{nk_n}x_{k_n} = t_n$$

does not vanish. The x's may therefore be solved as certain linear combinations of the t's. Hence, for the t's sufficiently small, the x's will also be small. More than this, we may find a $\delta > 0$ such that $|t_i| < \delta$, $i = 1, 2, \ldots, n$ implies

$$|a_{jk_1}x_{k_1}| + |a_{jk_2}x_{k_2}| + \cdots + |a_{jk_n}x_{k_n}| < \frac{\varepsilon}{2} \quad j = 1, 2, \ldots, n. \quad (5.3.18)$$

Now in view of condition (B), we can determine an integer $q' > k_n$ such that

$$|b| \left| \frac{a_{jq'}}{a_{nq'}} \right| < \min(\delta, \varepsilon/2) \quad \text{for } j = 1, 2, \ldots, n - 1. \quad (5.3.19)$$

Now, set $x_k = 0$ if $q < k < q'$ and $k \neq k_1, k_2, \ldots, k_n$. Determine $x_{k_1}, x_{k_2}, \ldots, x_{k_n}$ from the system (5.3.17) wherein we have set

$$\begin{cases} t_j = -\dfrac{ba_{jq'}}{a_{nq'}} & j = 1, 2, \ldots, n - 1 \\ t_n = 0. \end{cases} \qquad (5.3.20)$$

In view of (5.3.19) and (5.3.20), we have $|t_j| < \delta$ for $j = 1, 2, \ldots, n$ and hence (5.3.18) holds. If we set

$$x_{q'} = \frac{b}{a_{nq'}} \quad \text{then} \quad |a_{jq'}x_{q'}| = \left|\frac{a_{jq'}}{a_{nq'}}\right| \, |b| < \frac{\varepsilon}{2}$$

by (5.3.19).

Finally,

$$a_{j,q+1}x_{q+1} + a_{j,q+2}x_{q+2} + \cdots + a_{j,q'}x_{q'}$$

$$= a_{jk_1}x_{k_1} + a_{jk_2}x_{k_2} + \cdots + a_{jk_n}x_{k_n} + a_{jq'}\frac{b}{a_{jq'}} = t_j - t_j = 0$$

$$j = 1, 2, \ldots, n-1$$

and

$$a_{n,q+1}x_{q+1} + a_{n,q+2}x_{q+2} + \cdots + a_{nq'}x_{q'} = a_{nk_1}x_{k_1} + a_{nk_2}x_{k_2} + \cdots$$

$$+ a_{nk_n}x_{k_n} + a_{nq'}\frac{b}{a_{nq'}} = t_n + b = b$$

Thus, if $x_{q+1}, \ldots, x_{q'}$ have been selected in the above manner, all the required conditions are fulfilled.

Proof of Theorem 5.3.1: We begin by dividing the unknowns into blocks in an inductive fashion. In view of condition (A), the sequence $a_{11}, a_{12}, \ldots,$ must contain infinitely many nonzero elements. Determine q_1 such that $a_{1q_1} \neq 0$. Now set $x_1 = 0$, $x_2 = 0, \ldots, x_{q_1-1} = 0$, $x_{q_1} = b_1/a_{1q_1}$. Conditions (5.3.10) imposed on the first block are satisfied. Suppose now that we have obtained integers

$$q_1 < q_2 < \cdots < q_{n-1} \text{ and values } x_1, \ldots, x_{q_{n-1}}$$

where $n \geq 2$. We take the next step as follows. Use Lemma 5.3.2 with $q = q_{n-1}$, $b = b_n{}'$ (Cf. (5.3.13)) and determine $q' = q_n$ and the values of the nth block of unknowns, $x_{q_{n-1}+1}, \ldots, x_{q_n}$ so that (5.3.15) is satisfied and so that

$$|a_{jq_{n-1}+1}x_{q_{n-1}+1}| + \cdots + |a_{jq_n}x_{q_n}| \leq \frac{1}{2^n}, \ 1 \leq j < n \qquad (5.3.21)$$

i.e., (5.3.16) with $\varepsilon = \frac{1}{2^n}$. Note that $b_n{}'$ involves only the values $x_1, \ldots,$ $x_{q_{n-1}}$ so that this step may be taken. This completes the inductive definition.

In view of (5.3.13), it is easy to see that $x_1, x_2, \ldots,$ satisfy the original system (5.3.4)—at least in a blockwise fashion. Let j be fixed, i.e., consider a fixed row. As soon as $n > j$, (5.3.21) holds, and this tells us that the jth row must be absolutely convergent. From (5.3.14), we see that when $n \geq j$, the jth row has partial sums $a_{j1}x_1 + \cdots + a_{jq_n}x_{q_n} = b_j$, for the infinite sequence of indices $q_n, q_{n+1}, \ldots$ Since $\sum\limits_{k=1}^{\infty} a_{jk}x_k$ converges, it follows that $\sum\limits_{k=1}^{\infty} a_{jk}x_k = b_j$. This completes the proof.

It should be emphasized that nothing has been said about the uniqueness of the solution. Quite the contrary. Under these conditions, there must be an infinity of solutions. (See Prob. 7.)

5.4   Applications of Pólya's Theorem.   Theorem 5.3.1 will now be applied to give a proof of a theorem similar to Theorem 5.2.1.

THEOREM 5.4.1.   *Let* $z_1, z_2, \ldots,$ *satisfy* $0 < |z_1| < |z_2| < \cdots, \lim\limits_{n \to \infty} |z_n| = \rho \le \infty$. *To each point* z_j *associate a nonnegative integer* m_j *and* $m_j + 1$ *arbitrary values* $w_j{}^0, w_j{}^1, \ldots, w_j{}^{m_j}$. *Then we can find a function* $f(z)$ *that is analytic in* $|z| < \rho$ *and satisfies the interpolation conditions:*

$$f(z_j) = w_j{}^0, f'(z_j) = w_j{}^1, \ldots, f^{(m_j)}(z_j) = w_j{}^{m_j} \quad j = 1, 2, \ldots. \quad (5.4.1)$$

Proof: Assuming we have $f(z) = \sum\limits_{k=0}^{\infty} a_k z^k$, with a_k to be determined, we have for $s \ge 1$,

$$f^{(s)}(z) = \sum_{k=0}^{\infty} a_k k(k-1) \cdots (k-s+1)z^{k-s}. \quad (5.4.2)$$

The conditions (5.4.1) therefore lead to the infinite system

$$a_0 + z_1 a_1 + z_1{}^2 a_2 + z_1{}^3 a_3 + \cdots = w_1{}^0$$
$$a_1 + 2z_1 a_2 + 3z_1{}^2 a_3 + \cdots = w_1{}^1$$
$$2a_2 + 6z_1 a_3 + \cdots = w_1{}^2$$
$$\cdot$$
$$\cdot \quad (5.4.3)$$
$$\cdot$$
$$(m_j)! \, a_{m_j} + \cdots = w_1{}^{m_j}$$
$$a_0 + z_2 a_1 + z_2{}^2 a_2 + z_2{}^3 a_3 + \cdots = w_2{}^0$$
$$a_1 + 2z_2 + 3z_2{}^2 a_3 + \cdots = w_2{}^1$$
$$\cdot \quad \cdot$$
$$\cdot \quad \cdot$$
$$\cdot \quad \cdot$$

Now condition (B), (5.3.6) is satisfied. For if the jth row refers to an sth derivative at z_p, $s \ge 1$, we have from (5.4.2)

$$\frac{a_{j-1,k+1}}{a_{j,k+1}} = \frac{z_p k(k-1) \cdots (k-s+2)}{k(k-1) \cdots (k-s+1)} = \frac{z_p}{k-s+1} \to 0.$$

However, if the jth row relates to a value of f at a point, then the condition $|z_{p-1}| < |z_p|$ assures that the limit of $a_{j-1,k}/a_{j,k}$ is zero. Consider the coefficients arising from the system (5.4.3). We shall prove that the determinant formed from the first n rows and any n adjacent columns cannot vanish. This will tell us that condition (A) is fulfilled. These determinants may be thought of as generalized Vandermonde determinants. To avoid losing the thread of the argument in a welter of indices, look at the specific case $n = 4$, $m_1 = 2$. The 4×4 determinant formed from the first 4 rows and 4 adjacent columns is

$$D = \begin{vmatrix} z_1^k & z_1^{k+1} & z_1^{k+2} & z_1^{k+3} \\ kz_1^{k-1} & (k+1)z_1^k & (k+2)z_1^{k+1} & (k+3)z_1^{k+2} \\ k(k-1)z_1^{k-2} & (k+1)kz_1^{k-1} & (k+2)(k+1)z_1^k & (k+3)(k+2)z_1^{k+1} \\ z_2^k & z_2^{k+1} & z_2^{k+2} & z_2^{k+3} \end{vmatrix}$$

$$(5.4.4)$$

Form the related system of linear homogeneous equations in 4 unknowns $v_1, \ldots, v_4$:

$$z_1^k v_1 + z_1^{k+1} v_2 + z_1^{k+2} v_3 + z_1^{k+3} v_4 = 0$$
$$kz_1^{k-1} v_1 + (k+1)z_1^k v_2 + (k+2)z_1^{k+1} v_3 + (k+3)z_1^{k+2} v_4 = 0$$
$$k(k-1)z_1^{k-2} v_1 + (k+1)kz_1^{k-1} v_2$$
$$+ (k+2)(k+1)z_1^k v_3 + (k+3)(k+2)z_1^{k+1} v_4 = 0 \quad\quad (5.4.5)$$
$$z_2^k v_1 + z_2^{k+1} v_2 + z_2^{k+2} v_3 + z_2^{k+3} v_4 = 0$$

If $D = 0$, then by Theorem 1.2.2, we can find $v_1, \ldots, v_4$, not all zero, satisfying (5.4.5). With these values, form the polynomial

$$P(z) = v_1 z^k + v_2 z^{k+1} + v_3 z^{k+2} + v_4 z^{k+3}.$$

P is of degree $\leq k + 3$ and does not vanish identically. It has a k-fold zero at $z = 0$ and in view of (5.4.5), a 3-fold zero at z_1 and a zero at z_2. That is, it has zeros of total multiplicity $k + 4$. This is impossible, and hence $D \neq 0$. (Cf. the argument used in 2.3, Ex. 6.)

We now employ Theorem 5.3.1 and obtain values $a_0, a_1, \ldots$, for which (5.4.3) holds, all series being absolutely convergent. Since, in particular, $\sum_{k=0}^{\infty} a_k |z_j|^k < \infty$, $f(z) = \sum_{k=0}^{\infty} a_k z^k$ is convergent in $|z| < \rho$ and the formal work of (5.4.3) is valid.

Can one construct an analytic function whose derivatives at a point have been prescribed in advance? Ex. 1 of 5.1 shows that this is not always possible if the point is interior to the region of analyticity. But, by moving the point to the boundary of this region, it becomes possible.

THEOREM 5.4.2 (Borel). *Given an arbitrary sequence of real numbers* $m_0, m_1, \ldots,$ *we can find a function which is analytic in* $(-1, 1)$ *and for which*

$$\lim_{x \to 1^-} f^{(n)}(x) = m_n, \quad n = 0, 1, \ldots \tag{5.4.6}$$

Proof: Write, tentatively, $f(x) = \sum_{n=0}^{\infty} a_n x^n$, $f'(x) = \sum_{n=0}^{\infty} n a_n x^{n-1}$, etc. On the basis of these assumed expansions, set up the following infinite system of linear equations for $a_0, a_1, \ldots.$

$$
\begin{aligned}
a_0 + a_1 + a_2 + \cdots &= m_0 \\
a_1 + 2a_2 + 3a_3 + \cdots &= m_1 \\
2a_2 + 6a_3 + \cdots &= m_2
\end{aligned}
\tag{5.4.7}
$$

.

.

.

A typical column of coefficients of this system is $1, k, k(k-1), k(k-1) \times (k-2), \ldots,$ so that condition (B) of Theorem 5.3.1 is immediate. Consider, next, any $n \times n$ determinant formed from the first n rows of the coefficient matrix of (5.4.7). By the addition of appropriate linear combinations of the rows, it may be converted into a determinant whose typical column is $1, k, \ldots, k^{n-1}$. This is a Vandermonde and does not vanish. Condition (A) of Theorem 5.3.1 is satisfied, and we can find numbers $a_0, a_1, \ldots,$ for which all the series in (5.4.7) converge to the right-hand side.

Form $f(x) = \sum_{n=0}^{\infty} a_n x^n$. By the first equation of (5.4.7), $f(x)$ is analytic in $|x| < 1$. By Abel's Theorem (see e.g., Titchmarsh, [1] p. 229) $\lim_{x \to 1^-} f(x) = m_0$. Moreover, $f'(x) = \sum_{n=0}^{\infty} n a_n x^{n-1}$, $|x| < 1$. In view of the second equation of (5.4.7) and Abel's Theorem, $\lim_{x \to 1^-} f'(x) = m_1$. In this way, we can establish that (5.4.6) holds generally.

NOTES ON CHAPTER V

5.3–5.4 G. Pólya [1]. R. G. Cooke [1]. Theorems 5.4.1 and 5.4.2 have attracted wide attention and many proofs and generalizations can be found. See, e.g., Pólya [2], Ritt [1], Franklin [1].

PROBLEMS

1. Given $0 \le x_1 < x_2 < \cdots < 1$, $\lim_{n \to \infty} x_n = 1$. Find necessary and sufficient conditions on a_n in order that the problem $f(x_n) = a_n$ $n = 1, 2, \ldots,$ have a solution $f(x)$ that is differentiable in $0 \le x \le 1$.

2. Construct a function $f(x)$ that satisfies

$$xf(x) = f(x + 1), \quad 0 < x < \infty, \quad f(n) = \Gamma(n), \quad n = 1, 2, \ldots,$$

is convex for $0 < x < \infty$, but is not $\Gamma(x)$.

3. The function $f(z) = e^{1/(z+i)} - 1$ has an infinity of zeros in $|z| < 1$. Does this contradict the uniqueness principle for analytic functions?

4. Prove the following generalization of the Lagrange interpolation formula. Let $z_0, z_1, \ldots, (\neq 0)$ be distinct points with $\lim\limits_{n \to \infty} z_n = \infty$. Let $w(z)$ be an entire function with simple zeros at $z_0, z_1, \ldots$. If

$$\sum_{k=0}^{\infty} \left| \frac{a_k}{z_k w'(z_k)} \right| < \infty,$$

then

$$\sum_{k=0}^{\infty} \frac{a_k w(z)}{w'(z_k)(z - z_k)}$$

converges absolutely and uniformly in every $|z| \leq R$ to an entire function $f(z)$ for which $f(z_k) = a_k, k = 0, 1, \ldots$.

5. Specialize the result of the previous exercise by writing $w(z) = \sin z$ and obtain a theorem for the cardinal series.

6. Let $M(r)$ be an arbitrary positive function $0 \leq r < \infty$. We can find an entire function $f(z)$ such that $\max\limits_{0 \leq \theta \leq 2\pi} |f(re^{i\theta})| \geq M(r), 0 \leq r < \infty$. In other words, we can find an entire function whose growth is uniformly arbitrarily rapid. Hint: write $f(z) = a_0 + \sum\limits_{k=1}^{\infty} a_k \left(\frac{z}{k}\right)^{\lambda_k}$ and select a_k and λ_k sufficiently large. (Poincaré.)

7. Let the system $\sum\limits_{k=1}^{\infty} a_{jk} x_k = b_j, j = 1, 2, \ldots$, satisfy (A) and (B) of Theorem 5.3.1. For any $m > 0$, show that we may obtain a solution with $x_1, x_2, \ldots, x_m$ prescribed arbitrarily, and hence there is an infinity of solutions.

8. Suppose that $f(z) = \sum\limits_{n=0}^{\infty} a_n z^n$ converges at $z = 1$. Show that $f(1) = 1$, $f\left(\frac{1}{2}\right) = f\left(\frac{2}{3}\right) = \cdots = f\left(\frac{n}{n+1}\right) = \cdots = 0$ is impossible. Hence in Theorem 5.3.1 condition (A) cannot itself guarantee the existence of a solution.

9. Let $m_0, m_1, \ldots,$ be an arbitrary sequence of real numbers. Show there exists a function f of class $C^{\infty}[-1, 1]$ for which $f^{(n)}(0) = m_n \quad n = 0, 1, \ldots$.

10. Let z_n be distinct complex numbers with $\lim\limits_{n \to \infty} z_n = \infty$. If c_n is completely arbitrary, we can find an entire function $f(z)$ such that

$$\int_{z_n}^{z_{n+1}} f(z) \, dz = c_n \quad n = 0, 1, \ldots .$$

11. Use a theorem on infinite interpolation to construct a function that is analytic in $|z| < 1$ and has $|z| = 1$ as a natural boundary.

12. Let $x_1, x_2, \ldots,$ be distinct real numbers that satisfy $\lim_{n \to \infty} x_n = \infty$. If $m_0, m_1, \ldots,$ are completely arbitrary real numbers, we can find $u_0, u_1, \ldots,$ such that $\sum_{i=1}^{\infty} u_i x_i^n = m_n$ $n = 0, 1, \ldots,$ each series converging absolutely. (R. P. Boas, Jr.)

13. The problems of Theorems 5.4.1 and 5.4.2 have an infinity of solutions.

14. There is no entire function that satisfies

$$f(1) = 1, \quad f(-1) = 0, \quad f^{(2n+1)}(0) = 0 \quad n = 0, 1, \ldots.$$

15. Construct a function of class C^{∞} that satisfies the conditions of Prob. 14.

CHAPTER VI

Uniform Approximation

6.1 The Weierstrass Approximation Theorem. We come now to the 2nd fundamental theorem of this book, the Weierstrass approximation theorem of 1885.

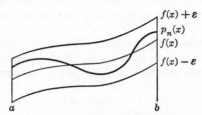

Figure 6.1.1.

THEOREM 6.1.1. *Let $f(x) \in C[a, b]$. Given an $\varepsilon > 0$ we can find a polynomial $p_n(x)$ (of sufficiently high degree) for which*

$$|f(x) - p_n(x)| \leq \varepsilon, \quad a \leq x \leq b. \tag{6.1.1}$$

Weierstrass' theorem asserts the possibility of *uniform* approximation by polynomials to continuous functions over a closed interval.

It is instructive to contrast this theorem with Taylor's theorem for analytic functions for the two are often confounded. Suppose that $f(z)$ is analytic in the circle $|z| \leq R$. Then we have $f(z) = \sum_{k=0}^{\infty} a_k z^k$, convergent uniformly in $|z| \leq R$. Hence it is clear that given an $\varepsilon > 0$, we can take sufficiently many terms of this power series and arrive at a polynomial $p_n(z) = \sum_{k=0}^{n} a_k z^k$ for which $|f(z) - p_n(z)| \leq \varepsilon$ for $|z| \leq R$. A fortiori, on the real segment $-R \leq x \leq R$ we have $|f(x) - p_n(x)| \leq \varepsilon$. But for functions that are not analytic, there is no expansion in power series. Yet Weierstrass' theorem assures us we can approximate uniformly functions which are merely continuous. Given a sequence $\varepsilon_1, \varepsilon_2, \ldots \to 0$, we can find polynomials

$$\begin{aligned}
p_{n_1}(x) &= a_{0n_1} + a_{1n_1}x + \cdots + a_{n_1 n_1}x^{n_1} \\
p_{n_2}(x) &= a_{0n_2} + a_{1n_2}x + \cdots + a_{n_2 n_2}x^{n_2}
\end{aligned} \tag{6.1.2}$$

$$\vdots$$

for which $|f(x) - p_{n_k}(x)| \leq \varepsilon_k$, $-R \leq x \leq R$, $k = 1, 2, \ldots$. Consequently $\lim\limits_{k \to \infty} p_{n_k}(x) = f(x)$ uniformly in $-R \leq x \leq R$.

From approximations, it is simple enough to go to expansions, for we can write the "collapsing" series

$$f(x) = p_{n_1}(x) + (p_{n_2}(x) - p_{n_1}(x)) + (p_{n_3}(x) - p_{n_2}(x)) + \cdots \quad (6.1.3)$$

which, evidently, converges to $f(x)$ uniformly on $-R \leq x \leq R$. Thus, briefly, an analytic function can be expanded in a uniformly convergent *power series*, and a continuous but nonanalytic function can be expanded in a uniformly convergent *series of general polynomials*, with no possibility of rearranging its terms so as to produce a convergent power series.

6.2   The Bernstein Polynomials. There are many proofs of the Weierstrass theorem, and we shall present S. Bernstein's proof. While it is not the simplest conceptually, it is easily the most elegant.

DEFINITION 6.2.1.   Let $f(x)$ be defined on $[0, 1]$. The nth ($n \geq 1$) Bernstein polynomial for $f(x)$ is given by

$$B_n(f; x) = \sum_{k=0}^{n} f\left(\frac{k}{n}\right) \binom{n}{k} x^k (1 - x)^{n-k}. \quad (6.2.1)$$

Notice that

$$B_n(f; 0) = f(0) \quad B_n(f; 1) = f(1). \quad (6.2.2)$$

It is clear that $B_n \in \mathscr{P}_n$. In certain cases, it may degenerate and become a polynomial of degree lower than n.

THEOREM 6.2.1.

$$B_n(f; x) = \sum_{t=0}^{n} \Delta^t f(0) \binom{n}{t} x^t \quad (6.2.3)$$

where the differences have been computed from the functional values at $0/n, 1/n, \ldots, (n-1)/n, n/n$.

Proof: $B_n(f; x) = \sum\limits_{k=0}^{n} f\left(\dfrac{k}{n}\right) \binom{n}{k} x^k (1 - x)^{n-k}$

$$= \sum_{k=0}^{n} f\left(\frac{k}{n}\right) \binom{n}{k} x^k \sum_{j=0}^{n-k} \binom{n-k}{j} (-1)^{n-k-j} x^{n-k-j}$$

$$= \sum_{k=0}^{n} \sum_{j=0}^{n-k} f\left(\frac{k}{n}\right) \binom{n}{k} \binom{n-k}{j} (-1)^{n-k-j} x^{n-j}.$$

Rearranging the summation, we obtain

$$\sum_{t=0}^{n} x^t \sum_{k=0}^{t} f\left(\frac{k}{n}\right)\binom{n}{k}\binom{n-k}{n-t}(-1)^{t-k} = \sum_{t=0}^{n} x^t \binom{n}{t} \sum_{k=0}^{t} f\left(\frac{k}{n}\right)\binom{t}{k}(-1)^{t-k}$$

$$= \sum_{t=0}^{n} \Delta^t f(0)\binom{n}{t} x^t.$$

The last equality follows from (2.7.5).

Ex. 1. If $f \in \mathscr{P}_m$ then $\Delta^t f(0) = 0$ for $t > m$. (6.2.3) then implies that $B_n(f; x) \in \mathscr{P}_m$ for all n.

Ex. 2. Useful identities may be derived by applying (6.2.3) to the functions $1, x, x^2$. For $f(x) = 1$, we have $\Delta^0 f(0) = 1, \Delta^1 f(0) = 0, \Delta^2 f(0) = 0$, etc. Hence

$$B_n(1; x) = \sum_{k=0}^{n} \binom{n}{k} x^k (1-x)^{n-k} = 1. \tag{6.2.4}$$

This, of course, is the binomial expansion for $1^n = (x + (1-x))^n$.

Ex. 3. For $f(x) = x$, we have $\Delta^0 f(0) = 0, \Delta^1 f(0) = \dfrac{1}{n}, \Delta^2 f(0) = 0, \ldots$. Hence,

$$B_n(x; x) = \sum_{k=0}^{n} \frac{k}{n}\binom{n}{k} x^k (1-x)^{n-k} = \frac{1}{n}\binom{n}{1} x = x. \tag{6.2.5}$$

Ex. 4. For $f(x) = x^2$, we have

$$\Delta^0 f(0) = 0, \quad \Delta^1 f(0) = \frac{1}{n^2}, \quad \Delta^2 f(0) = \frac{2}{n^2}, \quad \Delta^3 f(0) = 0, \ldots$$

Hence

$$B_n(x^2; x) = \sum_{k=0}^{n} \left(\frac{k}{n}\right)^2 \binom{n}{k} x^k (1-x)^{n-k} = \frac{1}{n^2}\binom{n}{1} x + \frac{2}{n^2}\binom{n}{2} x^2. \tag{6.2.6}$$

Ex. 5. $B_n(e^{\alpha x}; x) = (x e^{\alpha/n} + (1-x))^n$.

THEOREM 6.2.2 (Bernstein). *Let $f(x)$ be bounded on $[0, 1]$. Then*

$$\lim_{n \to \infty} B_n(f; x) = f(x) \tag{6.2.7}$$

at any point $x \in [0, 1]$ at which f is continuous. If $f \in C[0, 1]$, the limit (6.2.7) holds uniformly in $[0, 1]$.

Proof: A. Note the identity

$$\sum_{k=0}^{n} (k - nx)^2 \binom{n}{k} x^k (1-x)^{n-k} \equiv nx(1-x). \tag{6.2.8}$$

To prove this, expand this sum into

$$\sum_{k=0}^{n} k^2 \binom{n}{k} x^k (1-x)^{n-k} - 2nx \sum_{k=0}^{n} k \binom{n}{k} x^k (1-x)^{n-k}$$
$$+ n^2 x^2 \sum_{k=0}^{n} \binom{n}{k} x^k (1-x)^{n-k},$$

and combine this with identities (6.2.4), (6.2.5), and (6.2.6).

B. For a given $\delta > 0$ and for $0 \le x \le 1$, we have

$$\sum_{|k/n-x| \ge \delta} \binom{n}{k} x^k (1-x)^{n-k} \le \frac{1}{4n\delta^2}. \tag{6.2.9}$$

This notation means that we sum over those values of $k = 0, 1, \ldots, n$ for which $\left|\dfrac{k}{n} - x\right| \ge \delta$. To prove (6.2.9), note that $\left|\dfrac{k}{n} - x\right| \ge \delta$ implies $\dfrac{1}{\delta^2}\left(\dfrac{k}{n} - x\right)^2 \ge 1$. Hence

$$\sum_{|k/n-x| \ge \delta} \binom{n}{k} x^k (1-x)^{n-k} \le \frac{1}{\delta^2} \sum_{|k/n-x| \ge \delta} \left(\frac{k}{n} - x\right)^2 \binom{n}{k} x^k (1-x)^{n-k}$$

$$\le \frac{1}{\delta^2} \sum_{k=0}^{n} \left(\frac{k}{n} - x\right)^2 \binom{n}{k} x^k (1-x)^{n-k} = \frac{nx(1-x)}{\delta^2 n^2} \le \frac{1}{4n\delta^2}.$$

The last inequality follows since $x(1-x) \le \frac{1}{4}$ for all x.

C. We have $1 = \sum_{k=0}^{n} \binom{n}{k} x^k (1-x)^{n-k}$ from (6.2.4). Hence,

$$f(x) = \sum_{k=0}^{n} f(x) \binom{n}{k} x^k (1-x)^{n-k}$$

so that

$$f(x) - B_n(f; x) = \sum_{k=0}^{n} \left\{ f(x) - f\left(\frac{k}{n}\right) \right\} \binom{n}{k} x^k (1-x)^{n-k}$$

$$= \sum_{|k/n-x| < \delta} \left\{ f(x) - f\left(\frac{k}{n}\right) \right\} \binom{n}{k} x^k (1-x)^{n-k}$$

$$+ \sum_{|k/n-x| \ge \delta} \left\{ f(x) - f\left(\frac{k}{n}\right) \right\} \binom{n}{k} x^k (1-x)^{n-k}.$$

The function $f(x)$ is assumed bounded in $[0, 1]$. Hence for some $M > 0$, $|f(x)| \le M$ and for any two values $\alpha, \beta \in [0, 1]$, $|f(\alpha) - f(\beta)| \le 2M$. Let x be a point of continuity of f. Given an $\varepsilon > 0$, we can find a δ such that

$|f(x) - f(y)| < \varepsilon$ whenever $|y - x| < \delta$. Thus, using these estimates and using this δ in (6.2.9), we have

$$|f(x) - B_n(f; x)| \leq \sum_{|k/n-x|<\delta} \left|f(x) - f\left(\frac{k}{n}\right)\right| \binom{n}{k} x^k (1-x)^{n-k}$$

$$+ \sum_{|k/n-x|\geq\delta} \left|f(x) - f\left(\frac{k}{n}\right)\right| \binom{n}{k} x^k (1-x)^{n-k}$$

$$\leq \varepsilon \sum_{|k/n-x|<\delta} \binom{n}{k} x^k (1-x)^{n-k} + 2M \sum_{|k/n-x|\geq\delta} \binom{n}{k} x^k (1-x)^{n-k}$$

$$\leq \varepsilon \sum_{k=0}^{n} \binom{n}{k} x^k (1-x)^{n-k} + \frac{2M}{4n\delta^2} = \varepsilon + \frac{M}{2n\delta^2}.$$

From this inequality, we see that $|f(x) - B_n(f; x)| \leq 2\varepsilon$ for n sufficiently large. Since ε is arbitrary, (6.2.7) follows.

Suppose now that $f \in C[0, 1]$; then f is uniformly continuous there. Given an $\varepsilon > 0$, we can find a δ such that $|f(x) - f(y)| < \varepsilon$ for *all* x, y in $[0, 1]$ satisfying $|x - y| < \delta$. The above inequality holds independently of the x selected and the convergence to $f(x)$ is uniform in $[0, 1]$. We express this as a corollary.

COROLLARY 6.2.3. *If* $f(x) \in C[0, 1]$, *then given an* $\varepsilon > 0$, *we have for all sufficiently large* n,

$$|f(x) - B_n(f; x)| \leq \varepsilon, \quad 0 \leq x \leq 1. \tag{6.2.10}$$

Bernstein's Theorem not only proves the existence of polynomials of uniform approximation, but provides a simple explicit representation for them.

The results for $[0, 1]$ are easily transferred to $[a, b]$ by means of the linear transformation

$$y = \frac{x - a}{b - a} \tag{6.2.11}$$

that converts $[a, b]$ into $[0, 1]$.

COROLLARY 6.2.4 (Theorem 6.1.1). *Let* $f(x) \in C[a, b]$. *Then given* $\varepsilon > 0$, *we can find a polynomial* $p(x)$ *such that* $|f(x) - p(x)| \leq \varepsilon$ *for* $a \leq x \leq b$.

Proof: Consider $g(y) = f(a + (b - a)y)$. $g \in C[0, 1]$. Hence given an $\varepsilon > 0$ we can find a polynomial $r(y)$ such that $|g(y) - r(y)| \leq \varepsilon$, $0 \leq y \leq 1$.

Set $p(x) = r\left(\frac{x - a}{b - a}\right)$, which is a polynomial in x, and the required inequality follows.

6.3   Simultaneous Approximation of Functions and Derivatives.

In contrast to other modes of approximation—in particular to Tscheby-scheff or best uniform approximation which will be studied subsequently—the Bernstein polynomials yield smooth approximants. If the approximated function is differentiable, not only do we have $B_n(f; x) \to f(x)$ but $B_n'(f; x) \to f'(x)$. A corresponding statement is true for higher derivatives. The Bernstein polynomials therefore provide *simultaneous approximation* of the function and its derivatives.

In order to make the force of this result felt, we call attention to the following examples from real and complex analysis.

Ex. 1.   Uniform approximation does not automatically carry with it approximation of the derivatives. Consider $f_n(x) = \dfrac{1}{n} \sin nx$ on $[0, 2\pi]$. Since $|f_n(x)| \leq \dfrac{1}{n}$, the sequence f_n converges to 0 uniformly on $[0, 2\pi]$. On the other hand, $f_n'(x) = \cos nx$, so that f_n' does not approach $0' = 0$.

This phenomenon may be present in sequences of polynomials. Let $f_n(x) = \dfrac{1}{n} T_n(x)$, Cf. (3.3.2). Since $|T_n(x)| \leq 1$ on $[-1, 1]$, it follows that $f_n \to 0$ uniformly there. Now, $f_n'(x) = (1 - x^2)^{-\frac{1}{2}} \sin (n \arccos x)$, and if we set $x_n = \cos \dfrac{\pi}{2n}$, then $f_n'(x_n) = \csc \dfrac{\pi}{2n}$. The sequence of derivatives of f_n cannot approach any function of $C[-1, 1]$ uniformly.

Ex. 2.   Uniform approximation of analytic functions by analytic functions is totally different. Let R be a region bounded by a simple closed curve C. Let $f(z)$ and $p(z)$ be two functions analytic in R and on C. Suppose that $|f(z) - p(z)| \leq \varepsilon$ on C. By the Maximum Principle, this inequality, and hence uniform approximation, persists throughout R. Moreover, by Cauchy's Inequality (1.9.8) we have

$$|f^{(n)}(z) - p^{(n)}(z)| \leq \frac{n! \, L(C)}{2\pi \delta^{n+1}} \, \varepsilon \tag{6.3.1}$$

for z confined to a point set S in R the distance of whose points from C is no less than δ. For fixed S and n, allow $\varepsilon \to 0$ and (6.3.1) tells us that the nth derivative of the approximant is also a uniform approximation to the nth derivative of the approximee. In the complex analytic case, uniform approximation over regions carries with it the simultaneous uniform approximation, in the above sense, of all the derivatives.

LEMMA 6.3.1.   *Let $p \geq 0$ be an integer.* *Then*

$$B_{n+p}^{(p)}(f; x) = \frac{(n + p)!}{n!} \sum_{t=0}^{n} \Delta^p f\left(\frac{t}{n + p}\right)\binom{n}{t} x^t (1 - x)^{n-t}. \tag{6.3.2}$$

Proof: Apply Leibnitz' formula

$$(uv)^{(p)} = \sum_{j=0}^{p} \binom{p}{j} u^{(j)} v^{(p-j)} \tag{6.3.3}$$

to (6.2.1) and obtain

$$B_{n+p}^{(p)}(f; x) = \sum_{k=0}^{n+p} f\left(\frac{k}{n+p}\right)\binom{n+p}{k} \sum_{j=0}^{p} \binom{p}{j} (x^k)^{(j)}[(1-x)^{n+p-k}]^{(p-j)}. \tag{6.3.4}$$

Now, we have $(x^k)^{(j)} = k!\, x^{k-j}/(k-j)!$, $k - j \geq 0$ and

$$[(1-x)^{n+p-k}]^{(p-j)} = (-1)^{p-j}(n+p-k)!$$
$$\times (1-x)^{n+j-k}/(n+j-k)!,\ k - j \leq n.$$

Therefore (6.3.4) becomes

$$B_{n+p}^{(p)}(f; x) =$$

$$\sum_{k=0}^{n+p} \sum_{\substack{j=0 \\ 0 \leq k-j \leq n}}^{p} f\left(\frac{k}{n+p}\right) \frac{(n+p)!}{(k-j)!\,(n+j-k)!} \binom{p}{j}(-1)^{p-j} x^{k-j}(1-x)^{n+j-k}. \tag{6.3.5}$$

If we set $k - j = t$, $k = t + j$, we see that $0 \leq t \leq n$, $j = 0, 1, \ldots, p$, corresponds to the range of the sum in (6.3.5). We may write (6.3.5) as

$$B_{n+p}^{(p)}(f; x) = (n+p)! \sum_{t=0}^{n} \frac{x^t(1-x)^{n-t}}{t!\,(n-t)!} \sum_{j=0}^{p}(-1)^{p-j}\binom{p}{j} f\left(\frac{t+j}{n+p}\right) \tag{6.3.6}$$

(6.3.2) now follows from (2.7.4).

THEOREM 6.3.2. *Let* $f(x) \in C^p[0, 1]$. *Then*

$$\lim_{n \to \infty} B_n^{(p)}(f; x) = f^{(p)}(x)\ \textit{uniformly on}\ [0, 1]. \tag{6.3.7}$$

Proof: By (3.4.5) we have $\Delta^p f\left(\dfrac{t}{n+p}\right) = \dfrac{1}{(n+p)^p} f^{(p)}(\xi_t)$ for some ξ_t satisfying $\dfrac{t}{n+p} < \xi_t < \dfrac{t+p}{n+p}$, $t = 0, 1, \ldots, n$. Hence from Lemma 6.3.1,

$$B_{n+p}^{(p)}(f; x) = \frac{(n+p)!}{n!\,(n+p)^p} \sum_{t=0}^{n} f^{(p)}(\xi_t)\binom{n}{t} x^t(1-x)^{n-t}.$$

It follows that

$$\frac{n!\,(n+p)^p}{(n+p)!} B_{n+p}^{(p)}(f; x) = \sum_{t=0}^{n} f^{(p)}\left(\frac{t}{n}\right)\binom{n}{t} x^t(1-x)^{n-t}$$

$$+ \sum_{t=0}^{n}\left\{f^{(p)}(\xi_t) - f^{(p)}\left(\frac{t}{n}\right)\right\}\binom{n}{t} x^t(1-x)^{n-t}. \tag{6.3.8}$$

Since $\dfrac{t}{n+p} \leq \dfrac{t}{n} \leq \dfrac{t+p}{n+p}$, $t = 0, 1, \ldots, n$, it follows from the bounds on

ξ_t that $\left| \xi_t - \dfrac{t}{n} \right| < \dfrac{t+p}{n+p} - \dfrac{t}{n+p} = \dfrac{p}{n+p}$.

From the uniform continuity of $f^{(p)}(x)$, given an $\varepsilon > 0$, we can find an n_0 such that for all $n \geq n_0$ and all t, $|f^{(p)}(\xi_t) - f^{(p)}(t/n)| < \varepsilon$. As in the proof of Theorem 6.2.2, the second sum in (6.3.8) is less than ε in absolute value for $n \geq n_0$ and for all $x \in [0, 1]$. Furthermore, $\lim_{n \to \infty} \dfrac{n! \, (n+p)^p}{(n+p)!} = 1$, and by Theorem 6.2.2, the first sum approaches $f^{(p)}(x)$ uniformly. The theorem follows from this.

More general results may be established on the assumption that $f^{(p)}(x)$ exists at individual points of the interval.

THEOREM 6.3.3.   *Let p be a fixed integer with $0 \leq p \leq n$. If*

$$m \leq f^{(p)}(x) \leq M, \quad 0 \leq x \leq 1 \tag{6.3.9}$$

then

$$m \leq \frac{n^p}{n(n-1)\cdots(n-p+1)} \, B_n^{(p)}(f; x) \leq M, \, 0 \leq x \leq 1. \tag{6.3.10}$$

For $p = 0$, the multiplier of $B_n^{(p)}$ is to be interpreted as 1. If

$$f^{(p)}(x) \geq 0, \quad 0 \leq x \leq 1 \tag{6.3.11}$$

then

$$B_n^{(p)}(f; x) \geq 0, \quad 0 \leq x \leq 1. \tag{6.3.12}$$

If $f(x)$ is nondecreasing on $0 \leq x \leq 1$, then $B_n(f; x)$ is nondecreasing there. If $f(x)$ is convex on $0 \leq x \leq 1$ then $B_n(f; x)$ is convex there.

Proof: From (6.3.2) and (6.2.3) we have for $p = 1, 2, \ldots, n$,

$$B_n^{(p)}(f; x) = n(n-1)\cdots(n-p+1) \sum_{t=0}^{n-p} \Delta^p f\left(\frac{t}{n}\right) \binom{n-p}{t} x^t (1-x)^{n-p-t}. \tag{6.3.13}$$

By the extended mean value theorem, Cor. 3.4.4,

$$\Delta^p f\left(\frac{t}{n}\right) = \frac{1}{n^p} f^{(p)}(\xi_t), \quad \frac{t}{n} < \xi_t < \frac{t+p}{n}.$$

For $p = 0$, this equality obviously holds with $\xi_t = t/n$. Hence,

$$Q = \frac{n^p}{n(n-1)\cdots(n-p+1)} \, B_n^{(p)}(f; x)$$

$$= \sum_{t=0}^{n-p} f^{(p)}(\xi_t) \binom{n-p}{t} x^t (1-x)^{n-p-t}.$$

In view of (6.3.9) and the fact that $x^t(1-x)^{n-p-t} \geq 0$ on $[0, 1]$, it follows that

$$m = m \sum_{t=0}^{n-p} \binom{n-p}{t} x^t (1-x)^{n-p-t} \le Q \le M \sum_{t=0}^{n-p} \binom{n-p}{t} x^t (1-x)^{n-p-t} = M.$$

This demonstrates (6.3.10). (6.3.11) follows by setting $m = 0$.

If $f(x)$ is nondecreasing, $\Delta f(t/n) \ge 0$ and hence from (6.3.13) with $p = 1$, $B_n{}'(f; x) \ge 0$ on $[0, 1]$ and this implies that $B_n(f; x)$ is nondecreasing. Finally, if f is convex, then by (3.2.2), $\Delta^2 f(t/n) \ge 0$. From (6.3.13) with $p = 2$, this implies that $B_n{}''(f; x) \ge 0$. By Theorem 3.2.1, this, in turn, implies that B_n is convex in every closed subinterval of $(0, 1)$. Since B_n is continuous, it is convex in $[0, 1]$.

THEOREM 6.3.4. *Let $f(x)$ be convex in $[0, 1]$. Then, for $n = 2, 3, \ldots$,*

$$B_{n-1}(f; x) \ge B_n(f; x), \quad 0 < x < 1. \tag{6.3.14}$$

If $f \in C[0, 1]$, the strict inequality holds unless f is linear in each of the intervals $\left[\dfrac{j-1}{n-1}, \dfrac{j}{n-1} \right], j = 1, 2, \ldots, n-1$. *In this case, $B_{n-1}(f; x) = B_n(f; x)$.*

Proof: In (6.2.1) set $t = \dfrac{x}{1-x}$ and obtain

$$(1-x)^{-n}(B_{n-1}(f; x) - B_n(f; x))$$

$$= (1+t) \sum_{k=0}^{n-1} f\left(\frac{k}{n-1}\right) \binom{n-1}{k} t^k - \sum_{k=0}^{n} f\left(\frac{k}{n}\right) \binom{n}{k} t^k$$

$$= \sum_{k=1}^{n-1} f\left(\frac{k}{n-1}\right) \binom{n-1}{k} t^k + f(0) + \sum_{k=1}^{n-1} f\left(\frac{k-1}{n-1}\right) \binom{n-1}{k-1} t^k + f(1)t^n$$

$$- \sum_{k=1}^{n-1} f\left(\frac{k}{n}\right) \binom{n}{k} t^k - f(0) - f(1)t^n = \sum_{k=1}^{n-1} c_k t^k, \text{ where}$$

$$c_k = \frac{(n-1)!}{(k-1)!\,(n-k-1)!} \left\{ \frac{1}{k} f\left(\frac{k}{n-1}\right) \right.$$

$$\left. + \frac{1}{n-k} f\left(\frac{k-1}{n-1}\right) - \frac{n}{k(n-k)} f\left(\frac{k}{n}\right) \right\}. \tag{6.3.15}$$

Now $\dfrac{k-1}{n-1} < \dfrac{k}{n} < \dfrac{k}{n-1}$, and since f is convex, the bracketed quantity in (6.3.15) is ≥ 0 by Definition 3.2.1. Therefore $\sum_{k=1}^{n-1} c_k t^k \ge 0$ and (6.3.14) follows.

If f is linear in each of the intervals $\left[\dfrac{j-1}{n-1}, \dfrac{j}{n-1} \right]$, then all the c_j are 0 and hence $B_{n-1} \equiv B_n$. Conversely, if $B_{n-1} \equiv B_n$, then all the c_j are 0, and since $f \in C[0, 1]$ and is convex, (6.3.15) implies that f is linear in each interval.

The geometric interpretation of these theorems is this. The Bernstein approximant of a continuous function lies between the extreme values of the function itself, and its higher derivatives are bounded by (6.3.10). Monotonic and convex functions yield monotonic and convex approximants respectively. In a word—and this is reflected in Figure 6.3.1—the Bernstein approximants mimic the behavior of the function to a remarkable degree.

There is a price that must be paid for these beautiful approximation properties: the convergence of the Bernstein polynomials is very slow.

Ex. 3.   From (6.2.6) we have $B_n(x^2; x) - x^2 = \dfrac{x(1 - x)}{n}$. The convergence is like $1/n$.

It is far slower than what can be achieved by other means. If f is bounded, then at a point where $f''(x)$ exists and does not vanish, $B_n(f; x)$ converges to $f(x)$ precisely like C/n. (See Theorem 6.3.6.) This fact seems to have precluded any numerical application of Bernstein polynomials from having been made. Perhaps they will find application when the properties of the approximant in the large are of more importance than the closeness of the approximation.

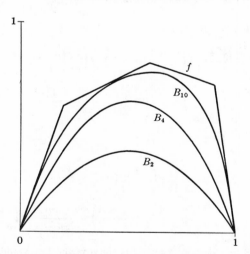

Figure 6.3.1   Illustrating the Approximation Properties of the Bernstein Polynomials of a Concave Function.

$B_2(f; x) = \frac{3}{2}(x - x^2)$
$B_4(f; x) = \frac{5}{2}x - 3x^2 + \frac{3}{2}x^3 - x^4$
$B_{10}(f; x) = 3x - 30x^3 + 105x^4 - 189x^5 + 210x^6 - 160x^7 + 90x^8 - 35x^9 + 6x^{10}$.
The graph of f is polygonal and joins $(0,0)$, $(.2,.6)$, $(.6,.8)$, $(.9,.7)$, $(1,0)$.

LEMMA 6.3.5. *There is a constant c independent of n such that for all x in* $[0, 1]$,

$$\sum_{|k/n - x| \geq n^{-\frac{1}{4}}} \binom{n}{k} x^k (1 - x)^{n-k} \leq \frac{c}{n^{\frac{3}{2}}}. \tag{6.3.16}$$

Proof: Consider the sums

$$S_m(x) = \sum_{k=0}^{n} (k - nx)^m \binom{n}{k} x^k (1 - x)^{n-k}. \tag{6.3.17}$$

We have already established by (6.2.4), (6.2.5), and (6.2.8), that $S_0(x) = 1$, $S_1(x) = 0$, $S_2(x) = nx(1 - x)$. Differentiating (6.3.17) we have

$$S_m'(x) = \sum_{k=0}^{n} \binom{n}{k} (k - nx)^{m-1} x^{k-1} (1 - x)^{n-k-1} [-mnx(1 - x) + (k - nx)^2]$$

$$= -mnS_{m-1}(x) + \frac{S_{m+1}(x)}{x(1 - x)}. \tag{6.3.18}$$

Hence,

$$S_{m+1}(x) = x(1 - x)[S_m'(x) + mnS_{m-1}(x)]. \tag{6.3.19}$$

We may conclude from this recurrence that each sum $S_m(x)$ is a polynomial in x and n. In particular, S_3 is of 1st, S_4 is of 2nd, S_5 is of 2nd, and S_6 is of 3rd degree in n. Hence, for some constant c, $|S_6(x)| \leq cn^3$ for x in $[0, 1]$.

Inasmuch as $\left| \dfrac{k}{n} - x \right| \geq n^{-\frac{1}{4}}$ implies $\dfrac{(k - nx)^6}{n^{\frac{9}{2}}} \geq 1$,

$$\sum_{|k/n - x| \geq n^{-\frac{1}{4}}} \binom{n}{k} x^k (1 - x)^{n-k} \leq \frac{1}{n^{\frac{9}{2}}} \sum_{k=0}^{n} (k - nx)^6 \binom{n}{k} x^k (1 - x)^{n-k}$$

$$= n^{-\frac{9}{2}} S_6(x) \leq \frac{c}{n^{\frac{3}{2}}}.$$

THEOREM 6.3.6 (Voronovsky). *Let* $f(x)$ *be bounded in* $[0, 1]$ *and let* x_0 *be a point of* $[0, 1]$ *at which* $f''(x_0)$ *exists. Then,*

$$\lim_{n \to \infty} n[B_n(f; x_0) - f(x_0)] = \tfrac{1}{2} x_0 (1 - x_0) f''(x_0). \tag{6.3.20}$$

Proof: From Theorem 1.6.6 we have

$$f(x) = f(x_0) + f'(x_0)(x - x_0) + \frac{f''(x_0)(x - x_0)^2}{2} + s(x)(x - x_0)^2$$

where $\lim_{x \to x_0} s(x) = 0$. Set $x = \dfrac{k}{n}$.

$$f\left(\frac{k}{n}\right) = f(x_0) + f'(x_0)\left(\frac{k}{n} - x_0\right) + \frac{f''(x_0)}{2}\left(\frac{k}{n} - x_0\right)^2 + s\left(\frac{k}{n}\right)\left(\frac{k}{n} - x_0\right)^2$$

$$\tag{6.3.21}$$

Multiply both sides of (6.3.21) by $\binom{n}{k}x_0^k(1-x_0)^{n-k}$ and sum from $k=0$ to $k=n$. In view of identities (6.2.4–6), we obtain

$$B_n(f;x_0) = f(x_0) + \frac{x_0(1-x_0)f''(x_0)}{2n} + \sum_{k=0}^{n} s\left(\frac{k}{n}\right)\left(\frac{k}{n}-x_0\right)^2\binom{n}{k}x_0^k(1-x_0)^{n-k}.$$
(6.3.22)

Designate the third term in (6.3.22) by S. Let $\varepsilon > 0$ be given. We can find n sufficiently large that $|x-x_0| < \frac{1}{n^{\frac{1}{4}}}$ implies $|s(x)| \leq \varepsilon$. Hence,

$$|S| \leq \sum_{|k/n-x_0|<n^{-\frac{1}{4}}}\left| s\left(\frac{k}{n}\right)\right|\left(\frac{k}{n}-x_0\right)^2\binom{n}{k}x_0^k(1-x_0)^{n-k}$$
(6.3.23)
$$+ \sum_{|k/n-x_0|\geq n^{-\frac{1}{4}}}\left| s\left(\frac{k}{n}\right)\right|\left(\frac{k}{n}-x_0\right)^2\binom{n}{k}x_0^k(1-x_0)^{n-k}.$$

Thus,

$$|S| \leq \varepsilon\sum_{k=0}^{n}\left(\frac{k}{n}-x_0\right)^2\binom{n}{k}x_0^k(1-x_0)^{n-k} + M\sum_{|k/n-x_0|\geq n^{-\frac{1}{4}}}\binom{n}{k}x_0^k(1-x_0)^{n-k}$$

where $M = \sup\limits_{0\leq x\leq 1} s(x)(x-x_0)^2$. By (6.2.8) and Lemma 6.3.5,

$$|S| \leq \frac{\varepsilon x_0(1-x_0)}{n} + \frac{MC}{n^{\frac{3}{2}}}.$$

It follows from (6.3.22) that

$$\left| n[B_n(f;x_0) - f(x_0)] - \frac{x_0(1-x_0)}{2}f''(x_0)\right| = |nS| \leq \varepsilon x_0(1-x_0) + \frac{MC}{n^{\frac{1}{2}}}.$$
(6.3.24)

Since ε is arbitrary, (6.3.20) follows.

6.4 Approximation by Interpolation: Fejér's Proof.

THEOREM 6.4.1.   *Let $x_1, x_2, \ldots, x_n$ be the zeros of the Tschebyscheff polynomial $T_n(x)$, $\left(x_j = \cos\frac{2j-1}{2n}\pi\right)$. Let $f(x) \in C[-1,1]$ and suppose that $H_{2n-1}(f;x)$ is that element of $\mathscr{P}_{2n-1}$ for which*

$$\left.\begin{array}{l} H_{2n-1}(f;x_k) = f(x_k)\\ H'_{2n-1}(f;x_k) = 0 \end{array}\right\} \quad k = 1, 2, \ldots, n.$$
(6.4.1)

Then, $\lim\limits_{n\to\infty} H_{2n-1}(f;x) = f(x)$ uniformly in $[-1,1]$.

Proof: We are confronted here with a problem of Hermite interpolation.

As already observed in (2.5.20), the general solution is given by

$$H_{2n-1}(f; x) = \sum_{k=1}^{n} y_k A_k(x) + \sum_{k=1}^{n} y_k' B_k(x) \qquad (6.4.2)$$

where

$$A_k(x) = \left[1 - \frac{w''(x_k)}{w'(x_k)} (x - x_k) \right] l_k^2(x)$$

$$B_k(x) = (x - x_k) l_k^2(x)$$

$$l_k(x) = \frac{w(x)}{w'(x_k)(x - x_k)}, \; w(x) = (x - x_1) \cdots (x - x_n). \qquad (6.4.3)$$

In the present construction, we select $y_k' = 0$, $k = 1, \ldots, n$, so that our interpolation polynomial reduces to

$$H_{2n-1}(f; x) = \sum_{k=1}^{n} f(x_k) A_k(x). \qquad (6.4.4)$$

We next compute $l_k(x)$, $w'(x_k)$, $w_k''(x)$. Now

$$w(x) = c_n T_n(x) = c_n \cos(n \text{ arc cos } x), \; c_n = \frac{1}{2^{n-1}},$$

and hence,

$$\frac{w'(x)}{c_n} = \frac{n \sin(n \text{ arc cos } x)}{\sqrt{1 - x^2}}, \text{ and}$$

$$\frac{w''(x)}{c_n} = n \left[\frac{x \sin(n \text{ arc cos } x)}{(1 - x^2)^{\frac{3}{2}}} - \frac{n(1 - x^2)^{\frac{1}{2}} \cos(n \text{ arc cos } x)}{(1 - x^2)^{\frac{3}{2}}} \right].$$

At $x = x_k$, $\cos(n \text{ arc cos } x) = 0$, and $\sin(n \text{ arc cos } x) = \sin((k - \frac{1}{2})\pi) = (-1)^{k-1}$.

Therefore, $\dfrac{w'(x_k)}{c_n} = \dfrac{(-1)^{k-1} n}{\sqrt{1 - x_k^2}}$, $\dfrac{w''(x_k)}{c_n} = \dfrac{n x_k (-1)^{k-1}}{(1 - x_k^2)^{\frac{3}{2}}}$,

$$l_k(x) = \frac{(-1)^{k-1} \sqrt{1 - x_k^2} \, T_n(x)}{n(x - x_k)}, \; 1 - \frac{w''(x_k)}{w'(x_k)} (x - x_k) = \frac{1 - x x_k}{1 - x_k^2}$$

$$A_k(x) = \left[1 - \frac{w''(x_k)}{w'(x_k)} (x - x_k) \right] l_k^2(x) = \frac{1 - x x_k}{1 - x_k^2} \frac{T_n^2(x)(1 - x_k^2)}{n^2(x - x_k)^2}$$

$$= (1 - x x_k) \left(\frac{T_n(x)}{n(x - x_k)} \right)^2.$$

Formula (6.4.4) can be rewritten as

$$H_{2n-1}(f; x) = \sum_{k=1}^{n} f(x_k)(1 - x x_k) \left(\frac{T_n(x)}{n(x - x_k)} \right)^2 = \sum_{k=1}^{n} f(x_k) A_k(x) \qquad (6.4.5)$$

where

$$A_k(x) = (1 - x x_k) \left(\frac{T_n(x)}{n(x - x_k)} \right)^2. \qquad (6.4.6)$$

Observe that since $|x_k| < 1$, $A_k(x) \geq 0$ for $-1 \leq x \leq 1$. Observe further that

$$H_{2n-1}(1; x) = \sum_{k=1}^{n} A_k(x) \tag{6.4.7}$$

is that unique polynomial $P_{2n-1}(x)$, of degree $\leq 2n - 1$, for which

$$P_{2n-1}(x_k) = 1, \quad P'_{2n-1}(x_k) = 0, \quad k = 1, 2, \ldots, n.$$

The polynomial 1 fits these requirements and hence

$$H_{2n-1}(1; x) = \sum_{k=1}^{n} A_k(x) \equiv 1. \tag{6.4.8}$$

After these algebraic preliminaries we can turn to the proof of convergence. Since $\sum_{k=1}^{n} A_k(x) \equiv 1$, it follows that $f(x) = \sum_{k=1}^{n} f(x)A_k(x)$, so that

$$f(x) - H_{2n-1}(f; x) = \sum_{k=1}^{n} (f(x) - f(x_k))A_k(x)$$

and

$$|f(x) - H_{2n-1}(f; x)| \leq \sum_{k=1}^{n} |f(x) - f(x_k)| \, A_k(x). \tag{6.4.9}$$

Since $f(x) \in C[-1, 1]$, it is uniformly continuous there. This means that if $\varepsilon > 0$ is given, we can find a $\delta > 0$ such that

$$|x_1 - x_2| \leq \delta \text{ implies } |f(x_1) - f(x_2)| \leq \varepsilon, \, -1 \leq x_1, x_2 \leq 1. \tag{6.4.10}$$

For a given x in $[-1, 1]$, split the indices $k = 1, 2, \ldots, n$ into two sets: I: $|x - x_k| < \delta$. II: $|x - x_k| \geq \delta$. Then,

$$|f(x) - H_{2n-1}(f; x)| \leq \sum_{k \in \mathrm{I}} |f(x) - f(x_k)| \, A_k(x) + \sum_{k \in \mathrm{II}} |f(x) - f(x_k)| \, A_k(x).$$

We now estimate each of these sums. In view of (6.4.10) and (6.4.8),

$$\sum_{k \in \mathrm{I}} |f(x) - f(x_k)| \, A_k(x) \leq \varepsilon \sum_{k \in \mathrm{I}} A_k(x) \leq \varepsilon \sum_{k=1}^{n} A_k(x) = \varepsilon.$$

Consider next $A_k(x)$ for $|x - x_k| \geq \delta$, $-1 \leq x \leq 1$.

$$A_k(x) = (1 - xx_k)\left[\frac{T_n(x)}{n(x - x_k)}\right]^2.$$

Now, $0 < 1 - xx_k < 2$, $|T_n(x)| \leq 1$, $|x - x_k| \geq \delta$. Hence $A_k(x) \leq \dfrac{2}{n^2\delta^2}$. Since $f \in C[-1, 1]$, it is bounded on $[-1, 1]$ by some constant M:

$$|f(x)| \leq M, \, -1 \leq x \leq 1.$$

Thus, $|f(x) - f(x_k)| \leq 2M,$

$$\sum_{k \in \text{II}} |f(x) - f(x_k)| A_k(x) \leq \sum_{k \in \text{II}} 2M \cdot \frac{2}{n^2 \delta^2} \leq \frac{4M}{n^2 \delta^2} \sum_{k=1}^{n} 1 = \frac{4M}{n\delta^2} .$$

Combining these estimates, $|f(x) - H_{2n-1}(f; x)| \leq \varepsilon + \dfrac{4M}{n\delta^2}$. Having been

given an ε, δ is determined. Select n so large that $\dfrac{4M}{n\delta^2} \leq \varepsilon$ Thus, for n

sufficiently large and for all $-1 \leq x \leq 1$, $|f(x) - H_{2n-1}(f; x)| \leq 2\varepsilon$.

6.5 Simultaneous Interpolation and Approximation. If $f \in C[a, b]$, it may be approximated uniformly by a polynomial. We know also that we may interpolate to f at a set of points in $[a, b]$. Can these processes be combined? Given n points $x_1, x_2, \ldots, x_n$ in $[a, b]$ and given $\varepsilon > 0$, can we find a polynomial $p(x)$ such that $|f(x) - p(x)| \leq \varepsilon$, $x \in [a, b]$, and $p(x_i) = f(x_i)$, $i = 1, 2, \ldots, n$? Such approximations may be very desirable.

Figure 6.5.1.

THEOREM 6.5.1 (Walsh).   *Let S be a closed bounded point set in the complex plane. Let $z_1, \ldots, z_n$ be n distinct points of S. Suppose that $f(z)$ is defined on S and is uniformly approximable by polynomials there. Then it is uniformly approximable by polynomials p that satisfy the auxiliary conditions*

$$p(z_i) = f(z_i), \quad i = 1, 2, \ldots, n.$$

Proof: Given an $\varepsilon > 0$, select a polynomial $p(z)$ such that

$$|f(z) - p(z)| \leq \varepsilon, z \in S.$$

Set

$$q(z) = \sum_{k=1}^{n} (f(z_k) - p(z_k)) l_k(z)$$

$$l_k(z) = \frac{w(z)}{(z - z_k) w'(z_k)}$$

$$w(z) = \prod_{k=1}^{n} (z - z_k). \tag{6.5.2}$$

Then $q(z)$ is the unique element of $\mathscr{P}_{n-1}$ with $q(z_k) = f(z_k) - p(z_k)$, $k = 1, 2, \ldots, n$. Now

$$\max_{z \in S} |q(z)| \leq \sum_{k=1}^{n} |f(z_k) - p(z_k)| \max_{z \in S} |l_k(z)| \leq \varepsilon M$$

where

$$M = \sum_{k=1}^{n} \max_{z \in S} |l_k(z)|. \tag{6.5.3}$$

Note that M depends only upon S and $z_1, \ldots, z_n$. Set

$$p_1(z) = p(z) + q(z). \tag{6.5.4}$$

Then

$$p_1(z_k) = p(z_k) + q(z_k) = f(z_k), \quad k = 1, 2, \ldots, n.$$

Moreover,

$$|f(z) - p_1(z)| \leq |f(z) - p(z)| + |q(z)| \leq \varepsilon + M\varepsilon, \quad z \in S.$$

This inequality proves the theorem.

Ex. 1.   By Weierstrass' Theorem, any $f(x) \in C[a, b]$ is uniformly approximable by polynomials. Hence the answer to the question in the introductory paragraph of 6.5 is "yes."

Ex. 2.   Let $f(z)$ be analytic in $|z| \leq R$. Since $f(z)$ may be expanded in a power series which is uniformly convergent there, it is uniformly approximable there by polynomials. Let $z_1, \ldots, z_n$ be distinct points in $|z| \leq R$. Then we can find a polynomial $p(z)$ with $|f(z) - p(z)| \leq \varepsilon$, $|z| \leq R$ and $p(z_k) = f(z_k)$, $k = 1, 2, \ldots, n$.

6.6   Generalizations of the Weierstrass Theorem.

The Weierstrass Theorem has been generalized in many different directions. We shall meet some of the results in Chapter XI where closure and completeness are studied. Here, we shall look at generalizations to functions of N real variables. If a real function of N real variables is continuous on a closed bounded set of R_N, it may be approximated uniformly by polynomials in the N variables. There are many proofs of this fact. One proof—an extension of Theorem 6.2.2—makes use of generalized Bernstein polynomials: if $f(x_1, x_2, \ldots, x_N)$ is continuous on the hypercube $C: 0 \leq x_j \leq 1, j = 1, 2, \ldots, N$, then the *generalized Bernstein polynomial*

$$B(f; x_1, x_2, \ldots, x_N) = \sum_{k_1=0}^{n_1} \cdots \sum_{k_N=0}^{n_N} \binom{n_1}{k_1}\binom{n_2}{k_2} \cdots \binom{n_N}{k_N}$$

$$\times f\left(\frac{k_1}{n_1}, \frac{k_2}{n_2}, \ldots, \frac{k_N}{n_N}\right) x_1^{k_1}(1 - x_1)^{n_1-k_1} \cdots x_N^{k_N}(1 - x_N)^{n_N-k_N}, \tag{6.6.1}$$

converges uniformly in C to f as $\min_j n_j \to \infty$.

In order to provide an alternate approach, and to present a result with a more contemporary flavor, we shall prove the Stone-Weierstrass Theorem and derive the N-dimensional Weierstrass Theorem as a consequence. Stone's Theorem was inspired, in part, by an elementary proof of Weierstrass' Theorem given by Lebesgue in 1908.

We shall limit our discussion to real functions that are defined on a finite interval I of R_n: $-\infty < a_i \le x_i \le b_i < \infty$. We designate points of I by P, Q, etc.

LEMMA 6.6.1.   *Let F be a family of functions that are real and continuous on I and such that*

$$f_1, f_2 \in F \text{ implies } \max[f_1, f_2] \in F \text{ and } \min[f_1, f_2] \in F. \qquad (6.6.2)$$

In order for a function f that is continuous on I to be uniformly approximable by members of F, it is necessary and sufficient that for any two points P_1 and P_2 of I and for any $\varepsilon > 0$, there be a function $t(P) \in F$ such that

$$|f(P_i) - t(P_i)| < \varepsilon, \quad i = 1, 2. \qquad (6.6.3)$$

Proof: If uniform approximation is possible, then given $\varepsilon > 0$ we can find a $t(P) \in F$ such that

$$|f(P) - t(P)| < \varepsilon, \quad P \in I$$

and so (6.6.3) follows trivially.

Conversely, suppose that (6.6.3) holds. Select a fixed $Q \in I$ and a fixed $\varepsilon > 0$. Then, for any point R, we can find a function $t(P)(= t(P; Q, R, \varepsilon))$ such that $|f(Q) - t(Q)| < \varepsilon$ and $|f(R) - t(R)| < \varepsilon$. In particular,

$$t(R) < f(R) + \varepsilon. \qquad (6.6.4)$$

By continuity of t and f, this inequality must persist in a certain neighborhood N_R of R. As R runs over all the points of I, the corresponding neighborhoods must cover I. Hence by the Heine-Borel Theorem, we can find a finite number of them $N_{R_1}, N_{R_2}, \ldots, N_{R_k}$ that cover I. The corresponding functions $t(P; Q, R_i)$ satisfy

$$t(P; Q, R_i) < f(P) + \varepsilon, \quad P \in N_i, \quad i = 1, 2, \ldots, k. \qquad (6.6.5)$$

Define

$$t^-(P; Q) = \min\{t(P; Q, R_1), t(P; Q, R_2), \ldots, t(P; Q, R_k)\}. \qquad (6.6.6)$$

By (6.6.2), iterated, $t^- \in F$ and by (6.6.5),

$$t^-(P, Q) < f(P) + \varepsilon, \quad P \in I. \qquad (6.6.7)$$

Again, for each i we have

$$|f(Q) - t(Q; Q, R_i)| < \varepsilon$$

so that

$$t(Q; Q, R_i) > f(Q) - \varepsilon. \qquad (6.6.8)$$

It follows from (6.6.6) that

$$t^-(Q; Q) > f(Q) - \varepsilon. \tag{6.6.9}$$

By continuity, (6.6.9) must persist in a neighborhood O of Q:

$$t^-(P; Q) > f(P) - \varepsilon. \tag{6.6.10}$$

Now let Q run over I. These neighborhoods cover I, and we may find a finite number of them $O_1, O_2, \ldots, O_r$ corresponding to $Q_1, \ldots, Q_r$ that cover I. Since

$$t^-(P; Q_i) > f(P) - \varepsilon, \quad P \in O_i, \quad i = 1, 2, \ldots, r, \tag{6.6.11}$$

and since the O_i cover I, for every $P \in I$, the inequality

$$t^-(P; Q_i) > f(P) - \varepsilon \tag{6.6.12}$$

must hold for some i.

If we set

$$s(P) = \max \{t^-(P, Q_1), \ldots, t^-(P, Q_r)\} \tag{6.6.13}$$

then by what we have just said,

$$s(P) > f(P) - \varepsilon, \quad \text{for all } P \in I. \tag{6.6.14}$$

On the other hand, by (6.6.7), $t^-(P; Q) < f(P) + \varepsilon$, $P \in I$ for all Q. Hence, $s(P) < f(P) + \varepsilon$, $P \in I$. Combining this with (6.6.14),

$$|f(P) - s(P)| < \varepsilon, \quad P \in I. \tag{6.6.15}$$

Finally, by (6.6.2) iterated, $s(P) \in F$.

Ex. 1. Let I be $-\infty < a \le x \le b < \infty$ and F be the set of all piecewise linear functions defined on I. It is easy to verify that F satisfies (6.6.2). Condition (6.6.3) can be satisfied with $\varepsilon = 0$ by means of a linear function. *Conclusion:* Every continuous function can be approximated uniformly on I by continuous piecewise linear functions.

DEFINITION 6.6.1. Let F be a family of real valued functions. By the *lattice hull L* of F is meant the intersection of all families of functions that contain F and contain the functions $\max (f_1, f_2)$ and $\min (f_1, f_2)$ whenever they contain f_1 and f_2.

Note that $F \subseteq L$. If all the functions of F are continuous, then the functions of L must also be continuous.

DEFINITION 6.6.2. An *algebra* $\mathscr{A}$ of real valued continuous functions defined on I is a set of such functions that possesses the following property.

$$f_1, f_2 \in \mathscr{A}, c \text{ real, implies } f_1 + f_2 \in \mathscr{A}, cf_1 \in \mathscr{A}, f_1 f_2 \in \mathscr{A}. \tag{6.6.16}$$

Note that (6.6.16) implies that any polynomial in f_1 with real coefficients and of the form $a_1 f_1 + a_2 f_1^2 + \cdots + a_n f_1^n \in \mathscr{A}$.

LEMMA 6.6.2. *Let $\mathscr{A}$ be an algebra of continuous real valued functions defined on I. Let L designate its lattice hull. Then, the elements of L are uniformly approximable by elements of $\mathscr{A}$.*

Proof: Designate by $\mathscr{B}$ the set of functions that are uniformly approximable by elements of $\mathscr{A}$. We shall show: (1) $\mathscr{B}$ is an algebra and (2) if $f \in \mathscr{B}$ then $|f| \in \mathscr{B}$.

Since we have

$$\max (f_1, f_2) = \tfrac{1}{2}[(f_1 + f_2) + |f_1 - f_2|]$$
$$\min (f_1, f_2) = \tfrac{1}{2}[(f_1 + f_2) - |f_1 - f_2|], \tag{6.6.17}$$

it follows from (1) and (2) that if f_1, $f_2 \in \mathscr{B}$ then $\max (f_1, f_2) \in \mathscr{B}$, and $\min (f_1, f_2) \in \mathscr{B}$. Since obviously $\mathscr{A} \subseteq \mathscr{B}$, it will follow from Definition 6.6.1 that $L \subseteq \mathscr{B}$. This will establish the lemma.

As far as (1) is concerned, we need only imitate the familiar proofs of the elementary theorems on limits of sums and products. To prove (2), observe that since the elements of $\mathscr{A}$ are continuous on I, they are bounded there. Since an $f \in \mathscr{B}$ is uniformly approximable by elements of $\mathscr{A}$, it, too, must be bounded there. Set $M = \sup_{P \in I} |f(P)|$. By Theorem 6.5.1, let $P(t)$ be a polynomial with $P(0) = 0$ and with $|P(t) - |t|| \leq \varepsilon$ for $|t| \leq M$. Since the values of f lie in $[-M, M]$, we have $|P(f) - |f|| \leq \varepsilon$. But from (1) and the remark following Definition 6.6.2, $P(f) \in \mathscr{B}$. Hence we can find a $g \in \mathscr{A}$ such that $|P(f) - g| \leq \varepsilon$. Combining these inequalities, $||f| - g| \leq 2\varepsilon$ on I. Therefore $|f| \in \mathscr{B}$.

THEOREM 6.6.3 (Stone-Weierstrass). *Let $\mathscr{A}$ be an algebra of real valued continuous functions defined on I. In order that an arbitrary continuous real valued function f be uniformly approximable on I by elements of $\mathscr{A}$, it is necessary and sufficient that for any two points P_1, $P_2 \in I$, and any $\varepsilon > 0$, we can find a $g \in \mathscr{A}$ such that $|f(P_i) - g(P_i)| \leq \varepsilon$, $i = 1, 2$.*

Proof: If f is uniformly approximable on I by elements in $\mathscr{A}$, then this condition obviously holds. Suppose, conversely, that the condition holds. Let $L(\mathscr{A})$ denote the lattice hull of $\mathscr{A}$ and let f be continuous on I. For any two points P_1, $P_2 \in I$, and any $\varepsilon > 0$, we can find a $g \in \mathscr{A}$ (and a fortiori $\in L(\mathscr{A})$) such that $|f(P_i) - g(P_i)| \leq \varepsilon$, $i = 1, 2$. Hence, by Lemma 6.6.1 with $F = L(\mathscr{A})$, we can approximate f uniformly on I by elements of $L(\mathscr{A})$. On the other hand, the elements of $L(\mathscr{A})$ can, by Lemma 6.6.2, be uniformly approximated by the elements of $\mathscr{A}$ itself. Combining these two approximations, we can approximate f by elements of $\mathscr{A}$.

COROLLARY 6.6.4. *Let $f(x_1, \ldots, x_N)$ be real and continuous on I. Then it can be approximated uniformly on I by polynomials in $x_1, x_2, \ldots, x_N$.*

Proof: For the algebra $\mathscr{A}$ take the set of polynomials in $x_1, x_2, \ldots, x_N$.

Let $P_i: (x_1^{(i)}, x_2^{(i)}, \ldots, x_N^{(i)})$, $i = 1, 2$, be distinct points and consider

$$g(x_1, x_2, \ldots, x_N) = f(P_1) + \frac{f(P_2) - f(P_1)}{\sum\limits_{i=1}^{N} (x_i^{(2)} - x_i^{(1)})^2} \sum_{i=1}^{N} (x_i - x_i^{(1)})(x_i^{(2)} - x_i^{(1)}).$$

(6.6.18)

This is a polynomial in $x_1, \ldots, x_N$ and $g(P_i) = f(P_i)$, $i = 1, 2$. The conditions of Theorem 6.6.3 are satisfied with $\varepsilon = 0$.

NOTES ON CHAPTER VI

6.2–6.3   Bernstein polynomials are described in Gontscharoff [1], Natanson [1], pp. 1–7, 174–182. Lorentz [1] is a penetrating study of these interesting polynomials and includes a discussion of their behavior in the complex plane, applications to moment problems, and generalizations.

For Theorem 6.3.4 and for applications of the Bernstein polynomials to variation reducing approximations, see Schoenberg [1].

For a deeper study of the rate of convergence of the Fejér scheme in Theorem 6.4.1, see Shisha et al. [1].

An interesting and unifying approach to Theorems 6.2.2, 6.4.1, and 12.2.8 (Bernstein and Fejér) is provided by the theory of positive linear functionals as developed by Korovkin in [2].

6.5   Walsh [2], p. 310.

6.6   The Stone-Weierstrass theorem can be found in McShane and Botts [1], Dieudonné [1], pp. 131–134. Dunford and Schwartz [1], pp. 272, 383–385. The chapter by Stone in Buck [6] is highly recommended.

PROBLEMS

1.   Let $f(x) \in C^1[a, b]$. If $p(x)$ is a polynomial that approximates f' to within ε on $[a, b]$ then $q(x) = \int_a^x p(x)\, dx + f(a)$ is a polynomial that approximates f to within $(b - a)\varepsilon$ on $[a, b]$. Extend to higher derivatives.

2.   Let $f(x) \in C^\infty[a, b]$. Show (without using Bernstein polynomials) that we can find a sequence of polynomials $p_n(x)$ such that $\lim\limits_{n \to \infty} p_n^{(j)}(x) = f^{(j)}(x)$ uniformly on $[a, b]$ $j = 0, 1, \ldots$.

3.   Let $p_{nk}(x) = \binom{n}{k} x^k (1 - x)^{n-k}$, $0 \le x \le 1$. Prove that the maximum value occurs at $x = \dfrac{k}{n}$. If $\dfrac{k}{n} \to x$ as $n \to \infty$, $p_{nk}\left(\dfrac{k}{n}\right)$ is asymptotically equal to $(2\pi n x (1 - x))^{-\frac{1}{2}}$.

4.   Bernstein polynomials over the interval $[a, b]$ may be defined by

$$B_n(f, a, b; x) = \frac{1}{h^n} \sum_{k=0}^{n} f\left(a + \frac{k}{n} h\right) \binom{n}{k} (x - a)^k (b - x)^{n-k}, \qquad h = b - a.$$

Prove a theorem analogous to Theorem 6.2.2.

5. Compare $B_4(\sqrt{x}; \frac{1}{2})$ with $\sqrt{\frac{1}{2}}$.

6. Let $S(x) = |2x - 1|, 0 \leq x \leq 1$. Compute $B_4(S(x); x)$, $B_6(S(x); x)$. Show that $B_{2n}\left(S(x); \frac{1}{2}\right) = \frac{1}{2^{2n}}\binom{2n}{n}$ and study the rapidity of the approach to 0 of $S(\frac{1}{2}) - B_{2n}(S(x); \frac{1}{2})$.

7. Let $f(x) = 0$ for $0 \leq x \leq \frac{1}{2}, f(x) = x - \frac{1}{2}$ for $\frac{1}{2} \leq x \leq 1$. Show that $B_{2n}(f; x)$ has nth order contact with f at both 0 and 1.

8. Verify Voronovsky's Theorem for $f(x) = e^x$ by a direct computation.

9. Obtain an explicit expression for $B_n(x^3; x)$ and show directly that

$$\lim_{n \to \infty} n[B_n(x^3; x) - x^3] = 3x^2(1 - x).$$

10. Let $f(x) \in C[a, b]$. It is uniformly approximable on $[a, b]$ by polynomials with rational coefficients.

11. If $f(x) \in C[a, b]$, uniform approximation by polynomials with integer coefficients is not necessarily possible.

12. Prove Ex. 1, 6.6 directly.

CHAPTER VII

Best Approximation

7.1   What is Best Approximation?   In Chapter VI, we have studied several situations in which functions can be approximated arbitrarily closely by polynomials. It goes without saying that in order to achieve more and more accuracy in the approximations, the approximants will (in general) have to be of higher and higher degree. But it is of considerable importance both for theory as well as for numerical practice to accomplish as much as possible with polynomials of a fixed degree. For instance, how well can the function x^4 be approximated over $0 \le x \le 1$ by a straight line? In order to answer such a question, the notion of *closeness of approximation* must be defined. Frequently, we measure the closeness of approximation over the interval by taking the maximum deviation between the function and its approximant. At other times, we may wish to use alternate definitions. The maximum deviation considered over a finite set of points, or the integral of the square of the deviations are frequently employed.

Once a criterion of closeness of approximation has been decided upon, we may begin to answer specific questions. We may, for instance, look into the problem of whether, among the elements of $\mathscr{P}_n$, there is one whose closeness to a given function $f(x)$ is not exceeded by any other element of $\mathscr{P}_n$. If there is, it is known as a *best approximation* to $f(x)$. Change the criterion of closeness of approximation and the best approximation will change.

Ex. 1.   Approximate $y = x^4$ over $[0, 1]$ by a straight line $l(x)$ so that

$$\text{(a)} \quad \int_0^1 (x^4 - l(x))^2 \, dx = \text{minimum}$$

$$\text{(b)} \quad \int_0^1 (x^4 - l(x))^2 \, dx + \int_0^1 (d/dx(x^4 - l(x)))^2 \, dx = \text{minimum}$$

$$\text{(c)} \quad \max_{0 \le x \le 1} |x^4 - l(x)| = \text{minimum}$$

The answers are given by

$$\text{(a)} \; l(x) = \tfrac{4}{5}x - \tfrac{1}{5}$$
$$\text{(b)} \; l(x) = \tfrac{54}{55}x - \tfrac{16}{55}$$
$$\text{(c)} \; l(x) = x - \tfrac{3}{16}\sqrt[3]{2} = x - .236 \ldots$$

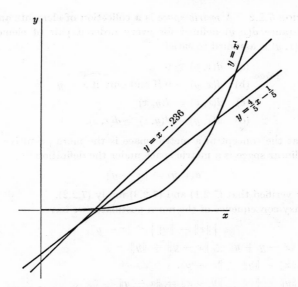

Figure 7.1.1   Least Square and Best Uniform Linear
Approximations to x^4 on $[0, 1]$.

The major investigations in the theory of best approximation concern themselves with (a) Under what circumstances is there a best approximation? If it exists, is it unique? (b) How can best approximations be characterized analytically or geometrically? (c) How can the best approximants be computed numerically? (d) What are the asymptotic properties of best approximation? We shall prove theorems in all these categories.

It would be good to have an abstract mathematical structure to describe properties of best approximation independently of the specific criterion of closeness of approximation. Such a structure is furnished by the theory of *Normed Linear Spaces,* and it is to this theory that we now turn.

7.2 Normed Linear Spaces

DEFINITION 7.2.1.   A linear space X is called a *normed linear space* if for each element x of the space there is defined a real number designated by $\|x\|$ with the following properties:

(a) $\|x\| \geq 0$ (positivity)

(b) $\|x\| = 0$ if and only if $x = 0$ (definiteness)

(c) $\|\alpha x\| = |\alpha|\,\|x\|$ for every scalar α (homogeneity)   (7.2.1)

(d) $\|x + y\| \leq \|x\| + \|y\|$ (triangle inequality)

The quantity $\|x\|$ is known as the *norm of x.*

DEFINITION 7.2.2.   A *metric space* is a collection of elements and a measure of distance $d(x, y)$ defined for every ordered pair of elements. The function $d(x, y)$ is assumed to satisfy

$$
\begin{align}
&\text{(a)} \quad d(x, y) \geq 0 \\
&\text{(b)} \quad d(x, y) = 0 \text{ if and only if } x = y \\
&\text{(c)} \quad d(x, y) = d(y, x) \\
&\text{(d)} \quad d(x, y) + d(y, z) \geq d(x, z).
\end{align}
\tag{7.2.2}
$$

Note that the concept of a metric space is the more primitive one since a normed linear space is a metric space under the definition

$$d(x, y) = \|x - y\|. \tag{7.2.3}$$

It is easily verified that (7.2.1) and (7.2.3) imply (7.2.2).

As an easy consequence of the norm postulates we have

$$\big| \|x\| - \|y\| \big| \leq \|x - y\|, \tag{7.2.4}$$

for $\|x\| = \|x - y + y\| \leq \|x - y\| + \|y\|$.

Therefore $\|x\| - \|y\| \leq \|x - y\|$.

Similarly $\|y\| - \|x\| \leq \|y - x\| = \|x - y\|$

and (7.2.4) follows.

The following examples of normed linear spaces find frequent application.

Ex. 1.   The real line $-\infty < x < \infty$ with $\|x\| = |x|$.

Ex. 2.   The real n-dimensional Cartesian space R_n of elements

$$x = (x_1, x_2, \ldots, x_n)$$

with the definition $\|x\| = (x_1{}^2 + x_2{}^2 + \cdots + x_n{}^2)^{\frac{1}{2}}$. This is known as the "square norm."

Ex. 3. R_n or C_n with the definition $\|x\| = (|x_1|^p + \cdots + |x_n|^p)^{1/p}$, $p \geq 1$. This is known as the "p norm." Properties (7.2.1)(a)–(c) are easy to verify. Property (d) is the *Minkowski Inequality* and takes a number of steps to reach.

LEMMA 7.2.1.   *If* $x, y \geq 0$, $a, b > 0$ *and* $a + b = 1$ *then*

$$x^a y^b \leq ax + by. \tag{7.2.5}$$

Equality holds if and only if $x = y$.

Proof: Let $t > 1$, $m < 1$. Let $f(t) = t^m$.

From Theorem 1.6.2,

$$f(t) = f(1) + (t - 1)f'(\xi) \qquad 1 < \xi < t.$$

Hence $t^m = 1 + (t - 1)m\xi^{m-1}$.

Since $\xi^{m-1} < 1$,

$$t^m < 1 + m(t - 1). \tag{7.2.6}$$

Assume, for the moment that $x > y > 0$. Set $t = x/y > 1$, $m = a$, so that $1 - m = b$. Then, from (7.2.6),

$$(x/y)^a < 1 + a(x/y - 1)$$

or

$$x^a y^{1-a} < y + a(x - y) = ax + (1 - a)y.$$

Hence

$$x^a y^b < ax + by.$$

If $0 < x < y$, we may interchange the roles of a and b and arrive at the same inequality. If $x = y$, $x^a y^b = x^{a+b} = x = ax + bx$.

LEMMA 7.2.2 (Hölder's Inequality). *Let x_k and y_k be complex. If $p > 1$, and $1/p + 1/q = 1$,*

$$\left| \sum_{k=1}^{n} x_k y_k \right| \leq \left(\sum_{k=1}^{n} |x_k|^p \right)^{1/p} \left(\sum_{k=1}^{n} |y_k|^q \right)^{1/q}. \tag{7.2.7}$$

Note that $1/p$, $1/q > 0$.

Proof: From Lemma 7.2.1, with $x = \dfrac{|x_k|^p}{\sum\limits_{k=1}^{n} |x_k|^p}$, $y = \dfrac{|y_k|^q}{\sum\limits_{k=1}^{n} |y_k|^q}$,

$a = 1/p$, $b = 1 - 1/p = 1/q$, we have

$$\left(\frac{|x_k|^p}{\sum\limits_{k=1}^{n} |x_k|^p} \right)^{1/p} \left(\frac{|y_k|^q}{\sum\limits_{k=1}^{n} |y_k|^q} \right)^{1/q} \leq 1/p \, \frac{|x_k|^p}{\sum\limits_{k=1}^{n} |x_k|^p} + 1/q \, \frac{|y_k|^q}{\sum\limits_{k=1}^{n} |y_k|^q}. \tag{7.2.8}$$

Summing (7.2.8) from $k = 1$ to $k = n$, and multiplying by $\left(\sum\limits_{k=1}^{n} |x_k|^p \right)^{1/p}$ $\times \left(\sum\limits_{k=1}^{n} |y_k|^q \right)^{1/q}$ we obtain

$$\sum_{k=1}^{n} |x_k y_k| \leq \left(\sum_{k=1}^{n} |x_k|^p \right)^{1/p} \left(\sum_{k=1}^{n} |y_k|^q \right)^{1/q}. \tag{7.2.9}$$

Since $\left| \sum\limits_{k=1}^{n} x_k y_k \right| \leq \sum\limits_{k=1}^{n} |x_k y_k|$, (7.2.7) follows. For equality, in (7.2.8) and hence in (7.2.7) we must have

$$\frac{|x_k|^p}{\sum\limits_{k=1}^{n} |x_k|^p} = \frac{|y_k|^q}{\sum\limits_{k=1}^{n} |y_k|^q} \quad k = 1, 2, \ldots, n$$

i.e.,

$$|x_k|^p = \text{constant } |y_k|^q, \quad k = 1, 2, \ldots, n. \tag{7.2.10}$$

Moreover,
$$\left| \sum_{k=1}^{n} x_k y_k \right| < \sum_{k=1}^{n} |x_k y_k|$$
unless
$$x_k y_k = |x_k y_k| e^{i\theta}, \quad \theta = \text{constant}, \quad k = 1, 2, \ldots, n. \tag{7.2.11}$$

Hence (7.2.10) and (7.2.11) together are necessary and sufficient for equality in (7.2.7).

THEOREM 7.2.3 (Minkowski's Inequality). If $p \geq 1$,

$$\left\{ \sum_{i=1}^{n} |x_i + y_i|^p \right\}^{1/p} \leq \left(\sum_{i=1}^{n} |x_i|^p \right)^{1/p} + \left(\sum_{i=1}^{n} |y_i|^p \right)^{1/p} \tag{7.2.12}$$

or, in the notation of Ex. 3., $\|x + y\| \leq \|x\| + \|y\|$.

Proof: If $p = 1$, (7.2.12) reduces to the triangle inequality. If $p > 1$,

$$\sum_{i=1}^{n} |x_i + y_i|^p = \sum_{i=1}^{n} |x_i + y_i| \, |x_i + y_i|^{p-1}$$
$$\leq \sum_{i=1}^{n} |x_i| \, |x_i + y_i|^{p-1} + \sum_{i=1}^{n} |y_i| \, |x_i + y_i|^{p-1}$$

By Lemma 7.2.2,

$$\sum_{i=1}^{n} |x_i| \, |x_i + y_i|^{p-1} \leq \left(\sum_{i=1}^{n} |x_i|^p \right)^{1/p} \left(\sum_{i=1}^{n} |x_i + y_i|^{(p-1)q} \right)^{1/q}$$
$$= \left(\sum_{i=1}^{n} |x_i|^p \right)^{1/p} \left(\sum_{i=1}^{n} |x_i + y_i|^p \right)^{1/q}.$$

Similarly, $\quad \sum_{i=1}^{n} |y_i| \, |x_i + y_i|^{p-1} \leq \left(\sum_{i=1}^{n} |y_i|^p \right)^{1/p} \left(\sum_{i=1}^{n} |x_i + y_i|^p \right)^{1/q}.$

Combining these inequalities and dividing by $\left(\sum_{i=1}^{n} |x_i + y_i|^p \right)^{1/q}$ we obtain (7.2.12).

Suppose that $p > 1$. If there is equality in (7.2.12), we must have

$$\left. \begin{aligned} \sum_{i=1}^{n} |x_i| \, |x_i + y_i|^{p-1} &= \left(\sum_{i=1}^{n} |x_i|^p \right)^{1/p} \left(\sum_{i=1}^{n} |x_i + y_i|^p \right)^{1/q} \\ \sum_{i=1}^{n} |y_i| \, |x_i + y_i|^{p-1} &= \left(\sum_{i=1}^{n} |y_i|^p \right)^{1/p} \left(\sum_{i=1}^{n} |x_i + y_i|^p \right)^{1/q} \end{aligned} \right\} \tag{7.2.13}$$

and

$$\sum_{i=1}^{n} |x_i + y_i| \, |x_i + y_i|^{p-1} = \sum_{i=1}^{n} |x_i| \, |x_i + y_i|^{p-1} + \sum_{i=1}^{n} |y_i| \, |x_i + y_i|^{p-1}. \tag{7.2.14}$$

By the remark following Lemma 7.2.2, (7.2.13) implies that

$$|x_i|^p = c_1 |x_i + y_i|^p, \ |y_i|^p = c_2 |x_i + y_i|^p, \quad i = 1, 2, \ldots, n.$$

Assuming that neither the x_i nor the y_i are all zero,

$$|x_i| = \text{constant } |y_i|, \quad i = 1, 2, \ldots, n.$$

Since $|\alpha + \beta| < |\alpha| + |\beta|$ unless α and β have the same direction, we conclude from (7.2.14) that

$$x_i = |x_i|\, e^{i\phi}, \, y_i = |y_i|\, e^{i\phi}, \quad i = 1, 2, \ldots, n.$$

Hence,

$$x_i = cy_i \quad i = 1, 2, \ldots, n, \tag{7.2.15}$$

with $c > 0$.

Ex. 4.   The same linear space as in Ex. 2, with the definition $\|x\| = \max\limits_{1 \leq j \leq n} |x_j|$. This is known as the *uniform norm*. It is also called the $p = \infty$ *norm* in virtue of the identity $\lim\limits_{p \to \infty} (|x_1|^p + |x_2|^p + \cdots + |x_n|^p)^{1/p} = \max\limits_{1 \leq j \leq n} |x_j|$. (Prove.) One sometimes expresses this as $\lim\limits_{p \to \infty} \|x\|_p = \|x\|_\infty$.

Ex. 5.   If $\{x_i\}$ and $\{y_i\}$ are two sequences of real or complex numbers such that $\sum\limits_{i=1}^{\infty} |x_i|^p < \infty, \sum\limits_{i=1}^{\infty} |y_i|^p < \infty, p \geq 1$, then

$$\left\{ \sum_{i=1}^{\infty} |x_i + y_i|^p \right\}^{1/p} \leq \left(\sum_{i=1}^{\infty} |x_i|^p \right)^{1/p} + \left(\sum_{i=1}^{\infty} |y_i|^p \right)^{1/p}. \tag{7.2.16}$$

This implies that the set of all infinite sequences $\{x_i\}$ with $\sum\limits_{i=1}^{\infty} |x_i|^p < \infty$ and addition and scalar multiplication defined analogously to Ex. 1, 1.3, form a linear space with $\|\{x_i\}\| = \left(\sum\limits_{i=1}^{\infty} |x_i|^p \right)^{1/p}$. This space is called the ℓ^p space.

Ex. 6.   Let $B[a, b]$ be the set of all bounded functions defined in $[a, b]$. Define

$$\|f\| = \sup_{a \leq x \leq b} |f(x)|.$$

Then we have a normed linear space. To prove (7.2.1)(d), observe that

$$|f(x)| \leq \sup |f(x)|, \, |g(x)| \leq \sup |g(x)|.$$

Hence,

$$|f(x) + g(x)| \leq |f(x)| + |g(x)| \leq \sup |f(x)| + \sup |g(x)|.$$

Since this is true for all x, it follows that

$$\sup |f(x) + g(x)| \leq \sup |f(x)| + \sup |g(x)|.$$

Ex. 7. $C[a, b]$ with the definition $\|f\| = \max\limits_{a \leq x \leq b} |f(x)|$ is a normed linear space.

Ex. 8.   Let R be a region of the complex plane. The set of functions analytic in R and bounded there forms a linear space $B(R)$. It may be normed by defining $\|f\| = \sup\limits_{z \in R} |f(z)|$.

Ex. 9.   Let $w(x) \in C[a, b]$ and $w(x) > 0$. We may norm $C[a, b]$ by defining $\|f\| = \left(\int_a^b w(x) f^2(x) \, dx \right)^{\frac{1}{2}}$ (weighted square norm).

Ex. 10. $L^p[a, b]$ (Cf. Def. 1.4.0) is a normed linear space for $p \geq 1$. We define $\|f\| = \left(\int_a^b |f(x)|^p \, dx \right)^{1/p}$. The triangle inequality is the Minkowski inequality for integrals.

$$\left(\int_a^b |f(x) + g(x)|^p \, dx \right)^{1/p} \leq \left(\int_a^b |f(x)|^p \, dx \right)^{1/p} + \left(\int_a^b |g(x)|^p \, dx \right)^{1/p}. \quad (7.2.17)$$

For $p > 1$ this may be defined from the Hölder inequality for integrals. This states that if $f \in L^p[a, b]$ and $g \in L^q[a, b]$, $1/p + 1/q = 1$, $p > 1$, then $fg \in L[a, b]$ and

$$\left| \int_a^b f(x) g(x) \, dx \right| \leq \left(\int_a^b |f(x)|^p \, dx \right)^{1/p} \left(\int_a^b |g(x)|^q \, dx \right)^{1/q}. \quad (7.2.18)$$

The proof runs parallel to that of Lemma 7.2.2. The particular case, $p = 2$,

$$\left| \int_a^b f(x) g(x) \, dx \right| \leq \left(\int_a^b |f(x)|^2 \, dx \right)^{\frac{1}{2}} \left(\int_a^b |g(x)|^2 \, dx \right)^{\frac{1}{2}} \quad (7.2.19)$$

is the very important *Schwarz inequality*.

Ex. 11.   Let $a \leq x_1 < x_2 < \ldots < x_n \leq b$. The linear space $\mathscr{P}_{n-1}$ may be normed as follows

$$\|p\| = \max_{1 \leq j \leq n} |p(x_j)|.$$

7.3   Convex Sets.

DEFINITION 7.3.1.   Let X be a linear space. If x_1 and x_2 are two distinct elements of X, the set of all elements of the form

$$x = tx_1 + (1 - t)x_2, \quad 0 \leq t \leq 1 \quad (7.3.1)$$

is called the *line segment joining x_1 and x_2*.

DEFINITION 7.3.2.   Let X be a linear space. A subset C of X is called *convex* if C contains all the elements on the line segment joining any two of its elements. That is, if $x_1, x_2 \in C$ then so does $tx_1 + (1 - t)x_2$, $0 \leq t \leq 1$.

Ex. 1.   In R_2, a line segment, the quadrant $x \geq 0$, $y \geq 0$, the interior of an ellipse are all convex sets. (Proofs?)

Ex. 2.   The set of all polynomials with nonnegative coefficients is convex.

Ex. 3.   Let X be a normed linear space. The ball $\|x\| \leq r$ is convex. For suppose $\|x\| \le r$, $\|y\| \le r$, and $0 \le t \le 1$. Then

$$\|tx + (1 - t)y\| \le |t|\,\|x\| + |1 - t|\,\|y\| \le tr + (1 - t)r = r.$$

Hence $tx + (1 - t)y$ belongs to the set.

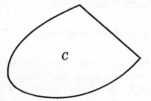

Figure 7.3.1 A Convex Set.

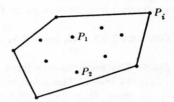

Figure 7.3.2 The Convex Hull.

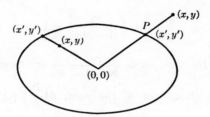

Figure 7.3.3.

DEFINITION 7.3.3.   The *convex hull* of a given set S is the intersection of all convex sets containing S.

Ex. 4.   The convex hull of the point set $\frac{1}{2} \le x^2 + y^2 \le 1, y \ge 0$ is the semicircle $x^2 + y^2 \le 1, y \ge 0$.

Ex. 5.   The convex hull of the points $P_1, P_2, \ldots, P_n$ lying in a plane may be "found" by driving nails in at P_i and wrapping a string around the configuration.

An elegant example of a normed linear space is furnished by the *Minkowski plane*. Let there be given in the x, y plane a bounded, convex set S with boundary C. We will suppose it contains the origin in its interior and that it is symmetric with respect to the origin; i.e., if $(x, y) \in S$ then $(-x, -y) \in S$. If $P = (x, y)$ is a point other than the origin, the directed line extending from $(0, 0)$ to (x, y) can be shown to intersect C in precisely one point (x', y').

Define

$$\|P\| = \|(x, y)\| = \frac{\sqrt{x^2 + y^2}}{\sqrt{x'^2 + y'^2}}, \quad \|(0, 0)\| = 0. \qquad (7.3.2)$$

It should be clear that the points of the plane interior to C have norm < 1, those exterior to C have norm > 1, while those on C have norm $= 1$.

THEOREM 7.3.1. R_2 *normed by* (7.3.2) *is a normed linear space.*

Proof: The requirements for a norm are (7.2.1)(a)–(d). The first three are easy to check. The triangle inequality requires some ingenuity. We wish to show that $\|P_1 + P_2\| \le \|P_1\| + \|P_2\|$. If either $\|P_1\| = 0$ or $\|P_2\| = 0$, the inequality is trivial. Assume that neither is 0 and set

$$\|P_1\| + \|P_2\| = t.$$

We can then write $\|P_1\| = \theta t$, $\|P_2\| = (1 - \theta)t$ for some $0 < \theta < 1$.

By homogeneity, $$\left\| \frac{P_1}{\theta t} \right\| = \left\| \frac{P_2}{(1 - \theta)t} \right\| = 1.$$

This tells us that the two points $\dfrac{P_1}{\theta t}$, $\dfrac{P_2}{(1 - \theta)t}$ are located on C. By the convexity of the closure of S, the point $\theta\left(\dfrac{P_1}{\theta t}\right) + (1 - \theta)\dfrac{P_2}{(1 - \theta)t} = \dfrac{1}{t}(P_1 + P_2)$ is also in the closure of S and hence

$$\left\| \frac{1}{t}(P_1 + P_2) \right\| = \frac{1}{t}\|P_1 + P_2\| \le 1.$$

Therefore,

$$\|P_1 + P_2\| \le \|P_1\| + \|P_2\|.$$

The same definition can be introduced in spaces of higher dimension. C is called the *gauge curve* of the normed plane.

Ex. 6.   If C is the unit circle, then $\|P\| = (x^2 + y^2)^{\frac{1}{2}}$, and we have the Euclidean norm.

Ex. 7.   If C is the square with sides $x = \pm 1, y = \pm 1$, then

$$\|P\| = \max(|x|, |y|).$$

Ex. 8. C is the square with sides $x \pm y = \pm 1$. Then $\|P\| = |x| + |y|$.

7.4   The Fundamental Problem of Linear Approximation.   Let X be a normed linear space. Select n linearly independent elements $x_1, \ldots, x_n$. Let y be an additional element. We wish to approximate y by an appropriate linear combination of the $x_1, \ldots, x_n$. The closeness of two elements will be defined as the norm of their difference. We therefore would like to make $\|y - (a_1 x_1 + a_2 x_2 + \cdots + a_n x_n)\|$ as small as possible. The element

$$y - (a_1 x_1 + \cdots + a_n x_n)$$

is called the *error* or *discrepancy*.

DEFINITION 7.4.1.   A *best approximation* to y by linear combinations of $x_1, \ldots, x_n$ is an element $a_1 x_1 + \cdots + a_n x_n$ for which

$$\|y - (a_1 x_1 + \cdots + a_n x_n)\| \leq \|y - (b_1 x_1 + \cdots + b_n x_n)\|$$

for every choice of constants $b_1, \ldots, b_n$.

A best approximation solves the problem of minimizing the error norm.

Ex. 1.   Let X be $C[0, 1]$ normed by $\|f\| = \max\limits_{0 \leq x \leq 1} |f(x)|$. Take $n = 1$, x_1 as the function 1, and y the function e^x. A best approximation to e^x by constants is the constant a that minimizes $\max\limits_{0 \leq x \leq 1} |e^x - a|$. The (unique) solution is $a = \frac{1}{2}(e + 1)$ and the error norm is $\frac{1}{2}(e - 1)$.

Ex. 2.   Let X be $C[0, 1]$ normed by $\|f\| = \left(\int_0^1 f^2 \, dx \right)^{\frac{1}{2}}$. Take $n = 1$, x_1 the function 1, and y the function e^x. A best approximation to e^x by constants is the constant a which minimizes $\left(\int_0^1 (e^x - a)^2 \, dx \right)^{\frac{1}{2}}$. The (unique) solution is given by $e - 1$ and the error norm is $(\frac{1}{2}(4e - e^2 - 3))^{\frac{1}{2}}$.

Ex. 3.   Let X be R_3 normed by $\|x\| = \max\limits_{1 \leq i \leq 3} |x_i|$. Take $n = 2$, $x_1 = (1, 0, 0)$, $x_2 = (0, 1, 0)$ and $y = (3, 5, 2)$. The minimum error norm is 2 and can be achieved with any coefficients a_1, a_2, for which $|3 - a_1| \leq 2$, $|5 - a_2| \leq 2$. Though there is a best approximation, it is not unique.

Ex. 4.   Let $X = \mathscr{P}_1$ normed by $\|f\| = |f(0)| + |f(1)|$. What constant is a best approximant to the polynomial x? We have $\|x - a\| = |a| + |1 - a|$. Hence, as a varies, the minimum value is 1 and is assumed for every $0 \leq a \leq 1$.

The problem of finding best approximations can be pictured geometrically. The set of all linear combinations $a_1 x_1 + \cdots + a_n x_n$ form a linear subspace of dimension n. We can picture this as a plane. The element y will not, in general, lie in this plane, and we would like to locate the point of the plane closest to y.

THEOREM 7.4.1.   *Given y and n linearly independent elements $x_1, \ldots, x_n$. The problem of finding $\min\limits_{a_i} \|y - (a_1 x_1 + \cdots + a_n x_n)\|$ has a solution.*

Proof: Consider the error norm

$$d(a_1, a_2, \ldots, a_n) = \|y - (a_1x_1 + a_2x_2 + \cdots + a_nx_n)\| \qquad (7.4.1)$$

as a function of the n real or complex variables $a_1, \ldots, a_n$. This function is continuous in the variables a_i, for, the difference

$$|d(a_1', a_2', \ldots, a_n') - d(a_1, a_2, \ldots, a_n)|$$
$$= \big| \, \|y - (a_1'x_1 + \cdots + a_n'x_n)\| - \|y - (a_1x_1 + \cdots + a_nx_n)\| \, \big|$$
$$\leq \|(a_1' - a_1)x_1 + \cdots + (a_n' - a_n)x_n\|$$
$$\leq |a_1' - a_1| \, \|x_1\| + \cdots + |a_n' - a_n| \, \|x_n\|. \qquad (7.4.2)$$

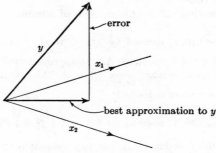

Figure 7.4.1.

The first inequality follows from (7.2.4) and the second from (7.2.1) (d), (c). Since the x's are fixed, (7.4.2) implies that the difference of the d's must be small if the difference of the a's is small. In a similar way, the function

$$h(a_1, \ldots, a_n) = \|a_1x_1 + \cdots + a_nx_n\| \qquad (7.4.3)$$

is a continuous function of the a's.

Let S designate the spherical surface

$$S: |a_1|^2 + |a_2|^2 + \cdots + |a_n|^2 = 1 \qquad (7.4.4)$$

in R_n (or C_n). S is closed and bounded and hence h must take on a minimum value $m > 0$ there. The possibility $m = 0$ is ruled out for we should have for some nonidentically vanishing a_i, $\|a_1x_1 + \cdots + a_nx_n\| = 0$. This implies that $a_1x_1 + \cdots + a_nx_n = 0$ and contradicts the assumption that the x's are linearly independent. Now, writing r for $(|a_1|^2 + |a_2|^2 + \cdots + |a_n|^2)^{\frac{1}{2}}$,

$$h(a_1, \ldots, a_n) = r \left\| \frac{a_1}{r} x_1 + \cdots + \frac{a_n}{r} x_n \right\| \qquad (7.4.5)$$

so that

$$h(a_1, \ldots, a_n) \geq mr. \qquad (7.4.6)$$

Moreover,

$$d = \|y - (a_1 x_1 + \cdots + a_n x_n)\| \geq \|a_1 x_1 + \cdots + a_n x_n\| - \|y\|$$
$$\geq mr - \|y\|. \tag{7.4.7}$$

This inequality means that if we select coefficients that are very large, d must be very large for it increases without limit as $r \to \infty$. This in turn means that in looking for a minimum value for d, we can confine our attentions to a certain sphere in the a-space. More precisely, set

$$\rho = \inf d(a_1, \ldots, a_n) \quad \text{and} \quad R = \frac{1 + \rho + \|y\|}{m}. \tag{7.4.8}$$

If $|a_1|^2 + \cdots + |a_n|^2 > R^2$, (7.4.7) implies

$$d > mR - \|y\| = 1 + \rho > \rho. \tag{7.4.9}$$

Hence

$$\inf_{\mathrm{I}} d(a_1, \ldots, a_n) = \inf_{\mathrm{II}} d(a_1, \ldots, a_n) \tag{7.4.10}$$

where I refers to the whole space of the a's and II to the portion $|a_1|^2 + \cdots + |a_n|^2 \leq R^2$. Since d is continuous, the value of the right hand of (7.4.10) is assumed in $|a_1|^2 + \cdots + |a_n|^2 \leq R^2$, and this completes the proof.

COROLLARY 7.4.2. *Let $f(x) \in C[a, b]$ and n be a fixed integer. The problem of finding $\min_{a_0, \ldots, a_n} \max_{a \leq x \leq b} |f(x) - (a_0 + a_1 x + \cdots + a_n x^n)|$ has a solution. As we shall prove subsequently, the solution is unique. It is called the Tschebyscheff approximation of degree $\leq n$ to $f(x)$. We shall designate it by $T_n(f; x)$.*

COROLLARY 7.4.3. *Let $f(x) \in C[a, b]$ and n be a fixed integer. Let $p \geq 1$. The problem of finding $\min_{a_0, a_1, \ldots, a_n} \int_a^b |f(x) - (a_0 + a_1 x + \cdots + a_n x^n)|^p \, dx$ has a solution. Such a solution yields a best approximation to $f(x)$ in the sense of least pth powers. We need only assume that $f \in L^p[a, b]$.*

COROLLARY 7.4.4. *Let B be a bounded region in the complex z plane. Let $f(z)$ be analytic in B and remain continuous in $\bar{B}$, the closure of B. The problem of finding $\min_{a_0, \ldots, a_n} \max_{z \in \bar{B}} |f(z) - (a_0 + a_1 z + \cdots + a_n z^n)|$ has a solution. This polynomial will be proven unique in Theorem 7.5.6 and will be designated by $T_n(f(z); z)$.*

COROLLARY 7.4.5. *Let $x_0, \ldots, x_k$ be $k + 1$ distinct points. Let $k \geq n$. The problem of determining $\min_{a_0, \ldots, a_n} \max_{0 \leq i \leq k} |f(x_i) - (a_0 + a_1 x_i + \cdots + a_n x_i^n)|$ has a solution.*

COROLLARY 7.4.6. *Let* $x_0, \ldots, x_k$ *be* $k + 1$ *distinct points,* $k \geq n$. *The problem of determining* $\min\limits_{a_0,\ldots,a_n} \sum\limits_{i=0}^{k} (f(x_i) - (a_0 + a_1 x_i + \cdots + a_n x_i^n))^2$ *possesses a solution. This is the common problem of least squares data fitting by polynomials.*

COROLLARY 7.4.7. *Let values* a_{ij}, y_i *be given for* $1 \leq i \leq p$, $1 \leq j \leq n$, $p > n$. *The problem of finding* $\min\limits_{x_j} \max\limits_{1 \leq i \leq p} |y_i - (a_{i1} x_1 + a_{i2} x_2 + \cdots + a_{in} x_n)|$ *has a solution.*

This is the "solution" of an over-determined system of linear equations, accepting as the answer those values that render minimum the maximum of the individual discrepancies. Other norms may be used. The most frequently used norm is the square norm.

COROLLARY 7.4.8. *Let* $C_p[-\pi, \pi]$ *designate the linear space of functions which are continuous on* $[-\pi, \pi]$ *and are periodic:* $f(\pi) = f(-\pi)$. *Then, there is a trigonometric polynomial of order* $\leq n$,

$$T_n(x) = \sum_{k=0}^{n} a_k \cos kx + \sum_{k=1}^{n} b_k \sin kx$$

for which $\max\limits_{-\pi \leq x \leq \pi} |f(x) - T_n(x)|$ *is minimum.*

DEFINITION 7.4.2. For a given $y; x_1, \ldots, x_n$ set

$$\min_{a_i} \|y - (a_1 x_1 + \cdots + a_n x_n)\| = E_n(y; x_1, \ldots, x_n) = E_n(y). \quad (7.4.11)$$

$E_n(y)$ is the measure of best approximation that can be achieved when y is approximated by linear combinations of the x's. Geometrically, it may be thought of as the distance from y to the subspace spanned by $x_1, \ldots, x_n$.

Evidently we have

$$E_1(y) \geq E_2(y) \geq E_3(y) \geq \cdots \qquad (7.4.12)$$

This is true since linear combinations of $x_1, x_2, \ldots, x_k$ are also linear combinations of $x_1, x_2, \ldots, x_k, x_{k+1}$.

7.5 Uniqueness of Best Approximation.

We have observed that under the hypothesis of Theorem 7.4.1 there is always one best approximant. But there may be more than one. In fact, the best approximants form a convex set.

THEOREM 7.5.1. *Let* S *designate the set of best approximants to* y *in the situation of Theorem 7.4.1. Then* S *is convex.*

Proof: Let x and w be two best approximants to y. Then, $\|y - x\| = E_n(y)$, $\|y - w\| = E_n(y)$. Suppose that $\alpha \geq 0$, $\beta \geq 0$ and $\alpha + \beta = 1$. Then

$$\|y - \alpha x - \beta w\| = \|\alpha(y - x) + \beta(y - w)\| \leq \alpha \|y - x\| + \beta \|y - w\|$$
$$= (\alpha + \beta)E_n(y) = E_n(y).$$

But $\alpha x + \beta w$ is also a linear combination of $x_1, x_2, \ldots, x_n$ and hence $\alpha x + \beta w$ must also be a best approximant.

COROLLARY 7.5.2. *The set of best approximants consists either of one element or of infinitely many elements.*

Proof: For if it contains two distinct elements, it must contain the whole line of elements joining the two.

Ex. 1. In Ex. 3, 7.4, the totality of best approximations is given by $|3 - a_1| \leq 2$, $|5 - a_2| \leq 2$. The points (a_1, a_2) lie in a square, a convex figure.

A fairly extensive sufficient condition can be given which assures the uniqueness of the best approximation.

DEFINITION 7.5.1. A normed linear space X is called *strictly convex* if $\|x\| \leq r$, $\|y\| \leq r$ imply $\|x + y\| < 2r$ unless $x = y$.

Ex. 2. The space C_n of complex sequences $x = (x_1, \ldots, x_n)$ with

$$\|x\| = \left(\sum_{i=1}^{n} |x_i|^p \right)^{1/p}, p > 1,$$

is strictly convex. For if

$$\|x\| \leq r, \|y\| \leq r \quad \text{and} \quad \|x + y\| = 2r,$$

then

$$2r = \|x + y\| \leq \|x\| + \|y\| \leq 2r.$$

Therefore,

$$\|x + y\| = \|x\| + \|y\|.$$

Hence

$$\|x\| = \|y\| = r.$$

By (7.2.15),

$$x_i/y_i = \lambda > 0 \quad i = 1, 2, \ldots, n.$$

Since $\|x\| = \|y\|$, $\lambda = 1$, and so $x = y$.

Ex. 3. The normed linear spaces ℓ^p and $L^p[a, b]$, $1 < p < \infty$, are strictly convex.

Ex. 4. The space $C[-1, 1]$ with $\|f\| = \max_{-1 \leq x \leq 1} |f(x)|$ is not strictly convex. For if $f(x) = 1 - x^2$, $g(x) = 1 - x^4$, $\|f\| = \|g\| = 1$, $\|f + g\| = 2$ but

$$f(x) \not\equiv g(x).$$

Ex. 5. R_2 with $\|(x_1, x_2)\| = \max(|x_1|, |x_2|)$ is not strictly convex for

$$\|(1, 0) + (1, 1)\| = \|(1, 0)\| + \|(1, 1)\|.$$

THEOREM 7.5.3.   *In a normed linear space X with a strictly convex norm, the problem of best approximation (posed in Theorem 7.4.1) has a unique solution.*

Proof: Suppose there are two distinct best approximants to y, u_1 and u_2. Then $\|y - u_1\| = \|y - u_2\| = E_n(y)$. Now $y - u_1$ and $y - u_2$ are also distinct. Hence by strict convexity,

$$\|(y - u_1) + (y - u_2)\| < 2E_n(y).$$

This is equivalent to $\|y - \frac{1}{2}(u_1 + u_2)\| < E_n(y)$. But this would mean that the element $\frac{1}{2}(u_1 + u_2)$, which is also a linear combination of $x_1, \ldots, x_n$, is closer to y than the minimum possible distance. This is a contradiction.

COROLLARY 7.5.4.   *Best approximation in the spaces $L^p[a, b]$, ℓ^p, $1 < p < \infty$ is unique.*

The important case of best uniform approximation is, unfortunately, not covered by the general result of Theorem 7.5.3 and must be treated by its own methods.

We begin by establishing a geometric lemma whose utility will become clear during the course of the proof of Theorem 7.5.6.

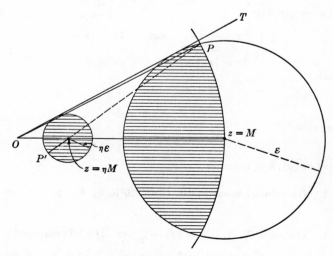

Figure 7.5.1.

LEMMA 7.5.5.   *Let M and $\varepsilon < M/2$ be fixed. Designate by* I *the set of points* $|z - \eta M| \le \eta \varepsilon$ *and by* II *the set of points common to* $|z| \le M$ *and* $|z - M| \le \varepsilon$. *Set* $\rho(\eta) = \max\limits_{\substack{z_1 \in \mathrm{I}, \\ z_2 \in \mathrm{II}}} |z_1 - z_2|$. *Then, for any* $0 < \eta < \frac{1}{3}$, $\rho(\eta) < M$.
The same conclusion obviously holds if the whole figure is rotated about 0.

Proof: Since $\varepsilon < M/2$, $\eta < \frac{1}{3} < \dfrac{M - \varepsilon}{M + \varepsilon}$. The sets I and II have no common point and appear as drawn in the figure. If $\theta = \measuredangle\, TOM$, then $\sin \theta = \dfrac{\eta \varepsilon}{\eta M} < \frac{1}{2}$. Hence $0 < \theta < \pi/6$.

A moment's consideration leads one to the conclusion that $\rho = \overline{PP'}$. Hence,

$$\rho(\eta) = (M^2 + \eta^2 M^2 - 2M^2 \eta \cos \theta')^{\frac{1}{2}} + \eta \varepsilon$$

where $\theta' = \measuredangle\, POM$. Since $\theta' < \theta < \pi/6$, $-\cos \theta' < -\dfrac{\sqrt{3}}{2}$. Hence,

$$\rho(\eta) < M((1 + \eta^2 - \sqrt{3}\eta)^{\frac{1}{2}} + \eta/2).$$

Let $f(\eta) = (1 + \eta^2 - \sqrt{3}\eta)^{\frac{1}{2}} + \eta/2$. Then $f(\eta) = 1$ if and only if $\eta = 0$ or $\eta = \frac{4}{3}(\sqrt{3} - 1) = .976$. Furthermore, $f(\frac{1}{2}) = (\frac{5}{4} - \frac{1}{2}\sqrt{3})^{\frac{1}{2}} + \frac{1}{4} = .87 < 1$. Therefore when $0 < \eta < .976$, $f(\eta) < 1$. Consequently $\rho(\eta) < M$ for $0 < \eta < \frac{1}{3}$.

THEOREM 7.5.6 (Tonelli).   *Let S be a closed and bounded set in the complex plane that contains more than $n + 1$ points. Let $f(z)$ be continuous on S and set*

$$M = \min_{p \in \mathscr{P}_n} \max_{z \in S} |f(z) - p(z)|. \tag{7.5.1}$$

Let $p_n(z)$ be any polynomial that realizes this extreme value and set

$$r(z) = f(z) - p_n(z). \tag{7.5.2}$$

Then,
   A. *The number of distinct points of S at which $|r(z)|$ takes on its maximum value is greater than $n + 1$.*
   B. *There is a unique solution to the problem* (7.5.1).
   Proof: There is first the trivial case in which $M = 0$. Then

$$\max_{z \in S} |f(z) - p_n(z)| = 0$$

so that

$$p_n(z) = f(z)$$

throughout S. This implies A. Since a minimizing polynomial must agree with f in more than $n + 1$ points, it is uniquely determined for it is in $\mathscr{P}_n$.

Assume next that $M > 0$. We first show that A implies B. Suppose that $p_n, q_n \in \mathscr{P}_n$ are two distinct polynomials for which

$$\max_{z \in S} |f(z) - p_n(z)| = \max_{z \in S} |f(z) - q_n(z)| = M.$$

Now,
$$\max_{z \in S} |f(z) - \tfrac{1}{2}(p_n(z) + q_n(z))|$$

$$\leq \max_{z \in S} \tfrac{1}{2} |f(z) - p_n(z)| + \max_{z \in S} \tfrac{1}{2} |f(z) - q_n(z)| = M/2 + M/2 = M.$$

Since $\tfrac{1}{2}(p_n + q_n) \in \mathscr{P}_n$, the definition of M implies that

$$\max_{z \in S} |f(z) - \tfrac{1}{2}(p_n(z) + q_n(z))| = M. \tag{7.5.3}$$

This means that $\tfrac{1}{2}(p_n + q_n)$ is a minimizing polynomial.

Let z' be a point of S for which the maximum in (7.5.3) is achieved. Then

$$M = |f(z') - \tfrac{1}{2}(p_n(z') + q_n(z'))| \leq \tfrac{1}{2}|f(z') - p_n(z')| + \tfrac{1}{2}|f(z') - q_n(z')|$$
$$\leq M/2 + M/2 = M.$$

This implies that

$$|f(z') - \tfrac{1}{2}(p_n(z') + q_n(z'))| = \tfrac{1}{2} |f(z') - p_n(z')| + \tfrac{1}{2} |f(z') - q_n(z')|$$
and $$|f(z') - p_n(z')| = M \quad \text{and} \quad |f(z') - q_n(z')| = M.$$

Since, moreover, $|a + b| = |a| + |b|$ implies that $\arg a = \arg b$ (or $ab = 0$), it follows that

$$f(z') - p_n(z') = f(z') - q_n(z') \quad \text{or} \quad p_n(z') - q_n(z') = 0.$$

According to A, there are at least $n + 2$ distinct points of type z'. Since $p_n - q_n \in \mathscr{P}_n$, it follows that $p_n \equiv q_n$.

To prove A, suppose the contrary, that $|r(z)| = M$ only at $z_1, z_2, \ldots, z_m$ and $m \leq n + 1$. Let $q(z)$ be an element of $\mathscr{P}_n$ for which

$$q(z_i) = r(z_i) \qquad i = 1, 2, \ldots, m.$$

We will show that for sufficiently small η,

$$\max_{z \in S} |r(z) - \eta q(z)| < M. \tag{7.5.4}$$

Thus,
$$\max_{z \in S} |f(z) - p_n(z) - \eta q(z)| < M. \tag{7.5.5}$$

Since $p_n + \eta q \in \mathscr{P}_n$, this would contradict the definition of M in (7.5.1). To prove (7.5.4),

(a) Select $\varepsilon < M/2$.

(b) In view of the uniform continuity of r and q over S, determine δ so that

$$|r(z_1) - r(z_2)| \leq \varepsilon, |q(z_1) - q(z_2)| \leq \varepsilon \tag{7.5.6}$$
for $$|z_1 - z_2| \leq \delta, z_1, z_2 \in S.$$

(c) Let C_δ designate the set of points of S lying in at least one of the circles

$$|z - z_k| < \delta \qquad k = 1, 2, \ldots, m.$$

Let $T_\delta = S - C_\delta$. Then $S = C_\delta \cup T_\delta$.

(d) Let $M_\delta = \max\limits_{z \in T_\delta} |r(z)|$. (If T_δ is empty, set $M_\delta = 0$. Similarly below.) In view of the fact that $|r(z)| = M$ only at $z_1, \ldots, z_m$, we have $M_\delta < M$.

(e) Set $\max\limits_{z \in S} |q(z)| = Q$ and select an η:

$$0 < \eta < \min\left(\tfrac{1}{3}, \frac{M - M_\delta}{2Q}\right).$$

Then,

$$\max_{z \in T_\delta} |r(z) - \eta q(z)| \le \max_{z \in T_\delta} |r(z)| + \eta \max_{z \in T_\delta} |q(z)|$$

$$< M_\delta + \frac{M - M_\delta}{2Q} Q = \frac{M + M_\delta}{2} < M. \qquad (7.5.7)$$

Now for C_δ. Let C_k designate the set of points common to S and to

$$|z - z_k| < \delta.$$

Then, $C_\delta = \bigcup\limits_{k=1}^{m} C_k$. If $z \in C_k$, then $|r(z)| \le M = |r(z_k)|$, and by (b)

$$|r(z) - r(z_k)| \le \varepsilon < M/2.$$

Thus, the values of $r(z)$ lie in a region II_k as explained in the lemma. If $z \in C_k$, then by (b), $|q(z) - q(z_k)| \le \varepsilon$; but since $q(z_k) = r(z_k)$, we have $|\eta q(z) - \eta r(z_k)| \le \eta\varepsilon$. Thus, for $z \in C_k$, the values of $\eta q(z)$ lie in a region I_k. Under the assumption on η,

$$|r(z) - \eta q(z)| \le \rho(\eta) = \max_{\substack{z_1 \in \mathrm{I}_k \\ z_2 \in \mathrm{II}_k}} |z_1 - z_2| < M.$$

This conclusion is independent of k and hence

$$\max_{z \in C_\delta} |r(z) - \eta q(z)| < M. \qquad (7.5.8)$$

Combining (7.5.8) with (7.5.7) we obtain (7.5.4). This completes the proof of the Theorem. For further elaboration of a similar argument carried out in the real domain, see the proof of Theorem 7.6.2.

We remark that if S contains $n + 1$ points, then $M = 0$ and B holds but not A. If S contains fewer than $n + 1$ points, $M = 0$ and the solution is not unique.

Ex. 6. Let S be a closed bounded point set in the complex plane containing more than $n - 1$ points. The problem of finding

$$\min_{a_i} \max_{z \in S} |z^n - (a_{n-1}z^{n-1} + a_{n-2}z^{n-2} + \cdots + a_0)| \qquad (7.5.9)$$

has a unique solution. The total expression above is a polynomial of degree n with leading coefficient 1 whose maximum modulus over S is minimum. We will designate it by $T_n(S; z)$. We note that it is frequently called the *Tscheby-scheff polynomial of degree n for S.* This adds greatly to the ambiguity of the expression "Tschebyscheff polynomial."

Ex. 7.   Let S consist of n distinct points $z_1, \ldots, z_n$. Then

$$T_n(S; z) = (z - z_1)(z - z_2) \cdots (z - z_n).$$

Ex. 8.   Let $S: [-1, 1]$. Then $T_n(S; z) = \tilde{T}_n(z)$ (Cf. Def. 3.3.2).

Ex. 9.   If C designates the unit circle, $T_n(C; z) = z^n$. We shall prove that if $p_n(z) = z^n + a_1 z^{n-1} + \cdots + a_n \not\equiv z^n$, then $\max\limits_{z \in C} |p_n(z)| > 1$. Consider

$$q(z) = \frac{p_n(z)}{z^n} = 1 + \frac{a_1}{z} + \cdots + \frac{a_n}{z^n}.$$

The function $q(z)$ is analytic in the closed exterior of C (including ∞), and $q(\infty) = 1$. By the Maximum Principle, since q is not a constant,

$$\max_{|z|=1} |q(z)| > |q(\infty)| = 1.$$

Therefore, $\max\limits_{|z|=1} |p_n(z)| = \max\limits_{|z|=1} |z^n| \, |q(z)| > 1$.

7.6   Best Uniform (Tschebyscheff) Approximation of Continuous Functions.

Let $f \in C[a, b]$. We know by Theorem 7.5.6 that the problem of finding $\min\limits_{p \in \mathscr{P}_n} \max\limits_{a \leq x \leq b} |f(x) - p(x)|$ has a unique solution. Designate the solution by $p_n(x)$ and set $E_n(f) = \max\limits_{a \leq x \leq b} |f(x) - p_n(x)|$. (The polynomial $p_n(x)$ is frequently called the *Tschebyscheff approximation of degree* $\leq n$ *to* $f(x)$.)

THEOREM 7.6.1.   *If* $f \in C[a, b]$, *then*

$$E_0(f) \geq E_1(f) \geq \cdots \quad and \quad \lim_{n \to \infty} E_n(f) = 0. \tag{7.6.1}$$

Proof: We have already noted the monotonicity in (7.4.12).
Let $\varepsilon > 0$ be given. By Weierstrass' theorem we can find a polynomial of degree m, $q_m(x)$, such that $|f(x) - q_m(x)| \leq \varepsilon, a \leq x \leq b$. Hence,

$$E_m(f) = \max_{a \leq x \leq b} |f(x) - p_m(x)| \leq \max_{a \leq x \leq b} |f(x) - q_m(x)| \leq \varepsilon.$$

Thus, $E_n(f) \leq \varepsilon$ for all $n \geq m$, establishing the second assertion.

We shall now characterize the behavior of the best uniform approximants. An examination of the cases $n = 0$ and $n = 1$ will provide insight to the general theorem and will help us to understand a simple, but fussy, proof.

Let $f(x) \in C[a, b]$. We are interested in solving the problem of finding $\min\limits_c \max\limits_{a \leq x \leq b} |f(x) - c|$. A glance at Figure 7.6.1 leads us to the following answer. Let $m = \min\limits_{a \le x \le b} f(x)$, $M = \max\limits_{a \le x \le b} f(x)$. Then the minimizing constant is

$$c = \tfrac{1}{2}(m + M) \tag{7.6.2}$$

and

$$E_0(f) = \tfrac{1}{2}(M - m). \tag{7.6.3}$$

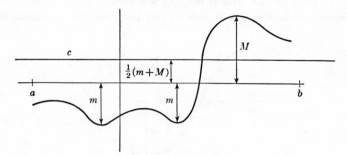

Figure 7.6.1.

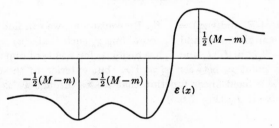

Figure 7.6.2.

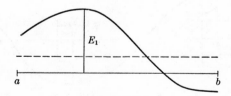

Figure 7.6.3.

Notice that when the error curve $\varepsilon(x) = f(x) - c$ is drawn, the value $\pm E_0$ is assumed by it at least *twice*: once with a plus sign and once with a minus sign. Suppose next that $p_1(x) = a_0 + a_1 x$ solves the problem of finding $\min\limits_{P \in \mathscr{P}_1} \max\limits_{a \le x \le b} |f(x) - p(x)|$. Consider the error curve $\varepsilon(x) = f(x) - p_1(x)$. Set $E_1 = \max\limits_{a \le x \le b} |f(x) - p_1(x)|$. Since $|\varepsilon(x)|$ is in $C[a, b]$, this maximum is assumed at least once. If it were taken on only once (Fig. 7.6.3) then by the addition of an appropriate constant to $p_1(x)$ we could lower E_1. This would contradict the definition of E_1.

Suppose E_1 is taken on only two times. It must be taken on with opposite signs, otherwise we can argue as above and lower E_1. But even if it were taken on with opposite signs, we could subtract from $\varepsilon(x)$, and hence from the original $p_1(x)$, an appropriate linear function which would have the effect of reducing the size of the maximum values without raising other values in excess of this. This is more or less evident geometrically (look at the dashed line in Fig. 7.6.4), but we can formalize the argument in this way.

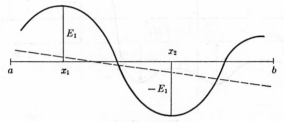

Figure 7.6.4.

Let $\varepsilon(x_1) = +E_1$ and $\varepsilon(x_2) = -E_1$. By continuity, we can find two closed intervals, I_1 containing x_1, and I_2 containing x_2 such that $\varepsilon(x) > E_1/2$ in I_1 and $\varepsilon(x) < -E_1/2$ in I_2. I_1 and I_2 are disjoint, for $\varepsilon(x)$ is of opposite sign in them. Pick a point x_0 between x_1 and x_2, but exterior to these intervals, and let $\ell(x)$ be a fixed linear function that passes through x_0, is positive in I_1, and negative in I_2. (Fig. 7.6.5)

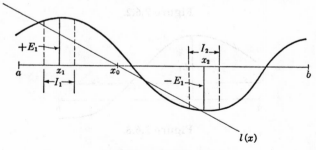

Figure 7.6.5.

Let J designate the closure of the set $[a, b] - I_1 - I_2$, and write $E_1' = \max_{x \in J} |\varepsilon(x)|$. We can obviously arrange matters so that J does not contain x_1 or x_2. Since the maximum of $|\varepsilon(x)|$ is assumed only at x_1 and x_2 we have $E_1' < E_1$. Finally, select a quantity η that satisfies

$$0 < \eta < (E_1 - E_1')/2 \max_{a \leq x \leq b} |\ell(x)|. \qquad (7.6.4)$$

The claim is now made that

$$\max_{a \leq x \leq b} |f(x) - p_1(x) - \eta\ell(x)| = \max_{a \leq x \leq b} |\varepsilon(x) - \eta\ell(x)| < E_1 \quad (7.6.5)$$

and this contradicts the definition of E_1. To show this: on I_1, by (7.6.4)

$$0 < \eta\ell(x) < \frac{\ell(x)(E_1 - E_1')}{2 \max\limits_{a \leq x \leq b} |\ell(x)|} \leq \frac{E_1}{2} < \varepsilon(x).$$

Hence $\varepsilon(x) > \eta\ell(x)$ and

$$|\varepsilon(x) - \eta\ell(x)| = \varepsilon(x) - \eta\ell(x)$$
$$\leq E_1 - \min_{x \in I_1} \eta\ell(x) = E_1 - \text{something positive} < E_1.$$

A similar argument holds for I_2. Now for J. Using (7.6.4),

$$\max_{x \in J} |\varepsilon(x) - \eta\ell(x)| \leq \max_{x \in J} |\varepsilon(x)| + \eta \max_{x \in J} |\ell(x)|$$

$$< E_1' + \frac{\max\limits_{x \in J} |\ell(x)|}{\max\limits_{a \leq x \leq b} |\ell(x)|} \frac{(E_1 - E_1')}{2} < E_1' + E_1 - E_1' = E_1.$$

Therefore (7.6.5) holds.

It follows that there must be at least three points $x_1 < x_2 < x_3$ where $\varepsilon(x_i) = \pm E_1$. The error must alternate in signs at these three points: E_1, $-E_1$, E_1 or $-E_1$, E_1, $-E_1$. For one alternation has already been established, and if we had, say, E_1, $-E_1$, $-E_1$, the same argument could be used to show that E_1 could be lowered.

THEOREM 7.6.2 (The Tschebyscheff Equioscillation Theorem).

Let $f(x) \in C[a, b]$ and $p(x)$ be the best uniform approximant to f of degree n. Let $E_n = \max\limits_{a \leq x \leq b} |f(x) - p(x)|$ and $\varepsilon(x) = f(x) - p(x)$. There are at least $n + 2$ points $a \leq x_1 < x_2 < \cdots < x_{n+2} \leq b$ where $\varepsilon(x)$ assumes the values $\pm E_n$ and with alternating signs:

$$\begin{aligned} \varepsilon(x_i) &= \pm E_n & i &= 1, 2, \ldots, n + 2 \\ \varepsilon(x_i) &= -\varepsilon(x_{i+1}) & i &= 1, 2, \ldots, n + 1. \end{aligned} \quad (7.6.6)$$

Proof: Select an ε so small that $|x_1 - x_2| < \varepsilon$ implies

$$|\varepsilon(x_1) - \varepsilon(x_2)| \leq E_n/2.$$

This is possible by the uniform continuity of $\varepsilon(x)$. Divide $[a, b]$ into consecutive closed intervals of width $\leq \varepsilon$. Call the intervals on which $|\varepsilon(x)|$ assumes its maximum value $I_1, I_2, \ldots, I_m$. Since $\varepsilon(x)$ can vary at most $E_n/2$ in any of these intervals, we must have

$$\varepsilon(x) > E_n/2 \quad \text{or} \quad \varepsilon(x) < -E_n/2$$

there. Let $u_1, \ldots, u_m (= \pm 1)$ be the sign of $\varepsilon(x)$ over these intervals. We must prove that in this sequence there are at least $n + 1$ changes of sign.

We do this by showing that if there were fewer than $n + 1$ changes, we could find a polynomial whose E_n is less than that of $p(x)$.

If all the u's were the same, by adding an appropriate constant to $p(x)$ we could get a better approximation. Suppose then, we go through the intervals $I_1, \ldots, I_m$ and group them into consecutive groups where the u's have the same sign:

1st group $I^{(1)}, I^{(2)},$ $\ldots, I^{(j_1)}$ First sign

2nd group $I^{(j_1+1)}, I^{(j_1+2)}, \ldots, I^{(j_2)}$ Second sign

.   First sign

.   .

.   .

kth group $I^{(j_{k-1}+1)},$ $\ldots, I^{(j_k)}$

(Here we have written $I^{(1)} = I_1, \ldots, I^{(j_k)} = I_m$, etc.)

This scheme displays $k - 1$ changes of sign so let us assume that

$$k - 1 < n + 1 \quad \text{or} \quad k < n + 2.$$

Consider the intervals $I^{(j_1)}, I^{(j_1+1)}$. These intervals cannot be adjacent for $\varepsilon(x)$ is 0 in neither and yet it has opposite signs there. Hence (with an obvious notation) we can find an x_1:

$$I^{(j_1)} < x_1 < I^{(j_1+1)}$$

Similarly, $I^{(j_2)} < x_2 < I^{(j_2+1)}$

.

.

$$I^{(j_{k-1})} < x_{k-1} < I^{(j_{k-1}+1)}.$$

Form the polynomial $q(x) = (x_1 - x)(x_2 - x) \cdots (x_{k-1} - x)$.

Since we have assumed $k < n + 2$, it follows that $k - 1 \le n$, so that $q(x) \in \mathscr{P}_n$. $q(x)$ vanishes only at x_i. Since the x_i are between the intervals $I_1, \ldots, I_m$, q must have constant sign over each of these intervals.

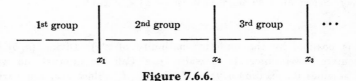

Figure 7.6.6.

Over the first group of intervals $q(x) = (x_1 - x) \cdots (x_{k-1} - x)$ is positive, for all factors are positive. Over the second group $q(x)$ is negative, for all but the first factor are positive, etc. By selecting $u = \pm 1$ appropriately, $uq(x)$ will coincide with $\varepsilon(x)$ in sign over all the intervals $I_1, \ldots, I_m$.

We now claim that for η sufficiently small, $p(x) + \eta u q(x)$ will be a better approximant to f than p is. This would be a contradiction. Let J be the closure of $[a, b] - I_1 - I_2 - \cdots - I_m$. Write $E_n' = \max_{x \in J} |\varepsilon(x)|$. Since the maxima of $|\varepsilon(x)|$ are assumed only on the I's we have $E_n' < E_n$. Select an η that satisfies

$$0 < \eta < (E_n - E_n')/2 \max_{a \leq x \leq b} |q(x)|. \qquad (7.6.7)$$

The rest of the argument parallels exactly the discussion following (7.6.5).

In one case it is possible to give an explicit construction of the best uniform approximant.

COROLLARY 7.6.3. *Let $f(x) \in C^2 [a, b]$ and let $f''(x) > 0$ there. If $a_0 + a_1 x$ is the best uniform linear approximant to f, then*

$$a_1 = \frac{f(b) - f(a)}{b - a}$$

$$a_0 = \tfrac{1}{2}(f(a) + f(c)) - \frac{f(b) - f(a)}{b - a} \frac{a + c}{2} \qquad (7.6.8)$$

where c is the unique solution of $f'(c) = \dfrac{f(b) - f(a)}{b - a}$.

Proof: One solution to the equation $f'(c) = \dfrac{f(b) - f(a)}{b - a}$ is guaranteed by the mean value theorem. Since $f'' > 0$, f' is increasing, and hence this solution is unique.

Now set $\varepsilon(x) = f(x) - (a_0 + a_1 x).$

By our theorem, there are at least 3 distinct points $x_1 < x_2 < x_3$ where $\varepsilon(x)$ reaches its maximum absolute value. One point, x_2, is interior to the interval and hence $\varepsilon'(x_2) = 0$. Since $\varepsilon'(x) = f'(x) - a_1$, ε' is also strictly increasing. The other two extreme points of $\varepsilon(x)$ must therefore be at a and b. Now with $x_1 = a$, $x_2 = c$, $x_3 = b$, it follows that

$$f(a) - (a_0 + a_1 a) = -(f(c) - (a_0 + a_1 c)) = f(b) - (a_0 + a_1 b)$$

and

$$\varepsilon'(c) = f'(c) - a_1 = 0.$$

These conditions, rearranged, lead to (7.6.8).

The best uniform (Tschebyscheff) approximant is completely characterized by the property of equioscillation at $n + 2$ points. This property is frequently the basis of numerical schemes for computing the approximant.

THEOREM 7.6.4. *Let $f(x) \in C[a, b]$. Given a $q(x) \in \mathscr{P}_n$ with*

$$\max_{a \leq x \leq b} |f(x) - q(x)| = \delta.$$

Let there be $n + 2$ points $a \leq x_1 < x_2 < \cdots < x_{n+2} \leq b$ such that

$$f(x_i) - q(x_i) = \pm\delta, \qquad i = 1, 2, \ldots, n + 2,$$

in an alternating fashion. Then,

$$\delta = E_n(f) \tag{7.6.9}$$

and q is the best uniform approximant to f in $\mathscr{P}_n$.

Proof: By definition, $E_n(f) \leq \delta$. Assume $E_n(f) < \delta$. Let $p(x)$ be the best uniform approximant. Then

$$q(x_i) - p(x_i) = q(x_i) - f(x_i) - (p(x_i) - f(x_i))$$

Since $\max\limits_{a \leq x \leq b} |f(x) - p(x)| = E_n(f) < \delta$, then writing $\tau_i = p(x_i) - f(x_i)$, we have $|\tau_i| < \delta$, $q(x_i) - p(x_i) = \pm\delta - \tau_i$. The function $q(x) - p(x) \in \mathscr{P}_n$ and has $n + 2$ points of alternation. It therefore has $n + 1$ zeros and consequently must be identically zero by Theorem 1.11.3. Conclusion: $q(x) \equiv p(x)$.

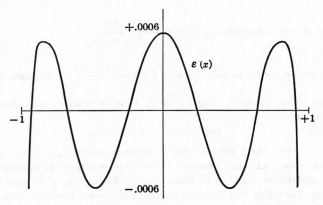

Figure 7.6.7 Tschebyscheff's Equioscillation Theorem.

The best uniform approximant to $\cos\dfrac{\pi}{2}x$ over $[-1, 1]$ out of $\mathscr{P}_5$ is

$$p(x) = 0.9994032 - 1.2227967x^2 + 0.2239903x^4.$$

$\varepsilon(x) = \cos\dfrac{\pi}{2}x - p(x)$ assumes its extreme values at $5 + 2$ points and with alternating sign.

7.7 Best Approximation by Nonlinear Families. The situation here is more complicated than in the case of linear families. A few examples will suffice to show what can happen.

Ex. 1.   Consider the problem $\min\limits_{a} \max\limits_{1 \leq x \leq 2} |0 - e^{ax}|$. By selecting a very large and negative, the maximum can be made as small as desired. The problem is, so to speak, solved by $a = -\infty$.

Ex. 2.   Consider the problem $\min\limits_{a} \max\limits_{0 \leq x \leq 1} |\frac{1}{2} - x^{a^2}|$. For any value a,

$$\max_{0 \leq x \leq 1} |\tfrac{1}{2} - x^{a^2}| = \tfrac{1}{2}.$$

Thus, while no proper value solves the first problem, any value at all solves the second.

One of the most familiar nonlinear families is fortunate enough to contain best uniform approximations.

THEOREM 7.7.1.   *Let $f(x) \in C[a, b]$ and let m and n be fixed integers ≥ 0. The problem of finding* $\min\limits_{a_i, b_i} \max\limits_{a \leq x \leq b} \left| f(x) - \dfrac{a_0 x^n + a_1 x^{n-1} + \cdots + a_n}{b_0 x^m + b_1 x^{m-1} + \cdots + b_m} \right|$ *has a solution.*

Proof: There is some redundancy in the coefficients of the rational function. We can adjust them so that $b_0^2 + b_1^2 + \cdots + b_m^2 = 1$. As the b's vary, we will certainly obtain some polynomials that do not vanish in $[a, b]$. If we set

$$\Delta = \inf_{a_i, b_i} \max_{a \leq x \leq b} \left| f(x) - \frac{a_0 x^n + \cdots}{b_0 x^m + \cdots} \right|,$$

then $0 \leq \Delta < \infty$. By the definition of Δ, we can find a sequence of rational functions

$$R_k(x) = \frac{A_k(x)}{B_k(x)}, \; A_k(x) = \sum_{i=0}^{n} a_i^{(k)} x^{n-i}, \; B_k(x) = \sum_{j=0}^{m} b_j^{(k)} x^{m-j}$$

so that if

$$\Delta_k = \max_{a \leq x \leq b} |f(x) - R_k(x)|, \tag{7.7.1}$$

then

$$\lim_{k \to \infty} \Delta_k = \Delta.$$

The coefficients $b_j^{(k)}$ are bounded due to the normalizing condition. The coefficients $a_i^{(k)}$ are also bounded. This can be seen as follows. From (7.7.1),

$$-\Delta_k \leq f(x) - R_k(x) \leq \Delta_k.$$

Hence

$$|R_k(x)| \leq \Delta_k + \max_{a \leq x \leq b} |f(x)| \leq M \tag{7.7.2}$$

for some constant M. This means that

$$|A_k(x)| \leq M |B_k(x)|. \tag{7.7.3}$$

Since the $b_j^{(k)}$ are bounded, the polynomials $B_k(x)$ are bounded on $[a, b]$, and therefore $A_k(x)$ are bounded. Now if a family of polynomials of bounded degrees are bounded, then its coefficients are bounded. (See Problem 40.)

Consider the points P_k: $(a_0^{(k)}, a_1^{(k)}, \ldots, a_n^{(k)}, b_0^{(k)}, b_1^{(k)}, \ldots, b_m^{(k)})$ in the space R_{m+n+2}. They lie in a bounded portion of that space. Hence, we may select a subsequence P_{k_l} that converges to a point

$$P' = (a_0', a_1', \ldots, a_n', b_0', b_1', \ldots, b_m').$$

Consider the rational functions corresponding to this subsequence, and reindex the subsequence so that we have

$$\lim_{k \to \infty} a_i^{(k)} = a_i' \qquad i = 0, 1, \ldots, n$$

$$\lim_{k \to \infty} b_j^{(k)} = b_j' \qquad j = 0, 1, \ldots, m. \tag{7.7.4}$$

Form

$$R'(x) = \frac{a_0'x^n + a_1'x^{n-1} + \cdots + a_n'}{b_0'x^m + b_1'x^{m-1} + \cdots + b_m'}. \tag{7.7.5}$$

If we can show that

$$\max_{a \le x \le b} |f(x) - R'(x)| = \Delta \tag{7.7.6}$$

then R' will be a best approximant and our proof is complete.

R', being rational, can have at most a finite number of infinities. Let $D(x)$ be the denominator of $R'(x)$ and select an x in $[a, b]$ such that $D(x) \ne 0$. At such a point we must have $\lim_{k \to \infty} R_k(x) = R'(x)$. Since

$$R'(x) = f(x) + R_k(x) - f(x) + R'(x) - R_k(x),$$
$$|R'(x)| \le |f(x)| + |f(x) - R_k(x)| + |R'(x) - R_k(x)|;$$

hence

$$|R'(x)| \le \max_{a \le x \le b} |f(x)| + \Delta_k + \varepsilon_k, \quad \varepsilon_k \to 0.$$

Setting $\mu = \sup_k \Delta_k$, and allowing $k \to \infty$,

$$|R'(x)| \le \max_{a \le x \le b} |f(x)| + \mu. \tag{7.7.7}$$

The bound (7.7.7) holds uniformly for any x in $[a, b]$ for which $D(x) \ne 0$. This, in turn, means that $R'(x)$ cannot have any infinities on $[a, b]$, for if it did, there would be values of x in a neighborhood of the infinity where the bound would be exceeded.

Let x be any point of $[a, b]$. Suppose first that $D(x) \ne 0$. Then for $k = 1, 2, \ldots,$

$$|f(x) - R'(x)| \le |f(x) - R_k(x)| + |R_k(x) - R'(x)|$$
$$\le \Delta_k + \eta_k \quad \text{where} \quad \eta_k \to 0.$$

Thus,

$$|f(x) - R'(x)| \le \Delta. \tag{7.7.8}$$

Suppose that $D(x) = 0$. Then we may find a sequence of points of $[a, b]$, x_1, $x_2, \ldots$, such that $x_i \to x$, $D(x_i) \neq 0$. Then by (7.7.8), $|f(x_i) - R'(x_i)| \leq \Delta$, $i = 1, 2, \ldots$, and by continuity, $|f(x) - R'(x)| \leq \Delta$. We have therefore shown that $|f(x) - R'(x)| \leq \Delta$ throughout $[a, b]$. By the definition of Δ, this implies (7.7.6).

NOTES ON CHAPTER VII

Works on functional analysis that have been found useful include Banach [1], Riesz-Sz. Nagy [1], Ljusternik and Sobolew [1], Kolmogorov and Fomin [1], Taylor [3], Dunford and Schwartz [1], Zaanen [1].

7.2   The convexity of x^r, $r > 1$, can be used to prove (7.2.5). See Boas [4], p. 148.

7.3   Eggleston [1] is a fine presentation of the theory of convex bodies.

7.4   Achieser [1].

7.5   For further results on uniqueness, see Hirschfeld [1], [2]. An extension of Theorem 7.5.6 to rational functions can be found in Walsh [2], p. 363.

7.6   de la Vallée Poussin [1], Natanson [1]. The numerical side of best uniform approximation by polynomials and rationals had to wait for the development of electronic computing equipment. For part of the vast literature that has developed around this problem, see Remez [1], Stiefel [1], Maehly and Witzgall [1], [2], Murnaghan and Wrench [1], [2]. Ward [1] expounds the problem from the point of view of linear programming.

For a more abstract approach to problems of Tschebyscheff type see Rivlin and Shapiro [1].

7.7   Approximation by nonlinear families is currently under investigation. See Motzkin [1], Rice [1], [2], [3]. See also Young [1].

PROBLEMS

1. Let n be fixed. S consists of the 2^n n-tuples A, B, $\ldots$, whose elements are either 0 or 1. If we set $d(A, B) = $ the number of places in which A and B differ, then S becomes a metric space. This is the *Hamming distance*.

2. Let S be a collection of sets A, B, $\ldots$, each of which contains a finite number of objects. If we set $d(A, B) = $ number of objects in $(A \cup B) - (A \cap B)$ then S is a metric space. This is the *Silverman distance*.

3. Is the following a norm in R_2: $\|(x, y)\| = \max(3|x| + 2|y|, 2|x| + 3|y|)$?

4. In R_n, define $\|(a_1, \ldots, a_n)\| = \max\limits_{0 \leq x \leq 1} \left| \sum\limits_{k=0}^{n-1} a_{k+1}x^k \right|$. Is this a norm?

5. Does the following expression define a norm in $C^1[a, b]$:

$$\|f\| = \max_{a \leq x \leq b} [|f(x)|, |f'(x)|]?$$

6. Show that R_n can be normed as follows: Let $w_1, w_2, \ldots, w_n$ be a fixed sequence of positive constants. Set $\|x\| = \max_{1 \le j \le n} w_j |x_j|$.

7. Prove that the postulate (7.2.1)(a) follows from (7.2.1)(c) and (7.2.1)(d).

8. Describe the gauge curve that gives rise to the norm $\|(x, y)\| = 3|x| + 2|y|$ in R_2.

9. In the space $C[-1, 1]$, is the norm $\|f\| = \int_{-1}^{+1} |f(x)|\, dx$ strictly convex?

10. Let the ellipse $\dfrac{x^2}{a^2} + \dfrac{y^2}{b^2} = 1$ act as the gauge curve for a Minkowski plane. Find the equation of the "Minkowski circle" of radius r and center (x_0, y_0).

11. Let C be a Minkowski gauge curve no part of which is a line segment. Prove that the resulting norm is strictly convex.

12. If X is strictly convex, then $\|x\| = \|y\| = r$, $x \ne y$ implies

$$\|tx + (1 - t)y\| < r \quad \text{for} \quad 0 < t < 1.$$

Geometrically, the surface $\|x\| = r$ does not contain a line segment.

13. If a linear space X satisfies (7.2.1)(a), (c), (d) but if $\|x\| = 0$ does not necessarily mean that $x = 0$, then X is called a *seminormed space*. Show that $C^1[a, b]$ with $\|f\|^2 = \int_a^b |f'(x)|^2\, dx$ is a seminormed space.

14. Prove that Theorem 7.4.1 holds in a seminormed linear space. Formulate several concrete examples.

15. Let X be a linear space and let $\mathscr{L}$ be a family of linear functionals taken from X^*. Show that $\|x\| = \sup_{L \in \mathscr{L}} |L(x)|$ defines a seminorm on X. When is it a norm?

16. Show that the fundamental theorem 7.4.1 holds if the a_i are allowed to vary only over a preassigned closed set.

17. Interpret Hölder's inequality (7.2.7) for $p = 1$ in the light of Ex. 4, 7.2.

18. Prove Young's criterion: Let y be a fixed element of a normed linear space and let the variable element x be a function of n real or complex parameters $a_1, \ldots, a_n$ defined for $|a_i| < \infty$: $x = x(a_1, \ldots, a_n)$. Suppose that (a) $\|x(a_1, \ldots, a_n)\|$ is a continuous function of its parameters (b) $\|x(a_1, \ldots, a_n)\| \le M$ implies there is an N such that $|a_i| \le N$ $i = 1, 2, \ldots, n$. Then the problem $\min_{a_i} \|y - x(a_1, \ldots, a_n)\|$ has a solution.

19. Show that Young's criterion is not necessary for the existence of minimizing parameters.

20. Solve the problem $\max_{0 \le x \le 1} |e^x - ax - b| = \text{minimum}$.

21. Solve the problem $\int_0^1 |e^x - a|\, dx = \text{minimum}$.

22. Solve the problem $\min_a \max_{0 \le x \le 1} |x^4 - ax|$. Is the solution unique?

23. The problem of finding $\min_c \int_{-1}^{1} |x - cx^2|\, dx$ does not have a unique solution.

24. Let $\mathscr{P}_2$ be normed by $\|p\| = |p(0)| + |p(1)| + |p(2)|$. Determine the best approximation to x^2 by a constant.

25. Referring to Figure 7.6.7, compare the accuracy of the best uniform approximant to $\cos \dfrac{\pi}{2} x$ with that of the Taylor approximant.

26. Let $f(x) = 0$, $-1 \leq x \leq 0$, $f(x) = 1$, $0 < x \leq 1$. Compute

$$\min_{g(x) \in C[-1,1]} \ \max_{-1 \leq x \leq 1} \ |f(x) - g(x)|.$$

27. Given n points in the plane P_i: (x_i, y_i) $i = 1, 2, \ldots, n$. P_0 is one additional point. Show that there exists a straight line through P_0: $y = a_0 x + a_1$ such that $\max\limits_{1 \leq i \leq n} |y_i - (a_0 x_i + a_1)| = $ minimum. Generalize.

28. Given n points in the plane $P_1, \ldots, P_n$. Let $d(P_i, \ell)$ designate the perpendicular distance from P_i to a straight line ℓ. Show that there is a line ℓ such that $\max\limits_{1 \leq i \leq n} d(P_i, \ell) = $ minimum.

29. Let $f(x) \in C[a, b]$ and consist of two linear portions. Determine the best uniform linear approximant to $f(x)$ over $[a, b]$. Interpret geometrically.

30. Let $B[a, b]$ designate the space of functions that are bounded over $[a, b]$. The problem of finding $\min\limits_{p \in \mathscr{P}_n} \ \max\limits_{a \leq x \leq b} |f(x) - p(x)|$ has a solution, but it is not necessarily unique.

31. Discuss the problem of minimizing $\max\limits_{-1 \leq x \leq 1} |1 - f(x)|$, where the approximants f satisfy $f \in C[-1, 1]$ and $\displaystyle\int_{-1}^{+1} f(x)\,dx = 0$.

32. Let $f(x) \in C^2[a, b]$ and be concave. P is a variable point on the curve and $T(x)$ is linear between $(a, f(a))$ and P, and between P and $(b, f(b))$. Show that there is a position of P that minimizes $\displaystyle\int_a^b f(x) - T(x)\,dx$. Interpret the position geometrically.

33. Characterize S in Theorem 7.5.1 as the intersection of two convex sets.

34. The best uniform approximation to $\sqrt{1 + x^2}$ on $[0, 1]$ by a linear function is $.955 + .414\,x$.

35. Derive the approximation $\sqrt{x^2 + y^2} \approx .955\,x + .414\,y$ $x \geq y > 0$, and determine the error incurred.

36. Find the best uniform approximation to $\sqrt{1 + x^3}$ by a straight line over $[0, 1]$.

37. Let $w(x), f(x) \in C[-1, 1]$. Prove there is one and only one polynomial in $\mathscr{P}_n$ for which $\max\limits_{-1 \leq x \leq 1} w(x) |f(x) - p(x)| = $ minimum.

38. Let $p(x) = z^k + a_1 z^{k-1} + \cdots + a_n$. If Γ is the lemniscate $|p(z)| = a$, the Tschebyscheff polynomial of degree nk for Γ is $[p(z)]^n$.

39. Discuss the problem of finding $\min\limits_{a,b} \ \max\limits_{0 \leq x \leq 1} |x - a \sin bx|$.

40. A family of polynomials of bounded degree whose values on $[a, b]$ are bounded must have bounded coefficients.

CHAPTER VIII

Least Square Approximation

8.1 Inner Product Spaces. We come now to the approximation process most commonly employed and most highly developed: least squares. An abstract vantage point from which it is convenient to survey the common features of various least square approximations is provided by the theory of *inner product spaces*. If the subjects of algebra, geometry, and analysis can be said to have a "center of gravity," it surely lies in this theory.

DEFINITION 8.1.1. A real linear space X with elements x will be called an *inner product space*, if there has been defined for each two elements x_1, x_2 a real number designated by (x_1, x_2) with the following properties

(a) $\quad (x_1 + x_2, x_3) = (x_1, x_3) + (x_2, x_3)$ $\qquad$ (Linearity)

(b) $\qquad (x_1, x_2) = (x_2, x_1)$ $\qquad\qquad$ (Symmetry)

(c) $\qquad (\alpha x_1, x_2) = \alpha(x_1, x_2), \quad \alpha$ real $\qquad$ (Homogeneity) $\quad$ (8.1.1)

(d) $\quad (x, x) \geq 0, \quad (x, x) = 0$ if and only if $x = 0$ $\quad$ (Positivity)

The quantity (x_1, x_2) is called the *inner product* of x_1 and x_2.

A similar definition can be made for complex linear spaces. The inner product (x_1, x_2) will be a complex number and (8.1.1)(b) must be replaced by the condition

$$\text{(b')} \quad (x_1, x_2) = \overline{(x_2, x_1)} \quad \text{(Hermitian Symmetry)} \qquad (8.1.1)$$

The bar in the above line designates the complex conjugate.

Ex. 1. $X = R_n$. $x = (x_1, x_2, \ldots, x_n), y = (y_1, y_2, \ldots, y_n)$. Let w_i be n fixed positive numbers. Define $(x, y) = \sum\limits_{i=1}^{n} w_i x_i y_i$.

Ex. 2. X is the complex Euclidean space, C_n with elements

$$x = (x_1, x_2, \ldots, x_n), x_i \text{ complex.}$$

Define $\qquad\qquad\qquad (x, y) = \sum\limits_{i=1}^{n} w_i x_i \overline{y_i}.$

Ex. 3. $X = C[a, b]$. If $x = f(t)$, $y = g(t)$, we define

$$(x, y) = (f, g) = \int_a^b f(t)\overline{g(t)}\, dt.$$

Ex. 4.   Let B designate a bounded two dimensional region. Let X designate the space of functions $f(x, y)$ that are continuous in the closure of B. $w(x, y)$ is a fixed positive continuous weight function. $(f, g) = \iint_B f(x, y)\overline{g(x, y)}w(x, y)\, dx\, dy$.

Ex. 5.   Let B be a bounded simply connected region of the complex plane with a simple, rectifiable boundary C. Let X be the complex linear space composed of all functions analytic in B and continuous in $B \cup C$. The line integral $(f, g) = \int_C f(z)\overline{g(z)}\, ds$, $ds^2 = dx^2 + dy^2$, is an inner product for X.

Ex. 6.   Let $X = L^2[a, b]$, and write $(f, g) = \int_a^b f(x)\overline{g(x)}\, dx$. (Cf. 7.2, Ex. 10, with $p = 2$.)

Note several simple consequences of (8.1.1). From (c), $(0, x) = 0$. From (c) and (b'), $(x_1, \alpha x_2) = \bar{\alpha}(x_1, x_2)$. From (a) and (b) or (b'), $(x_1, x_2 + x_3) = (x_1, x_2) + (x_1, x_3)$.

THEOREM 8.1.1 (The Schwarz Inequality).
In an inner product space,

$$|(x_1, x_2)|^2 \le (x_1, x_1)(x_2, x_2). \tag{8.1.2}$$

The equality sign holds if and only if x_1 and x_2 are dependent.

Proof: If $x_2 = 0$, the theorem reduces to the trivial inequality $0 \le 0$. Assume then that $x_2 \neq 0$. Let λ be an arbitrary complex number. We have from (8.1.1)(d),

$$(x_1 + \lambda x_2, x_1 + \lambda x_2) \ge 0.$$

This is equivalent to

$$(x_1, x_1) + \lambda(x_2, x_1) + \bar{\lambda}(x_1, x_2) + \lambda\bar{\lambda}(x_2, x_2) \ge 0.$$

This is true, in particular, for the number

$$\lambda = -(x_1, x_2)/(x_2, x_2).$$

Hence,

$$(x_1, x_1) - \frac{(x_1, x_2)(x_2, x_1)}{(x_2, x_2)} - \frac{(x_1, x_2)(x_2, x_1)}{(x_2, x_2)}$$

$$+ \frac{(x_1, x_2)(x_2, x_1)}{(x_2, x_2)^2} \cdot (x_2, x_2) \ge 0.$$

Then, $$(x_1, x_1) - \frac{(x_1, x_2)(x_2, x_1)}{(x_2, x_2)} \geq 0.$$

Therefore

$$(x_1, x_2)(x_2, x_1) \leq (x_1, x_1)(x_2, x_2),$$

or by (8.1.1)(b'), $$|(x_1, x_2)|^2 \leq (x_1, x_1)(x_2, x_2).$$

Suppose that equality holds. The case $x_2 = 0$ is trivial, so take $x_2 \neq 0$. By the above work we have

$$(x_1 + \lambda x_2, x_1 + \lambda x_2) = 0 \quad \text{with} \quad \lambda = -(x_1, x_2)/(x_2, x_2).$$

Hence by (8.1.1)(d), $$x_1 + \lambda x_2 = 0 \quad \text{and} \quad x_1 = \frac{(x_1, x_2)}{(x_2, x_2)} x_2.$$

Conversely if $x_1 = \alpha x_2$, then $|(x_1, x_2)|^2 = |\alpha|^2 (x_2, x_2)^2 = (x_1, x_1)(x_2, x_2)$.

THEOREM 8.1.2.   *If X is an inner product space, the equation*

$$\|x\| = \sqrt{(x, x)} \tag{8.1.3}$$

defines a norm in X, and X becomes a normed linear space.

Proof: The quantity $\sqrt{(x, x)}$ satisfies the requirements for a norm given in (7.2.1). The only requirement that is not immediately evident is the triangle inequality

$$\|x + y\| \leq \|x\| + \|y\|. \tag{8.1.4}$$

This is equivalent to

$$\|x + y\|^2 \leq \|x\|^2 + 2 \|x\| \|y\| + \|y\|^2$$

or $$(x + y, x + y) \leq (x, x) + 2\sqrt{(x, x)}\sqrt{(y, y)} + (y, y).$$

Since $$(x + y, x + y) = (x, x) + (y, x) + (x, y) + (y, y),$$

we must show $$(x, y) + (y, x) \leq 2\sqrt{(x, x)}\sqrt{(y, y)}.$$

But $$|(x, y) + (y, x)| \leq |(x, y)| + |(y, x)| \leq 2 |(x, y)|.$$

Hence, it suffices to show $|(x, y)| \leq \sqrt{(x, x)}\sqrt{(y, y)}$. But this is precisely the Schwarz Inequality.

In view of Theorem 8.1.2, we can make every inner product space into a normed linear space in a natural and automatic way.

EX. 7.   For ℓ^2, the Schwarz inequality (8.1.2) coincides with the Hölder inequality for infinite sequences. (Cf. (7.2.7) with $p = 2$.) The triangle inequality (8.1.4) coincides with the Minkowski inequality (7.2.16).

EX. 8.   A similar observation holds for $L^2[a, b]$.

THEOREM 8.1.3 (The Parallelogram Theorem). *For any elements x and y in an inner product space X we have*

$$\|x + y\|^2 + \|x - y\|^2 = 2 \|x\|^2 + 2 \|y\|^2. \tag{8.1.5}$$

Proof: Replace the norms by inner products and expand.

8.2 Angle Geometry for Inner Product Spaces. For two nonzero elements in a *real* inner product space, we have, from (8.1.2),

$$-1 \le \frac{(x_1, x_2)}{\|x_1\| \, \|x_2\|} \le 1.$$

There is consequently a unique value of θ in the range $0 \le \theta \le \pi$ that satisfies $\cos \theta = \dfrac{(x_1, x_2)}{\|x_1\| \, \|x_2\|}$.

DEFINITION 8.2.1. The angle θ between elements x_1, x_2 in a real inner product space is defined by

$$\cos \theta = \frac{(x_1, x_2)}{\|x_1\| \, \|x_2\|} \qquad 0 \le \theta \le \pi. \tag{8.2.1}$$

The justification of this definition lies in the fact that it extends the usual formulas of Euclidean geometry.

Ex. 1.   Let $X = R_3$. For two elements $x = (x_1, x_2, x_3)$, $y = (y_1, y_2, y_3)$, use the inner product $(x, y) = \sum\limits_{i=1}^{3} x_i y_i$. This leads to the norm (or distance from the origin) $\|x\|^2 = x_1{}^2 + x_2{}^2 + x_3{}^2$. Then

$$\cos \theta = \frac{x_1 y_1 + x_2 y_2 + x_3 y_3}{(x_1{}^2 + x_2{}^2 + x_3{}^2)^{\frac{1}{2}}(y_1{}^2 + y_2{}^2 + y_3{}^2)^{\frac{1}{2}}} .$$

This is familiar from analytic geometry.

In the case of a complex inner product space, the definition

$$\cos \theta = \frac{|(x, y)|}{\|x\| \, \|y\|} \qquad 0 \le \theta \le \pi, \tag{8.2.2}$$

is frequently employed, though this is not completely satisfactory. (See Problem 13.)

Two special cases are particularly noteworthy.

A. $\theta = 0$. In this case, $\cos \theta = 1$ and $|(x, y)| = \|x\| \, \|y\|$. According to Theorem 8.1.1, the elements x_1 and x_2 are dependent: $\alpha x_1 = \beta x_2$. We may say that x_1 and x_2 are *parallel*.

B. $\theta = \pi/2$. In this case, the elements are *perpendicular* or *orthogonal*. Since $\cos \theta = 0$, $|(x, y)|/\|x\| \, \|y\| = 0$, and this implies that $(x, y) = 0$. We sometimes write $x \perp y$ to express orthogonality and make the following definition.

DEFINITION 8.2.2. $x \perp y$ if and only if $(x, y) = 0$.

Ex. 2.   0 is the only self orthogonal element.

Ex. 3. $x \perp y$ implies $y \perp x$.

Ex. 4. $y \perp x_1, x_2, \ldots, x_n$ implies $y \perp a_1 x_1 + a_2 x_2 + \cdots + a_n x_n$.

Ex. 5.   Pythagoras' Theorem. $x \perp y$ implies

$$\|x + y\|^2 = \|x\|^2 + \|y\|^2$$

Ex. 6.   The Law of Cosines. In a real inner product space,

$$\|x + y\|^2 = \|x\|^2 + \|y\|^2 + 2 \, \|x\| \, \|y\| \cos \theta.$$

The inner product has a geometric interpretation as a projection. This is suggested by the accompanying two dimensional figure.

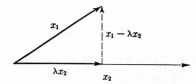

Figure 8.2.1.

Let x_1 and x_2 be nonzero elements. Select a scalar λ so that λx_2 is the projection of x_1 on x_2. Then

$$\lambda x_2 \perp x_1 - \lambda x_2.$$

This means

$$(\lambda x_2, x_1 - \lambda x_2) = 0.$$

Therefore $\lambda(x_2, x_1) - \lambda \bar{\lambda} \, (x_2, x_2) = 0$ and $\lambda = \dfrac{(x_1, x_2)}{(x_2, x_2)}$. This means that if $x_2 \neq 0$,

$$\text{projection of } x_1 \text{ on } x_2 = \frac{(x_1, x_2)}{(x_2, x_2)} \, x_2. \tag{8.2.3}$$

This equation serves to define *projection* in the abstract case.

When the element x_2 has unit length: $\|x_2\|^2 = (x_2, x_2) = 1$, then

$$\text{projection of } x_1 \text{ on } x_2 = (x_1, x_2)x_2. \tag{8.2.4}$$

8.3 Orthonormal Systems

DEFINITION 8.3.1. A set S of elements of an inner product space is called *orthonormal* if

$$(x, y) = \begin{cases} 0 & x \neq y \\ 1 & x = y \end{cases} \quad x, y \in S. \tag{8.3.1}$$

If we have merely

$$(x, y) = 0 \quad \text{for} \quad x \neq y, \tag{8.3.2}$$

the set is called *orthogonal*.

Ex. 1. In R_n with $(x, y) = \sum_{i=1}^{n} x_i y_i$, the unit vectors

$$(1, 0, \ldots, 0), (0, 1, 0, \ldots, 0), \ldots, (0, 0, \ldots, 1)$$

form an orthonormal set.

Ex. 2. Let the $n \times n$ matrix Q be orthogonal, i.e., $QQ' = I$ where Q' is the transpose of Q and I is the unit matrix. The rows (or columns) of Q form an orthonormal set in the space of Ex. 1.

Ex. 3. In $C[-\pi, \pi]$ or $L^2[-\pi, \pi]$, with $(f, g) = \int_{-\pi}^{\pi} f(x)g(x)\, dx$, the functions $(2\pi)^{-\frac{1}{2}}$, $\pi^{-\frac{1}{2}} \cos x$, $\pi^{-\frac{1}{2}} \sin x$, $\pi^{-\frac{1}{2}} \cos 2x$, $\pi^{-\frac{1}{2}} \sin 2x, \ldots$, form an orthonormal set.

Ex. 4. In $C[-1, 1]$ with $(f, g) = \int_{-1}^{1} \frac{f(x)g(x)\, dx}{\sqrt{1 - x^2}}$, the polynomials $(\pi)^{-\frac{1}{2}}T_0$, $\left(\frac{\pi}{2}\right)^{-\frac{1}{2}} T_n(x)$, $n = 1, 2, \ldots$, $T_n(x) = \cos (n \text{ arc cos } x)$, form an orthonormal set. These are the Tschebyscheff Polynomials (cf. Def. 3.3.1).

Ex. 5. In $C[-1, 1]$ or $L^2[-1, 1]$, with $(f, g) = \int_{-1}^{+1} \sqrt{1 - x^2} f(x)g(x)\, dx$ the functions $U_m(x) = \dfrac{\sin [(m + 1) \text{ arc cos } x]}{\sqrt{1 - x^2}}$ $m = 0, 1, 2, \ldots$, form an orthogonal set. For

$$\int_{-1}^{1} \sqrt{1 - x^2}\, U_m(x)U_n(x)\, dx = \int_{0}^{\pi} \sin (m + 1)\theta \sin (n + 1)\theta\, d\theta \begin{array}{l} = 0\ , \ m \neq n. \\[2mm] = \dfrac{\pi}{2}\ , \ m = n. \end{array}$$

The functions $\sqrt{\dfrac{2}{\pi}}\, U_n(x)$ are orthonormal. Actually, $U_m(x)$ is a polynomial of degree m in $[-1, 1]$; for, $\cos n\theta = T_n(\cos \theta)$. Now

$$\sin (n + 1)\theta = \sin n\theta \cos \theta + \cos n\theta \sin \theta = \sin n\theta \cos \theta + T_n(\cos \theta) \sin \theta.$$

Hence,

$$\frac{\sin (n + 1)\theta}{\sin \theta} = \frac{\sin n\theta}{\sin \theta} \cos \theta + T_n(\cos \theta).$$

By induction, therefore, $\dfrac{\sin (n + 1)\theta}{\sin \theta}$ is a polynomial of degree n in $\cos \theta$. $U_m(x)$ are the *Tschebyscheff polynomials of the second kind.*

Ex. 6.   In $C[a, b]$ with $(f, g) = \displaystyle\int_a^b A(x)f(x)g(x)\, dx$, $(A(x) \in C[a, b]$ and $\geq 0)$, the eigenfunctions of the self-adjoint differential problem

$$\begin{cases} y'' + \lambda A(x)y = 0 \\ y(a) = y(b) = 0 \end{cases}$$

corresponding to distinct eigenvalues are orthogonal.

Let y_j and y_k be two solutions of this problem corresponding to distinct values λ_j, λ_k. Then

$$(\lambda_k - \lambda_j)A(x)y_k(x)y_j(x) = y_jy_k'' - y_ky_j'' = \frac{d}{dx}[y_jy_k' - y_ky_j'].$$

Hence, $(\lambda_k - \lambda_j)\displaystyle\int_a^b A(x)y_k(x)y_j(x)\, dx = [y_jy_k' - y_ky_j']_a^b = 0.$

THEOREM 8.3.1 (Pythagorean Theorem).   *If* $x_1, \ldots, x_n$ *are orthogonal then*

$$\|x_1 + x_2 + \cdots + x_n\|^2 = \|x_1\|^2 + \|x_2\|^2 + \cdots + \|x_n\|^2. \qquad (8.3.3)$$

Proof: The cross terms in the expanded inner product vanish.

THEOREM 8.3.2.   *Any finite set of nonzero orthogonal elements* $x_1, x_2, \ldots,$ x_n *is linearly independent.*

Proof: Suppose $a_1x_1 + a_2x_2 + \cdots + a_nx_n = 0$ where the a's are not all 0. Then,

$$0 = (0, x_k) = (a_1x_1 + \cdots + a_nx_n, x_k) = a_k(x_k, x_k).$$

This implies that $a_k = 0$, $k = 1, 2, \ldots, n$, a contradiction.

The previous theorem has a partial converse which is of great importance. An independent set, of course, is not necessarily orthogonal, but it can be *orthogonalized.* That is, we can find linear combinations which are orthogonal. From among the many proofs, we select one that leads to the *Gram-Schmidt orthonormalizing process.*

THEOREM 8.3.3.   *Let* $x_1, x_2, \ldots$, *be a finite or infinite sequence of elements such that any finite number of elements* $x_1, x_2, \ldots, x_k$ *are linearly independent. Then, we can find constants*

$$a_{11}$$
$$a_{21} \quad a_{22}$$
$$a_{31} \quad a_{32} \quad a_{33}$$
$$.$$
$$.$$
$$.$$

such that the elements

$$x_1{}^* = a_{11}x_1$$
$$x_2{}^* = a_{21}x_1 + a_{22}x_2$$
$$x_3{}^* = a_{31}x_1 + a_{32}x_2 + a_{33}x_3 \qquad (8.3.4)$$
$$.$$
$$.$$
$$.$$

are orthonormal:

$$(x_i{}^*, x_j{}^*) = \delta_{ij} \qquad i, j = 1, 2, \ldots . \qquad (8.3.5)$$

Proof: Set, recursively,

$$y_1 = x_1 \quad \text{and} \quad x_1{}^* = y_1/\|y_1\|$$
$$y_2 = x_2 - (x_2, x_1{}^*)x_1{}^* \quad \text{and} \quad x_2{}^* = y_2/\|y_2\|$$
$$.$$
$$.$$

$$(8.3.6)$$

$$y_{n+1} = x_{n+1} - \sum_{k=1}^{n} (x_{n+1}, x_k{}^*)x_k{}^* \quad \text{and} \quad x_{n+1}^* = y_{n+1}/\|y_{n+1}\|$$
$$.$$
$$.$$

It is clear from the structure of this recursion that y_{n+1}, and hence, x_{n+1}, is a linear combination of $x_1, x_2, \ldots, x_{n+1}$. $\|y_i\|$ cannot vanish inasmuch as this would imply that $y_i = 0$. But $y_i = x_i + $ "lower" x's, and this would contradict the assumption that the x's are independent. The $x_i{}^*$ are normal:

$$(x_i{}^*, x_i{}^*) = \left(\frac{y_i}{\|y_i\|}, \frac{y_i}{\|y_i\|} \right) = \frac{1}{\|y_i\|^2} (y_i, y_i) = 1.$$

Finally we must prove that x_{n+1}^* (or y_{n+1}) is orthogonal to $x_n^*, x_{n-1}^*, \ldots,$ x_1^*. A simple computation verifies that $(y_2, x_1^*) = 0$. Assume that for $i \leq n, j < i$ we have proved $(y_i, x_j^*) = 0$. Then for $j \leq n$,

$$(y_{n+1}, x_j^*) = (x_{n+1} - \sum_{k=1}^{n} (x_{n+1}, x_k^*) x_k^*, x_j^*)$$

$$= (x_{n+1}, x_j^*) - \sum_{k=1}^{n} (x_{n+1}, x_k^*)(x_k^*, x_j^*)$$

$$= (x_{n+1}, x_j^*) - (x_{n+1}, x_j^*) = 0.$$

COROLLARY 8.3.4.   *The "leading coefficients" a_{ii} are positive.*
    For $a_{ii} = (\|y_i\|)^{-1}$.

COROLLARY 8.3.5.   *We can find constants*

$$b_{11}$$
$$b_{21} \quad b_{22}$$
$$b_{31} \quad b_{32} \quad b_{33}$$
$$\cdot$$
$$\cdot$$
$$\cdot$$

with $b_{ii} > 0$
such that

$$x_1 = b_{11} x_1^*$$
$$x_2 = b_{21} x_1^* + b_{22} x_2^*$$
$$\cdot$$
$$\cdot \qquad\qquad\qquad\qquad\qquad (8.3.7)$$
$$\cdot$$

$$x_n = b_{n1} x_1^* + b_{n2} x_2^* + \cdots + b_{nn} x_n^*.$$

Proof: $\qquad\qquad x_1 = \dfrac{1}{a_{11}} x_1^* \qquad$ so that $b_{11} = \dfrac{1}{a_{11}}$.

$$x_2 = \frac{-a_{21}}{a_{22}} x_1 + \frac{1}{a_{22}} x_2^*$$

$$= -\frac{a_{21}}{a_{22}a_{11}} x_1^* + \frac{1}{a_{22}} x_2^*$$

so that $\qquad\qquad b_{21} = \dfrac{-a_{21}}{a_{22}a_{11}}, \quad b_{22} = \dfrac{1}{a_{22}}$.

It is clear that we may proceed step by step in this way, since $a_{jj} > 0$, $j = 1, 2, \ldots$. Note that $b_{ii} = \dfrac{1}{a_{ii}} = \|y_i\| > 0$.

COROLLARY 8.3.6. $x_n{}^* \perp x_1, x_n{}^* \perp x_2, \ldots, x_n{}^* \perp x_{n-1}$.

Proof: $x_k = b_{k1}x_1{}^* + b_{k2}x_2{}^* + \cdots + b_{kk}x_k{}^*$ so that

$$(x_n{}^*, x_k) = \left(x_n{}^*, \sum_{i=1}^{k} b_{ki}x_i{}^*\right) = \sum_{i=1}^{k} b_{ki}(x_n{}^*, x_i{}^*) = 0 \quad \text{if} \quad k < n.$$

Note: In the subsequent portions of this book, we shall use the asterisk * on symbols of elements to designate orthonormal elements, and on symbols of spaces to designate conjugate spaces.

The following observation should be made. If $x_1, \ldots, x_n$ and $x_1{}^*, \ldots, x_n{}^*$ are related by (8.3.4), and we require that the latter are orthonormal and $a_{kk} > 0$, $k = 1, 2, \ldots, n$, then the constants a_{ij} are determined uniquely. (Prove!) The Gram-Schmidt process is merely one scheme for determining them.

On the other hand, if we allow

$$x_k{}^* = \alpha_{k1}x_1 + \alpha_{k2}x_2 + \cdots + \alpha_{kn}x_n, \qquad k = 1, 2, \ldots, n,$$

then there is much more freedom in the choice of the constants α_{ij}. Whenever we speak of orthogonalizing a sequence of elements $x_1, \ldots, x_n$, the reader should decide whether the statements made hold for the triangular scheme (8.3.4) only, or whether they are valid for the more general scheme.

Ex. 7.   The powers $1, x, x^2, \ldots$, are independent in $C[a, b]$. For, if

$$a_0 + a_1x + \cdots + a_nx^n \equiv 0, \qquad a \leq x \leq b,$$

then $a_i = 0$, $i = 0, 1, \ldots, n$. If $w(x)$ is a fixed positive, integrable function defined on $[a, b]$ then the integral

$$(f, g) = \int_a^b w(x)f(x)g(x)\,dx \tag{8.3.8}$$

forms an inner product in $C[a, b]$. The powers may therefore be orthogonalized with respect to this inner product and we obtain a set of polynomials

$$p_n{}^*(x) = k_nx^n + \cdots, \qquad n = 0, 1, 2, \ldots, \qquad k_n > 0$$

which are orthonormal in the sense that

$$\int_a^b w(x)p_m{}^*(x)p_n{}^*(x)\,dx = \delta_{mn}, \qquad m, n = 0, 1, \ldots. \tag{8.3.9}$$

In the case of a semi-infinite or an infinite interval $[a, b]$, we must assume that the weight factor $w(x)$ is such that the integrals $\int_a^b w(x)x^n\,dx$, $n = 0, 1, \ldots$, all exist.

The following special selections of $[a, b]$ and $w(x)$ have been studied extensively, and the resulting orthonormal polynomials constitute the "classical" orthonormal polynomials.

1. $a = -1$ $b = 1$ $w(x) = 1$ Legendre Polynomials
2. $a = -1$ $b = 1$ $w(x) = (1 - x^2)^{-\frac{1}{2}}$ Tschebyscheff Polynomials
                                                         (of the First Kind)
3. $a = -1$ $b = 1$ $w(x) = (1 - x^2)^{\frac{1}{2}}$ Tschebyscheff Polynomials
                                                         of the Second Kind.
4. $a = -1$ $b = 1$ $w(x) = (1 - x)^\alpha (1 + x)^\beta$;
                          $\alpha, \beta > -1$ Jacobi Polynomials
5. $a = 0$ $b = \infty$ $w(x) = x^\alpha e^{-x}, \alpha > -1$,   Laguerre Polynomials
6. $a = -\infty$ $b = \infty$ $w(x) = e^{-x^2}$ Hermite Polynomials

Ex. 8.   We compute the first 3 Legendre polynomials using the scheme of Theorem 8.3.3.

$$x_1 = 1, x_2 = t, x_3 = t^2 \, . \quad y_1 = 1 \, . \quad \|y_1\| = \left(\int_{-1}^{+1} dt \right)^{\frac{1}{2}} = \sqrt{2}.$$

$$x_1{}^* = \frac{1}{\sqrt{2}} \, . \quad y_2 = t - \left(t, \frac{1}{\sqrt{2}} \right) \frac{1}{\sqrt{2}} \, . \quad \left(t, \frac{1}{\sqrt{2}} \right) = \frac{1}{\sqrt{2}} \int_{-1}^{+1} t \, dt = 0$$

$$y_2 = t . \quad \|y_2\| = \left(\int_{-1}^{+1} t^2 \, dt \right)^{\frac{1}{2}} = \sqrt{\frac{2}{3}} \, . \quad x_2{}^* = \sqrt{\frac{3}{2}} \, t.$$

$$y_3 = t^2 - \left(t^2, \sqrt{\frac{3}{2}} \, t \right) \sqrt{\frac{3}{2}} \, t - \left(t^2, \frac{1}{\sqrt{2}} \right) \frac{1}{\sqrt{2}} = t^2 - \frac{1}{3} \, .$$

$$\|y_3\| = \left(\int_{-1}^{+1} \left(t^2 - \frac{1}{3} \right)^2 dt \right)^{\frac{1}{2}} = \frac{2}{15} \sqrt{10} \quad x_3{}^* = \frac{3}{4} \sqrt{10} \left(t^2 - \frac{1}{3} \right).$$

Though the Gram-Schmidt process may be employed, the Legendre polynomials of higher degree are more expeditiously computed via recurrence. (See Chapter X.)

Ex. 9.   Let $a \leq x_1 < x_2 < \cdots < x_{n+1} \leq b$ be $n + 1$ distinct points and $w_1, w_2, \ldots, w_{n+1}$ be $n + 1$ positive weights. The expression

$$(f, g) = \sum_{i=1}^{n+1} w_i f(x_i) g(x_i) \tag{8.3.10}$$

is an inner product for $\mathscr{P}_n$ (but not for $\mathscr{P}_m, m > n$, or for $C[a, b]$). We may therefore orthonormalize the powers and arrive at a set of polynomials

$$p_0{}^*(x), p_1{}^*(x), \ldots, p_n{}^*(x)$$

for which

$$\sum_{k=1}^{n+1} w_k p_i{}^*(x_k) p_j(x_k) = \delta_{ij} \qquad 0 \leq i, j \leq n. \tag{8.3.11}$$

These orthonormal polynomials are important in least square approximations on discrete sets of points.

Ex. 10. Consider the situation of Ex. 5 of 8.1. The complex powers $1, z,$ $z^2, \ldots,$ are independent elements of X. Hence, they may be orthonormalized with respect to the inner product $(f, g) = \int_C f(z)\overline{g(z)} \, ds$ to arrive at a set of polynomials $p_0{}^*(z), p_1{}^*(z), \ldots,$ for which

$$\int_C p_n{}^*(z)\overline{p_m{}^*(z)} \, ds = \delta_{mn} \qquad 0 \leq m, n < \infty.$$

8.4 Fourier (or Orthogonal) Expansions

DEFINITION 8.4.1. Let $x_1{}^*, x_2{}^*, \ldots,$ be a finite or infinite sequence of orthonormal elements. Let y be an arbitrary element. The series $\sum_{n=1}^{\infty} (y, x_n{}^*)x_n{}^*$ is the *Fourier series* for y. (If the sequence is finite we use a finite sum.) The constants $(y, x_n{}^*)$ are known as the *Fourier coefficients of y.*

One frequently writes

$$y \sim \sum_{n=1}^{\infty} (y, x_n{}^*)x_n{}^* \tag{8.4.1}$$

to indicate that the right-hand sum is associated in a formal way with the left-hand side. The relation between an element and its Fourier series has been the object of vast investigations and theories.

In view of (8.2.4) we may write (8.4.1) in the form

$$y \sim \sum_{n=1}^{\infty} (\text{Projection of } y \text{ on } x_n{}^*) \tag{8.4.2}$$

and hence the Fourier series of an element is merely the sum of the projections of the element on a system of orthonormal elements.

If $x_1, x_2, \ldots, \neq 0$ are orthogonal, but not necessarily normal, then

$$x_k{}^* = x_k/\|x_k\| \qquad k = 1, 2, \ldots, \tag{8.4.3}$$

are orthonormal so that (8.4.1) becomes

$$y \sim \sum_{n=1}^{\infty} \left(y, \frac{x_k}{\|x_k\|}\right) \frac{x_k}{\|x_k\|} = \sum_{k=1}^{\infty} \frac{(y, x_k)}{(x_k, x_k)} x_k. \tag{8.4.4}$$

Again, by (8.2.3) this may be interpreted as

$$y \sim \sum_{n=1}^{\infty} (\text{Projection of } y \text{ on } x_n). \tag{8.4.5}$$

Ex. 1.   In R_3 with $(x, y) = \sum\limits_{i=1}^{3} x_i y_i$, select

$$x_1{}^* = (1, 0, 0), \quad x_2{}^* = (0, 1, 0), \quad x_3{}^* = (0, 0, 1).$$

For a given $y = (a, b, c)$ we have $(y, x_1{}^*) = a$, $(y, x_2{}^*) = b$, $(y, x_3{}^*) = c$. The summation $(a, b, c) = a(1, 0, 0) + b(0, 1, 0) + c(0, 0, 1)$ is the Fourier expansion of y.

Ex. 2.   Take $C[-\pi, \pi]$ or $L^2[-\pi, \pi]$ with $(f, g) = \int_{-\pi}^{\pi} f(x)g(x)\, dx$. Orthonormal system: $(2\pi)^{-\frac{1}{2}}$, $(\pi)^{-\frac{1}{2}} \sin x$, $(\pi)^{-\frac{1}{2}} \cos x$, $(\pi)^{-\frac{1}{2}} \sin 2x$, ...

$$f(x) \sim \frac{a_0}{2} + \sum_{k=1}^{\infty} a_k \cos kx + b_k \sin kx, \quad a_k = \frac{1}{\pi} \int_{-\pi}^{\pi} f(x) \cos kx\, dx$$

$$b_k = \frac{1}{\pi} \int_{-\pi}^{\pi} f(x) \sin kx\, dx. \quad (8.4.6)$$

This is *the* Fourier Series.

Ex. 3. $C[-1, 1]$. $(f, g) = \int_{-1}^{1} \frac{f(x)g(x)\, dx}{\sqrt{1 - x^2}}$. Orthonormal system

$$(\pi)^{-\frac{1}{2}} T_0(x), \quad \left(\frac{\pi}{2}\right)^{-\frac{1}{2}} T_1(x), \quad \left(\frac{\pi}{2}\right)^{-\frac{1}{2}} T_2(x), \ldots,$$

$$f(x) \sim \frac{a_0}{2} + \sum_{k=1}^{\infty} a_k T_k(x), \quad a_k = \frac{2}{\pi} \int_{-1}^{+1} \frac{f(x) T_k(x)\, dx}{\sqrt{1 - x^2}} . \quad (8.4.7)$$

This is the Tschebyscheff-Fourier Series.

In the simple case of finite dimensional spaces, the Fourier expansion of an element coincides with the element. More precisely, the following theorem holds.

THEOREM 8.4.1.   *Let $x_1, \ldots, x_n$ be independent and let $x_i{}^*$ be the x's ortho-normalized. If $w = a_1 x_1 + \cdots + a_n x_n$, then*

$$w = \sum_{k=1}^{n} (w, x_k{}^*) x_k{}^*. \quad (8.4.8)$$

Proof: From Corollary 8.3.5, we have,

$$w = a_1(b_{11}x_1{}^*) + a_2(b_{21}x_1{}^* + b_{22}x_2{}^*) + \cdots + a_n(b_{n1}x_1{}^* + \cdots + b_{nn}x_n{}^*)$$
$$= c_1 x_1{}^* + c_2 x_2{}^* + \cdots + c_n x_n{}^*.$$

Now, for $1 \leq k \leq n$,

$$(w, x_k{}^*) = (c_1 x_1{}^* + \cdots + c_n x_n{}^*, x_k{}^*)$$
$$= c_1(x_1{}^*, x_k{}^*) + \cdots + c_k(x_k{}^*, x_k{}^*) + \cdots + c_n(x_n{}^*, x_k{}^*) = c_k,$$

and (8.4.8) follows.

Ex. 4.   If $p_n{}^*(x) = \sum_{j=0}^{n} k_{nj}x^j$, $k_{nn} \neq 0$, are polynomials that are orthonormal with respect to the inner product (f, g) then

$$p(x) = \sum_{k=0}^{n} (p, p_k{}^*)p_k{}^*(x) \tag{8.4.9}$$

for all $p \in \mathscr{P}_n$.

Ex. 5.   Let $x_1, \ldots, x_n$ ($\neq 0$) be an orthogonal set in R_n or C_n. Any element in the space is equal to the sum of its projections on $x_1, \ldots, x_n$.

8.5   Minimum Properties of Fourier Expansions.   Truncated Fourier expansions have the following minimum property.

THEOREM 8.5.1.   *Let $x_1{}^*, x_2{}^*, \ldots,$ be an orthonormal system and let y be arbitrary. Then,*

$$\left\| y - \sum_{i=1}^{N} (y, x_i{}^*)x_i{}^* \right\| \leq \left\| y - \sum_{i=1}^{N} a_i x_i{}^* \right\| \tag{8.5.1}$$

for any selection of constants $a_1, a_2, \ldots, a_N$.

Proof:

$$\left\| y - \sum_{i=1}^{N} a_i x_i{}^* \right\|^2 = \left(y - \sum_{i=1}^{N} a_i x_i{}^*, y - \sum_{i=1}^{N} a_i x_i{}^* \right)$$

$$= (y, y) - \sum_{i=1}^{N} a_i(x_i{}^*, y) - \sum_{i=1}^{N} \bar{a}_i(y, x_i{}^*) + \sum_{i,j=1}^{N} a_i \bar{a}_j(x_i{}^*, x_j{}^*)$$

$$= (y, y) - \sum_{i=1}^{N} a_i(x_i{}^*, y) - \sum_{i=1}^{N} \bar{a}_i(y, x_i{}^*) + \sum_{i=1}^{N} |a_i|^2$$

$$+ \sum_{i=1}^{N} (x_i{}^*, y)(y, x_i{}^*) - \sum_{i=1}^{N} (x_i{}^*, y)(y, x_i{}^*)$$

$$= (y, y) - \sum_{i=1}^{N} |(y, x_i{}^*)|^2 + \sum_{i=1}^{N} |a_i - (y, x_i{}^*)|^2.$$

Since the first two terms of the last member are independent of the a's, it is clear that the minimum of $\left\| y - \sum_{i=1}^{N} a_i x_i{}^* \right\|^2$ is achieved when and only when

$$a_i = (y, x_i{}^*) \qquad i = 1, 2, \ldots, N; \tag{8.5.2}$$

i.e., when the a's are the Fourier coefficients of y.

Least square problems of numerical analysis can be formulated in terms of finding $\min_{a_i} \left\| y - \sum_{i=1}^{N} a_i x_i \right\|$ in an appropriate inner product space. (Cf., e.g., Ex. 3.) The next corollary gives the solution to such problems.

COROLLARY 8.5.2. *Let* $x_1, \ldots, x_N$ *be an independent set of elements. The problem of finding that linear combination of* $x_1, \ldots, x_N$ *which minimizes*

$$\left\| y - \sum_{i=1}^{N} a_i x_i \right\| \text{ is solved by } \sum_{i=1}^{N} (y, x_i^*) x_i^*.$$

The x_i^*'s are orthonormalized x's. The solution is unique. This tells us that every least square problem is solved by an appropriate truncated Fourier series.

COROLLARY 8.5.3. $$\min_{a_i} \left\| y - \sum_{i=1}^{N} a_i x_i \right\|^2 = \|y\|^2 - \sum_{i=1}^{N} |(y, x_i^*)|^2.$$

Proof: Insert $a_i = (y, x_i^*)$ in the last equality of the proof of Theorem 8.5.1.

Since this minimum value is ≥ 0 we have

COROLLARY 8.5.4 (Bessel Inequality). *If* x_i^* *are orthonormal, then*

$$\sum_{i=1}^{N} |(y, x_i^*)|^2 \leq \|y\|^2. \tag{8.5.3}$$

COROLLARY 8.5.5. *If* x_i^* *are an infinite sequence of orthonormal elements then*

$$\sum_{i=1}^{\infty} |(y, x_i^*)|^2 \leq \|y\|^2. \tag{8.5.4}$$

COROLLARY 8.5.6. *If* x_i^* *are an infinite sequence of orthonormal elements then*

$$\lim_{i \to \infty} (y, x_i^*) = 0 \tag{8.5.5}$$

i.e., *the Fourier coefficients of any element approach zero.*

COROLLARY 8.5.7 (Minimum Property of Orthogonal Elements). *Let* x_1, $x_2, \ldots, x_n$ *be independent. Let* $x_1^*, x_2^*, \ldots, x_n^*$ *be the* x_k's *orthonormalized according to the triangular scheme of Theorem 8.3.3. Then, for all selections of constants* $a_1, \ldots, a_{n-1}$, *we have*

$$\|y_n\| = \left\| \frac{x_n^*}{a_{nn}} \right\| \leq \|a_1 x_1 + a_2 x_2 + \cdots + a_{n-1} x_{n-1} + x_n\|.$$

The notation of (8.3.6) is employed.

Proof: By Corollary 8.5.2, the problem

$$\min_{b_i} \|x_n - (b_1 x_1 + \cdots + b_{n-1} x_{n-1})\|$$

is solved by $\sum_{k=1}^{n-1} (x_n, x_k{}^*) x_k{}^*$. But from (8.3.6) this is precisely $x_n - y_n$.

Least square approximations (i.e., best approximations in an inner product space) of an element y by a combination of given independent elements $x_1, x_2, \ldots, x_n$ can be expressed in several ways: (1) as a linear combination $a_1 x_1 + \cdots + a_n x_n$ of the given elements, (2) as a linear combination $b_1 x_1{}^* + \cdots + b_n x_n{}^*$ of the orthonormalized x's. Although (1) may be more convenient, (2) possesses the advantage of *permanence*. That is to say, suppose we add an additional element x_{n+1} to our list and ask for best approximations to y by linear combinations of $x_1, \ldots, x_n, x_{n+1}$. Expressed in form (1), the answer will be some $a_1' x_1 + \cdots + a_n' x_n + a_{n+1}' x_{n+1}$ where the a_k''s may bear no simple relation to the a_k. Expressed in form (2), the answer retains the first n coefficients and merely adds one more:

$$b_1 x_1{}^* + b_2 x_2{}^* + \cdots + b_n x_n{}^* + b_{n+1} x_{n+1}^*.$$

Ex. 1. If $f \in C[-\pi, \pi]$ or even of $L^2[-\pi, \pi]$, then

$$\lim_{n \to \infty} \int_{-\pi}^{\pi} f(x) \sin nx \, dx = \lim_{n \to \infty} \int_{-\pi}^{\pi} f(x) \cos nx \, dx = 0.$$

This is Riemann's Theorem and is a consequence of Corollary 8.5.5. It holds under more general circumstances than demonstrated here.

Ex. 2. Let $p_n{}^*(x)$ be the Legendre polynomials. If

$$f \in C[-1, 1] \quad \text{then} \quad \lim_{n \to \infty} \int_{-1}^{+1} f(x) p_n{}^*(x) \, dx = 0.$$

Ex. 3. If $f \in L^2[a, b]$, the problem of finding

$$\min_{a_i} \int_a^b \left(f(x) - \sum_{i=0}^{n} a_i x^i \right)^2 dx$$

has a unique solution.

Ex. 4. Solve the problem

$$\min_{a_i} \int_{-1}^{+1} (e^x - a_0 - a_1 x - a_2 x^2)^2 \, dx.$$

Use the Legendre polynomials

$$x_1{}^* = \sqrt{\tfrac{1}{2}}, \quad x_2{}^* = \sqrt{\tfrac{3}{2}} \, x, \quad x_3{}^* = \tfrac{3}{4}\sqrt{10}(x^2 - \tfrac{1}{3}).$$

The Fourier coefficients of e^x are

$$b_0 = \int_{-1}^{+1} \sqrt{\tfrac{1}{2}} e^x \, dx = \sqrt{\tfrac{1}{2}}(e - e^{-1}),$$

$$b_1 = \int_{-1}^{+1} \sqrt{\tfrac{3}{2}} x e^x \, dx = \sqrt{\tfrac{3}{2}}(2e^{-1}),$$

$$b_2 = \int_{-1}^{+1} \tfrac{3}{4}\sqrt{10}(x^2 - \tfrac{1}{3})e^x \, dx = \frac{3\sqrt{10}}{4}\left(\frac{2e - 14e^{-1}}{3}\right)$$

The minimizing polynomial is therefore

$$p(x) = \tfrac{1}{2}(e - e^{-1}) + 3e^{-1}x + \frac{90}{16}\left(\frac{2e - 14e^{-1}}{3}\right)(x^2 - \tfrac{1}{3})$$

$$= \tfrac{15}{4}(e - 7e^{-1})x^2 + 3e^{-1}x + \tfrac{33}{4}e^{-1} - \tfrac{3}{4}e.$$

$$\approx .537x^2 + 1.104x + .996.$$

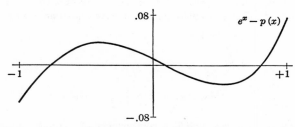

Figure 8.5.1   Error in Least Square Approximation of e^x by a Parabola.

Ex. 5.   Let $p_n(x) = k_{nn}x^n + \cdots$ be the Legendre polynomials. Then,

$$\int_{-1}^{+1}\left(\frac{p_n(x)}{k_{nn}}\right)^2 dx = \frac{2^{2n+1}(n!)^4}{(2n)!\,(2n+1)!} \le \int_{-1}^{+1}(x^n + a_1 x^{n-1} + \cdots + a_n)^2\, dx \quad (8.5.6)$$

for all selections of a's. (Use Theorem 10.3.5 and Theorem 10.3.4 for the equality.)

Ex. 6.   Let $T_n(x) = 2^{n-1}x^n + \cdots$ be the Tschebyscheff polynomial of the first kind. Then,

$$\frac{1}{2^{2n-2}}\int_{-1}^{+1}\frac{(T_n(x))^2}{\sqrt{1 - x^2}}\, dx = \frac{\pi}{2^{2n-1}} \le \int_{-1}^{+1}\frac{(x^n + a_1 x^{n-1} + \cdots + a_n)^2}{\sqrt{1 - x^2}}\, dx \quad (8.5.7)$$

for all selections of a's. Here $n \ge 1$.

Ex. 7.   Lanczos Economization.

If $\dfrac{a_0}{2} + \displaystyle\sum_{k=1}^{\infty} a_k T_k(x)$ is the Tschebyscheff-Fourier series of a continuous $f(x)$ (cf. 8.4.7), then the partial sum $\dfrac{a_0}{2} + \displaystyle\sum_{k=1}^{n} a_k T_k(x)$ solves the problem of finding

$$\min_{b_i} \int_{-1}^{+1}\left(f(x) - \sum_{k=0}^{n} b_k x^k\right)^2 (1 - x^2)^{-\frac{1}{2}}\, dx.$$ But this partial sum is very nearly the

solution to the problem $\min\limits_{b_i} \max\limits_{-1 \le x \le 1} \left| f(x) - \sum\limits_{k=0}^{n} b_k x^k \right|$. For suppose that we write $f(x) = \dfrac{a_0}{2} + a_1 T_1(x) + \cdots + a_n T_n(x) + a_{n+1} T_{n+1}(x)$ plus a remainder which we neglect. Then, $f(x) - \left(\dfrac{a_0}{2} + a_1 T_1(x) + \cdots + a_n T_n(x) \right) = a_{n+1} T_{n+1}(x)$. Since $a_{n+1} T_{n+1}(x)$ has $n + 2$ equal maxima and minima alternating in sign, Theorem 7.6.4 tells us that the contents of the parenthesis is the best uniform approximation to $f(x)$ from $\mathscr{P}_n$. For this reason, the partial Tschebyscheff-Fourier series are sometimes used as a starting point in determining best uniform approximations.

If f is a polynomial, its Tschebyscheff-Fourier expansion can be obtained by using the table, given in the Appendix, of powers as combinations of Tschebyscheff polynomials.

Ex. 7A. Economize $f(x) = 1 + \dfrac{x}{2} + \dfrac{x^2}{3} + \dfrac{x^3}{4} + \dfrac{x^4}{5} + \dfrac{x^5}{6}$ on the interval $[-1, 1]$ allowing a tolerance of $\varepsilon = .05$. We have,

$$
\begin{aligned}
f(x) &= T_0 + \tfrac{1}{2}T_1 + \tfrac{1}{3} \cdot \tfrac{1}{2}(T_0 + T_1) + \tfrac{1}{4} \cdot \tfrac{1}{4}(3T_1 + T_3) \\
&\quad + \tfrac{1}{5} \cdot \tfrac{1}{8}(3T_0 + 4T_2 + T_4) + \tfrac{1}{6} \cdot \tfrac{1}{16}(10T_1 + 5T_3 + T_5) \\
&= \tfrac{149}{120}T_0 + \tfrac{76}{96}T_1 + \tfrac{32}{120}T_2 + \tfrac{11}{96}T_3 + \tfrac{3}{120}T_4 + \tfrac{1}{96}T_5.
\end{aligned}
$$

Since $|T_n(x)| = |\cos(n \operatorname{arc\,cos} x)| \le 1$, we can delete the last two terms and we incur an error of at most $\tfrac{3}{120} + \tfrac{1}{96} < .05$. Hence

$$
\tfrac{149}{120}T_0 + \tfrac{76}{96}T_1 + \tfrac{32}{120}T_2 + \tfrac{11}{96}T_3
$$

is in $\mathscr{P}_3$ and approximates $f(x)$ to within .05 on $[-1, 1]$.

Ex. 7B. The Tschebyscheff-Fourier coefficients of $\cos \dfrac{\pi x}{2}$ are

$$
\begin{aligned}
\frac{1}{\pi} \int_{-1}^{+1} \frac{\cos \dfrac{\pi x}{2} \cos(n \operatorname{arc\,cos} x)}{\sqrt{1 - x^2}} \, dx & \\
= \frac{1}{\pi} \int_{0}^{\pi} \cos\left(\frac{\pi}{2} \cos y \right) \cos ny \, dy &= \begin{cases} (-1)^k 2 J_{2k}(\pi/2) & n = 2k \\ 0 & n = 2k + 1 \end{cases}
\end{aligned}
$$

where $J_n(x)$ is the Bessel Function of order n. (See, for instance, G. N. Watson [1] p. 21.) Hence, $\cos \dfrac{\pi x}{2} \sim J_0(\pi/2) - 2J_2(\pi/2)T_2(x) + 2J_4(\pi/2)T_4(x) - \cdots$. The partial sum of order 4 is $= 0.9993966 - 1.2227432x^2 + 0.2239366x^4$, and this may be compared with the best uniform approximation given in Fig. 7.6.7.

8.6 The Normal Equations

THEOREM 8.6.1. *Let* $x_1, x_2, \ldots, x_n$ *be independent elements and let* $x_1^*,$ $x_2^*, \ldots, x_n^*$ *be the* x's *orthonormalized. Then, for any element* $y,$

$$
\left(y - \sum_{k=1}^{n} (y, x_k^*) x_k^* \right) \perp x_j^*.
$$

Proof:

$$\left(y - \sum_{k=1}^{n} (y, x_k{}^*)x_k{}^*, x_j{}^*\right)$$

$$= (y, x_j{}^*) - \sum_{k=1}^{n} (y, x_k{}^*)(x_k{}^*, x_j{}^*) = (y, x_j{}^*) - (y, x_j{}^*) = 0.$$

COROLLARY 8.6.2. *y minus its best approximation by linear combinations of $x_1, \ldots, x_n$ is orthogonal to each x_j.*

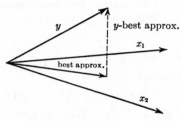

Figure 8.6.1

In geometric language, we speak of the set of all possible linear combinations $a_1x_1 + a_2x_2 + \cdots a_nx_n$ as constituting a *linear manifold*. A linear manifold is a natural generalization of the notion of a plane through the origin, and this corollary states that the shortest distance from a fixed element to a point of a linear manifold is the length of an element perpendicular to the manifold. (See also Def. 9.4.1.)

THEOREM 8.6.3. *Let $a_1x_1 + a_2x_2 + \cdots + a_nx_n$ be the best approximation to y from among the linear combinations of $x_1, \ldots, x_n$ (assumed independent). Then, the coefficients a_i are the solution of the following system of equations.*

$$a_1(x_1, x_1) + a_2(x_2, x_1) + \cdots + a_n(x_n, x_1) = (y, x_1)$$

$$\vdots \qquad\qquad \vdots \qquad\qquad \vdots \qquad\qquad (8.6.1)$$

$$a_1(x_1, x_n) + a_2(x_2, x_n) + \cdots + a_n(x_n, x_n) = (y, x_n).$$

These equations are known as *the normal equations.*

Proof: By the previous corollary, $(y - a_1x_1 - \cdots - a_nx_n, x_j) = 0$. When expanded, this is the jth equation of system (8.6.1).

8.7 Gram Matrices and Determinants

DEFINITION 8.7.1. Given a sequence of elements $x_1, x_2, \ldots, x_n$ in an inner product space. The $n \times n$ matrix

$$G = ((x_i, x_j)) = \begin{bmatrix} (x_1, x_1) & (x_1, x_2), \ldots, (x_1, x_n) \\ \cdot & \cdot \\ \cdot & \cdot \\ \cdot & \cdot \\ (x_n, x_1) & (x_n, x_2), \ldots, (x_n, x_n) \end{bmatrix} \qquad (8.7.1)$$

is known as *the Gram matrix* of $x_1, x_2, \ldots, x_n$. Its determinant

$$g = g(x_1, \ldots, x_n) = |(x_i, x_j)| = |(x_j, x_i)| \qquad (8.7.2)$$

is known as the *Gram determinant* of the elements.

The Gram matrix is the transpose of the coefficient matrix of the normal equations. It is also the matrix of the bilinear form

$$(a_1 x_1 + a_2 x_2 + \cdots + a_n x_n, b_1 x_1 + b_2 x_2 + \cdots + b_n x_n) = \sum_{i,j=1}^{n} a_i \bar{b}_j (x_i, x_j). \qquad (8.7.3)$$

Notice that $g(x_1, \ldots, x_n)$ is a symmetric function of its arguments. For, consider $g(x_1, \ldots, x_i, \ldots, x_j, \ldots, x_n)$ and suppose that x_i and x_j have been interchanged yielding $g(x_1, \ldots, x_j, \ldots, x_i, \ldots, x_n)$. In the determinant expression for the latter, interchange the ith and jth columns and the ith and jth rows and obtain the determinant for the former.

LEMMA 8.7.1. *Let* $y_i = \sum_{j=1}^{n} a_{ij} x_j$, $i = 1, 2, \ldots, n$. *Let* A *designate the matrix* (a_{ij}) *and* $\tilde{A}$ *be its conjugate transpose* $(\bar{a}_{ji})$. *Then*

$$G(y_1, y_2, \ldots, y_n) = AG(x_1, x_2, \ldots, x_n)\tilde{A} \qquad (8.7.4)$$

and

$$g(y_1, \ldots, y_n) = |\det A|^2 \, g(x_1, \ldots, x_n.) \qquad (8.7.5)$$

Proof:

$$\begin{bmatrix} (x_1, y_1) & (x_1, y_2) & \cdots & (x_1, y_n) \\ \cdot & & & \cdot \\ \cdot & & & \cdot \\ \cdot & & & \cdot \\ (x_n, y_1) & (x_n, y_2) & \cdots & (x_n, y_n) \end{bmatrix}$$

$$= \begin{bmatrix} (x_1, x_1) & (x_1, x_2) & \cdots & (x_1, x_n) \\ \cdot & & & \cdot \\ \cdot & & & \cdot \\ \cdot & & & \cdot \\ (x_n, x_1) & (x_n, x_2) & \cdots & (x_n, x_n) \end{bmatrix} \begin{bmatrix} \bar{a}_{11} & \bar{a}_{21} & \cdots & \bar{a}_{n1} \\ \cdot & & & \cdot \\ \cdot & & & \cdot \\ \cdot & & & \cdot \\ \bar{a}_{1n} & \bar{a}_{2n} & \cdots & \bar{a}_{nn} \end{bmatrix}$$

$$= G(x_1, x_2, \ldots, x_n)\tilde{A}.$$

Furthermore,

$$
\begin{bmatrix}
a_{11} & a_{12} & \cdots & a_{1n} \\
\cdot & & & \cdot \\
\cdot & & & \cdot \\
\cdot & & & \cdot \\
a_{n1} & a_{n2} & \cdots & a_{nn}
\end{bmatrix}
\begin{bmatrix}
(x_1, y_1) & (x_1, y_2) & \cdots & (x_1, y_n) \\
\cdot & & & \\
\cdot & & & \\
\cdot & & & \\
(x_n, y_1) & (x_n, y_2) & \cdots & (x_n, y_n)
\end{bmatrix}
$$

$$
=
\begin{bmatrix}
(y_1, y_1) & (y_1, y_2) & \cdots & (y_1, y_n) \\
\cdot & & & \cdot \\
\cdot & & & \cdot \\
\cdot & & & \cdot \\
(y_n, y_1) & (y_n, y_2) & \cdots & (y_n, y_n)
\end{bmatrix}
= G(y_1, \ldots, y_n).
$$

Combining these two equations we obtain the first identity of the lemma. The second comes from taking determinants and observing that $|\tilde{A}| = \overline{|A|}$. As a special case,

$$
g(\sigma_1 x_1, \sigma_2 x_2, \ldots, \sigma_n x_n) = |\sigma_1|^2 \, |\sigma_2|^2 \cdots |\sigma_n|^2 \, g(x_1, x_2, \ldots, x_n). \quad (8.7.6)
$$

THEOREM 8.7.2. *Let $x_1, \ldots, x_n \neq 0$. Then,*

$$
0 \leq g(x_1, x_2, \ldots, x_n) \leq \|x_1\|^2 \, \|x_2\|^2 \cdots \|x_n\|^2. \quad (8.7.7)
$$

The lower extreme $g = 0$ occurs if and only if the elements x_i are dependent. The upper extreme occurs if and only if the elements are orthogonal.

If the elements have been normalized: $\|x_i\| = 1$, then we have

$$
0 \leq g \leq 1. \quad (8.7.8)
$$

Proof: Suppose first that the x's are dependent. Then, we can find constants $a_1, \ldots, a_n$ not all zero such that $a_1 x_1 + a_2 x_2 + \cdots + a_n x_n = 0$. Suppose that $a_j \neq 0$, and consider the transformation

$$
y_1 = x_1
$$
$$
\cdot
$$
$$
\cdot
$$
$$
\cdot
$$
$$
y_{j-1} = x_{j-1}
$$
$$
y_j = a_1 x_1 + a_2 x_2 + \cdots + a_n x_n = 0 \quad (8.7.9)
$$
$$
y_{j+1} = \qquad\qquad x_{j+1}
$$
$$
\cdot
$$
$$
\cdot
$$
$$
\cdot
$$
$$
y_n = \qquad\qquad\qquad x_n.
$$

Since $(y_j, y_i) = (0, y_i) = 0$, $g(y_1, \ldots, y_n) = 0$.

Now

$$|A| = \begin{vmatrix} 1 & 0 & \cdots & 0 & \cdots & 0 \\ & & & & & \\ & & & & & \\ 0 & 1 & \cdots & 0 & \cdots & 0 \\ \cdot & & & & & \\ a_1 & a_2 & \cdots & a_j & \cdots & a_n \\ \cdot & & & & & \\ \cdot & & & & & \\ 0 & 0 & \cdots & 0 & \cdots & 1 \end{vmatrix}.$$

Expanding this according to minors of the jth row, we find

$$|A| = 0 + 0 + \cdots + a_j \cdot 1 + 0 + \cdots + 0 = a_j \neq 0.$$

It follows from Lemma 8.7.1 that $g(x_1, x_2, \ldots, x_n) = 0$.

Next, suppose that the x's are independent. Then by Theorem 8.3.3 we can find constants a_{ij} such that the elements

$$x_k{}^* = a_{k1}x_1 + a_{k2}x_2 + \cdots + a_{kk}x_k, \quad a_{kk} > 0$$

are orthonormal. By our lemma,

$$1 = |\delta_{ij}| = g(x_1{}^*, x_2{}^*, \ldots, x_n{}^*) = g(x_1, x_2, \ldots, x_n) |A|^2$$

where

$$|A| = \begin{vmatrix} a_{11} & 0 & \cdots & 0 \\ a_{21} & a_{22} & \cdots & 0 \\ \cdot & & & \cdot \\ \cdot & & & \cdot \\ \cdot & & & \cdot \\ a_{n1} & a_{n2} & \cdots & a_{nn} \end{vmatrix} = a_{11}a_{22} \cdots a_{nn}.$$

Hence,

$$g(x_1, x_2, \ldots, x_n) = \frac{1}{a_{11}{}^2} \cdot \frac{1}{a_{22}{}^2} \cdots \frac{1}{a_{nn}{}^2} > 0 . \qquad (8.7.10)$$

Hence $g = 0$ occurs when and only when the x's are dependent.

We show next that

$$\frac{1}{a_{kk}{}^2} \leq \|x_k\|^2 . \qquad (8.7.11)$$

From (8.3.6), $\dfrac{1}{a_{kk}^{\,2}} = \|y_k\|^2 = \left\| x_k - \sum\limits_{j=1}^{k-1} (x_k, x_j{}^*)x_j{}^* \right\|^2 \le \|x_k\|^2$ (from Theorem 8.5.1 with $a_i = 0$).

If the x_i are orthogonal, then G is a matrix with $\|x_i\|^2$ on the diagonal and 0's elsewhere. In this case, $g(x_1, x_2, \ldots, x_n) = \|x_1\|^2 \|x_2\|^2 \cdots \|x_n\|^2$. Suppose, conversely, $g(x_1, x_2, \ldots, x_n) = \|x_1\|^2 \|x_2\|^2 \cdots \|x_n\|^2$. We have from (8.7.10) $g(x_1, \ldots, x_n) = \dfrac{1}{a_{11}^{\,2}} \cdots \dfrac{1}{a_{nn}^{\,2}}$. Now since $\dfrac{1}{a_{kk}^{\,2}} \le \|x_k\|^2$, it follows that $\|y_k\|^2 = \dfrac{1}{a_{kk}^{\,2}} = \|x_k\|^2$, $k = 1, 2, \ldots, n$. But from (8.3.6) and Corollary 8.5.3,

$$\|y_k\|^2 = \|x_k\|^2 - \sum_{i=1}^{k-1} |(x_k, x_i{}^*)|^2. \text{ Hence } \sum_{i=1}^{k-1} |(x_k, x_i{}^*)|^2 = 0 \text{ for } k = 1,$$

$2, \ldots, n$. This implies the orthogonality of the vectors $x_1, x_2, \ldots, x_n$.

COROLLARY 8.7.3 (Hadamard's Determinant Inequality). *Let* $D = (a_{ij})$ *be an* $n \times n$ *matrix with complex elements. Then,*

$$|D|^2 \le \prod_{k=1}^{n} (|a_{k1}|^2 + |a_{k2}|^2 + \cdots + |a_{kn}|^2). \qquad (8.7.12)$$

If the elements a_{ij} *satisfy* $|a_{ij}| \le M$, $i, j = 1, 2, \ldots, n$, *then*

$$|D| \le M^n n^{n/2}. \qquad (8.7.13)$$

Proof: Let x_i designate the vector $(a_{i1}, a_{i2}, \ldots, a_{in})$. Use the Hermitian inner product $(x_i, x_j) = \sum\limits_{k=1}^{n} a_{ik}\bar{a}_{jk}$ in C_n.

If $\tilde{D}$ designates the conjugate transpose of D, $|\bar{a}_{ji}|$, then

$$D\tilde{D} = \begin{vmatrix} a_{11} & a_{12} & \cdots & a_{1n} \\ a_{21} & a_{22} & \cdots & a_{2n} \\ \cdot & & & \\ \cdot & & & \\ \cdot & & & \\ a_{n1} & a_{n2} & \cdots & a_{nn} \end{vmatrix} \begin{vmatrix} \bar{a}_{11} & \bar{a}_{21} & \cdots & \bar{a}_{n1} \\ \bar{a}_{12} & \bar{a}_{22} & \cdots & \bar{a}_{n2} \\ \cdot & & & \\ \cdot & & & \\ \cdot & & & \\ \bar{a}_{1n} & \bar{a}_{2n} & \cdots & \bar{a}_{nn} \end{vmatrix}$$

$$= |(x_i, x_j)| = g(x_1, x_2, \ldots, x_n) \le \|x_1\|^2 \|x_2\|^2 \cdots \|x_n\|^2.$$

Now $\|x_i\|^2 = \sum\limits_{k=1}^{n} a_{ik}\bar{a}_{ik} = \sum\limits_{k=1}^{n} |a_{ik}|^2$. And, since, $|\tilde{D}| = \overline{|D|}$, (8.7.12) follows. If $|a_{ij}| \le M$, then $|a_{k1}|^2 + |a_{k2}|^2 + \cdots + |a_{kn}|^2 \le nM^2$, so that $|D|^2 \le n^n M^{2n}$.

Ex. 1. (Gram's Criterion). Let $f_i(t) \in C[a, b]$ $i = 1, 2, \ldots, n$. These functions are linearly independent if and only if the Gram determinant

$$\left| \int_a^b f_i(t) f_j(t) \, dt \right| > 0.$$

A similar result holds for $f_i(t) \in L^2[a, b]$.

THEOREM 8.7.4. *Let $x_1, x_2, \ldots, x_n$ be independent. If*

$$\delta = \min_{a_i} \| y - (a_1 x_1 + a_2 x_2 + \cdots + a_n x_n) \|, \tag{8.7.14}$$

then,

$$\delta^2 = \frac{g(x_1, x_2, \ldots, x_n, y)}{g(x_1, x_2, \ldots, x_n)}. \tag{8.7.15}$$

Proof: Let the minimizing element $a_1 x_1 + a_2 x_2 + \cdots + a_n x_n$ be called s. Then

$$\delta^2 = \| y - s \|^2 = (y - s, y - s) = (y - s, y) - (y - s, s).$$

By Theorem 8.6.1, $(y - s, s) = 0$ so that

$$\delta^2 = (y - s, y) = (y, y) - (s, y)$$

and

$$(s, y) = (y, y) - \delta^2. \tag{8.7.16}$$

Write the normal equations in the following form and append to them the expanded version of (8.7.16):

$$a_1(x_1, x_1) + a_2(x_2, x_1) + \cdots + a_n(x_n, x_1) - \qquad (y, x_1) \quad = 0$$
$$\cdot \qquad\qquad\qquad\qquad\qquad\qquad\qquad\qquad \cdot$$
$$\cdot \qquad\qquad\qquad\qquad\qquad\qquad\qquad\qquad \cdot$$
$$\cdot \qquad\qquad\qquad\qquad\qquad\qquad\qquad\qquad \cdot \qquad (8.7.17)$$
$$a_1(x_n, x_1) + a_2(x_n, x_2) + \cdots + a_n(x_n, x_n) - \qquad (y, x_n) \quad = 0$$
$$a_1(x_1, y) + a_2(x_2, y) + \cdots + a_n(x_n, y) + [\delta^2 - (y, y)] = 0$$

If we introduce a dummy value $a_{n+1} = 1$ as a coefficient of the elements of the last column, then (8.7.17) becomes a system of $n + 1$ homogeneous linear equation in $n + 1$ variables $a_1, \ldots, a_n, a_{n+1} (= 1)$, which possesses a nontrivial solution. The determinant of this system must therefore vanish:

$$\begin{vmatrix} (x_1, x_1) & (x_2, x_1) & \cdots & (x_n, x_1) & 0 - (y, x_1) \\ \cdot & & & & \cdot \\ \cdot & & & & \cdot \\ \cdot & & & & \cdot \\ (x_1, x_n) & (x_2, x_n) & \cdots & (x_n, x_n) & 0 - (y, x_n) \\ (x_1, y) & (x_2, y) & \cdots & (x_n, y) & \delta^2 - (y, y) \end{vmatrix} = 0.$$

Therefore

$$
\begin{vmatrix}
(x_1, x_1) & (x_2, x_1) & \cdots & (x_n, x_1) & 0 \\
\cdot & & & & \cdot \\
\cdot & & & & \cdot \\
\cdot & & & & \cdot \\
(x_1, x_n) & (x_2, x_n) & \cdots & (x_n, x_n) & 0 \\
(x_1, y) & (x_2, y) & \cdots & (x_n, y) & \delta^2
\end{vmatrix}
$$

$$
=
\begin{vmatrix}
(x_1, x_1) & (x_2, x_1) & \cdots & (x_n, x_1) & (y, x_1) \\
\cdot & & & & \cdot \\
\cdot & & & & \cdot \\
\cdot & & & & \cdot \\
(x_1, x_n) & (x_2, x_n) & \cdots & (x_n, x_n) & (y, x) \\
(x_1, y) & (x_2, y) & \cdots & (x_n, y) & (y, y)
\end{vmatrix}
$$

and

$$
\delta^2 g(x_1, x_2, \ldots, x_n) = g(x_1, x_2, \ldots, x_n, y).
$$

THEOREM 8.7.5. *Let $x_1, \ldots, x_n$ be independent. The solution s to the minimum problem*

$$
\min_{a_i} \| y - (a_1 x + \cdots + a_n x_n) \|
$$

is given by

$$
s = -
\begin{vmatrix}
(x_1, x_1) & (x_2, x_1) & \cdots & (x_n, x_1) & (y, x_1) \\
\cdot & & & \cdot & \cdot \\
\cdot & & & \cdot & \cdot \\
\cdot & & & \cdot & \cdot \\
(x_1, x_n) & (x_2, x_n) & \cdots & (x_n, x_n) & (y, x_n) \\
x_1 & x_2 & \cdots & x_n & 0
\end{vmatrix}
\div g(x_1, \ldots, x_n).
\tag{8.7.18}
$$

The remainder or error, $y - s$, is given by

$$
y - s =
\begin{vmatrix}
(x_1, x_1) & (x_2, x_1) & \cdots & (x_n, x_1) & (y, x_1) \\
\cdot & & & & \cdot \\
\cdot & & & & \cdot \\
\cdot & & & & \cdot \\
(x_1, x_n) & (x_2, x_n) & \cdots & (x_n, x_n) & (y, x_n) \\
x_1 & x_2 & \cdots & x_n & y
\end{vmatrix}
\div g(x_1, \ldots, x_n).
\tag{8.7.19}
$$

Proof: From the normal equations (8.6.1) and Cramer's rule (1.2.2) we have

$$a_1 = \begin{vmatrix} (y, x_1) & (x_2, x_1) & \cdots & (x_n, x_1) \\ (y, x_2) & (x_2, x_2) & \cdots & (x_n, x_2) \\ \cdot & \cdot & & \cdot \\ \cdot & \cdot & & \cdot \\ \cdot & \cdot & & \cdot \\ (y, x_n) & (x_2, x_n) & \cdots & (x_n, x_n) \end{vmatrix} \div g(x_1, \ldots, x_n), \quad (8.7.20)$$

with similar formulas for the other a's. If we expand the determinant in (8.7.18) according to the minors of the last row we obtain expressions for the coefficients of the x's which coincide with those just mentioned.

Since

$$y = \begin{vmatrix} (x_1, x_1) & (x_2, x_1) & \cdots & (x_n, x_1) & 0 \\ \cdot & & & \cdot & \cdot \\ \cdot & & & \cdot & \cdot \\ \cdot & & & \cdot & \cdot \\ (x_1, x_n) & (x_2, x_n) & \cdots & (x_n, x_n) & 0 \\ x_1 & x_2 & \cdots & x_n & y \end{vmatrix} \div g(x_1, \ldots, x_n), \quad (8.7.21)$$

adding (8.7.18) to (8.7.21) yields (8.7.19).

COROLLARY 8.7.6. *Let the elements* $x_1, x_2, \ldots,$ *be linearly independent and be orthonormalized according to the Gram-Schmidt scheme yielding* $x_1{}^*, x_2{}^*, \ldots$ *. Then,*

$$x_n{}^* = \frac{1}{\sqrt{g(x_1, \ldots, x_{n-1})g(x_1, \ldots, x_n)}} \begin{vmatrix} (x_1, x_1) & (x_2, x_1) & \cdots & (x_n, x_1) \\ \cdot & & & \cdot \\ \cdot & & & \cdot \\ \cdot & & & \cdot \\ (x_1, x_{n-1}) & (x_2, x_{n-1}) & \cdots & (x_n, x_{n-1}) \\ x_1 & x_2 & \cdots & x_n \end{vmatrix}$$

$$(8.7.22)$$

$n > 1,$

$x_1{}^* = x_1/\sqrt{g(x_1)}.$

The "leading coefficient" in $x_n{}^*$, a_{nn}, *is given by*

$$a_{nn} = \sqrt{\frac{g(x_1, \ldots, x_{n-1})}{g(x_1, \ldots, x_n)}}, \quad n > 1. \quad (8.7.23)$$

Proof: Consider the minimum problem

$$\min_{a_i} \|x_n - (a_1 x_1 + \cdots + a_{n-1} x_{n-1})\|.$$

According to Corollary 8.5.7, the solution is given by

$$\frac{x_n{}^*}{a_{nn}} = x_n - (a_1 x_1 + \cdots + a_{n-1} x_{n-1}), \quad a_{nn} > 0. \qquad (8.7.24)$$

According to (8.7.19),

$$x_n - (a_1 x_1 + \cdots + a_{n-1} x_{n-1})$$

$$= \begin{vmatrix} (x_1, x_1) & (x_2, x_1) & \cdots & (x_{n-1}, x_1) & (x_n, x_1) \\ & & & & \\ & & & & \\ (x_1, x_{n-1}) & (x_2, x_{n-1}) & \cdots & (x_{n-1}, x_{n-1}) & (x_n, x_{n-1}) \\ x_1 & x_2 & \cdots & x_{n-1} & x_n \end{vmatrix} \div g(x_1, \ldots, x_{n-1}) \qquad (8.7.25)$$

From Corollary 8.5.7, $\dfrac{1}{a_{nn}} = \left\| \dfrac{x_n{}^*}{a_{nn}} \right\| = \min_{a_i} \|x_n - (a_1 x_1 + \cdots + a_{n-1} x_{n-1})\|$, and from Theorem 8.7.4, this minimum $= \sqrt{\dfrac{g(x_1, \ldots, x_n)}{g(x_1, \ldots, x_{n-1})}}$. Hence (8.7.22) follows from (8.7.24) and (8.7.23).

Ex. 2.   The Legendre polynomials are given by

$$p_0{}^*(x) = \frac{1}{\sqrt{2}}, \quad p_1{}^*(x) = \frac{1}{\sqrt{2}\sqrt{\frac{4}{3}}} \begin{vmatrix} 2 & 0 \\ 1 & x \end{vmatrix},$$

$$p_2{}^*(x) = \frac{1}{\sqrt{\frac{4}{3}}\sqrt{\frac{32}{135}}} \begin{vmatrix} 2 & 0 & \frac{2}{3} \\ 0 & \frac{2}{3} & 0 \\ 1 & x & x^2 \end{vmatrix}, \ldots.$$

8.8 Further Properties of the Gram Determinant

THEOREM 8.8.1.   *The Gram determinant* $g(x_1, x_2, \ldots, x_n)$ *has the following properties.*

(a) *g is a symmetric function of its arguments.*

(b) $g(x_1, \ldots, \sigma x_j, \ldots, x_n) = |\sigma|^2 \, g(x_1, \ldots, x_n)$

(c) $g(x_1, \ldots, x_j + \sigma x_k, \ldots, x_n) = g(x_1, \ldots, x_n), \quad j \neq k$

(d) $g^{\frac{1}{2}}(x_1' + x_1'', x_2, \ldots, x_n) \leq g^{\frac{1}{2}}(x_1', x_2, \ldots, x_n)$
$\qquad\qquad\qquad\qquad\qquad + g^{\frac{1}{2}}(x_1'', x_2, \ldots, x_n)$

(e) $g(x_1, \ldots, x_n) \leq g(x_1, \ldots, x_p) g(x_{p+1}, \ldots, x_n), \quad 1 \leq p < n.$

$$(8.8.1)$$

Equality in (e) *holds if and only if*

$$(x_i, x_j) = 0 \quad 1 \le i \le p, \quad p + 1 \le j \le n.$$

(Compare Theorem 13.1.2.)

Proof: Statements (a) and (b) have already been proved. (c). Write the left hand member of (8.8.1)(c) as a determinant of inner products. Expand the inner products in the jth row and use the elementary properties of determinants. Then expand the inner products in the jth column and do likewise. (d). We may assume that $x_2, \ldots, x_n$ are independent. Otherwise both members of (8.8.1) (d) vanish and the inequality holds trivially. Orthonormalize $x_2, \ldots, x_n$ and call the orthonormal vectors $x_2{}^*, \ldots, x_n{}^*$. Then, from Theorem 8.5.1 and (8.7.15),

$$\frac{g^{\frac{1}{2}}(x_1' + x_1'', x_2, \ldots, x_n)}{g^{\frac{1}{2}}(x_2, \ldots, x_n)} = \left\| x_1' + x_1'' - \sum_{k=2}^{n} (x_1' + x_1'', x_k{}^*)x_k{}^* \right\|$$

$$= \left\| x_1' - \sum_{k=2}^{n} (x_1', x_k{}^*)x_k{}^* + x_1'' - \sum_{k=2}^{n} (x_1'', x_k{}^*)x_k{}^* \right\|$$

$$\le \left\| x_1' - \sum_{k=2}^{n} (x_1', x_k{}^*)x_k{}^* \right\| + \left\| x_1'' - \sum_{k=2}^{n} (x_1'', x_k{}^*)x_k{}^* \right\|$$

$$= \frac{g^{\frac{1}{2}}(x_1', x_2, \ldots, x_n)}{g^{\frac{1}{2}}(x_2, \ldots, x_n)} + \frac{g^{\frac{1}{2}}(x_1'', x_2, \ldots, x_n)}{g^{\frac{1}{2}}(x_2, \ldots, x_n)}.$$

Multiplying this inequality by the denominator, we arrive at (d).

(e) Let k satisfy $1 \le k < p$. Then, since there is more competition on the left-hand side,

$$\min_{a_j} \|x_k - (a_{k+1}x_{k+1} + \cdots + a_n x_n)\|^2 \le \min_{b_j} \|x_k - (b_{k+1}x_{k+1} + \cdots + b_p x_p)\|^2.$$

Similarly,

$$\min_{c_j} \|x_p - (c_{p+1}x_{p+1} + \cdots + c_n x_n)\|^2 \le \|x_p\|^2.$$

Therefore by Theorem 8.7.4,

$$\frac{g(x_k, x_{k+1}, x_{k+2}, \ldots, x_n)}{g(x_{k+1}, \ldots, x_n)} \le \frac{g(x_k, x_{k+1}, \ldots, x_p)}{g(x_{k+1}, \ldots, x_p)} \tag{8.8.2}$$

and

$$\frac{g(x_p, x_{p+1}, \ldots, x_n)}{g(x_{p+1}, \ldots, x_n)} \le g(x_p). \tag{8.8.3}$$

In particular, writing $k = 1, k = 2, \ldots, p - 1$ in (8.8.2) we have

$$
\left.
\begin{aligned}
\frac{g(x_1, \ldots, x_n)}{g(x_2, \ldots, x_n)} &\leq \frac{g(x_1, \ldots, x_p)}{g(x_2, \ldots, x_p)} \\[2mm]
\frac{g(x_2, \ldots, x_n)}{g(x_3, \ldots, x_n)} &\leq \frac{g(x_2, \ldots, x_p)}{g(x_3, \ldots, x_p)} \\
&\ \ \vdots \\
\frac{g(x_p, \ldots, x_n)}{g(x_{p+1}, \ldots, x_n)} &\leq g(x_p).
\end{aligned}
\right\}
\qquad (8.8.4)
$$

Multiplying these inequalities together,

$$
\frac{g(x_1, \ldots, x_n)}{g(x_{p+1}, \ldots, x_n)} \leq g(x_1, \ldots, x_p),
$$

or

$$
g(x_1, \ldots, x_n) \leq g(x_1, \ldots, x_p)g(x_{p+1}, \ldots, x_n). \qquad (8.8.5)
$$

Now, equality in (8.8.5) can hold if and only if it holds in each of the relations (8.8.4); i.e., if and only if

$$
\min_{c_j} \|x_p - (c_{p+1}x_{p+1} + \cdots + c_n x_n)\|^2 = \|x_p\|^2 \qquad (8.8.6)
$$

and

$$
\min_{a_j} \|x_k - (a_{k+1}x_{k+1} + \cdots + a_n x_n)\|^2 = \min_{b_j} \|x_k - (b_{k+1}x_{k+1} + \cdots + b_p x_p)\|^2
$$
$$
k = 1, 2, \ldots, p - 1. \qquad (8.8.7)
$$

Now, by Corollary 8.5.3, (8.8.6) holds if and only if

$$
(x_p, x_j) = 0 \quad j = p + 1, \ldots, n.
$$

Now (8.8.7) with $k = p - 1$ reads

$$
\min_{a_j} \|x_{p-1} - (a_p x_p + \cdots + a_n x_n)\|^2 = \min_{b_p} \|x_{p-1} - b_p x_p\|^2.
$$

By the same principle, this holds if and only if $x_{p-1} \perp x_{p+1}, x_{p+2}, \ldots, x_n$. Considering $k = p - 2, p - 3, \ldots, 1$ successively, we arrive at the stated orthogonality conditions.

The Gram determinant has a very striking geometrical interpretation. Let there be given n vectors in $R_n \colon x_i = (x_{i1}, x_{i2}, \ldots, x_{in})$. These vectors are the edges of a certain n-dimensional parallelotope (the generalization of a parallelogram) whose volume will be designated by $V = V(x_1, x_2, \ldots, x_n)$.

It can be shown that

$$V = \text{absolute value of} \begin{vmatrix} x_{11} & x_{12} & \cdots & x_{1n} \\ x_{21} & x_{22} & \cdots & x_{2n} \\ \cdot & \cdot & & \cdot \\ \cdot & \cdot & & \cdot \\ \cdot & \cdot & & \cdot \\ x_{n1} & x_{n2} & \cdots & x_{nn} \end{vmatrix}. \tag{8.8.8}$$

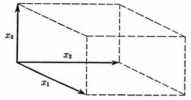

Figure 8.8.1   Parallelotope in R_3.

(Though elementary, the derivation of this formula is far from trivial. An axiomatic derivation can be found in Schreier and Sperner [1] Ch. II.)

Multiply the determinant in (8.8.8) by its transpose and obtain

$$V^2 = \begin{vmatrix} (x_1, x_1) & (x_1, x_2) & \cdots & (x_1, x_n) \\ \cdot & \cdot & & \cdot \\ \cdot & \cdot & & \cdot \\ \cdot & \cdot & & \cdot \\ (x_n, x_1) & (x_n, x_2) & \cdots & (x_n, x_n) \end{vmatrix} = g(x_1, \ldots, x_n). \tag{8.8.9}$$

Hence,

$$V(x_1, \ldots, x_n) = \sqrt{g(x_1, x_2, \ldots, x_n)}. \tag{8.8.10}$$

A derivation of (8.8.8) can be given via Theorem 8.7.4 if we assume by way of analogy to the situation in 2 and 3 dimensions that the volume of the parallelotope can be found by multiplying altitudes. That is, assume

$$V(x_1, x_2, \ldots, x_n) = \|x_1\| \, d(x_2; x_1) \, d(x_3; x_1, x_2) \cdots d(x_n; x_1, x_2, \ldots, x_{n-1}) \tag{8.8.11}$$

where $d(x_i; x_1, x_2, \ldots, x_{i-1})$ designates the distance from x_i to the linear manifold spanned by $x_1, \ldots, x_{i-1}$. Since

$$d(x_i; x_1, x_2, \ldots, x_{i-1}) = \sqrt{\frac{g(x_1, x_2, \ldots, x_i)}{g(x_1, x_2, \ldots, x_{i-1})}}, \tag{8.8.12}$$

$$V^2 = g(x_1) \frac{g(x_1, x_2)}{g(x_1)} \frac{g(x_1, x_2, x_3)}{g(x_1, x_2)} \cdots \frac{g(x_1, x_2, \ldots, x_n)}{g(x_1, x_2, \ldots, x_{n-1})} = g(x_1, \ldots, x_n). \tag{8.8.13}$$

8.9   Closure and Its Consequences

DEFINITION 8.9.1.   A finite or infinite system of elements, $x_1, x_2, \cdots$, in a normed linear space X is called *closed* if every element $x \in X$ can be approximated arbitrarily closely by finite linear combinations of the x_i. That is, given $x \in X$ and $\varepsilon > 0$, we can find constants $a_1, \ldots, a_n$ such that

$$\|x - (a_1 x_1 + a_2 x_2 + \cdots + a_n x_n)\| \leq \varepsilon. \qquad (8.9.1)$$

Ex. 1.   Any set of n independent vectors $x_1, \ldots, x_n$ in R_n or C_n is closed. In this case, the approximation can be made perfect (Theorem 1.3.1). But one can also argue as follows. Since x_i are independent, $g(x_1, \ldots, x_n) > 0$. If

$$x_i = (x_{i1}, x_{i2}, \ldots, x_{in}),$$

then

$$g(x_1, \ldots, x_n) = \begin{vmatrix} x_{11} & \cdots & x_{1n} \\ \cdot & & \cdot \\ \cdot & & \cdot \\ \cdot & & \cdot \\ x_{n1} & \cdots & x_{nn} \end{vmatrix} \begin{vmatrix} \bar{x}_{11} & \cdots & \bar{x}_{n1} \\ \cdot & & \cdot \\ \cdot & & \cdot \\ \cdot & & \cdot \\ \bar{x}_{1n} & \cdots & \bar{x}_{nn} \end{vmatrix} > 0$$

and therefore $|x_{ij}| \neq 0$. Given any y, the system $a_1 x_1 + a_2 x_2 + \cdots + a_n x_n = y$ may be solved for the a_i.

Ex. 2.   Let X be $C[a, b]$ with $\|f\| = \max\limits_{a \leq x \leq b} |f(x)|$. The powers $1, x, x^2, \ldots$, are closed in X. This is a reformulation of the Weierstrass Theorem 6.1.1.

Ex. 3.   Let X be $C[a, b]$ with $\|f\|^2 = \int_a^b |f(x)|^2 \, dx$. The powers $1, x, x^2, \ldots$, are closed in X.

Ex. 4.   Let X be $L[a, b]$ with $\|f\| = \int_a^b |f(x)| \, dx$. The powers $1, x, x^2, \ldots$, are closed in X. This is a generalization of the Weierstrass Theorem.

According to Theorem 1.4.0(e), given an $\varepsilon > 0$, we can find an absolutely continuous function $g(x)$ such that $\int_a^b |f(x) - g(x)| \, dx \leq \varepsilon/2$. Since g is continuous we can find a polynomial p such that $|g(x) - p(x)| \leq \dfrac{\varepsilon}{2(b - a)}$, $a \leq x \leq b$ and hence, $\int_a^b |g(x) - p(x)| \, dx \leq \varepsilon/2$. Therefore by (7.2.17)

$$\int_a^b |f(x) - g(x) + g(x) - p(x)| \, dx \leq \varepsilon/2 + \varepsilon/2 = \varepsilon.$$

Ex. 5.   Let X be the set of analytic functions that are continuous in $|z| \leq 1$. Set

$$\|f\| = \max_{|z| \leq 1} |f(z)|.$$

The powers $z, z^2, z^3, \ldots$, are not closed in X. If they were, given an $\varepsilon > 0$ we could find constants $a_1, \ldots, a_n$ such that

$$\max_{|z| \leq 1} |1 - (a_1 z + a_2 z^2 + \cdots + a_n z^n)| \leq \varepsilon.$$

Setting $z = 0$, we would have, in particular, $1 \leq \varepsilon$.

Before studying the implications of closure, it will be important to recall a number of topological concepts. Let X be a metric space with a distance function $d(x, y)$. (Definition 7.2.2.) If $x_0 \in X$, the set $U(x_0, r)$ consisting of all elements $x \in X$ for which $d(x, x_0) < r$ is called an *open ball*. An element x of a subset S is called an *interior* element of S if there is an $r > 0$ such that $U(x, r) \subseteq S$. In a metric space and hence in a normed linear space, the notion of convergence can be defined:

DEFINITION 8.9.2.   A sequence of elements $\{x_n\}$ of a metric space is said to converge to an element $x \in X$ if

$$\lim_{n \to \infty} d(x, x_n) = 0. \tag{8.9.2}$$

In a normed linear space, (8.9.2) is equivalent to

$$\lim_{n \to \infty} \|x - x_n\| = 0. \tag{8.9.2'}$$

A convergent sequence cannot converge to two different elements; for suppose $\lim_{n \to \infty} d(x, x_n) = 0$ and $\lim_{n \to \infty} d(y, x_n) = 0$. By (7.2.2)(d), $0 \leq d(x, y) \leq d(x, x_n) + d(y, x_n)$. Allowing $n \to \infty$, we obtain $d(x, y) = 0$ and hence $x = y$. We may speak of x as the *limit* of the sequence $\{x_n\}$ and write

$$\lim_{n \to \infty} x_n = x. \tag{8.9.3}$$

Convergence of type (8.9.2') is sometimes called *convergence in norm* or, if the norm happens to be given by an integral expression, *convergence in the mean*. In the case of normed linear spaces of functions, this serves to distinguish it from other types of convergence (pointwise, uniform, etc.). One must always make this distinction, for a sequence in a normed linear space of functions may converge in norm without converging in the pointwise sense.

Ex. 6.   Let X be $C[-1, 1]$ with $\|f\|^2 = \int_{-1}^{+1} (f(x))^2 \, dx$. Let $f_n(x) = \left[\dfrac{n}{1 + n^4 x^2}\right]^{\frac{1}{2}}$. Then,

$$\|0 - f_n\|^2 = \int_{-1}^{+1} \frac{n}{1 + n^4 x^2} \, dx = \frac{2}{n} \arctan n^2 \to 0.$$

However, $f_n(0) \to \infty$.

Let X be a metric space and $S \subseteq X$. The *closure* $\bar{S}$ of S is defined as the set of all limits of convergent sequences of S. Obviously $S \subseteq \bar{S}$. If $S = \bar{S}$, S is called *closed*. A set S is *dense* in X if $\bar{S} = X$. X is *separable* if there is in it a countable dense set.

If $\lim_{n \to \infty} x_n = x$, then from $d(x_m, x_n) \leq d(x, x_m) + d(x, x_n)$ it follows that we can make $d(x_m, x_n) \leq \varepsilon$ for all $m, n > N(\varepsilon)$. But as in the case of the metric space of rational numbers (with $d(x, y) = |x - y|$), the converse is not true. It is important to distinguish those spaces in which it is true.

DEFINITION 8.9.3.   A sequence of elements of X, $\{x_n\}$ is called a *Cauchy sequence*, if for every $\varepsilon > 0$, there is an integer $N(\varepsilon)$ such that $d(x_m, x_n) \leq \varepsilon$ for all $m, n \geq N(\varepsilon)$. A space X is called *complete* if every Cauchy sequence has a limit in X.

Specifically, if X is a complete normed linear space, and if for any $\varepsilon > 0$, we can find $N(\varepsilon)$ such that

$$\|x_m - x_n\| \leq \varepsilon, \quad m, n \geq N(\varepsilon), \tag{8.9.4}$$

then there is an $x \in X$ such that

$$\lim_{n \to \infty} \|x - x_n\| = 0. \tag{8.9.5}$$

A complete normed linear space is often called a *Banach Space*.

Ex. 7.   The complex Euclidian space C_n is complete and hence is a Banach Space. Let $x_m = (x_1^{(m)}, x_2^{(m)}, \ldots, x_n^{(m)})$. If $\{x_m\}$ is a Cauchy sequence, then

$$\sum_{i=1}^{n} |x_i^{(m)} - x_i^{(p)}|^2 \leq \varepsilon \quad \text{for} \quad m, p > N(\varepsilon).$$

Hence, for any particular i, $|x_i^{(m)} - x_i^{(p)}|^2 \leq \varepsilon$ for $m, p \geq N(\varepsilon)$. Thus, for each i, $x_i^{(m)}$ is a Cauchy sequence and has a limit x_i: $\lim_{m \to \infty} |x_i - x_i^{(m)}| = 0$. If we set $x = (x_1, \ldots, x_n)$ then $\|x - x_m\|^2 = \sum_{i=1}^{n} |x_i - x_i^{(m)}|^2 \leq \varepsilon^2$ for $m \geq N'(\varepsilon)$.

Ex. 8.   Let $C[a, b]$ be normed by $\|f\| = \max_{a \leq x \leq b} |f(x)|$. This space is complete. For if $\max_{a \leq x \leq b} |f_m(x) - f_n(x)| \leq \varepsilon, m, n > N(\varepsilon)$ then the sequence $f_m(x)$ is uniformly convergent on $[a, b]$. Hence there is a function $f(x) \in C[a, b]$ for which

$$|f(x) - f_n(x)| \leq \varepsilon, a \leq x \leq b, n \geq N'(\varepsilon),$$

and this implies that f_n converges to f in the norm considered.

Ex. 9.   On the other hand, if X is $C[a, b]$ normed by $\|f\|^2 = \int_a^b |f(x)|^2 \, dx$, then X is not complete. This can be shown by exhibiting a Cauchy sequence in X which does not converge to an element of X.

For simplicity take $a = -1$, $b = 1$ and let $f_n(x)$ be the continuous function

$$f_n(x) = \begin{cases} -1 & -1 \leq x \leq -\dfrac{1}{n} \\ nx & -\dfrac{1}{n} \leq x \leq \dfrac{1}{n} \\ 1 & \dfrac{1}{n} \leq x \leq 1 \end{cases}.$$

Let $f(x)$ be the discontinuous function

$$f(x) = \begin{cases} -1 & -1 \leq x \leq 0 \\ 1 & 0 < x \leq 1 \end{cases}.$$

Now

$$\|f(x) - f_n(x)\|^2 = \int_{-1/n}^{0} (-1 - nx)^2 \, dx + \int_{0}^{1/n} (1 - nx)^2 \, dx = \frac{2}{3n}.$$

And therefore $\lim_{n \to \infty} \|f - f_n\|^2 = 0$. f_n converges (in norm) to f and is a fortiori a Cauchy sequence. But it cannot converge in norm to a continuous function $g(x)$, for

$$\|f - g\| = \|f - f_n + f_n - g\| \leq \|f - f_n\| + \|g - f_n\|.$$

If therefore $\|g - f_n\| \to 0$ and $\|f - f_n\| \to 0$, then allowing $n \to \infty$, we obtain

$$\|f - g\| = 0. \text{ Thus, } \int_{-1}^{0} (1 + g(x))^2 \, dx = 0 \quad \text{and} \int_{0}^{1} (g(x) - 1)^2 \, dx = 0.$$

This means that $g(x) = -1$ for $-1 < x < 0$ and $g(x) = 1$ for $0 < x < 1$.

We now come to the fundamental theorem of orthonormal (Fourier) expansions.

THEOREM 8.9.1. *Let x_1^*, x_2^*, . . . , be a sequence of orthonormal elements in an inner product space X. The sequence may consist of only a finite number of elements. Appropriate changes are then to be made below. Consider the following seven statements.*

(A) *The x_i^* are closed in X.*

(B) *The Fourier series of any element $y \in X$ converges in norm to y; i.e.,*

$$\lim_{n \to \infty} \left\| y - \sum_{k=1}^{n} (y, x_k^*) x_k^* \right\| = 0. \tag{8.9.6}$$

(C) *Parseval's identity holds. That is, for any $y \in X$,*

$$\|y\|^2 = (y, y) = \sum_{n=1}^{\infty} |(y, x_n^*)|^2. \tag{8.9.7}$$

(C') *The extended Parseval identity holds. That is, for any* $x, y \in X$,

$$(x, y) = \sum_{n=1}^{\infty} (x, x_n{}^*)(x_n{}^*, y). \qquad (8.9.8)$$

(D) *There is no strictly larger orthonormal system containing* $x_1{}^*, x_2{}^*, \ldots$.

(E) *The elements* $x_1{}^*, x_2{}^*, \ldots$, *have the completeness property. That is,* $y \in X$ *and* $(y, x_k{}^*) = 0, k = 1, 2, \ldots$, *implies* $y = 0$.

(F) *An element of* X *is determined uniquely by its Fourier coefficients. That is, if* $(w, x_k{}^*) = (y, x_k{}^*)$ $k = 1, 2, \ldots$, *then* $w = y$.

Then

$$A \leftrightarrow B \leftrightarrow C \leftrightarrow C' \to D \leftrightarrow E \leftrightarrow F. \qquad (8.9.9)$$

If X *is a complete inner produce space,* $D \to C$ *and all seven statements are equivalent:*

$$A \leftrightarrow B \leftrightarrow C \leftrightarrow C' \leftrightarrow D \leftrightarrow E \leftrightarrow F. \qquad (8.9.10)$$

We have used "$\to$" to mean "implies" and "$\leftrightarrow$" to mean "implies and is implied by."

Proof: Assume A. Now

$$\left\| y - \sum_{k=1}^{n} (y, x_k{}^*)x_k{}^* \right\| \leq \left\| y - \sum_{k=1}^{n} a_k x_k{}^* \right\| \qquad \text{by (8.5.1).}$$

By A, the last expression can be made $\leq \varepsilon$. If B holds, we can approximate any element y by its Fourier segments; hence $x_k{}^*$ is closed. Thus $A \leftrightarrow B$.

By orthogonality,

$$\left(x - \sum_{k=1}^{n} (x, x_k{}^*)x_k{}^*, \ y - \sum_{k=1}^{n} (y, x_k{}^*)x_k{}^* \right) = (x, y) - \sum_{k=1}^{n} (x, x_k{}^*)(x_k{}^*, y).$$

By the Schwarz inequality,

$$\left| (x, y) - \sum_{k=1}^{n} (x, x_k{}^*)(x_k{}^*, y) \right| \leq \left\| x - \sum_{k=1}^{n} (x, x_k{}^*)x_k{}^* \right\| \cdot \left\| y - \sum_{k=1}^{n} (y, x_k{}^*)x_k{}^* \right\|.$$

If B holds, then the right-hand members both approach zero and hence $B \to C'$.

By selecting $x = y$ in C' it is clear that C holds. Hence $C' \to C$.

By Corollary 8.5.3,

$$0 \leq \left\| y - \sum_{k=1}^{n} (y, x_k{}^*)x_k{}^* \right\|^2 = \|y\|^2 - \sum_{k=1}^{n} |(y, x_k{}^*)|^2.$$

Hence $C \to B$, and thus $A \leftrightarrow B \leftrightarrow C \leftrightarrow C'$.

Assume A and suppose that $x_1{}^*, x_2{}^*, \ldots, w, (w \neq x_i{}^*)$, is also an orthonormal system. This augmented system is also closed in X. Since $A \to C'$, we have both

$$\|w\|^2 = \sum_{k=1}^{\infty} |(w, x_k{}^*)|^2 + (w, w),$$

$$\|w\|^2 = \sum_{k=1}^{\infty} |(w, x_k{}^*)|^2$$

Hence $(w, w) = 0$ and this is a contradiction since $\|w\| = 1$. This means that $A \to D$.

Suppose there is a $y \in X$, $y \neq 0$ such that $(y, x_k{}^*) = 0$, $k = 1, 2, \ldots$. Then, $x_1{}^*, x_2{}^*, \ldots, y/\|y\|$ would be an orthonormal system strictly larger than $x_1{}^*, x_2{}^*, \ldots$. Thus, $D \leftrightarrow E$.

Suppose $(w, x_k{}^*) = (y, x_k{}^*)$ $k = 1, 2, \ldots$. Then $(w - y, x_k{}^*) = 0$ $k = 1, 2, \ldots$. Assuming E, $w - y = 0$. Therefore $E \to F$.

If E were false, we could find a $z \neq 0$ with $(z, x_k{}^*) = 0$ $k = 1, 2, \ldots$. For any y, $(y, x_k{}^*) = (y + z, x_k{}^*)$ $k = 1, 2, \ldots$. So y and $y + z$ would be two distinct elements with the same Fourier coefficients. F would then be false. Therefore $F \to E$. This completes the chain of implications (8.9.9).

Assume that X is complete. We will show that $F \to B$ and this will establish the implications (8.9.10). Let $w \in X$ and consider the elements

$$s_n = \sum_{k=1}^{n} (w, x_k{}^*)x_k{}^*. \qquad (8.9.11)$$

For $n > m$, we have $s_n - s_m = \sum_{k=m+1}^{n} (w, x_k{}^*)x_k{}^*$ so that

$$\|s_n - s_m\|^2 = \sum_{k=m+1}^{n} |(w, x_k{}^*)|^2. \qquad (8.9.12)$$

By (8.5.3), $\sum_{k=1}^{\infty} |(w, x_k{}^*)|^2 < \infty$. Therefore given an ε, we can find an $N(\varepsilon)$ such that $\sum_{k=m+1}^{n} |(w, x_k{}^*)|^2 \leq \varepsilon$ for all $m, n \geq N(\varepsilon)$. Thus $\{s_n\}$ is a Cauchy sequence, and by the assumed completeness of X converges to an element $s \in X$:

$$\lim_{n \to \infty} \|s - s_n\| = 0. \qquad (8.9.13)$$

Let v be fixed and $n \geq v$. Then

$$(s - s_n, x_v{}^*) = (s, x_v{}^*) - (s_n, x_v{}^*) = (s, x_v{}^*) - (w, x_v{}^*).$$

By the Schwarz inequality,

$$|(s, x_v{}^*) - (w, x_v{}^*)| = |(s - s_n, x_v{}^*)| \leq \|s - s_n\| \, \|x_v{}^*\| = \|s - s_n\|.$$

In view of (8.9.13), we find that

$$(s, x_v{}^*) = (w, x_v{}^*) \quad v = 1, 2, \ldots.$$

By F, this implies that $s = w$, so that (8.9.13) reads

$$\lim_{n \to \infty} \left\| w - \sum_{k=1}^{n} (w, x_k{}^*) x_k{}^* \right\| = 0.$$

But this is precisely B.

We remark in passing that Parseval's identity is a generalization of the Theorem of Pythagoras 8.3.1.

The completeness property in (E) may be defined for any set of elements:

DEFINITION 8.9.4.   A set of elements S in an inner product space X is *complete* if

$$(y, x) = 0 \quad \text{for all } x \in S \tag{8.9.14}$$

implies $y = 0$.

As we have seen, in a complete inner product space, completeness and closure are equivalent concepts and some authors use these words with interchanged meaning. In Chapter XI, the notion of complete sequence is extended to normed linear spaces and the relation between closure and completeness is probed further.

Ex. 10.   Let X be $C[a, b]$ with $(f, g) = \int_a^b f(x) g(x)\, dx$. Given an $f \in C[a, b]$, and $\varepsilon > 0$, we can find constants a_k such that

$$|f(x) - (a_0 + a_1 x + \cdots + a_n x^n)| \leq \varepsilon \qquad a \leq x \leq b.$$

By integration,

$$\int_a^b (f(x) - (a_0 + \cdots + a_n x^n))^2\, dx \leq \varepsilon^2 (b - a).$$

It follows that the powers $1, x, x^2, \ldots$, are closed in X. All statements (A)–(F) now follow with the elements $x_k{}^*$ interpreted as certain modified Legendre polynomials. In particular, if $f \in C[a, b]$ and if

$$\int_a^b f(x) x^n\, dx = 0 \qquad n = 0, 1, 2, \ldots, \quad \text{then} \quad f(x) \equiv 0.$$

Ex. 11.   If X is $L^2[a, b]$ with the same inner product then the same conclusion holds (See Theorems 11.2.1 and 9.2.2).

Ex. 12.   Fourier expansions of continuous functions in terms of Jacobi polynomials converge in norm.

Parseval's identity holds for such expansions. If $w(x) \geq 0$ and $\int_a^b w(x)\, dx < \infty$ then these results may be extended to the class $L^2[a, b; w]$ of measurable functions f for which $\int_a^b w(x) |f|^2\, dx < \infty$.

THEOREM 8.9.2 (Riesz).   *Let X be a complete inner product space and let*

a_k *be constants such that* $\sum\limits_{k=1}^{\infty} |a_k|^2 < \infty$. *Let* $\{x_k{}^*\}$ *be a complete orthonormal sequence. Then there is a* $y \in X$ *such that*

$$(y, x_k{}^*) = a_k \quad k = 1, 2, \ldots. \tag{8.9.15}$$

Proof: Consider the elements $s_n = \sum\limits_{k=1}^{n} a_k x_k{}^*$.
Now $\|s_n - s_m\|^2 = \sum\limits_{k=m+1}^{n} |a_k|^2$. In view of $\sum\limits_{k=1}^{\infty} |a_k|^2 < \infty$, $\{s_n\}$ is a Cauchy sequence, and there is a y such that $\lim\limits_{n\to\infty} \|y - s_n\| = 0$. With k fixed and $n \geq k$,

$$\|y - s_n\| = \|y - s_n\| \, \|x_k{}^*\| \geq |(y - s_n, x_k{}^*)| = |(y, x_k{}^*) - a_k|.$$

Allowing $n \to \infty$, we obtain (8.9.15).

8.10 Further Geometrical Properties of Complete Inner Product Spaces. We have seen in Theorem 8.5.1 that there is a minimum distance from a given element to a linear manifold. How can this be extended to more general subsets? Theorem 8.10.1 provides a sufficient condition of great importance.

THEOREM 8.10.1. *Let* X *be a complete inner product space. Let* M *be a closed (i.e., topologically), convex, and nonempty subset of* X. *Let* $y \in X$ *and set*

$$d = \inf_{x \in M} \|y - x\|. \tag{8.10.1}$$

Then there is a unique x_0 *in* M *such that*

$$\|y - x_0\| = d. \tag{8.10.2}$$

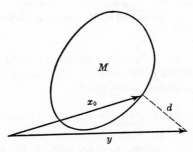

Figure 8.10.1.

Proof: By (8.10.1) we can find a sequence of elements x_n in M such that

$$\lim_{n\to\infty} \|y - x_n\| = d. \tag{8.10.3}$$

The parallelogram theorem (Theorem 8.1.3) tells us that

$$\|x_m - x_n\|^2 = 2\|x_m - y\|^2 + 2\|x_n - y\|^2 - \|2y - x_m - x_n\|^2$$
$$= 2\|x_m - y\|^2 + 2\|x_n - y\|^2 - 4\|y - \tfrac{1}{2}(x_m + x_n)\|^2.$$

M is convex; hence $\tfrac{1}{2}(x_m + x_n)$ is in M and $\|y - \tfrac{1}{2}(x_m + x_n)\|^2 \geq d^2$. Therefore,

$$\|x_m - x_n\|^2 \leq 2\|y - x_m\|^2 + 2\|y - x_n\|^2 - 4d^2.$$

In view of (8.10.3), $\|x_m - x_n\|^2 \to 0$ as m, $n \to \infty$. This means that $\{x_m\}$ is a Cauchy sequence. Since X is complete, there is an x_0 in X such that $\|x_m - x_0\| \to 0$. Since M is closed, x_0 is in M. Now

$$\|y - x_0\| \leq \|y - x_n\| + \|x_n - x_0\| \to d + 0 = d.$$

On the other hand, from (8.10.1), $\|y - x_0\| \geq d$. Therefore $\|y - x_0\| = d$.

Suppose we have x_0 and x_1 with $\|y - x_0\| = \|y - x_1\| = d$. Since M is convex, $\tfrac{1}{2}(x_0 + x_1)$ is in M. Hence,

$$d \leq \|y - \tfrac{1}{2}(x_0 + x_1)\| = \|\tfrac{1}{2}y - \tfrac{1}{2}x_0 + \tfrac{1}{2}y - \tfrac{1}{2}x_1\|$$

$$\leq \tfrac{1}{2}\|y - x_0\| + \tfrac{1}{2}\|y - x_1\| = \frac{d}{2} + \frac{d}{2} = d.$$

Therefore $\|y - \tfrac{1}{2}(x_0 + x_1)\| = d$. By the parallelogram law,

$$\|x_0 - x_1\|^2 = 2\|y - x_0\|^2 + 2\|y - x_1\|^2 - 4\|y - \tfrac{1}{2}(x_0 + x_1)\|^2$$
$$= 2d^2 + 2d^2 - 4d^2 = 0.$$

Therefore $x_0 = x_1$.

Ex. 1. In the plane, let M designate a nonempty, closed, convex set of points. If P is a point not contained in M, there is a unique segment of minimum length connecting P and M.

THEOREM 8.10.2. *Let X be a complete inner product space and let M be a closed linear subspace that is not the whole of X. Then there exists a nonzero element $z \perp M$, i.e., $(z, y) = 0$ for all $y \in M$.*

Proof: Let $w \notin M$. Set $d = \inf_{y \in M} \|w - y\|$. By Theorem 8.10.1, we can find a y_0 in M with $\|w - y_0\| = d$. Let $z = w - y_0$. Now $z \neq 0$, otherwise $w = y_0 \in M$. Since M is linear, $y_0 + Cy$ is in M for all $y \in M$ and all C. Hence, $d \leq \|w - y_0 - Cy\| = \|z - Cy\|$. Then, $\|z - Cy\|^2 - \|z\|^2 \geq 0$. This means that $|C|^2 \|y\|^2 - C(y, z) - \bar{C}(z, y) \geq 0$. In particular, select $C = \sigma(z, y)$ where σ is real.
Then,

$$|(z, y)|^2 \{\sigma^2 \|y\|^2 - 2\sigma\} \geq 0 \quad \text{for all } \sigma.$$

But for $\sigma > 0$ and sufficiently small, $\sigma^2 \|y\|^2 - 2\sigma$ is negative. Hence $(z, y) = 0$.

COROLLARY 8.10.3. *The minimal element extending from a given element to a closed linear subspace is perpendicular to the subspace.*

Here is a second application of Theorem 8.10.1 more directly related to questions of approximation theory.

THEOREM 8.10.4. *Let M designate the set of polynomials of degree $\leq n$ that are convex on $[a, b]$. Let $f(x) \in L^2[a, b]$. Then the problem*

$$\min_{p \in M} \|f - p\|, \quad \|f\|^2 = \int_a^b |f|^2 \, dx, \qquad (8.10.4)$$

possesses a unique solution.

Proof: A polynomial $p(x)$ is convex on $[a, b]$ if and only if $p''(x) \geq 0$ there. If p and q are convex on $[a, b]$, then $tp(x) + (1 - t)q(x)$ is also convex on $[a, b]$ for $0 \leq t \leq 1$, inasmuch as $tp'' + (1 - t)q'' \geq 0$. The set M is therefore convex.

We show next that M is a closed subset of $L^2[a, b]$. Let $p_k(x) \in M$ converge to $f(x) \in L^2[a, b]$; i.e., let $\lim_{k \to \infty} \int_a^b |f(x) - p_k(x)|^2 \, dx = 0$. Let $q(x)$ be the best approximation to $f(x)$ in $\mathscr{P}_n$ (Cf. Cor. 8.5.2). Then,

$$0 \leq \int_a^b |f(x) - q(x)|^2 \, dx \leq \int_a^b |f(x) - p_k(x)|^2 \, dx.$$

Allowing $k \to \infty$ we have $\int_a^b |f(x) - q(x)|^2 \, dx = 0$, so that $f(x) = q(x)$ is in $\mathscr{P}_n$.

Let $P_0{}^*(x), P_1{}^*(x), \ldots, P_n{}^*(x)$ designate the orthonormal polynomials for $[a, b]$. Then,

$$f(x) - p_k(x) \equiv a_{0k}P_0{}^*(x) + \cdots + a_{nk}P_n{}^*(x), \quad k = 1, 2, \ldots,$$

for some constants a_{ik} and hence,

$$\int_a^b |f(x) - p_k(x)|^2 \, dx = \sum_{i=0}^n |a_{ik}|^2 \to 0.$$

Thus,
$$\lim_{k \to \infty} a_{ik} = 0, \quad i = 0, 1, \ldots, n.$$

Now $|f(z) - p_k(z)| \leq \sum_{i=0}^n |a_{ik}| \, |P_i{}^*(z)|$; hence over any bounded region in the complex plane $p_k(z) \to f(z)$ uniformly. Hence $p_k''(z) \to f''(z)$ uniformly there and since $p_k''(x) \geq 0$ on $[a, b]$, it follows that $f''(x) \geq 0$ there. Thus M is closed. The Theorem now follows by an application of Theorem 8.10.1. We refer to Theorem 9.2.2 for the completeness of $L^2[a, b]$.

NOTES ON CHAPTER VIII

General works that have been found useful include Halmos [1] and Aronszajn [1]. Also the references cited in Chapter 7.

8.1   For spaces that do not satisfy the positivity requirement (8.1.1)(d), see Synge [1], Part III.

8.2   The Euclidean geometry of n dimensions is developed in Sommerville [1].

8.5   For additional examples of Tschebyscheff expansions, see Murnaghan and Wrench [2], Clenshaw [1]. Davis and Rabinowitz [2] presents an extensive survey of the applications of least square methods to numerical analysis.

8.7–8.8   Kowalewski [2] pp. 223–229. Gantmacher [1], Vol. I. Kaczmarz and Steinhaus [1] pp. 73–78, Szegö [1] pp. 23–27.

8.9–8.10   Banach [1] Chap. IV, Kaczmarcz and Steinhaus [1], Chap. II. Further discussion of Theorem 8.9.1 is in Olmstead [1].

PROBLEMS

1. If $-\infty < \lambda, \mu < \infty$, then $\displaystyle\lim_{T\to\infty} \frac{1}{2T} \int_{-T}^{T} \cos \lambda x \cos \mu x \, dx = 0$, $\lambda \neq \mu$. What is the value of the limit for $\lambda = \mu$?

2. If X is a real inner product space $\|x + y\| = \|x - y\|$ if and only if $(x, y) = 0$. What if X is a complex inner product space?

3. In a real inner product space $\|x\| = \|y\|$ implies that

$$(x, x - y) = (y, y - x).$$

Give a geometric interpretation.

4. If X is a real inner product space, $(x, y) = 0$ if and only if $\|ax + y\| \geq \|y\|$ for all real a.

5. If x_1 and x_2 are orthonormal, then $x_1 + x_2$ and $x_1 - x_2$ are orthogonal. Geometrical interpretation?

6. Let $\|x_i\| = 1$, $i = 1, 2, \ldots, n$. $\|x_i - x_j\| = 1$, $i \neq j$. Determine the angle between x_i and x_j. Interpretation in R_2 and R_3?

7. Prove that an inner product space is strictly convex.

8. In a real inner product space $(x, y) = \frac{1}{4}[\|x + y\|^2 - \|x - y\|^2]$. In a complex inner product space, $\mathrm{Re}\,(x, y) = \frac{1}{4}[\|x + y\|^2 - \|x - y\|^2]$ $\mathrm{Im}\,(x, y) = -\frac{1}{4}[\|ix + y\|^2 - \|ix - y\|^2]$.

9. Give an example of a normed linear space in which the parallelogram theorem fails.

10. In $C^1[a, b]$ define

$$(1)\ (f, g) = \int_a^b f'(x)g'(x)\, dx$$

$$(2)\ (f, g) = \int_a^b f'(x)g'(x)\, dx + f(a)g(a).$$

Which of these is an inner product?

11. Let X be a real Minkowski plane with points P. It is an inner product space if and only if the locus $\|P\| = 1$ is an ellipse.

12. Estimate $\displaystyle\int_0^1 \frac{x^n\,dx}{1+x}$ from above by the Schwarz inequality. Estimate from below and above by using the mean value theorem for integrals. Compare with the exact answer for $n = 6$.

13. In the case of a complex inner product space two definitions of angle are possible: (1) $\cos\phi = \dfrac{|(x, y)|}{\|x\|\,\|y\|}$, (2) $\cos\phi = \dfrac{\mathrm{Re}\,(x, y)}{\|x\|\,\|y\|}$. According to (1), $\phi = \pi/2$ if and only if $(x, y) = 0$, but the law of cosines does not hold. According to (2), the reverse is true.

14. Let X be an inner product space and $x_1, \ldots, x_n$ be n independent elements. Introduce $L_k(x) = (x, x_k)$ and use the LU decomposition of $((x_i, x_j))$ (2.6.19) to derive the orthonormalized $x_1{}^*, x_2{}^*, \ldots, x_n{}^*$.

15. Let $n > 2$. "Solve" the overdetermined system of real equations in x, y

$$a_1 x + b_1 y = c_1$$
$$\cdot \qquad \cdot$$
$$\cdot \qquad \cdot$$
$$\cdot \qquad \cdot$$
$$a_n x + b_n y = c_n$$

by minimizing $\displaystyle\sum_{i=1}^n (c_i - a_i x - b_i y)^2$.

16. Find the least square polynomial approximations of degrees 0, 2, and 4 to $|x|$ over $[-1, 1]$.

17. Approximate x^2 in $L^2[0, 1]$ by a combination of $1, x$. By a combination of x^{100}, x^{101}. Compare the answers.

18. In an inner product space, minimize

$$\|x_1 - x\|^2 + \|x_2 - x\|^2 + \cdots + \|x_n - x\|^2$$

where x is a linear combination of $x_1, \ldots, x_n$. Interpret geometrically.

19. Show that the value of y that minimizes the sum of the squares of the "relative errors," $\displaystyle\sum_{j=1}^n \left[\frac{a_j - y}{y}\right]^2$, is $y = \displaystyle\sum_{j=1}^n a_j{}^2 \Big/ \sum_{j=1}^n a_j$.

20. Let a triangle T have sides a, b, c. If P is a point in the plane, denote by x, y, z, the distance from P to the sides. What position of P minimizes $x + y + z$?

21. With the notation of Problem 20, show that there is a point P_L such that $\dfrac{x}{a} = \dfrac{y}{b} = \dfrac{z}{c}$. This is known as the *Lemoine Point* of the triangle. Show that this point minimizes $x^2 + y^2 + z^2$. Hint: Use the Lagrange identity

$$(a^2 + b^2 + c^2)(x^2 + y^2 + z^2) = (ax + by + cz)^2$$
$$+ (ay - bx)^2 + (bz - cy)^2 + (cx - az)^2.$$

22. In R_n, $\|y - Ax\|^2$ is minimized for $x = x_1$ if and only if $A'Ax_1 = A'y$.

23. Given $0 < x_1 < x_2 < \cdots < x_n \le 1$; $f_k(x) = 1, 0 \le x \le x_k$; $f_k(x) = 0$, $x_k < x \le 1$; $k = 1, 2, \ldots, n$. Compute $G(f_1, f_2, \ldots, f_n)$.

24. If x_i are normal: $\|x_i\| = 1$, prove that the sequence $\mu_n = g(x_1, x_2, \ldots, x_n)$ is nonincreasing.

25. Let the matrix $G(x_1, x_2, \ldots, x_n)$ have rank r. Then r of the elements $x_1, \ldots, x_n$ are linearly independent and $n - r$ elements are linear combinations of these.

26. Prove that Hadamard's determinant inequality is an equality if and only if the rows or columns of D are orthogonal.

27. Prove that the n-dimensional volume, V_n, of the regular n-simplex of side 1 (generalization to n dimensions of the equilateral triangle) is given by

$$V_n = \frac{1}{n!} \sqrt{\frac{n+1}{2^n}}.$$

28. Let X be an inner product space and $x_1, \ldots, x_n$ be n independent elements. Then, $g(y, x_1, \ldots, x_n) = |(y, z)|^2$ for all y, and where z depends only on $x_1, \ldots, x_n$, if and only if the dimension of $X \leq n + 1$. The z is unique. If the dimension of $X < n + 1$, $z = 0$. The element z is known as the *Grassman outer product of* $x_1, \ldots, x_n$.

29. Use the result of the last problem to generalize the formula of analytic geometry for the distance from a point to a line. In R_{n+1}, let $x_1, \ldots, x_n$ be independent. Then, the distance d from y to the hyperplane spanned by $x_1, \ldots, x_n$ is given by $d = \dfrac{|(y, z)|}{\|z\|}$. Check the case $n = 2$.

30. Suppose there is a real valued function of n elements of an inner product space $V(x_1, x_2, \ldots, x_n)$ such that (1) $V(x_1, \ldots, x_n) = \|x_1\| \|x_2\| \cdots \|x_n\|$ whenever $x_1, \ldots, x_n$ are orthogonal. (2) $V(x_1, \ldots, x_j, \ldots, x_k, \ldots, x_n) = V(x_1, \ldots, x_j + \sigma x_k, \ldots, x_k, \ldots, x_n)$ for all constants σ and all j, k $(j \neq k)$. Prove that $V(x_1, \ldots, x_n) = g^{\frac{1}{2}}(x_1, \ldots, x_n)$.

31. Give a geometric interpretation of Hadamard's Inequality and of inequality (8.8.1)(e).

32. Let $g(x) \in C^2[0, 1]$ and suppose that $g''(x) < 0$ at some point $x \in [0, 1]$. Suppose that a polynomial $p(x)$ is such that $|g(x) - p(x)| \leq \varepsilon$ for $0 \leq x \leq 1$. For ε sufficiently small, there is a point η in $[0, 1]$ where $p''(\eta) < 0$.

33. What is the implication of the previous exercise about the possibility of uniform convex polynomial approximations?

34. Let $\mathscr{P}_n$ designate a normed linear space of polynomials of degree $\leq n$. If $p_k(z)$ is a sequence of elements of $\mathscr{P}_n$ for which $\|p_k\| \to 0$, then $p_k(z) \to 0$ uniformly on any bounded set.

35. If X satisfies all conditions for an inner product space, except that $(x, x) = 0$ does not necessarily imply $x = 0$, X is called an *indefinite inner product space*. Give examples of such a space. How much of the present chapter is valid for such spaces?

CHAPTER IX

Hilbert Space

9.1 Introduction. Hilbert space is the natural generalization to an infinite number of dimensions of the real or complex Euclidean spaces R_n and C_n. There are many advantages to be gained from working in a Hilbert space. In the first place, our spatial intuition acquired in 1, 2, and 3 dimensions carries over to some extent, and theorems and processes can be "seen" geometrically as well as analytically. In the second place, the norm in the space is associated with a quadratic expression, so that the processes of minimization lead to linear problems. Finally, all (separable) Hilbert spaces are abstractly equivalent to one another. This means that the theorems established have wide application.

DEFINITION 9.1.1. A complete inner product space will be called a *Hilbert space*, H, if the following additional requirements are fulfilled

(a) H is infinite dimensional; that is, given any integer n, we can find n independent elements. $\hspace{2cm}$ (9.1.1)

(b) There is a closed (or complete) sequence of elements in H.

We have already observed in Chapter VIII that inner product spaces have a good bit of geometry associated with them. The requirement that H be complete means that all the conditions A–F of Theorem 8.9.1 are equivalent.

Condition (9.1.1)(a) provides H with more dimensions than any R_n while condition (b) restricts the number of dimensions to being countably infinite. These conditions are largely a matter of convenience, and the practice of authors with respect to them varies.

A Hilbert space is, at the very first level, a linear space. This linear space may be either real or complex. Accordingly, the Hilbert Space is spoken of as real or complex.

If X is an inner product space and the sequence $x_1, x_2, \ldots$, is complete, we can find a subsequence $x_{k_1}, x_{k_2}, \ldots$ (possibly finite) that is both complete and independent. For, beginning with the first nonzero element, inspect the sequence and strike out the first element that depends upon the previous elements. Inspect the subsequence that remains and do the same. Proceeding in this way we obtain a subsequence $x_{k_1}, x_{k_2}, \ldots$, of independent elements. Moreover, any element x_q struck from the list is a linear combination of a certain number of elements of the subsequence: $x_q = a_1 x_{k_1} + \cdots + a_p x_{k_p}$.

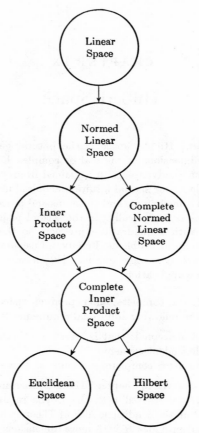

Figure 9.1.1   The Hierarchy of Linear Spaces.

Hence, $(y, x_{k_i}) = 0,\ i = 1, 2, \dots,$ implies $(y, x_i) = 0,\ i = 1, 2, \dots.$ This means that the subsequence is itself complete.

If X is an infinite dimensional complete inner product space, a finite sequence $x_1, \dots, x_n$ cannot be complete. For, otherwise we may suppose it to be independent, as above, and orthonormalize it to obtain a complete orthonormal sequence $x_1{}^*, \dots, x_n{}^*.$ In view of (8.9.6), any element y equals $\sum_{k=1}^{n} (y, x_k{}^*)x_k{}^*$ so that there cannot be $n + 1$ independent elements.

In view of this discussion and Theorem 8.3.3, condition (9.1.1) may be replaced by

$$\text{There is a complete orthonormal infinite} \atop \text{sequence of elements in } H. \qquad (9.1.2)$$

Recall that a metric space S is *separable* if there is a denumerable dense set of elements. Condition (9.1.1)(b) implies that H is separable. For let $\{x_i{}^*\}$ be a complete orthonormal sequence. The finite linear combinations of the $x_i{}^*$ with rational (real or complex) coefficients is a denumerable set. Given $y \in H$ and $\varepsilon > 0$, then by Theorem 8.9.1(B), for sufficiently large n,

$$\left\| y - \sum_{k=1}^{n} (y, x_k{}^*)x_k{}^* \right\| < \varepsilon/2.$$ If rationals r_k are selected so that

$$\sum_{k=1}^{n} |r_k - (y, x_k{}^*)|^2 < \varepsilon^2/4,$$

then

$$\left\| \sum_{k=1}^{n} (y, x_k{}^*)x_k{}^* - \sum_{k=1}^{n} r_k x_k{}^* \right\|^2 < \varepsilon^2/4.$$

Hence $\left\| y - \sum_{k=1}^{n} r_k x_k{}^* \right\| < \varepsilon$. This approximant to y is one of the denumerable set.

Hilbert spaces as defined here are sometimes called *separable Hilbert spaces*, but we shall omit the qualifying term.

The following lemma is of occasional use.

LEMMA 9.1.1. *Let $x_1{}^*, \ldots, x_n{}^*$ be n orthonormal elements in a Hilbert space. We can augment these elements by $x_{n+1}^*, x_{n+2}^*, \ldots$, such that $x_1{}^*, x_2{}^*, \ldots$, is complete and orthonormal.*

Proof: The Hilbert space has a complete sequence $y_1, y_2, \ldots$. The sequence $x_1{}^*, \ldots, x_n{}^*, y_1, y_2, \ldots$, is obviously complete (since $(y, y_i) = 0$, $i = 1, 2, \ldots$, already implies $y = 0$). By the process described above, we can extract a subsequence that is complete and independent. The subsequence will begin with $x_1{}^*, \ldots, x_n{}^*$ since these elements are independent. Orthonormalizing this subsequence, we obtain $x_1{}^*, \ldots, x_n{}^*$ over again, plus additional elements that we call $x_{n+1}^*, x_{n+2}^*, \ldots$.

9.2 Three Hilbert Spaces. There are many known examples of Hilbert spaces, but we limit our presentation to three.

I

THEOREM 9.2.1. *The set of all infinite sequences $\{a_i\}$ for which*

$$\sum_{i=1}^{\infty} |a_i|^2 < \infty, \tag{9.2.1}$$

augmented by the usual definitions for addition and scalar products and by

$$(a, b) = \sum_{i=1}^{\infty} a_i \overline{b_i}, \quad a = \{a_i\}, \ b = \{b_i\}, \tag{9.2.2}$$

as the definition of an inner product, constitutes a Hilbert space.

It is called the *sequential Hilbert space* and is designated by ℓ^2.

Proof: We have already observed in Ex. 5, 7.2, that ℓ^2 is a normed linear space. For any two elements a, b of ℓ^2, we have by (7.2.9)

$$\sum_{i=1}^{n} |a_i \overline{b}_i| \leq \left(\sum_{i=1}^{n} |a_i|^2 \right)^{\frac{1}{2}} \left(\sum_{i=1}^{n} |b_i|^2 \right)^{\frac{1}{2}}.$$

Hence, $\sum_{i=1}^{\infty} |a_i \overline{b}_i| < \infty$ and $\sum_{i=1}^{\infty} a_i \overline{b}_i$ converges. The expression (a, b) is therefore defined for all a, $b \in \ell^2$, and it is now a trivial computation to show that it satisfies the requirements (8.1.1) (a)–(d) for an inner product.

We prove next that the space is complete. Let $a^{(n)} = \{a_i^{(n)}\}$ be a Cauchy sequence. That is, let

$$\lim_{m,n \to \infty} \|a^{(m)} - a^{(n)}\|^2 = \lim_{m,n \to \infty} \sum_{i=1}^{\infty} |a_i^{(m)} - a_i^{(n)}|^2 = 0. \tag{9.2.3}$$

For each fixed subscript i, then, we have $\lim_{m,n \to \infty} |a_i^{(m)} - a_i^{(n)}| = 0$. Each of the sequences $a_i^{(1)}, a_i^{(2)}, \ldots$, is therefore a Cauchy sequence of real or complex numbers, and by the completeness of these spaces, possesses a limit:

$$\lim_{m \to \infty} a_i^{(m)} = a_i, \quad i = 1, 2, 3 \cdots. \tag{9.2.4}$$

For each integer k, set

$$\sigma_k^2 = \sup_{m,n \geq k} \|a^{(m)} - a^{(n)}\|^2. \tag{9.2.5}$$

By (9.2.3), we have

$$\lim_{k \to \infty} \sigma_k = 0. \tag{9.2.6}$$

Now for N arbitrary and for all $m \geq n$, we have

$$\sum_{i=1}^{N} |a_i^{(m)} - a_i^{(n)}|^2 \leq \sum_{i=1}^{\infty} |a_i^{(m)} - a_i^{(n)}|^2 = \|a^{(m)} - a^{(n)}\|^2 \leq \sigma_n^2. \tag{9.2.7}$$

In (9.2.7) keep N and n fixed and let $m \to \infty$. Since we are dealing with a finite sum, we have

$$\sum_{i=1}^{N} |a_i - a_i^{(n)}|^2 \leq \sigma_n^2. \tag{9.2.8}$$

By Minkowski's inequality,

$$\left(\sum_{i=1}^{N} |a_i|^2 \right)^{\frac{1}{2}} \leq \left(\sum_{i=1}^{N} |a_i^{(n)}|^2 \right)^{\frac{1}{2}} + \left(\sum_{i=1}^{N} |a_i - a_i^{(n)}|^2 \right)^{\frac{1}{2}} \tag{9.2.9}$$
$$\leq \|a^{(n)}\| + \sigma_n.$$

This is true for all N. Keeping n fixed, allow $N \to \infty$ and obtain

$$\left(\sum_{i=1}^{\infty} |a_i|^2 \right)^{\frac{1}{2}} \leq \|a^{(n)}\| + \sigma_n. \tag{9.2.10}$$

This shows us that the sequence $\{a_i\}$ is an element of ℓ^2.

We next show that

$$\lim_{n \to \infty} \|a - a^{(n)}\| = 0 \tag{9.2.11}$$

where $a = \{a_i\}$.

In (9.2.8) allow $N \to \infty$. This yields

$$\sum_{i=1}^{\infty} |a_i - a_i^{(n)}|^2 \leq \sigma_n^2.$$

In view of (9.2.6), (9.2.11) follows. The Cauchy sequence $a^{(n)}$ approaches the element a as a limit.

The unit elements $u_1 = (1, 0, \ldots)$, $u_2 = (0, 1, 0, \ldots)$, $\ldots$, are independent and orthonormal. They are complete, for if $a = (a_1, a_2, \ldots) \in \ell^2$, then $a_i = (a, u_i) = 0$, $i = 1, 2, \ldots$, implies $a = 0$.

II

The second Hilbert space is $L^2[a, b]$, consisting of all functions defined on $[a, b]$ which are measurable and for which $|f(x)|^2$ is integrable. The inner product is defined by

$$(f, g) = \int_a^b f(x)\overline{g(x)} \, dx \tag{9.2.12}$$

and the norm by

$$\|f\|^2 = (f, f) = \int_a^b |f(x)|^2 \, dx. \tag{9.2.13}$$

We may consider functions that are real valued or that are complex valued. This leads to two separate spaces, but the proofs are the same for each.

It should be recalled that two functions differing only on a set of zero measure have the same Lebesgue integral. Hence, according to (9.2.13), there are functions not identically zero with zero norm. In order to avoid this difficulty, we treat as identical any two functions whose values differ on a set of zero measure at most. This means that the elements of our space should not be the functions themselves but equivalence classes of functions. To set up the work in this form is a nicety, and we shall not insist upon it.

THEOREM 9.2.2. $L^2[a, b]$ *with inner product* (9.2.12) *and the identification of functions discussed above is a Hilbert space.*

Proof: We have seen (Ex. 10, 7.2) that $L^2[a, b]$ is a normed linear space and that the inner product expression (9.2.12) has meaning for $f, g \in L^2[a, b]$. By Theorem 1.4.0, and simple properties of the integral, $L^2[a, b]$ is an inner product space as well.

We next show that $L^2[a, b]$ is complete. That is, we show that every

Cauchy sequence converges to an element of the space. In symbols,

$$\lim_{m,n\to\infty} \int_a^b |f_m(x) - f_n(x)|^2\, dx = \lim_{m,n\to\infty} \|f_m - f_n\|^2 = 0 \qquad (9.2.14)$$

implies the existence of an $f(x) \in L^2[a, b]$ for which

$$\lim_{n\to\infty} \int_a^b |f(x) - f_n(x)|^2\, dx = \lim_{n\to\infty} \|f - f_n\|^2 = 0. \qquad (9.2.15)$$

In view of (9.2.14), we can find a strictly increasing sequence of positive integers $n_1, n_2, \ldots,$ such that

$$\|f_{n_k} - f_{n_{k+1}}\| \le \frac{1}{2^k}, \quad k = 1, 2, \ldots. \qquad (9.2.16)$$

If g is an arbitrary function of $L^2[a, b]$, then by the Schwarz inequality, $\int_a^b |g(x)|\, |f_{n_k}(x) - f_{n_{k+1}}(x)|\, dx \le \|g\|\, \|f_{n_k} - f_{n_{k+1}}\| \le \|g\|\, 2^{-k}$. Hence,

$$\sum_{k=1}^\infty \int_a^b |g(x)|\, |f_{n_k}(x) - f_{n_{k+1}}(x)|\, dx \le \|g\|\, (\tfrac{1}{2} + \tfrac{1}{4} + \cdots) = \|g\|. \qquad (9.2.17)$$

Thus, interchanging summation and integration,

$$\int_a^b |g(x)| \sum_{k=1}^\infty |f_{n_k}(x) - f_{n_{k+1}}(x)|\, dx \le \|g\| < \infty.$$

This tells us that

$$|g(x)| \sum_{k=1}^\infty |f_{n_k}(x) - f_{n_{k+1}}(x)| < \infty \qquad (9.2.18)$$

almost everywhere on $[a, b]$, and hence that $\sum_{k=1}^\infty |f_{n_k}(x) - f_{n_{k+1}}(x)| < \infty$ almost everywhere. This last statement is true because if the series diverged on a set of positive measure, we could take a test function g that was nonzero on this set, and obtain a contradiction to (9.2.18).

Now $\sum_{k=1}^\infty (f_{n_{k+1}}(x) - f_{n_k}(x))$ must also converge almost everywhere. Its partial sums are $f_{n_{k+1}}(x) - f_{n_1}(x)$. Hence, for an appropriate function $f(x)$ defined almost everywhere,

$$\lim_{k\to\infty} f_{n_k}(x) = f(x). \qquad (9.2.19)$$

We next show that $f(x) \in L^2[a, b]$. In view of (9.2.19), for fixed j, we have almost everywhere $\lim_{k\to\infty} |f_{n_k}(x) - f_{n_j}(x)|^2 = |f(x) - f_{n_j}(x)|^2$. Hence by Fatou's lemma,

$$\int_a^b |f(x) - f_{n_j}(x)|^2\, dx \le \liminf_{k\to\infty} \int_a^b |f_{n_k}(x) - f_{n_j}(x)|^2\, dx. \qquad (9.2.20)$$

Now

$$\|f_{n_k} - f_{n_j}\| \leq \|f_{n_j} - f_{n_{j+1}}\| + \|f_{n_{j+1}} - f_{n_{j+2}}\| + \cdots + \|f_{n_{k-1}} - f_{n_k}\|$$
$$\leq \frac{1}{2^j} + \frac{1}{2^{j+1}} + \cdots = \frac{1}{2^{j-1}}.$$

Hence,

$$\|f_{n_k} - f_{n_j}\|^2 \leq \frac{1}{2^{2j-2}}, \text{ so that } \liminf_{k \to \infty} \int_a^b |f_{n_k}(x) - f_{n_j}(x)|^2 \, dx \leq \frac{1}{2^{2j-2}} < \infty.$$

This means that $\int_a^b |f(x) - f_{n_j}(x)|^2 \, dx < \infty$, so that $f - f_{n_j} \in L^2[a, b]$. Since $f = (f - f_{n_j}) + f_{n_j}$, and each of these is in $L^2[a, b]$, their sum, f, must be in $L^2[a, b]$. From (9.2.20) and the last inequality for the lim inf,

$$\lim_{j \to \infty} \|f - f_{n_j}\| = 0.$$

Now $\|f - f_n\| \leq \|f - f_{n_j}\| + \|f_{n_j} - f_n\|$. The first term on the right can be made arbitrarily small, as we have seen. The second can also be made arbitrarily small in virtue of (9.2.14). Hence (9.2.15) follows.

To wind up the proof of the theorem, we must show that $L^2[a, b]$ is infinite dimensional and contains a complete (or closed) sequence. The functions $1, x, x^2, \ldots$, are in $L^2[a, b]$ and are independent. Moreover, they are complete. This will be established in Theorem 11.2.1.

The proof of completeness is capable of wide generalization. In the first place, completeness holds with the norm, $\|f\|^p = \int_a^b |f(x)|^p \, dx$, $p \geq 1$. Secondly, positive weighting functions may be used. For $p = 2$, each weight leads to a corresponding Hilbert space.

III

The third Hilbert space to be studied here is comprised of certain single valued analytic functions. It has a totally different flavor from the two previous spaces for the reason that convergence in norm now implies uniform convergence. A certain part of the discussion that follows could have been abridged by employing the Lebesgue integral, but it is of some interest to see the theory built up with only the Riemann integral.

Let B designate a fixed region (open connected set) lying in the complex z plane. It is clear (intuitively at least for simple regions, and we shall not pursue the topological question further†) that we can find a sequence of closed bounded regions $B_1, B_2, \ldots$, with the following properties

   (a) B_n is contained in B, $n = 1, 2, \ldots$.

   (b) B_n is contained in the interior of B_{n+1}, $n = 1, 2, \ldots$.

   (c) The sequence B_n exhausts B in the sense that any point of B (with the exception of $z = \infty$, if it lies in B) ultimately belongs to some B_n and hence to all subsequent B's.

† See Walsh, [1], p. 10.

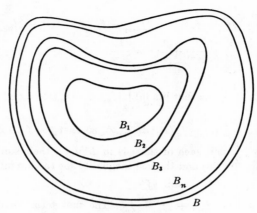

Figure 9.2.1.

If S in any closed bounded set in B then S is contained in B_n for all n sufficiently large. For suppose not. Then we shall be able to find a point z_1 in S but not in B_1, z_2 in S but not in B_2, ... The points $z_1, z_2, \ldots$ have a limit point $z^* \neq \infty$ which is in S. Now z^* lies in some B_m by hypothesis (c) and hence is interior to B_{m+1} by (b). A whole neighborhood of z^* therefore lies in B_{m+1}, and this means that an infinity of points of the sequence $z_1, z_2, \ldots$, lie in B_{m+1}. Let $q \, (> m + 1)$ be such that $z_q \in B_{m+1}$. Then $z_q \in B_q$. This is a contradiction.

We shall deal with functions $w(z)$ defined on B and possessing Riemann integrals over B_n, $n = 1, 2, \ldots$, for all sequences B_n.

If $\lim\limits_{n \to \infty} \iint\limits_{B_n} w(z) \, dx \, dy$ exists for all sequences B_n and its value is independent of the particular sequence B_n assumed to satisfy (a)–(c), we shall write

$$\iint\limits_{B} w(z) \, dx \, dy = \lim_{n \to \infty} \iint\limits_{B_n} w(z) \, dx \, dy. \tag{9.2.21}$$

If $w(z) \geq 0$, and if for some sequence B_n we set $I_n = \iint\limits_{B_n} w(z) \, dx \, dy$, then

$B_n \subset B_{n+1}$ implies that I_n is nondecreasing. Hence, $\lim\limits_{n \to \infty} I_n$ exists (or is $+\infty$). Suppose that $\lim\limits_{n \to \infty} I_n = I < \infty$. Let D_n be a second sequence satisfying (a)–(c). A fixed D_n is contained in some B_k for k sufficiently large. Hence

$$\iint\limits_{D_n} w(z) \, dx \, dy \leq \iint\limits_{B_k} w(z) \, dx \, dy \leq I. \tag{9.2.22}$$

Thus $\lim\limits_{n \to \infty} \iint\limits_{D_n} w(z)\, dx\, dy = L \leq I.$

We may reverse the roles of the D's and the B's and obtain $I \leq L$. Hence $I = L$, and this tells us that the limit is independent of the particular selection of regions B_n.

DEFINITION 9.2.1.   Let B be a region. The set of functions $f(z)$ which are single valued and analytic in B and for which

$$\iint\limits_{B} |f(z)|^2\, dx\, dy < \infty. \qquad (9.2.23)$$

will be designated by $L^2(B)$.

Ex. 1.   If B is bounded and $f(z)$ remains analytic or even continuous in the closure of B then $f \in L^2(B)$.

LEMMA 9.2.3.   *If $f(z)$ and $g(z) \in L^2(B)$ then the linear combination $af(z) + bg(z) \in L^2(B)$ for all complex constants a, b. $L^2(B)$ is therefore a linear space.*

Proof: First of all, the combination $af(z) + bg(z)$ is single valued and analytic in B. Now,

$$|af(z) + bg(z)|^2 + |af(z) - bg(z)|^2 \equiv 2(|a|^2\, |f(z)|^2 + |b|^2\, |g(z)|^2).$$

Hence $|af(z) + bg(z)|^2 \leq 2(|a|^2\, |f(z)|^2 + |b|^2\, |g(z)|^2)$. Therefore

$$\iint\limits_{B_n} |af(z) + bg(z)|^2\, dx\, dy \leq 2\, |a|^2 \iint\limits_{B_n} |f(z)|^2\, dx\, dy + 2\, |b|^2 \iint\limits_{B_n} |g(z)^2|\, dx\, dy.$$

Allowing $n \to \infty$, the last two integrals possess finite limits by hypothesis.

Therefore $\lim\limits_{n \to \infty} \iint\limits_{B_n} |af(z) + bg(z)|^2\, dx\, dy$ exists and is $< \infty$.

LEMMA 9.2.4.   *If f and $g \in L^2(B)$ then $\iint\limits_{B} f\bar{g}\, dx\, dy$ exists.*

Proof: The following is an algebraic identity:

$$f\bar{g} = \frac{1}{2}\, |f + g|^2 + \frac{i}{2}\, |f + ig|^2 - \frac{1 + i}{2}\, |f|^2 - \frac{1 + i}{2}\, |g|^2.$$

This expresses the function $f\bar{g}$ as a linear combination of nonnegative functions.

Integrating over B_n,

$$\iint\limits_{B_n} f\bar{g}\, dx\, dy = \frac{1}{2}\iint\limits_{B_n} |f+g|^2\, dx\, dy + \frac{i}{2}\iint\limits_{B_n} |f+ig|^2\, dx\, dy$$

$$- \frac{1+i}{2}\iint\limits_{B_n} |f|^2\, dx\, dy - \frac{1+i}{2}\iint\limits_{B_n} |g|^2\, dx\, dy.$$

Since f and g are in $L^2(B)$, the previous lemma tells us that both $f + g$ and $f + ig$ are in $L^2(B)$. The limits of each of the four integrals on the right, as $n \to \infty$ exist independently of the sequence B_n and are $< \infty$. This must also be true of the integral on the left.

LEMMA 9.2.5.   *For f, $g \in L^2(B)$, write*

$$(f, g) = \iint\limits_{B} f\bar{g}\, dx\, dy. \tag{9.2.24}$$

The expression (f, g) is an inner product for $L^2(B)$. With it, $L^2(B)$ is an inner product space and therefore a normed linear space wherein

$$\|f\|^2 = \iint\limits_{B} |f|^2\, dx\, dy. \tag{9.2.25}$$

Proof: Since $(f+g)\bar{h} = f\bar{h} + g\bar{h}$, $(\alpha f)\bar{g} = \alpha(f\bar{g})$, $f\bar{g} = \overline{g\bar{f}}$, the algebraic properties required of an inner product are evident by integrating over B_n and passing to the limit. If $(f, f) = \iint\limits_{B} |f|^2\, dx\, dy = 0$, then $\iint\limits_{B_n} |f|^2\, dx\, dy = 0$ for all n. Since $|f|^2$ is continuous over B_n, it follows that $f(z) \equiv 0$ on B_n, and hence throughout B.

LEMMA 9.2.6.   *Let C_r designate the closed circle $|z - z_0| \le r$. Suppose that $f(z)$ is analytic in C_r. If*

$$f(z) = a_0 + a_1(z - z_0) + a_2(z - z_0)^2 + \dots, \tag{9.2.26}$$

then

$$\iint\limits_{C_r} |f(z)|^2\, dx\, dy = \pi \sum_{n=0}^{\infty} |a_n|^2 \frac{r^{2n+2}}{n+1}. \tag{9.2.27}$$

Proof: The series (9.2.26) converges uniformly and absolutely in C_r. For this reason,

$$\iint\limits_{C_r} |f(z)|^2\, dx\, dy = \iint\limits_{C_r} \left(\sum_{m=0}^{\infty} a_m(z-z_0)^m \right)\left(\sum_{m=0}^{\infty} \bar{a}_m(\bar{z}-\bar{z}_0)^m \right) dx\, dy$$

$$= \sum_{m,n=0}^{\infty} a_m\bar{a}_n \iint\limits_{C_r} (z-z_0)^m(\bar{z}-\bar{z}_0)^n\, dx\, dy.$$

Set $z - z_0 = \rho e^{i\theta}$. Then

$$\iint\limits_{C_r} (z - z_0)^m (\bar{z} - \bar{z}_0)^n \, dx \, dy = \int_0^{2\pi} d\theta \, e^{(m-n)i\theta} \int_0^r \rho^{m+n+1} \, d\rho.$$

If $m \neq n$, the inner integral vanishes. When $m = n$, we have $\pi \dfrac{r^{2(n+1)}}{n+1}$.

Thus, $\quad \displaystyle\iint\limits_{C_r} |f(z)|^2 \, dx \, dy = \pi \sum_{n=0}^{\infty} |a_n|^2 \frac{r^{2n+2}}{n+1}.$

THEOREM 9.2.7. *Let* $f(z) = \displaystyle\sum_{n=0}^{\infty} a_n(z - z_0)^n.$ *Then* $S = \pi \displaystyle\sum_{n=0}^{\infty} |a_n|^2 \dfrac{r^{2n+2}}{n+1} < \infty$ *if and only if* $f \in L^2(C_r)$. *If* $f \in L^2(C_r)$ *then* $S = \|f\|^2$.

Proof: Assume $S < \infty$. Then, for some constant M,

$$|a_n|^2 \frac{r^{2n+2}}{n+1} \leq M, \, n = 0, 1, \ldots.$$

Hence,

$$|a_n|^{1/n} \leq \frac{M^{1/2n}(n+1)^{1/2n}}{r \cdot r^{1/n}}.$$

If ρ designates the radius of convergence of $\displaystyle\sum_{n=0}^{\infty} a_n(z - z_0)^n$, then

$$\frac{1}{\rho} = \limsup_{n \to \infty} |a_n|^{1/n} \leq \frac{1}{r}.$$

Thus, $\rho \geq r$ and therefore $f(z)$ is analytic in $|z - z_0| < r$. Select an r' with $0 < r' < r$. If $C_{r'}$ is $|z - z_0| \leq r'$, then by Lemma 9.2.6,

$$\iint\limits_{C_{r'}} |f(z)|^2 \, dx \, dy = \pi \sum_{n=0}^{\infty} |a_n|^2 \frac{(r')^{2n+2}}{n+1} < S < \infty.$$

Therefore $\displaystyle\lim_{r' \to r} \iint\limits_{C_{r'}} |f(z)|^2 \, dx \, dy < \infty$ and therefore $f \in L^2(C_r)$.

Suppose, conversely, that $f \in L^2(C_r)$. Then, for any r', $0 < r' < r$, $f(z)$ is analytic in the closed circle $|z - z_0| \leq r'$. Hence by the lemma,

$$\iint\limits_{C_{r'}} |f(z)|^2 \, dx \, dy = \pi \sum_{n=0}^{\infty} |a_n|^2 \frac{(r')^{2n+2}}{n+1} = S(r') < \infty.$$

Moreover, $\displaystyle\lim_{r' \to r^-} S(r')$ exists and equals $\|f\|^2 \, (< \infty)$. For any N,

$$\pi \sum_{n=0}^{N} |a_n|^2 \frac{(r')^{2n+2}}{n+1} \leq S(r').$$

Allowing $r' \to r^-$,

$$\pi \sum_{n=0}^{N} |a_n|^2 \frac{r^{2n+2}}{n+1} \leq \|f\|^2.$$

This estimate is independent of N. Allowing $N \to \infty$, we obtain

$$S \leq \|f\|^2 < \infty.$$

Since $S(r') \leq S$, by allowing $r' \to r^-$ we have $\|f\|^2 \leq S$ and therefore $S = \|f\|^2$.

LEMMA 9.2.8.   *Let $f(z) \in L^2(B)$. Let $r(z_0)$ be the distance from a fixed point $z_0 \in B$ to the boundary of B. Then,*

$$|f(z_0)|^2 \leq \frac{\|f\|^2}{\pi r^2(z_0)}. \tag{9.2.28}$$

Proof: Let $0 < r' < r(z_0)$. Then the circle $C_{r'}$: $|z - z_0| \leq r'$ is contained in B. This implies that $\iint\limits_{C_{r'}} |f(z)|^2 \, dx \, dy \leq \iint\limits_{B} |f(z)|^2 \, dx \, dy = \|f\|^2$. But from (9.2.27), ignoring all but the first term,

$$\pi |f(z_0)|^2 r'^2 = \pi |a_0|^2 r'^2 \leq \iint\limits_{C_{r'}} |f(z)|^2 \, dx \, dy \leq \|f\|^2.$$

Then,

$$|f(z_0)|^2 \leq \frac{\|f\|^2}{\pi r'^2}.$$

This is true for all $0 < r' < r$ and hence (9.2.28) follows.

LEMMA 9.2.9. $L^2(B)$ *is a complete inner product space.*

Proof: Let $\{f_n(z)\}$ be a Cauchy sequence of functions in $L^2(B)$. Given an ε, we can find an $N(\varepsilon)$ such that

$$\|f_m(z) - f_n(z)\| \leq \varepsilon, \quad m, n \geq N(\varepsilon). \tag{9.2.29}$$

We wish to show the existence of an analytic function $f(z)$ which is in $L^2(B)$ and for which $\lim\limits_{n \to \infty} \|f(z) - f_n(z)\| = 0$. Select a fixed B_k. On B_k we have from the previous Lemma,

$$|f_m(z) - f_n(z)| \leq \frac{\|f_m - f_n\|}{\pi^{\frac{1}{2}} r(z)}. \tag{9.2.30}$$

If $\rho(> 0)$ designates the minimum distance from B_k to the boundary of B, then we have uniformly in B_k,

$$|f_m(z) - f_n(z)| \leq \frac{\|f_m - f_n\|}{\pi^{\frac{1}{2}} \rho}. \tag{9.2.31}$$

In view of (9.2.31) and (9.2.29), $\{f_m(z)\}$ is a Cauchy sequence of functions with respect to the norm $\|\phi\| = \max\limits_{z \in B_k} |\phi(z)|$. (Cf. Ex. 8, 7.2.) The sequence therefore converges uniformly in B_k to a function $f(z)$ which must be analytic in the interior of B_k. Since B_k is arbitrary, $f(z)$ must be analytic in the whole of B. Now, for n and k fixed, $\lim\limits_{m \to \infty} \iint\limits_{B_k} |f_m(z) - f_n(z)|^2 \, dx \, dy =$

$\iint\limits_{B_k} |f(z) - f_n(z)|^2 \, dx \, dy$. This is true because of the uniform convergence of

f_m to f on B_k. For $m, n \geq N(\varepsilon)$ we have for all k,

$$\iint\limits_{B_k} |f_m(z) - f_n(z)|^2 \, dx \, dy \leq \iint\limits_{B} |f_m(z) - f_n(z)|^2 \, dx \, dy \leq \varepsilon^2.$$

Allowing $m \to \infty$,

$$\iint\limits_{B_k} |f(z) - f_n(z)|^2 \, dx \, dy \leq \varepsilon^2.$$

This statement is independent of k; hence

$$\lim\limits_{k \to \infty} \iint\limits_{B_k} |f(z) - f_n(z)|^2 \, dx \, dy \leq \varepsilon^2.$$

This implies that $f(z) - f_n(z)$ is in $L^2(B)$. Since $f_n(z)$ is in $L^2(B)$, their sum $f(z)$ is in $L^2(B)$. Since $\iint\limits_{B} |f(z) - f_n(z)|^2 \, dx \, dy \leq \varepsilon^2$ for $n \geq N(\varepsilon)$, we have convergence in norm to $f(z)$.

Having established that $L^2(B)$ is a complete inner product space, it remains only to show that $L^2(B)$ is infinite dimensional, that it contains a closed sequence of elements, and we will have proved that $L^2(B)$ is a Hilbert space. As with $L^2[a, b]$, we could refer to Theorem 11.4.8 telling us that the complex powers are closed in $L^2(B)$ for certain types of region B. However, we shall prove a stronger result by means of the Fréchet-Riesz representation theorem whose proof will be given shortly.

THEOREM 9.2.10. $L^2(B)$ *contains a complete sequence of functions. If B is bounded, or can be mapped 1-1 conformally onto a bounded region, then $L^2(B)$ is a Hilbert space.*

Proof: Let t be a fixed point of B. As is shown in 9.3, Ex. 4, $L_n(f) = f^{(n)}(t)$, $n = 0, 1, 2, \ldots$, are bounded linear functionals over $L^2(B)$. By Theorem 9.3.3, there exists, for each n, an element $g_n(z) \in L^2(B)$ such that

$$L_n(f) = (f, g_n) = f^{(n)}(t), \qquad f \in L^2(B). \tag{9.2.32}$$

If now $(f, g_n) = 0$, $n = 0, 1, \ldots$, then $f^{(n)}(t) = 0$, $n = 0, 1, \ldots$, and this implies that $f \equiv 0$. Therefore g_n is a complete sequence of functions.

If B is bounded, then $1, z, z^2, \ldots$, are independent and are all in $L^2(B)$. Hence $L^2(B)$ is infinite dimensional. If B can be mapped 1-1 conformally onto a bounded region D, then we can find an infinite sequence of independent functions in $L^2(B)$ by a change of variable. For, let D lie in the w plane, $w = u + iv$, and suppose that

$$w = M(z), \qquad z = x + iy \in B, \tag{9.2.33}$$

maps B 1-1 conformally onto D. Let D_n be a sequence for D satisfying (a)–(c). Let B_n be the images of D_n under the inverse map of (9.2.33). The B_n will be a sequence for B satisfying (a)–(c), If $f(w) \in L^2(D)$, we have

$$\iint\limits_{D_n} |f(w)|^2 \, du \, dv = \iint\limits_{B_n} |f(M(z))|^2 \begin{vmatrix} \dfrac{\partial u}{\partial x} & \dfrac{\partial u}{\partial y} \\[2mm] \dfrac{\partial v}{\partial x} & \dfrac{\partial v}{\partial y} \end{vmatrix} dx \, dy \tag{9.2.34}$$

But by the Cauchy-Riemann equations,

$$\frac{\partial u}{\partial x} = \frac{\partial v}{\partial y}, \quad \frac{\partial u}{\partial y} = -\frac{\partial v}{\partial x},$$

so that

$$\begin{vmatrix} \dfrac{\partial u}{\partial x} & \dfrac{\partial u}{\partial y} \\[2mm] \dfrac{\partial v}{\partial x} & \dfrac{\partial v}{\partial y} \end{vmatrix} = \left(\frac{\partial u}{\partial x}\right)^2 + \left(\frac{\partial u}{\partial y}\right)^2 = \left| \frac{\partial u}{\partial x} - i \frac{\partial u}{\partial y} \right|^2 = |M'(z)|^2.$$

Thus, the rule for the transformation of our double integral under the conformal map is

$$\iint\limits_{D_n} |f(w)|^2 \, du \, dv = \iint\limits_{B_n} |f(M(z))|^2 \, |M'(z)|^2 \, dx \, dy. \tag{9.2.35}$$

The functions $f(w) = 1, w, w^2, \ldots,$ are clearly in $L^2(D)$. It follows from (9.2.35), by allowing $n \to \infty$, that their "images,"

$$(M(z))^n M'(z) \quad n = 0, 1, \ldots,$$

will be an infinite independent set of functions in $L^2(B)$.

9.3 Bounded Linear Functionals in Normed Linear Spaces and in Hilbert Spaces.
We may distinguish two types of linear functionals defined on normed linear spaces: the bounded and the unbounded.

DEFINITION 9.3.1. Let L be a linear functional defined over the elements of a normed linear space X. L is said to be *bounded* if there exists a constant M such that

$$|L(x)| \leq M \, \|x\|, \qquad \text{for all } x \in X. \tag{9.3.1}$$

If no such constant exists, the functional is called *unbounded*.

Ex. 1. Let $X = C[a, b]$, $\|f\| = \max\limits_{a \le x \le b} |f(x)|$. Let $L(f) = \int_a^b w(x)f(x)\,dx$ for a fixed $w(x) \in C[a, b]$. Then,

$$|L(f)| \le \int_a^b |w(x)| \max\limits_{a \le x \le b} |f(x)|\,dx \le \|f\| \int_a^b |w(x)|\,dx.$$

Inequality (9.3.1) is satisfied with $M = \int_a^b |w(x)|\,dx$ and L is bounded.

Ex. 2. Let

$$X = C[a, b], \quad \|f\|^2 = \int_a^b (f(x))^2\,dx.$$

Let $L(f) = f(x_0)$ where $a \le x_0 \le b$. L is unbounded. For we may construct a sequence of functions $f_n(x)$, $n = 1, 2, \ldots$, with $f_n(x_0) = 1$ and

$$\int_a^b (f_n(x))^2\,dx = \varepsilon_n{}^2 \to 0.$$

This can be done in many ways. Now if L were bounded we should have

$$1 = |L(f_n)| \le M\varepsilon_n,$$

and this is impossible.

Ex. 3. Let X be an inner product space and x_0 be a fixed element. Then $L(x) = (x, x_0)$ is a linear functional defined on X. It is bounded, for

$$|L(x)| = |(x, x_0)| \le \|x\|\,\|x_0\|.$$

Ex. 4. Let n be a fixed integer ≥ 0. If t is a fixed point in a region B, then the functional

$$L(f) = f^{(n)}(t)$$

is bounded over $L^2(B)$.

Proof: Since t is an interior point, we can find a circle C_r: $|z - t| \le r$ contained in B. Since

$$f(z) = \sum_{n=0}^{\infty} \frac{f^{(n)}(t)}{n!} (z - t)^n$$

in C_r, then by (9.2.27),

$$\pi \sum_{n=0}^{\infty} \frac{|f^{(n)}(t)|^2}{n!^2(n + 1)} r^{2n+2} = \iint\limits_{C_r} |f(z)|^2\,dx\,dy \le \iint\limits_B |f(z)|^2\,dx\,dy.$$

For a particular n,

$$\frac{\pi |f^{(n)}(t)|^2}{n!^2(n + 1)} r^{2n+2} \le \iint\limits_B |f(z)|^2\,dx\,dy = \|f\|^2.$$

Therefore

$$|L(f)| = |f^{(n)}(t)| \le \frac{n!\sqrt{n + 1}}{\sqrt{\pi}\,r^{n+1}} \|f\|.$$

Some authors use the term "linear functional" to mean a bounded linear functional. But in interpolation theory, the same formal functional may be bounded or unbounded depending upon what space it is considered in, and it is therefore better to stress the fact of boundedness whenever it occurs.

DEFINITION 9.3.2. Let F be a functional defined over a normed linear space. F is said to be *continuous* if

$$x_n \to x \quad \text{implies} \quad F(x_n) \to F(x).$$

THEOREM 9.3.1. *A linear functional L defined on a normed linear space X is bounded if and only if it is continuous.*

Proof: Let L be bounded. Then $|L(x)| \leq M \|x\|$ for all $x \in X$. If now $\|x_n - x\| \to 0$, then $|L(x_n) - L(x)| = |L(x_n - x)| \leq M \|x_n - x\|$. Therefore $|L(x_n) - L(x)| \to 0$.

Conversely, suppose that L is continuous and unbounded. Then we can find a sequence of elements x_n such that $|L(x_n)| > n \|x_n\|$. Set $y_n = \dfrac{x_n}{n \|x_n\|}$. Then $\|y_n\| = \dfrac{1}{n}$. Hence $\|y_n - 0\| \to 0$. Now $|L(y_n)| = \dfrac{L(x_n)}{n \|x_n\|} > 1$. Since L is continuous, $L(y_n) \to L(0) = 0$, and this is a contradiction.

A norm may be associated with each bounded linear functional. Let I designate the set of values M for which condition (9.3.1) holds. Let M' be the inf of the set I. We can find $M_1, M_2, \ldots, \in I$ such that $M_n \to M'$. We have $|L(x)| \leq M_n \|x\|$ for all x and for $n = 1, 2, \ldots$. Keep x fixed and allow $n \to \infty$. Then, $|L(x)| \leq M' \|x\|$. This is true for all x. Therefore (9.3.1) holds with $M = M'$, and the set I has a minimum.

DEFINITION 9.3.3. Let L be a bounded linear functional defined on a normed linear space X. Then $\|L\|$ is defined as the minimum value M for which (9.3.1) holds. We have, obviously,

$$|L(x)| \leq \|L\| \|x\|, \qquad x \in X, \qquad (9.3.2)$$

and for every $\varepsilon > 0$, we can find an $x_0 \in X$ for which

$$|L(x_0)| > (\|L\| - \varepsilon) \|x_0\|. \qquad (9.3.3)$$

An alternate formula for $\|L\|$ is given by

$$\|L\| = \sup_{x \in X} \frac{|L(x)|}{\|x\|}. \qquad (9.3.4)$$

For, $|L(x)| \leq \|L\| \|x\|$ so that $\dfrac{|L(x)|}{\|x\|} \leq \|L\|$ and hence $\sup_x \dfrac{|L(x)|}{\|x\|} \leq \|L\|$.

On the other hand, given an $\varepsilon > 0$ there is an x_0 with (9.3.3) holding. Therefore $\sup_x \dfrac{|L(x)|}{\|x\|} \geq \dfrac{|L(x_0)|}{\|x_0\|} > \|L\| - \varepsilon$. These two inequalities imply (9.3.4).

Ex. 5. $X = C[a, b]$, $\|f\| = \max\limits_{a \leq x \leq b} |f(x)|$. Let $L(f) = \sum\limits_{k=1}^{n} a_k f(x_k)$ where $a \leq x_k \leq b$. Then, $|L(f)| \leq \sum\limits_{k=1}^{n} |a_k| \, |f(x_k)| \leq \|f\| \sum\limits_{k=1}^{n} |a_k|$. This implies that $\|L\| \leq \sum\limits_{k=1}^{n} |a_k|$. On the other hand, construct an $f(x) \in C[a, b]$ such that $|f(x)| \leq 1$, $a \leq x \leq b$, and $f(x_k) = \operatorname{sgn} a_k$, $k = 1, 2, \ldots, n$. Then, $\|f\| = 1$ and $|L(f)| = \sum\limits_{k=1}^{n} |a_k| \leq \|L\| \, \|f\|$. This implies that $\|L\| \geq \sum\limits_{k=1}^{n} |a_k|$. Therefore $\|L\| = \sum\limits_{k=1}^{n} |a_k|$.

Ex. 6. $X = C[a, b]$, $\|f\| = \max\limits_{a \leq x \leq b} |f(x)|$. Let $L(f) = \int_a^b f(x)w(x)\,dx$ where $w(x)$ is a fixed function of $C[a, b]$. A similar argument shows that

$$\|L\| = \int_a^b |w(x)|\,dx.$$

THEOREM 9.3.2. *The set of all bounded linear functionals defined over a normed linear space X is a linear space. Introducing the quantity $\|L\|$ by (9.3.4) makes this linear space into a normed linear space.*

Proof: Let L_1 and L_2 be bounded linear functionals over X. Then, for any $x \in X$,

$$|(a_1 L_1 + a_2 L_2)(x)| = |a_1 L_1(x) + a_2 L_2(x)| \leq |a_1| \, \|L_1\| \, \|x\| + |a_2| \, \|L_2\| \, \|x\|$$
$$= (|a_1| \, \|L_1\| + |a_2| \, \|L_2\|) \, \|x\|.$$

This implies that $a_1 L_1 + a_2 L_2$ is a bounded linear functional.

Let $L \neq 0$. Then there is a $y \neq 0$ such that $L(y) \neq 0$. Hence $\|L\| = \sup_x \dfrac{|L(x)|}{\|x\|} \geq \dfrac{|L(y)|}{\|y\|} > 0$. Therefore $\|L\| = 0$ if and only if $L = 0$. Secondly,

$$\|aL\| = \sup_x \frac{|aL(x)|}{\|x\|} = |a| \sup_x \frac{|L(x)|}{\|x\|} = |a| \, \|L\|. \text{ Finally,}$$

$$\|L_1 + L_2\| = \sup_x \frac{|L_1(x) + L_2(x)|}{\|x\|} \leq \sup_x \frac{|L_1(x)|}{\|x\|} + \sup_x \frac{|L_2(x)|}{\|x\|} = \|L_1\| + \|L_2\|.$$

The postulates for a norm are therefore satisfied.

DEFINITION 9.3.4. The normed linear space of bounded linear functionals defined on a given normed linear space X by means of (9.3.4) is known as the *normed conjugate space of X* and is designated by X^*.

In a complete inner product space, bounded linear functionals possess a particularly simple representation.

THEOREM 9.3.3 (Fréchet-Riesz). *If L is a bounded linear functional over a complete inner product space X, then there exists a unique element $x_0 \in X$ such that*

$$L(x) = (x, x_0), \qquad x \in X. \tag{9.3.5}$$

Proof: Let M designate the set of elements x such that $L(x) = 0$. M is clearly a linear space. Moreover, it is closed. For suppose that $x_n \in M$ and $\|x_n - x\| \to 0$. Since L is bounded,

$$|L(x - x_n)| \le \|L\| \, \|x - x_n\| \to 0.$$

Therefore $L(x) - L(x_n) \to 0$. But $L(x_n) = 0$. Hence $L(x) = 0$ and $x \in M$.

Now there are two possibilities. (a) M is the whole space. In this case $L = 0$ and we may take $x_0 = 0$. (b) M is not the whole space. In this case, by Theorem 8.10.2, we can find an element $y_0 \ne 0$ which is $\perp M$. If we set

$$x_0 = \frac{\overline{L(y_0)} y_0}{\|y_0\|^2}, \tag{9.3.6}$$

then we can show that

$$(x, x_0) = \left(x, \frac{\overline{L(y_0)} y_0}{(y_0, y_0)} \right) = L(x), \qquad x \in X. \tag{9.3.7}$$

Now (9.3.7) is equivalent to

$$L(x)(y_0, y_0) = L(y_0)(x, y_0). \tag{9.3.8}$$

Consider the elements $L(x)y_0 - L(y_0)x$, $x \in X$. These elements are in M, for $L(L(x)y_0 - L(y_0)x) = L(x)L(y_0) - L(y_0)L(x) = 0$. Hence y_0 is $\perp$ to these elements. This means that $(L(x)y_0 - L(y_0)x, y_0) = 0$ and this is precisely (9.3.8).

The x_0 is unique, for if

$$L(x) = (x, x_0) = (x, x_1), \qquad x \in X,$$

then $\qquad\qquad\quad (x, x_0 - x_1) = 0, \qquad\qquad x \in X.$

If we select $x = x_0 - x_1$, this implies that $\|x_0 - x_1\|^2 = 0$ and hence $x_0 = x_1$.

DEFINITION 9.3.5. The element x_0 is known as the *representer* of the linear functional L.

COROLLARY 9.3.4. *Let L be a bounded linear functional over a complete* *inner product space. Let x_0 be its representer. Then,*

$$\|L\| = \|x_0\| \tag{9.3.9}$$

and

$$|L(x_0)| = \|L\| \, \|x_0\|. \tag{9.3.10}$$

Proof: $L(x) = (x, x_0)$. Hence $|L(x)| \le |(x, x_0)| \le \|x_0\| \, \|x\|$. This implies that $\|L\| \le \|x_0\|$. But $\|L\| = \sup_x \dfrac{|L(x)|}{\|x\|} \ge \dfrac{|L(x_0)|}{\|x_0\|} = \dfrac{|(x_0, x_0)|}{\|x_0\|} = \|x_0\|$. Therefore $\|L\| \ge \|x_0\|$ and hence $\|L\| = \|x_0\|$. Finally, $|L(x_0)| = (x_0, x_0) = \|x_0\|^2 = \|L\| \, \|x_0\|$.

Ex. 7. In $L^2[a, b]$, every bounded linear functional has the form

$$L(f) = \int_a^b f(x)g(x)\, dx$$

with $g(x) \in L^2[a, b]$. In $L^2(B)$ such a functional has the form $L(f) = \iint_B f(z)\overline{g(z)}\, dx\, dy$ where $g(z) \in L^2(B)$. Moreover, $\|L\|^2 = \int_a^b (g(x))^2\, dx$ or $\iint_B |g(z)|^2\, dx\, dy$.

The representer of a functional has a simple formula in terms of a complete orthonormal sequence.

THEOREM 9.3.5. *Let H be a Hilbert space and $x_1{}^*, x_2{}^*, \ldots,$ be a complete orthonormal sequence of elements. If L is a bounded linear functional on H then $L(x) = (x, y)$ where y has the Fourier expression*

$$y \sim \sum_{k=1}^{\infty} \overline{L(x_k{}^*)}x_k{}^*. \tag{9.3.11}$$

Moreover

$$L(x) = \sum_{k=1}^{\infty} (x, x_k{}^*)L(x_k{}^*), \qquad x \in H \tag{9.3.12}$$

and

$$\|L\|^2 = \sum_{k=1}^{\infty} |L(x_k{}^*)|^2. \tag{9.3.13}$$

Proof: Let y be the representer of L. Then

$$y \sim \sum_{k=1}^{\infty} (y, x_k{}^*)x_k{}^* = \sum_{k=1}^{\infty} \overline{(x_k{}^*, y)}x_k{}^* = \sum_{k=1}^{\infty} \overline{L(x_k{}^*)}x_k{}^*.$$

By (8.9.8),

$$L(x) = (x, y) = \sum_{k=1}^{\infty} (x, x_k{}^*)(x_k{}^*, y) = \sum_{k=1}^{\infty} (x, x_k{}^*)L(x_k{}^*).$$

Finally, $\|L\|^2 = \|y\|^2$. By (8.9.7),

$$\|y\|^2 = \sum_{k=1}^{\infty} |(y, x_k{}^*)|^2 = \sum_{k=1}^{\infty} |L(x_k{}^*)|^2.$$

LEMMA 9.3.6 (Abel-Dini). *Let $w_n > 0$ and let $\sum_{n=1}^{\infty} w_n = \infty$. Then if $W_n = w_1 + w_2 + \cdots + w_n$,*

$$\sum_{n=1}^{\infty} \frac{w_n}{W_n} = \infty, \tag{9.3.14}$$

while

$$\sum_{n=1}^{\infty} \frac{w_n}{W_n^2} < \infty. \tag{9.3.15}$$

Proof: W_n increases to $+\infty$.

$$\frac{w_{n+1}}{W_{n+1}} + \frac{w_{n+2}}{W_{n+2}} + \cdots + \frac{w_{n+p}}{W_{n+p}} > \frac{w_{n+1} + \cdots + w_{n+p}}{W_{n+p}}$$

$$= \frac{W_{n+p} - W_n}{W_{n+p}} = 1 - \frac{W_n}{W_{n+p}}$$

For every fixed n, this last fraction approaches 1. Hence, from some p on, the partial sums exceed, say, $\frac{1}{2}$. The tails of a convergent series cannot all exceed a fixed amount, and hence (9.3.14) diverges. We have

$$\sum_{n=2}^{N} \left(\frac{1}{W_{n-1}} - \frac{1}{W_n} \right) = \frac{1}{W_1} - \frac{1}{W_N}.$$

Therefore the series

$$\sum_{n=2}^{\infty} \left(\frac{1}{W_{n-1}} - \frac{1}{W_n} \right) = \sum_{n=2}^{\infty} \frac{W_n - W_{n-1}}{W_{n-1} W_n}$$

is convergent. Thus,

$$\sum_{n=2}^{\infty} \frac{w_n}{W_{n-1} W_n} < \infty$$

and a fortiori

$$\sum_{n=1}^{\infty} \frac{w_n}{W_n^2} < \infty.$$

LEMMA 9.3.7 (Landau). *Let $\{a_n\}$ be a fixed sequence of complex numbers and suppose that $\sum_{n=1}^{\infty} a_n b_n$ converges for all sequences $\{b_n\}$ for which*

$$\sum_{n=1}^{\infty} |b_n|^2 < \infty.$$

Then,

$$\sum_{n=1}^{\infty} |a_n|^2 < \infty.$$

Proof: Suppose that $\sum_{n=1}^{\infty} |a_n|^2 = \infty$. Set $b_n = \dfrac{\overline{a_n}}{|a_1|^2 + \cdots + |a_n|^2}$. Then by (9.3.15) with $w_n = |a_n|^2$, $\sum_{n=1}^{\infty} |b_n|^2 < \infty$. On the other hand, by (9.3.14), $\sum_{n=1}^{\infty} a_n b_n = \infty$, and this contradicts the hypothesis.

THEOREM 9.3.8. *Let H be a Hilbert space and $\{x_n{}^*\}$ a complete orthonormal sequence. Let L be a linear functional defined on H and suppose that for all $x \in H$ we have*

$$L(x) = \sum_{k=1}^{\infty} (x, x_k{}^*)L(x_k{}^*). \qquad (9.3.16)$$

Then L is bounded on H and

$$\|L\|^2 = \sum_{k=1}^{\infty} |L(x_k{}^*)|^2. \qquad (9.3.17)$$

Proof: By Theorems 8.9.1(C) and 8.9.2, the set of all sequences $b_k = (x, x_k{}^*)$, $x \in H$, is identical to the set of all sequences $\{b_k\}$ for which

$$\sum_{k=1}^{\infty} |b_k|^2 < \infty.$$

By Lemma 9.3.7,

$$\sum_{k=1}^{\infty} |L(x_k{}^*)|^2 < \infty.$$

Applying the Schwarz inequality to (9.3.16),

$$|L(x)|^2 \leq \sum_{k=1}^{\infty} |(x, x_k{}^*)|^2 \sum_{k=1}^{\infty} |L(x_k{}^*)|^2 = \|x\|^2 \sum_{k=1}^{\infty} |L(x_k{}^*)|^2.$$

L is therefore bounded, and (9.3.17) follows from (9.3.13).

For examples illustrating this theorem, see Corollary 12.5.5.

THEOREM 9.3.9. *Let H be a Hilbert space . Let H^* be its normed conjugate space. Then H^* can be made into a Hilbert space in such a way that H and H^* are essentially the same. More precisely, we can find a one to one correspondence ($\leftrightarrow$) between H and H^* such that*

(a) $x_1 \leftrightarrow L_1$, $x_2 \leftrightarrow L_2$ implies $a_1 x_1 + a_2 x_2 \leftrightarrow a_1 L_1 + a_2 L_2$.
(b) $x \leftrightarrow L$ implies $\|x\| = \|L\|$.
(c) An inner product can be introduced in H^* by writing
$$(L_1, L_2) = (x_1, x_2) \text{ where } x_1 \leftrightarrow L_1, x_2 \leftrightarrow L_2. \qquad (9.3.18)$$
(d) The norm arising from this inner product coincides
with the original norm in H^* $\left(\text{i.e., } \|L\| = \sup_{x \in H} \dfrac{|L(x)|}{\|x\|}\right)$.

Proof: Let $\{x_i{}^*\}$ be a complete orthonormal system in H. Let $L \in H^*$. By Theorem 9.3.3, we have $L(y) = (y, w)$ for a unique $w \in H$ and for all $y \in H$. The quantities $(w, x_i{}^*)$ are the Fourier coefficients of w and hence, $\|w\|^2 = \sum_{i=1}^{\infty} |(w, x_i{}^*)|^2 < \infty$. The quantities $(x_i{}^*, w) = \overline{(w, x_i{}^*)}$ satisfy the same inequality $\sum_{i=1}^{\infty} |(x_i{}^*, w)|^2 < \infty$ and hence by Theorems 8.9.2 and 8.9.1(F), they are the Fourier coefficients of a unique element $\in H$ which will be designated by $\bar{w}$. Note that $\|\bar{w}\| = \|w\|$.

Make the correspondence $L \leftrightarrow \bar{w}$. This correspondence is one to one between the whole of H and the whole of H^*. For, each $L \in H^*$ determines a $w \in H$ and each w determines a $\bar{w}$. If $L_1 \to w_1$ and $L_2 \to w_2$ then $L_1 \neq L_2$ implies $w_1 \neq w_2$. For we can find an $x \in H$ such that $L_1(x) \neq L_2(x)$. Therefore $0 \neq L_1(x) - L_2(x) = (x, w_1) - (x, w_2)$. Hence $w_1 \neq w_2$. Now $\overline{w_1} \neq \overline{w_2}$. For otherwise, $(\overline{w_1}, x_i{}^*) = (\overline{w_2}, x_i{}^*)$, $i = 1, 2, \dots$. Then $(w_1, x_i{}^*) = (w_2, x_i{}^*)$, $i = 1, 2, \dots$, implying $w_1 = w_2$.

Conversely, let $v \in H$. Consider $\bar{v}$ as above and define L by means of $L(x) = (x, \bar{v})$. By the above, the element $\bar{\bar{v}}$ corresponds to L. But $\bar{\bar{v}} = v$. Thus v corresponds to some L in H^*.

(a) Let $x_1 \leftrightarrow L_1$, $x_2 \leftrightarrow L_2$. Then,

$$L_1(x) = (x, \overline{x_1}), \quad L_2(x) = (x, \overline{x_2})$$

so that

$$(a_1 L_1 + a_2 L_2)(x) = a_1 L_1(x) + a_2 L_2(x) = (x, \bar{a}_1 \bar{x}_1 + \bar{a}_2 \bar{x}_2).$$

Now

$$\overline{a_1 x_1 + a_2 x_2} = \bar{a}_1 \bar{x}_1 + \bar{a}_2 \bar{x}_2.$$

Hence

$$(a_1 L_1 + a_2 L_2)(x) = (x, \overline{a_1 x_1 + a_2 x_2})$$

and therefore

$$a_1 L_1 + a_2 L_2 \leftrightarrow a_1 x_1 + a_2 x_2.$$

(b) If $w \leftrightarrow L$ then $L(x) = (x, \bar{w})$. Hence by (9.3.9), $\|L\| = \|\bar{w}\| = \|w\|$.

(c) The inner product properties of (L_1, L_2) follow from those in H:

$$(L_1 + L_2, L_3) = (x_1 + x_2, x_3) = (x_1, x_3) + (x_2, x_3) = (L_1, L_3) + (L_2, L_3).$$

$$(L_1, L_2) = (x_1, x_2) = \overline{(x_2, x_1)} = \overline{(L_2, L_1)}.$$

$$(\alpha L_1, L_2) = (\alpha x_1, x_2) = \alpha(x_1, x_2) = \alpha(L_1, L_2).$$

$$(L_1, L_1) = (x_1, x_1) = \|x_1\|^2 > 0 \text{ and } = 0 \text{ if and only if } x_1 = 0,$$

hence if and only if $L_1 = 0$.

(d) $(L, L)^{\frac{1}{2}} = (x, x)^{\frac{1}{2}} = \|x\| = \|\bar{x}\|$. Since $L(y) = (y, \bar{x})$, $y \in H$, $\|L\| = \|\bar{x}\|$. Hence $\|L\| = (L, L)^{\frac{1}{2}}$.

Thus, H^* is an inner product space. To show completeness we need to prove that $\|L_m - L_n\| \leq \varepsilon$, $m, n \geq N_\varepsilon$ implies the existence of an L with $\|L - L_n\| \to 0$. Let $L_n \leftrightarrow x_n$. Then $\|x_n - x_m\| = \|L_n - L_m\| \leq \varepsilon$ for $m, n \geq N_\varepsilon$. Thus, $\{x_n\}$ is a Cauchy sequence in H. Hence there is an $x : \|x - x_n\| \to 0$. If $x \leftrightarrow L$ then $\|L - L_n\| = \|x - x_n\| \to 0$.

Finally, there is a complete orthonormal sequence in H^*. For if $\{x_k{}^*\}$ is complete and orthonormal in H and $x_n{}^* \leftrightarrow L_n{}^*$ then $\{L_n{}^*\}$ is complete and orthonormal in H^*. H^* is therefore a Hilbert space.

In virtue of (9.3.18), the spaces H and H^* are known as *isomorphic* and *isometric*.

DEFINITION 9.3.6. Let X be an inner product space. Y is an arbitrary subset of X. The set of elements x that are orthogonal to all elements of Y

is known as the *orthogonal complement* of Y and is designated by $Y^\perp$. In symbols: $(x, y) = 0$ for all $x \in Y^\perp$ and all $y \in Y$.

Ex. 8. In R_2, the orthogonal complement of the x axis is the y axis.

Ex. 9. Orthogonal complements are closed sets.

The following decomposition theorem holds.

THEOREM 9.3.10. *Let X be a complete inner product space. Let M be a closed linear subspace and $M^\perp$ its orthogonal complement. Then any $x \in X$ may be written uniquely as*

$$x = m + m^\perp \tag{9.3.19}$$

where $m \in M$ and $m^\perp \in M^\perp$.

Proof: We show first that M is complete. If $\{m_k\}$ is a Cauchy sequence of elements of M, then by the completeness of X there is an element $x \in X$ such that $\lim\limits_{k \to \infty} \|x - m_k\| = 0$. But, by the closure of M, x must be in M. Hence every Cauchy sequence of elements of M has a limit in M.

Let $x \in X$ be a fixed element and consider (m, x) as m varies over M. By Theorem 9.3.3, there is an $m_1 \in M$ such that $(m, m_1) = (m, x)$ for all $m \in M$. Write

$$x = m_1 + (x - m_1). \tag{9.3.20}$$

Now if $m' \in M$, $(x - m_1, m') = (x, m') - (m_1, m') = 0$. Hence $x - m_1 \in M^\perp$. The decomposition is unique. For suppose $m_1 + m_1{}^\perp = m_2 + m_2{}^\perp$. Then $m_1 - m_2 = m_2{}^\perp - m_1{}^\perp$. But $m_1 - m_2 \in M$ and $m_2{}^\perp - m_1{}^\perp \in M^\perp$. Now the only element simultaneously in M and $M^\perp$ is 0 and hence $m_1 = m_2$, $m_1{}^\perp = m_2{}^\perp$.

DEFINITION 9.3.7. The unique element m determined from x is called the *projection of x on M*: $m = \operatorname{proj}(x)$.

LEMMA 9.3.11. *If X is a separable metric space and S is an arbitrary subset of X, S is also separable.*

Proof: Let $\{x_k\}$ be a sequence of elements that is dense in X. The set of points x of X satisfying $d(x, x_k) < r$ is a ball of radius r centered at x_k. Designate it by $U(x_k, r)$ and consider all the balls $U(x_k, r)$ where r runs through all the positive rational numbers. These form a denumerable set and hence can be listed as a sequence $U_1, U_2, \ldots$.

If x is an arbitrary element of X, and if $V(x, \rho)$ is any ball with center at x, then we can find an m such that $x \in U_m \subseteq V(x, \rho)$. For select an x_j with $d(x, x_j) < \rho/2$ and a rational ρ_1 with $d(x, x_j) < \rho_1 < \rho/2$. If $z \in U(x_j, \rho_1)$, then $d(z, x_j) < \rho_1$. But $d(z, x) \leq d(z, x_j) + d(x_j, x) < \rho/2 + \rho/2 = \rho$ and hence $z \in V(x, \rho)$.

Now consider the members of the sequence $U_1, U_2, \ldots$, that have points in common with S. These are also denumerable, and hence can be listed in a sequence. Call them $T_1, T_2, \ldots$.

For each k select a $y_k \in T_k \cap S$. The sequence $\{y_k\}$ is dense in S. For let $z \in S$ and $\varepsilon > 0$ be given. As we have seen, we can find a U_m such that $z \in U_m$ and $U_m \subset V(z, \varepsilon)$. U_m contains a point of S and is therefore a T_n. T_n contains y_n, and hence $d(y_n, z) < \varepsilon$.

THEOREM 9.3.12. *Let H be a Hilbert space and $\{x_k{}^*\}$ be an orthonormal sequence that is not complete. Then we can find a sequence of elements $\{y_k{}^*\}$ (finite or infinite) such that $\{x_k{}^*\}$ and $\{y_k{}^*\}$ together form a complete orthonormal set (cf. Lemma 9.1.1).*

Proof: Designate by M the closed linear subspace generated by $x_1{}^*$, $x_2{}^*, \ldots$. That is, M consists of all finite linear combinations of $x_i{}^*$ plus the limits of sequences of such combinations. $M^\perp$ is the orthogonal complement of M. Since H is separable (cf. 9.1), Lemma 9.3.11 tells us that $M^\perp$ is also separable. Let $\{z_k\}$ be a sequence of elements of $M^\perp$ that is dense in $M^\perp$. Go through the sequence $z_1, z_2, \ldots$, and strike out any element that is dependent on its predecessors. Call the independent sequence that remains $\{y_k\}$. Orthonormalize this sequence to yield $\{y_k{}^*\}$.

If $z \in M^\perp$, and if $\varepsilon > 0$ is given, we can find a linear combination $\sum\limits_{k=1}^{n} a_k y_k{}^*$ such that $\left\| z - \sum\limits_{k=1}^{n} a_k y_k{}^* \right\| \leq \varepsilon$. For, we can find a k such that $\|z - z_k\| \leq \varepsilon$. Now z_k is either a y_j or a linear combination of $y_1, y_2, \ldots$, and hence a linear combination of $y_1{}^*, y_2{}^*, \ldots$.

Since $y_k{}^* \in M^\perp$, $(x_j{}^*, y_k{}^*) = 0$ and hence the combined set $\{x_j{}^*\}, \{y_k{}^*\}$ is orthonormal. This combined set is closed in H. For let $x \in H$. By Theorem 9.3.10 we can write $x = m + m^\perp$. Now by the definition of M, for appropriate constants b_k, $\left\| m - \sum\limits_{k=1}^{p} b_k x_k{}^* \right\| \leq \varepsilon/2$. Furthermore, as we have just seen, for appropriate constants b_k', $\left\| m^\perp - \sum\limits_{k=1}^{q} b_k' y_k{}^* \right\| \leq \varepsilon/2$. Hence

$$\left\| x - \sum_{k=1}^{p} b_k x_k{}^* - \sum_{k=1}^{q} b_k' y_k{}^* \right\| \leq \varepsilon.$$

This completes the proof.

THEOREM 9.3.13. *Let H and K be two Hilbert spaces (either both real or both complex). Then H and K are isomorphic and isometric. That is, we can find a one to one mapping T of H onto K such that for all constants α, β*

$$T(\alpha x_1 + \beta x_2) = \alpha T(x_1) + \beta T(x_2), \qquad x_1, x_2 \in H \qquad (9.3.21)$$

and such that

$$(T(x_1), T(x_2)) = (x_1, x_2). \qquad (9.3.22)$$

Proof: It suffices to take K as ℓ^2 (real or complex as the case may be). For since isomorphisms and isometries are obviously transitive, the general case can be proved by going through ℓ^2 as an intermediate space.

Let $x_1^*, x_2^*, \ldots$, be a complete orthonormal sequence in H. For any $x \in H$, the sequence of constants $a_k = (x, x_k^*)$, $k = 1, 2, \ldots$, is in ℓ^2 by (8.9.7). Define T by $T(x) = \{a_k\}$. Conversely, by Theorem 8.9.2, to any sequence $\{a_k\} \in \ell^2$, there is a unique element $x \in H$ such that $T(x) = \{a_k\}$. The linearity of T is obvious. Property (9.3.22) follows from (8.9.8).

9.4   Linear Varieties and Hyperplanes; Interpolation and Approximation in Hilbert Space

DEFINITION 9.4.1.   Let $x_1, \ldots, x_n$ be n independent elements of a linear space. The set of all linear combinations

$$x_0 + \sum_{i=1}^{n} a_i x_i \tag{9.4.1}$$

is known as a *linear variety of dimension n.*

DEFINITION 9.4.2.   Let $x_1, x_2, \ldots, x_n$ be n independent elements of an inner product space and let $c_1, c_2, \ldots, c_n$ be n given constants. The set of elements y that simultaneously satisfy the n equations

$$(y, x_i) = c_i, \qquad i = 1, 2, \ldots, n, \tag{9.4.2}$$

is known as a *hyperplane of co-dimension n.*

Ex. 1.   Linear varieties and hyperplanes are convex sets.

If $x_1, \ldots, x_n$ are orthonormalized to produce $x_1^*, \ldots, x_n^*$, then by Corollary 8.3.5, we can write the variety in the form

$$x_0 + \sum_{i=1}^{n} a_i x_i^* \tag{9.4.3}$$

and the hyperplane in the form

$$(y, x_i^*) = d_i, \qquad i = 1, 2, \ldots, n. \tag{9.4.4}$$

In an inner product space of finite dimension n, the concepts of linear variety and hyperplane are equivalent. More precisely, a linear variety of dimension p, $1 \leq p < n$, is a hyperplane of co-dimension $n - p$ and vice versa. For let $y = x_0 + \sum_{i=1}^{p} a_i x_i^*$ be a variety V of dimension p. Then $(y, x_k^*) = (x_0, x_k^*)$ for $k = p + 1, p + 2, \ldots, n$ and hence y lies on a hyperplane of co-dimension $n - p$. Conversely, let y satisfy $(y, x_k^*) = (x_0, x_k^*)$, $k = p + 1, \ldots, n$. Since any element z has the expansion $z = \sum_{k=1}^{n} (z, x_k^*) x_k^*$, we have $y = \sum_{k=1}^{p} (y, x_k^*) x_k^* + \sum_{k=p+1}^{n} (x_0, x_k^*) x_k^* = x_0 + \sum_{k=1}^{p} [(y, x_k^*) - (x_0, x_k^*)] x_k^*$. Hence y

lies on V. In the same way we can show that every hyperplane of co-dimension p is a linear variety of dimension $n - p$.

We begin with the problem of finite interpolation. This is essentially an instance of Theorem 2.2.2.

THEOREM 9.4.1.   *Let X be an inner product space. Let $x_1, \ldots, x_n$ be n independent elements. Then given any set of n constants $c_1, \ldots, c_n$, we can find an element y such that*

$$(y, x_i) = c_i, \qquad i = 1, 2, \ldots, n. \tag{9.4.5}$$

Proof: We can find a solution among the linear combinations of the x_i. Set $y = a_1 x_1 + \cdots + a_n x_n$. Then (9.4.5) becomes

$$a_1(x_1, x_i) + \cdots + a_n(x_n, x_i) = c_i, \qquad i = 1, 2, \ldots, n. \tag{9.4.6}$$

The system (9.4.6) has determinant $g(x_1, \ldots, x_n) \neq 0$ in view of the independence of the x_i. Hence there is a solution for any assignment of c's.

When we consider an interpolation problem with infinitely many conditions, the situation is not so simple.

Ex. 2.   In the sequential Hilbert space ℓ^2, let $x_1 = (1, 0, 0, \ldots)$, $x_2 = (0, 1, 0, 0, \ldots)$, $\ldots$, and consider the problem

$$(y, x_i) = 1, \qquad i = 1, 2, \ldots. \tag{9.4.7}$$

These conditions obviously require a solution of the form $y = (1, 1, 1, \ldots)$. But this element is not in the Hilbert space.

THEOREM 9.4.2.   *Let $\{x_i\}$ be an infinite sequence of independent elements of a Hilbert space H and let constants $\{c_i\}$ be given. A necessary and sufficient condition that there exist an element $y \in H$ such that*

$$(y, x_i) = c_i, \qquad i = 1, 2, \ldots, \tag{9.4.8}$$

is that

$$\sum_{k=1}^{\infty} |a_k|^2 < \infty \tag{9.4.9}$$

where $\bar{a}_1 = \bar{c}_1 / \sqrt{g(x_1)},$

$$\bar{a}_n = \frac{1}{\sqrt{g(x_1, \ldots, x_{n-1}) g(x_1, \ldots, x_n)}} \begin{vmatrix} (x_1, x_1) & (x_2, x_1) & \cdots & (x_n, x_1) \\ \vdots & & & \vdots \\ \vdots & & & \vdots \\ \vdots & & & \vdots \\ (x_1, x_{n-1}) & (x_2, x_{n-1}) & \cdots & (x_n, x_{n-1}) \\ \bar{c}_1 & \bar{c}_2 & \cdots & \bar{c}_n \end{vmatrix}.$$

$$n > 1. \tag{9.4.10}$$

If there is a solution, it is unique if and only if $\{x_i\}$ is complete (or closed) in H.

Proof: Orthonormalize the x_k using Theorem 8.3.3 and obtain $x_k{}^*$. In view of (8.7.22), the conditions (9.4.8) imply

$$(y, x_i{}^*) = a_i \qquad i = 1, 2, \ldots . \tag{9.4.11}$$

Conversely, it is easily shown through (8.3.7), that (9.4.11) implies (9.4.8). Now if there is a $y \in H$ satisfying (9.4.11), then by Corollary 8.5.4, (9.4.9) follows. Conversely, if (9.4.9) holds, then by Theorem 8.9.2, there is an element y for which (9.4.11) and hence (9.4.8) holds.

Suppose that $(y_1, x_i) = c_i = (y_2, x_i)$. Then $(y_1 - y_2, x_i) = 0, i = 1, 2, \ldots .$ If $\{x_i\}$ is complete then $y_1 - y_2 = 0$ and $y_1 = y_2$. If $\{x_i\}$ is incomplete, then we can find an element $z \neq 0$ such that $(z, x_i) = 0, i = 1, 2, \ldots .$ Hence, $(y, x_i) = c_i = (y + z, x_i)$ so that the solution to (9.4.8) is not unique.

Ex. 3. Under what circumstances can we have $\int_{-1}^{+1} f(x)x^n \, dx = c_n$, $n = 0$, $1, \ldots,$ for $f \in L^2[-1, 1]$? This is the *moment problem* for the space $L^2[-1, 1]$. Let $p_n{}^*(x) = a_{n0} + a_{n1}x + \cdots + a_{nn}x^n$ be the Legendre polynomials. Then the moment conditions above are totally equivalent to

$$\int_{-1}^{+1} f(x)p_n{}^*(x) \, dx = a_{n0}c_0 + a_{n1}c_1 + \cdots + a_{nn}c_n.$$

Hence, the necessary and sufficient condition is that

$$\sum_{n=0}^{\infty} |a_{n0}c_0 + a_{n1}c_1 + \cdots + a_{nn}c_n|^2 < \infty.$$

Since the powers are complete in $L^2[-1, 1]$ (Theorem 11.2.1), there can be at most one solution.

Let X be an inner product space and $x_1, \ldots, x_n$ be n independent elements. Let S be the linear subspace spanned by the x's. That is, S consists of all linear combinations $y = a_1x_1 + a_2x_2 + \cdots + a_nx_n$, or alternatively, of all combinations $y = a_1x_1{}^* + \cdots + a_nx_n{}^*$ where the $x_i{}^*$ are the x_i orthonormalized (in any way). Take an element $z \in X$ which is not perpendicular to all the $x_i{}^*$, and let V consist of all the elements $y \in S$ for which $(y, z) = 1$. V is a linear variety of dimension $\leq n - 1$. For suppose that $(x_1{}^*, z) \neq 0$. If $y = a_1x_1{}^* + \cdots + a_nx_n{}^*$ and $(y, z) = 1$, then

$$a_1(x_1{}^*, z) + \cdots + a_n(x_n{}^*, z) = 1$$

and so

$$y = \frac{x_1{}^*}{(x_1{}^*, z)} + a_2\left(-\frac{(x_2{}^*, z)}{(x_1{}^*, z)} x_1{}^* + x_2{}^*\right) + \cdots + a_n\left(-\frac{(x_n{}^*, z)}{(x_1{}^*, z)} x_1{}^* + x_n{}^*\right). \tag{9.4.12}$$

Since $x_1^*, \ldots, x_n^*$ are independent, the elements $-\dfrac{(x_i^*, z)}{(x_1^*, z)} x_1^* + x_i^*$, $i = 2, \ldots, n$ are easily seen to be independent. Hence (9.4.12) is of form (9.4.1) and the elements of V lie on the variety (9.4.12). Conversely, all the elements of this variety are in S and satisfy $(y, z) = 1$. Hence they belong to V.

It is desired to find the element of V closest to the origin. That is, select $y \in V$ such that $\|y\| = $ minimum.

THEOREM 9.4.3. *The unique solution to the above problem is given by*

$$y = \sum_{i=1}^{n} (z, x_i^*) x_i^* / \sum_{i=1}^{n} |(z, x_i^*)|^2. \qquad (9.4.13)$$

The minimal distance is given by

$$\|y\|^2 = 1/\sum_{i=1}^{n} |(z, x_i^*)|^2. \qquad (9.4.14)$$

Proof: Set $\sum_{i=1}^{n} |(z, x_i^*)|^2 = s \neq 0$, and write $a_i = \dfrac{(z, x_i^*)}{s} + b_i$ where the b_i are now to be determined. Now, $1 = (y, z) = 1 + \sum_{i=1}^{n} b_i(x_i^*, z)$, so that $\sum_{i=1}^{n} b_i(x_i^*, z) = 0$. But

$$\|y\|^2 = \sum_{i=1}^{n} |a_i|^2 = \sum_{i=1}^{n} \left(\frac{(z, x_i^*)}{s} + b_i \right) \left(\frac{(x_i^*, z)}{s} + \overline{b_i} \right)$$

$$= \frac{s}{s^2} + \frac{1}{s} \sum_{i=1}^{n} b_i(x_i^*, z) + \frac{1}{s} \sum_{i=1}^{n} \overline{b_i}(z, x_i^*) + \sum_{i=1}^{n} |b_i|^2$$

$$= \frac{1}{s} + \sum_{i=1}^{n} |b_i|^2.$$

The selection leading to the minimum $\|y\|^2$ is uniquely given by $b_i \equiv 0$ and the minimum value is $\dfrac{1}{s}$.

We turn to a second problem of approximation under side conditions. Let $x_1, x_2, \ldots, x_n$ be n independent elements and $d_1, \ldots, d_n$ be n given constants.

Find

$$\min_{y} \|x - y\| \qquad (9.4.15)$$

subject to

$$(y, x_i) = d_i, \quad i = 1, 2, \ldots, n. \qquad (9.4.16)$$

Geometrically speaking, find the shortest distance from the element x to the hyperplane (9.4.16).

If we set $w = x - y$, our problem becomes that of finding

$$\min_w \|w\| \tag{9.4.17}$$

subject to

$$(w, x_i) = c_i = (x, x_i) - d_i, \quad i = 1, 2, \ldots, n. \tag{9.4.18}$$

If the x's are orthonormalized (by Theorem 8.3.3) yielding $x_1^*, x_2^*, \ldots, x_n^*$, then (9.4.18) is totally equivalent to

$$(w, x_i^*) = a_i \tag{9.4.19}$$

where a_i are given by (9.4.10).

THEOREM 9.4.4.   *Let H be a Hilbert space. Then*

$$w = \sum_{i=1}^n a_i x_i^* \tag{9.4.20}$$

solves the problem (9.4.17), (9.4.18). *Moreover,*

$$\min_w \|w\|^2 = \sum_{i=1}^n |a_i|^2. \tag{9.4.21}$$

Proof: The x_i^* may be augmented yielding a complete orthonormal sequence for H (Lemma 9.1.1). Then, for any element $w \in H$,

$$\|w\|^2 = \sum_{i=1}^\infty |(w, x_i^*)|^2.$$

Any element w that satisfies (9.4.19) must therefore satisfy

$$\|w\|^2 = \sum_{i=1}^n |a_i|^2 + \sum_{i=n+1}^\infty |(w, x_i^*)|^2.$$

This expression is minimized if $(w, x_i^*) = 0$ for $i = n + 1, n + 2, \ldots$, and we are led to (9.4.20).

COROLLARY 9.4.5.   *In a Hilbert space let $x_1 \neq 0$. The problem of finding* $\min \|w\|$ *subject to* $(w, x_1) = d_1$ *is solved by* $w = \dfrac{d_1}{\|x_1\|^2} x_1$. *Furthermore,*

$$\min \|w\| = \frac{|d_1|}{\|x_1\|}.$$

COROLLARY 9.4.6.   *The equation of any hyperplane of co-dimension 1 can be written in the "normal" form*

$$(y, x^*) = d \tag{9.4.22}$$

where x^* is an appropriate element of unit length and d is the distance from the origin to the hyperplane.

DEFINITION 9.4.3.   The portion of a Hilbert space common to the hyperplane $P: (x, x_i) = c_i$, $i = 1, 2, \ldots, n$ and to the ball $\|x\| \leq r$ is called a *hypercircle*, C_r.

As before, orthonormalize $x_1, \ldots, x_n$ by the Gram-Schmidt process, and complete the sequence of orthonormal elements yielding $x_1{}^*, x_2{}^*, \ldots$. The conditions $(x, x_i) = c_i$ are equivalent to $(x, x_i{}^*) = a_i$, $i = 1, 2, \ldots, n$, where the a_i are given by (9.4.10). Hence, any element x of the hyperplane can be written as $x = \sum_{k=1}^{n} a_k x_k{}^* + \sum_{k=n+1}^{\infty} (x, x_k{}^*) x_k{}^* = w + \sum_{k=n+1}^{\infty} (x, x_k{}^*) x_k{}^*$. The last equality follows from (9.4.20). Now, $\|x\|^2 = \|w\|^2 + \sum_{k=n+1}^{\infty} |(x, x_k{}^*)|^2$. If $x \in C_r$, $r^2 \geq \|x\|^2 \geq \|w\|^2$. The element w is determined solely by x_i and c_i, $i = 1, 2, \ldots, n$, and is independent of r. If $r < \|w\|$, the hypercircle contains no elements.

THEOREM 9.4.7 (The Hypercircle Inequality). *Let w be the element of the hyperplane P nearest to the origin. Then, for any $x \in C_r$ and any bounded linear functional L we have*

$$|L(x) - L(w)|^2 \leq (r^2 - \|w\|^2) \sum_{k=n+1}^{\infty} |L(x_k{}^*)|^2. \qquad (9.4.23)$$

If h is the representer of L, this may be written as

$$|L(x) - L(w)|^2 \leq (r^2 - \|w\|^2) \left(\|h\|^2 - \sum_{k=1}^{n} |L(x_k{}^*)|^2 \right). \qquad (9.4.24)$$

Moreover, if $\|w\| \leq r$, there is an element in C_r for which this inequality becomes an equality.

Proof: From the above remarks, $x - w = \sum_{k=n+1}^{\infty} (x, x_k{}^*) x_k{}^*$ and hence by (9.3.12), $L(x - w) = \sum_{k=n+1}^{\infty} (x, x_k{}^*) L(x_k{}^*)$. By the Schwarz inequality and (8.9.7), $|L(x - w)|^2 \leq \|x - w\|^2 \sum_{k=n+1}^{\infty} |L(x_k{}^*)|^2$. Since $x - w \perp w$,

$$\|x - w\|^2 = \|x\|^2 - \|w\|^2 \leq r^2 - \|w\|^2.$$

Combining, we obtain (9.4.23). By (9.3.5), $\sum_{k=n+1}^{\infty} |L(x_k{}^*)|^2 = \sum_{k=n+1}^{\infty} |(x_k{}^*, h)|^2 = \sum_{k=1}^{\infty} |(h, x_k{}^*)|^2 - \sum_{k=1}^{n} |(x_k{}^*, h)|^2$. Again by (8.9.7) and (9.3.5), this is $\|h\|^2 - \sum_{k=1}^{n} |L(x_k{}^*)|^2$, giving us (9.4.24).

Assume now that $\|w\| \leq r$. We shall exhibit an element in C_r for which the inequality (9.4.23) becomes an equality. Let $z = \sum_{k=n+1}^{\infty} \overline{L(x_k{}^*)} x_k{}^*$. If $z = 0$, $L(x_k{}^*) = 0$, $k = n + 1, \ldots$. In this case, the element $x_0 = w$ is in C_r and equality in (9.4.23) holds trivially.

If $z \neq 0$, set

$$x_0 = w + \lambda z \qquad (9.4.25)$$

where

$$|\lambda| = (r^2 - \|w\|^2)^{\frac{1}{2}} / \|z\|. \qquad (9.4.26)$$

Now $(x_0, x_i^*) = (w, x_i^*) = a_i$, $i = 1, 2, \ldots, n$, and so $x_0 \in P$. Moreover, $\|x_0\|^2 = \|w\|^2 + |\lambda|^2 \|z\|^2 = r^2$, so that $x_0 \in C_r$. Furthermore,

$$|L(x_0 - w)| = |\lambda| \, |L(z)| = |\lambda| \sum_{k=n+1}^{\infty} |L(x_k^*)|^2 = |\lambda| \, \|z\|^2$$

$$= (r^2 - \|w\|^2)^{\frac{1}{2}} \|z\| = (r^2 - \|w\|^2)^{\frac{1}{2}} \left(\sum_{k=n+1}^{\infty} |L(x_k^*)|^2 \right)^{\frac{1}{2}}.$$

Thus, equality holds in (9.4.23) for $x = x_0$.

It should be remarked that the element z has the alternate representation

$$z = h - \sum_{k=1}^{n} (h, x_k^*) x_k^* = h - \sum_{k=1}^{n} \overline{L(x_k^*)} x_k^*; \tag{9.4.27}$$

for it is readily verified that the Fourier coefficients (with respect to x_k^*) of z and of the middle member of (9.4.27) are identical.

The hypercircle inequality may be used to obtain bounds for $L(x)$ having been given certain information about x. This is illustrated by the following example.

Ex. 4. Estimate $\displaystyle\int_{-\frac{1}{2}}^{\frac{1}{2}} x(t) \, dt$ on the basis of the following information:

$$x(t) \in L^2[-1, 1], \qquad \|x(t)\| \le r, \tag{9.4.28}(a)$$

and

$$\int_{-1}^{1} x(t) \, dt = 1, \qquad \int_{-1}^{1} t^2 x(t) \, dt = 1. \tag{9.4.28}(b)$$

The relevant hyperplane is $(x(t), x_i(t)) = 1$, $i = 1, 2$, where $x_1(t) = 1$ and $x_2(t) = t^2$. Orthonormalizing we obtain

$$x_1^*(t) = \tfrac{1}{2}\sqrt{2}, \qquad x_2^*(t) = -\tfrac{1}{4}\sqrt{10} + \tfrac{3}{4}\sqrt{10}\, t^2.$$

The hyperplane equations can be written as

$$(x(t), x_1^*(t)) = \tfrac{1}{2}\sqrt{2} = a_1, \qquad (x(t), x_2^*(t)) = \tfrac{1}{2}\sqrt{10} = a_2.$$

Therefore,

$$w = w(t) = (\tfrac{1}{2}\sqrt{2})(\tfrac{1}{2}\sqrt{2}) + \tfrac{1}{2}\sqrt{10}(-\tfrac{1}{4}\sqrt{10} + \tfrac{3}{4}\sqrt{10}\, t^2) = \tfrac{3}{4}(5t^2 - 1),$$

and

$$\|w\|^2 = a_1^2 + a_2^2 = 3.$$

Now,

$$L(x_1^*) = \int_{-\frac{1}{2}}^{\frac{1}{2}} \tfrac{1}{2}\sqrt{2} \, dt = \tfrac{1}{2}\sqrt{2},$$

$$L(x_2^*) = \int_{-\frac{1}{2}}^{\frac{1}{2}} (-\tfrac{1}{4}\sqrt{10} + \tfrac{3}{4}\sqrt{10}\, t^2) \, dt = -\tfrac{3}{16}\sqrt{10},$$

and

$$L(w) = \int_{-\frac{1}{2}}^{\frac{1}{2}} \tfrac{3}{4}(5t^2 - 1) \, dt = -\tfrac{7}{16}.$$

By inspection, $h = h(t) = 1$ for $-\frac{1}{2} \le t \le \frac{1}{2}$ and $h(t) = 0$ elsewhere. Hence, $\|h(t)\| = 1$. Inequality (9.4.24) becomes

$$\left(\int_{-\frac{1}{2}}^{\frac{1}{2}} x(t)\, dt + \tfrac{7}{16} \right)^2 \le (r^2 - 3)(1 - \tfrac{1}{2} - \tfrac{90}{256}) = \tfrac{19}{128}(r^2 - 3). \qquad (9.4.29)$$

We may express this as

$$-\tfrac{7}{16} - \varepsilon \le \int_{-\frac{1}{2}}^{\frac{1}{2}} x(t)\, dt \le -\tfrac{7}{16} + \varepsilon \qquad (9.4.30)$$

with $\varepsilon = \sqrt{\tfrac{19}{128}}\,\sqrt{r^2 - 3}$. Inequality (9.4.30) holds for all functions $x(t)$ satisfying the hypercircle conditions (9.4.28)(a), (b). Moreover, since one equality in (9.4.30) occurs for some element in the hypercircle, the midpoint of the range, $-\tfrac{7}{16}$, can be taken as a "best" value of $\int_{-\frac{1}{2}}^{\frac{1}{2}} x(t)\, dt$ relative to the information available.

NOTES ON CHAPTER IX

See the references on normed linear spaces and Hilbert spaces listed under Chapters VII and VIII.

9.2   For interchange of summation and integration and Fatou's lemma used in the proof of Theorem 9.2.2, see, e.g., Rudin [1], pp. 209–217. The Hilbert space $L^2(B)$ can be found in Bergman [2], Chapter I, Behnke and Sommer [1], pp. 256–282, Nehari [1], pp. 239–260. Related Hilbert spaces formed by using line integrals as inner products are described in Walsh [2], Chapter 6. Bergman and Schiffer [1] discuss Hilbert spaces of solutions of elliptic partial differential equations.

9.4   For interpolation problems in $L^2(B)$ see Bergman [2], pp. 47–49, Walsh and Davis [1]. The hypercircle inequality (Th. 9.4.7) is given in Synge [1], Chapter 2, and in Golomb and Weinberger [1], p. 133, where many applications to numerical analysis will be found.

PROBLEMS

1. Let $w_i > 0$ and let ℓ_w^2 designate the set of all sequences $\{a_i\}$ such that $\sum_{i=1}^{\infty} w_i |a_i|^2 < \infty$. Set $(a, b) = \sum_{i=1}^{\infty} w_i a_i \bar{b}_i$. Then ℓ_w^2 is a Hilbert space.

2. Prove that all sequences of the form $(a_1, 0, a_2, 0, a_3, 0, \ldots)$ with

$$\sum_{i=1}^{\infty} |a_i|^2 < \infty$$

constitute a sub-Hilbert space of ℓ^2. Generalize.

3. Let M be a linear subspace of a Hilbert space H. Show by an example that there may be a sequence $x_1{}^*, x_2{}^*, \ldots,$ of orthonormal elements that are complete for M but not for H.

4. If C is the unit circle, there are functions that are analytic in C but are not in $L^2(C)$.

5. If f is in $L^2(B)$, is f' in $L^2(B)$? Is $\int^z f(z)\, dz$ in $L^2(B)$?

6. Let B be a finite region and $z_1, z_2, \ldots, z_n$ be n fixed points in B. Let H

designate the set of all functions $f(z)$ in $L^2(B)$ for which $f(z_k) = 0, k = 1, 2, \ldots, n$. Show that H is a sub-Hilbert space of $L^2(B)$. Generalize.

7. If C is the unit circle, find $\min\limits_{f \in L^2(C)} \iint\limits_C |1 - zf(z)|^2 \, dx \, dy$.

8. If $X = C[a, b]$ with $\|f\| = \max\limits_{a \leq x \leq b} |f(x)|$, are either $F(f) = \int_a^b f^2(x) \, dx$ or $F(f) = f^2(x_1) - f^2(x_2), a \leq x_1, x_2 \leq b$ continuous functionals?

9. $X = C[0, 1]$ with $\|f\| = \int_0^1 |f(x)| \, dx$. Let $F(f) = \int_0^1 f^2(x) \, dx$. Show that F is not a continuous functional.

10. In $L^2[-1, 1]$, set $L_h(f) = \dfrac{1}{2h} \int_{-h}^h f(x) \, dx, 0 < h \leq 1$. Compute $\|L_h\|$ and study $\lim\limits_{h \to 0} \|L_h\|$.

11. X is a real normed linear space. If L is an additive and continuous functional, it must be homogeneous and hence is a bounded linear functional.

12. Let X be a finite dimensional normed linear space. Then any linear functional L is bounded on X.

13. A linear transformation of one Hilbert space into another is a mapping T for which $T(a_1 x_1 + a_2 x_2) = a_1 T(x_1) + a_2 T(x_2)$. A linear transformation U is called *isometric* if $\|U(x)\| = \|x\|$ for all $x \in H$. Prove that if U is isometric, $(Ux, Uy) = (x, y)$ and hence an isometry sends orthogonal systems into orthogonal systems.

14. If U is an isometry that maps H onto (the whole of) itself, then $\{U(x_n)\}$ is complete if and only if $\{x_n\}$ is complete.

15. Let $\{x_i^*\}$ and $\{y_i^*\}$ be two complete orthonormal systems for a Hilbert space H. The transformation $U\left(\sum\limits_{i=1}^{\infty} a_i x_i^*\right) = \sum\limits_{i=1}^{\infty} a_i y_i^*$ is isometric.

16. Exhibit the isomorphism and isometry of R_n and R_n^*.

17. If X is an inner product space of infinite dimension, a hyperplane in X is also of infinite dimension.

18. Formulate an interpolation problem in Hilbert space that has infinitely many conditions and has infinitely many independent solutions.

19. Discuss the interpolation problem $(y, x_1) = 1, (y, x_k) = 0, k = 2, 3, \ldots$, in a Hilbert space.

20. Formulate Theorem 9.4.2 as a theorem about a system of infinitely many linear equations in infinitely many unknowns.

21. In ℓ^2, what are necessary and sufficient conditions that the elements $x_1 = (a_{11}, 0, 0, \ldots), x_2 = (a_{21}, a_{22}, 0, 0, \ldots), x_3 = (a_{31}, a_{32}, a_{33}, 0, 0, \ldots)$ be independent? Be complete? Orthogonalize them.

22. Discuss the solution of the system

$$
\begin{aligned}
x_1 \qquad\qquad &= a_1 \\
-x_1 + x_2 \qquad &= a_2 \\
- x_2 + x_3 &= a_3 \\
\vdots \qquad\qquad
\end{aligned}
$$

from the point of view of Theorem 9.4.2.

CHAPTER X

Orthogonal Polynomials

10.1 General Properties of Real Orthogonal Polynomials. Let $[a, b]$ be a finite or infinite interval and let $w(x)$ be a positive weight function defined there. We assume that the integrals $\int_a^b w(x)x^n \, dx$, $n = 0, 1, \ldots$, all exist. Employ an inner product

$$(f, g) = (g, f) = \int_a^b w(x)f(x)g(x) \, dx \qquad (10.1.1)$$

and orthonormalize (by means of Theorem 8.3.3) the sequence of powers $1, x, x^2, \ldots$, with respect to this inner product. Designate the polynomials obtained by

$$p_n{}^*(x) = k_n x^n + \cdots, \quad k_n > 0. \qquad (10.1.2)$$

Polynomials that are merely orthogonal without being necessarily normal will be designated by p_n throughout this chapter.

Observe that if $p \in \mathscr{P}_{n-1}$ then

$$(p, p_n) = 0. \qquad (10.1.3)$$

This follows from Corollary 8.3.5 and the definition of orthogonality.

Though determinant expressions for orthogonal polynomials can be obtained from Corollary 8.7.6, they appear to be of limited importance. The following theorem, however, is of great utility.

THEOREM 10.1.1. *Real orthonormal polynomials satisfy a three term recurrence relationship.*

$$p_n{}^*(x) = (a_n x + b_n)p_{n-1}^*(x) - c_n p_{n-2}^*(x) \quad n = 2, 3, \ldots. \qquad (10.1.4)$$

The following form is particularly convenient for machine computation

$$\left.\begin{aligned}
&p_{-1} = 0 \\
&p_0 = 1 \\
&\quad\cdot \\
&\quad\cdot \\
&\quad\cdot \\
&p_{n+1}(x) = xp_n{}^*(x) - (xp_n{}^*, p_n{}^*)p_n{}^*(x) - (p_n, p_n)^{\frac{1}{2}}p_{n-1}^*(x) \\
&\qquad\qquad\qquad\qquad\qquad\qquad\qquad n = 0, 1, 2, \ldots \\
&p_n{}^*(x) = p_n(x)/(p_n, p_n)^{\frac{1}{2}}. \qquad\qquad n = 0, 1, 2, \ldots
\end{aligned}\right\} \qquad (10.1.5)$$

Proof: It is clear that $p_n{}^*(x)$ (defined by (10.1.5)) is a polynomial of degree n and is normal. We shall prove by induction that they are orthogonal. Assume that we have proved $(p_m{}^*, p_j{}^*) = 0$ for $j = 0, 1, \ldots, m - 1$ and for $m = 0, 1, \ldots, n$. We wish to show that $(p_{n+1}^*, p_j{}^*) = 0$ for $j = 0, 1, \ldots, n$. Now,

$$(p_{n+1}, p_j{}^*) = (xp_n{}^* - (xp_n{}^*, p_n{}^*)p_n{}^* - (p_n, p_n)^{\frac{1}{2}}p_{n-1}^*, p_j{}^*)$$
$$= (xp_n{}^*, p_j{}^*) - (xp_n{}^*, p_n{}^*)(p_n{}^*, p_j{}^*) - (p_n, p_n)^{\frac{1}{2}}(p_{n-1}^*, p_j{}^*).$$

Also $(xp_n{}^*, p_j{}^*) = (p_n{}^*, xp_j{}^*)$ as can be seen by referring to (10.1.1). For $j = 0, 1, 2, \ldots, n - 2$ we have by our induction hypothesis

$$(p_n{}^*, p_j{}^*) = 0, \quad (p_{n-1}^*, p_j{}^*) = 0, \quad (p_n{}^*, xp_j{}^*) = 0$$

since $xp_j{}^*$ is a polynomial of degree $\leq n - 1$. Hence $(p_{n+1}, p_j{}^*) = 0$ for $j = 0, 1, 2, \ldots, n - 2$. For $j = n - 1$ we have

$$(p_{n+1}, p_{n-1}^*) = (xp_n{}^*, p_{n-1}^*) - 0 - (p_n, p_n)^{\frac{1}{2}} \cdot 1.$$

Now, $(xp_n, p_{n-1}^*) = (p_n, xp_{n-1}^*)$. By the recurrence,

$$xp_{n-1}^* = p_n + \alpha p_{n-1}^* + \beta p_{n-2}^*.$$

Hence, $(p_n, xp_{n-1}^*) = (p_n, p_n + \alpha p_{n-1}^* + \beta p_{n-2}^*) = (p_n, p_n)$ by our induction hypothesis. Hence $(xp_n{}^*, p_{n-1}^*) = (p_n, p_n)^{\frac{1}{2}}$ and therefore $(p_{n+1}, p_{n-1}^*) = 0$. Finally, $(p_{n+1}, p_n{}^*) = (xp_n{}^*, p_n{}^*) - (xp_n{}^*, p_n{}^*) - 0 = 0$. In this way, the induction is carried to $n + 1$. Equation (10.1.4) follows from (10.1.5).

Further identification of the coefficients of the recursion is often useful.

THEOREM 10.1.2. *Let $p_n{}^*(x) = k_n x^n + s_n x^{n-1} + \cdots$ be orthonormal polynomials. Then the coefficients in the recurrence*

$$p_n{}^* = (a_n x + b_n)p_{n-1}^* - c_n p_{n-2}^* \tag{10.1.4}$$

are given by

$$\left. \begin{array}{l} a_n = \dfrac{k_n}{k_{n-1}}, \quad b_n = a_n\left(\dfrac{s_n}{k_n} - \dfrac{s_{n-1}}{k_{n-1}}\right) \\[2ex] c_n = a_n \dfrac{k_{n-2}}{k_{n-1}} = \dfrac{k_n k_{n-2}}{k_{n-1}^2}. \end{array} \right\} \quad n = 2, 3, \ldots. \tag{10.1.6}$$

Proof: The first two identities are obtained by inserting $k_n x^n + s_n x^{n-1} + \cdots$ into (10.1.4) and comparing the coefficients of x^n and x^{n-1}. The third identity can be proved in this way:

$$0 = (p_n{}^*, p_{n-2}^*) = (a_n xp_{n-1}^* + b_n p_{n-1}^* - c_n p_{n-2}^*, p_{n-2}^*)$$
$$= a_n(xp_{n-1}^*, p_{n-2}^*) - c_n.$$

But

$$(xp^*_{n-1}, p^*_{n-2}) = (p^*_{n-1}, xp^*_{n-2}) = (p^*_{n-1}, k_{n-2}x^{n-1})$$

$$= \frac{k_{n-2}}{k_{n-1}}(p^*_{n-1}, k_{n-1}x^{n-1}) = \frac{k_{n-2}}{k_{n-1}}(p^*_{n-1}, p^*_{n-1}) = \frac{k_{n-2}}{k_{n-1}}.$$

The formula for c_n is now apparent.

THEOREM 10.1.3.   *The zeros of real orthogonal polynomials are real, simple, and are located in the interior of* $[a, b]$.

Proof: Let n (≥ 1) be fixed. If $p_n(x)$ were of constant sign in $[a, b]$, say positive, then $\int_a^b w(x)p_n(x)\,dx = (p_n, p_0) > 0$. But this contradicts orthogonality. Hence $p_n(x_1) = 0$ for some $x_1 \in (a, b)$. Suppose that there is a zero at x_1 which is multiple. Then $\frac{p_n(x)}{(x-x_1)^2}$ would be a polynomial of degree $n - 2$. Hence $0 = \left(p_n(x), \frac{p_n(x)}{(x-x_1)^2}\right) = \left(1, \left(\frac{p_n(x)}{(x-x_1)}\right)^2\right) > 0$ and this is impossible. Therefore every zero is simple. Suppose now that $p_n(x)$ has j zeros $x_1, x_2, \ldots, x_j$ and no others lying in (a, b). Then,

$$p_n(x)(x - x_1)(x - x_2) \cdots (x - x_j) = P_{n-j}(x)(x - x_1)^2(x - x_2)^2 \cdots (x - x_j)^2$$

where P_{n-j} is a polynomial of degree $n - j$ that does not change sign in (a, b). Hence, $(p_n(x), (x - x_1) \cdots (x - x_j)) = (P_{n-j}, (x - x_1)^2 \cdots (x - x_j)^2)$. The right-hand side cannot vanish. But the left vanishes if $j < n$, so that $j \geq n$. But $j > n$ is impossible, and therefore $j = n$.

THEOREM 10.1.4.   *Let* $f(x) \in C[a, b]$; *then the Fourier segment*

$$\sum_{k=0}^n (f, p_k^*)p_k^*(x)$$

must coincide with $f(x)$ *in at least* $n + 1$ *points of* (a, b).

Proof: Let $R_n(x) = f(x) - \sum_{k=0}^n (f, p_k^*)p_k^*(x)$. Then from Theorem 8.6.1 we know that $(R_n(x), p_k^*(x)) = 0$, $k = 0, 1, \ldots, n$. In particular,

$$(R_n(x), p_0^*) = 0 = (R_n(x), 1).$$

Hence $R_n(x)$ must vanish somewhere in (a, b). Suppose now that it changes sign at $a < x_1 < x_2 < \cdots < x_j < b$ and at no other points of (a, b). Then, $R_n(x)$ is of constant and alternating sign in the segments

$$(a, x_1), (x_1, x_2), \cdots, (x_j, b),$$

and this is true of the polynomial $(x - x_1) \cdots (x - x_j)$. Thus, the product

$R_n(x)(x - x_1) \cdots (x - x_j)$ has constant sign in (a, b) and

$$(R_n(x)(x - x_1) \cdots (x - x_j), 1) = (R_n(x), (x - x_1) \cdots (x - x_j))$$

cannot vanish. But by orthogonality it must vanish for $j \le n$. Hence $j > n$ and the theorem follows.

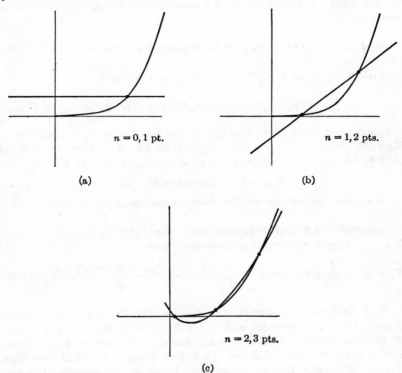

Figure 10.1.1 Coincidence of a Function and its Fourier Approximants.

DEFINITION 10.1.1. Let $p_n{}^*(x)$, $n = 0, 1, \ldots$, be a system of real orthonormal polynomials. The symmetric function

$$K_n(x, y) = \sum_{k=0}^{n} p_k{}^*(x) p_k{}^*(y) \tag{10.1.7}$$

is called the *kernel polynomial of order n of the orthonormal system*.

The kernel polynomial has the following *reproducing property*.

THEOREM 10.1.5. *For any polynomial* $P \in \mathscr{P}_n$,

$$(P(x), K_n(x, y))_x = P(y). \tag{10.1.8}$$

Conversely, if $K(x, y)$ is a polynomial of degree n at most in x and in y and if

$$(P(x), K(x, y))_x = P(y) \tag{10.1.9}$$

holds for all $P(x) \in \mathscr{P}_n$ then $K(x, y) = K_n(x, y)$.

The subscript x is placed outside the inner product to indicate the integration variable.

Proof: $P(x) = \sum_{k=0}^{n} (P, p_k^*)p_k^*(x)$ by Theorem 8.4.1. Hence,

$$(P(x), K_n(x, y))_x = \left(\sum_{k=0}^{n} (P, p_k^*)p_k^*(x), \sum_{k=0}^{n} p_k^*(x)p_k^*(y) \right)_x$$

$$= \sum_{m,k=0}^{n} (P, p_k^*)p_m^*(y)(p_k^*(x), p_m^*(x)) = \sum_{k=0}^{n} (P, p_k^*)p_k^*(y) = P(y).$$

Suppose now, that $(P(x), K(x, y))_x = P(y)$ for all $P \in \mathscr{P}_n$. Select $P(x) = K_n(x, w)$. Then, $(K_n(x, w), K(x, y))_x = K_n(y, w)$. But, also, in view of (10.1.8), $(K_n(x, w), K(x, y))_x = (K(x, y), K_n(x, w))_x = K(w, y)$. Hence,

$$K_n(y, w) = K_n(w, y) = K(w, y).$$

An alternate expression may be given for the kernel polynomial.

THEOREM 10.1.6 (Christoffel-Darboux). *Let $p_n^*(x) = k_n x^n + \cdots$, $n = 0, 1, \ldots$, be real orthonormal polynomials. Then,*

$$K_n(x, y) = \sum_{k=0}^{n} p_k^*(x)p_k^*(y) = \frac{k_n}{k_{n+1}} \frac{p_{n+1}^*(x)p_n^*(y) - p_n^*(x)p_{n+1}^*(y)}{x - y} \tag{10.1.10}$$

Proof: Designate the right-hand member of (10.1.10) by $K(x, y)$. Consider y fixed. Then the numerator of K is a polynomial of degree $\leq n + 1$ in x. Moreover it vanishes when $x = y$ and hence is divisible by $x - y$. Thus, $K(x, y)$ is a polynomial of degree $\leq n$ in both x and y. We shall show that if $p(x) \in \mathscr{P}_n$, $(p(x), K(x, y))_x = p(y)$ and hence by the previous theorem we will have $K(x, y) = K_n(x, y)$. Now,

$$(p(x), K(x, y))_x = \frac{k_n}{k_{n+1}} \left([p_{n+1}^*(x)p_n^*(y) - p_n^*(x)p_{n+1}^*(y)], \frac{p(x) - p(y)}{x - y} \right)_x$$

$$+ \frac{k_n}{k_{n+1}} p(y) \left(p_n^*(x), \frac{p_{n+1}^*(x) - p_{n+1}^*(y)}{x - y} \right)_x$$

$$+ \frac{k_n}{k_{n+1}} p(y) \left(p_{n+1}^*(x), \frac{p_n^*(y) - p_n^*(x)}{x - y} \right)_x.$$

Observe that $\dfrac{p(x) - p(y)}{x - y}$ and $\dfrac{p_n^*(x) - p_n^*(y)}{x - y}$ are polynomials of degree

$\leq n - 1$ in x (or equal 0 if $n = 0$). Hence by orthogonality, the first and third of the inner products on the right hand vanish. Then,

$$(p(x), K(x, y))_x = \frac{k_n}{k_{n+1}} p(y) \left(p_n{}^*(x), \frac{p_{n+1}^*(x) - p_{n+1}^*(y)}{x - y} \right)_x.$$

But,

$$\frac{k_n}{k_{n+1}} \frac{p_{n+1}^*(x) - p_{n+1}^*(y)}{x - y} = k_n \left[\frac{y^{n+1} - x^{n+1}}{x - y} + \cdots \right]$$

$$= k_n x^n + \cdots = p_n{}^*(x) + \text{polynomial of lower degree}.$$

Hence

$$\frac{k_n}{k_{n+1}} \left(p_n{}^*(x), \frac{p_{n+1}^*(y) - p_{n+1}^*(x)}{x - y} \right)_x = 1$$

and the theorem follows.

COROLLARY 10.1.7.

$$\sum_{k=0}^{n} \{p_n{}^*(x)\}^2 = \frac{k_n}{k_{n+1}} \{p_{n+1}^{*\prime}(x) p_n{}^*(x) - p_n{}^{*\prime}(x) p_{n+1}^*(x)\} \quad (10.1.11)$$

Proof: Allow $y \to x$ in 10.1.10 and evaluate the right-hand limit by de l'Hospital's rule.

10.2 Complex Orthogonal Polynomials. Let C be a rectifiable curve or arc lying in the plane of the complex variable $z = x + iy$. Consider the linear space of all polynomials with complex coefficients and for z on C. The complex powers $1, z, z^2, \ldots$, are independent elements, for if we had $a_0 + a_1 z + a_2 z^2 + \cdots + a_n z^n \equiv 0$ on C it would follow from the fundamental theorem of algebra that $a_0 = a_1 = \cdots = a_n = 0$. In this space,

$$(f, g) = \int_C f(z)\overline{g(z)} \, ds \qquad (10.2.1)$$

forms an inner product $\left(\int_C |f(z)|^2 \, ds = 0 \text{ implies } f(z) \equiv 0 \right)$. Hence, by Theorem 8.3.3 we may orthogonalize the powers and arrive at a set of polynomials

$$p_n{}^*(z) = k_n z^n + \cdots \quad ; \quad k_n > 0, \quad n = 0, 1, \ldots, \qquad (10.2.2)$$

that are orthonormal:

$$\int_C p_n{}^*(z) \overline{p_m{}^*(z)} \, ds = \delta_{mn}. \qquad (10.2.3)$$

The $p_n{}^*(z)$ are known as the *complex orthonormal polynomials corresponding to* C. If $w(z)$ is a positive function of the complex variable z defined on C, then, with appropriate integrability conditions,

$$(f, g) = \int_C f(z)\overline{g(z)}w(z)\, ds \qquad (10.2.4)$$

also constitutes an inner product and gives rise to a set of orthonormal polynomials. The polynomials arising from (10.2.3) or (10.2.4) are associated with the name of G. Szegö who first studied their properties extensively.

Complex orthogonal polynomials may also be constructed with double integrals. Let B designate a bounded region (open connected set) lying in the complex plane. If we introduce the inner product

$$(f, g) = \iint\limits_B f(z)\overline{g(z)}\, dx\, dy \qquad (10.2.5)$$

or

$$(f, g) = \iint\limits_B f(z)\overline{g(z)}w(z)\, dx\, dy \qquad (10.2.6)$$

for a suitable positive weight function $w(z)$, then sets of orthonormal polynomials can be generated. These orthogonal polynomials are associated with the names of T. Carleman and S. Bergman. (Cf. 9.2, III.)

Ex. 1.   The powers $\dfrac{1}{\sqrt{2\pi r^{2n+1}}}\, z^n$, $n = 0, 1, 2, \ldots$, are orthonormal on $|z| = r$. For

$$\int_{|z|=r} \frac{1}{\sqrt{2\pi r^{2n+1}}}\, z^n\, \frac{1}{\sqrt{2\pi r^{2m+1}}}\, \overline{z}^m\, ds = \frac{1}{2\pi r^{(m+n+1)}} \int_{|z|=r} z^n \overline{z}^m\, ds = \frac{1}{2\pi} \int_0^{2\pi} e^{i(m-n)\theta}\, d\theta$$

$$= 0 \text{ if } m \neq n \text{ and } 1 \text{ if } m = n.$$

Ex. 2.   The powers $\sqrt{\dfrac{n+1}{\pi}}\, \dfrac{z^n}{r^{n+1}}$, $n = 0, 1, 2, \ldots$, are orthonormal over the region $|z| < r$.

Ex. 3.   The Tschebyscheff polynomials of the first kind, $T_n(w)$, are orthogonal on every ellipse $\mathscr{E}_\rho$ (see 1.13) with respect to the weight function $|1 - w^2|^{-\frac{1}{2}}$. That is,

$$I_{m,n} = \int_{\mathscr{E}_\rho} \frac{T_m(w)\overline{T_n(w)}}{|1 - w^2|^{\frac{1}{2}}}\, |dw| = 0, \quad m \neq n; \quad |dw| = ds_w \qquad (10.2.7)$$

$$= \text{element of arc in the } w\text{-plane.}$$

Proof: Let

$$w = \tfrac{1}{2}(z + z^{-1}), \quad z = \rho e^{i\theta};$$

then by (4.4.2)

$$T_n(w) = \tfrac{1}{2}(\rho^n e^{in\theta} + \rho^{-n} e^{-in\theta}).$$

Now

$$dw/dz = \tfrac{1}{2}(1 - z^{-2}) \quad \text{so that} \quad |dw| = \frac{|z^2 - 1|}{|z^2|}\,|dz|$$

and

$$|1 - w^2|^{\frac{1}{2}} = \frac{1}{2}\frac{|z^2 - 1|}{|z|}\,.$$

Hence, by transforming to the z plane,

$$I_{m,n} = \frac{1}{4}\int_{|z|=\rho} \frac{(\rho^m e^{im\theta} + \rho^{-m} e^{-im\theta})(\rho^n e^{-in\theta} + \rho^{-n} e^{in\theta})}{|z|}\,|dz|$$

$$= \frac{1}{4}\int_0^{2\pi} (\rho^m e^{im\theta} + \rho^{-m} e^{-im\theta})(\rho^n e^{-in\theta} + \rho^{-n} e^{in\theta})\,d\theta = 0 \quad \text{when} \quad m \neq n.$$

Ex. 4. The Tschebyscheff Polynomials of the Second Kind

$$p_n{}^*(z) = 2\sqrt{\frac{n + 1}{\pi}}\,(\rho^{2n+2} - \rho^{-2n-2})^{-\frac{1}{2}}\,U_n(z),$$

$$U_n(z) = (1 - z^2)^{-\frac{1}{2}} \sin[(n + 1)\,\text{arc cos } z],$$

(10.2.8)

are orthonormal over the ellipse $\mathscr{E}_\rho$ with respect to the inner product

$$(f, g) = \iint_{\mathscr{E}_\rho} f(z)\overline{g(z)}\,dx\,dy.$$

Proof: Under the conformal map $z = \cos w, z = x + iy, w = u + iv$, the interior of the rectangle R with vertices at $\sigma i, \sigma i + \pi, -\sigma i + \pi, -\sigma i$ is mapped onto the ellipse $\mathscr{E}_\rho{}'$ consisting of the interior of $\mathscr{E}_\rho$ with the two segments $[-a, -1], [1, a]$ deleted. Now, $U_n(z) = (1 - z^2)^{-\frac{1}{2}} \sin[(n + 1)w]$ and

$$dx\,dy = \begin{vmatrix} \dfrac{\partial x}{\partial u} & \dfrac{\partial y}{\partial u} \\[2ex] \dfrac{\partial x}{\partial v} & \dfrac{\partial y}{\partial v} \end{vmatrix} du\,dv = \left|\frac{dz}{dw}\right|^2 dw\,dv = |1 - z^2|\,du\,dv.$$

Hence,

$$I_{m,n} = \iint_{\mathscr{E}_\rho} U_m(z)\overline{U_n(z)}\,dx\,dy = \iint_{\mathscr{E}_\rho{}'} U_m(z)\overline{U_n(z)}\,dx\,dy$$

$$= \iint_R \sin(m + 1)w\,\overline{\sin(n + 1)w}\,du\,dv.$$

With $p = m + 1, r = n + 1$, and since

$$\sin p(u + iv) = \sin pu \cosh pv + i \cos pu \sinh pv,$$

$$I_{m,n} = \int_{-\sigma}^{\sigma} \cosh pv \cosh rv \, dv \int_{0}^{\pi} \sin pu \sin ru \, du + \int_{-\sigma}^{\sigma} \sinh pv \sinh rv \, dv$$

$$\times \int_{0}^{\pi} \cos pu \cos ru \, du + i \int_{-\sigma}^{\sigma} \sinh pv \cosh rv \, dv \int_{0}^{\pi} \cos pu \sin ru \, du -$$

$$i \int_{-\sigma}^{\sigma} \cosh pv \sinh rv \, dv \int_{0}^{\pi} \sin pu \cos ru \, du$$

Now $\sinh pv \cosh rv$ and $\cosh pv \sinh rv$ are odd functions of v. Their integral over $[-\sigma, \sigma]$ therefore vanishes. Furthermore,

$$\int_{0}^{\pi} \sin pu \sin ru \, du = \int_{0}^{\pi} \cos pu \cos ru \, du = \begin{matrix} 0, & p \neq r \\ \dfrac{\pi}{2}, & p = r. \end{matrix}$$

Hence, $I_{m,n} = 0$ for $m \neq n$. Now

$$I_{p,p} = \frac{\pi}{2} \int_{-\sigma}^{\sigma} \cosh^2 pv + \sinh^2 pv \, dv = \frac{\pi}{2} \int_{-\sigma}^{\sigma} \cosh 2pv \, dv = \frac{\pi \sinh 2p\sigma}{2p}.$$

But $\rho = e^{\sigma}$. Hence, $I_{p,p} = \dfrac{\pi}{4p} (\rho^{2p} - \rho^{-2p})$.

Ex. 5. Let S designate the square with sides $x = \pm 1, y = \pm 1$, and write

$$(f, g) = \int_{S} f(z)\overline{g(z)} \, ds.$$

The following polynomials are orthonormal with respect to this inner product.

$$s_0(z) = .35355$$
$$s_1(z) = .30619z$$
$$s_2(z) = .25877z^2$$
$$s_3(z) = .21348z^3$$
$$s_4(z) = .18697z^4 + .14957$$
$$s_5(z) = .15811z^5 + .18070z$$
$$s_6(z) = .13396z^6 + .20049z^2$$
$$s_7(z) = .11372z^7 + .20905z^3$$
$$s_8(z) = .09656z^8 + .20627z^4 - .00666$$

Though in isolated cases there are recurrence formulas relating successive complex orthogonal polynomials, there does not appear to be a general theorem of this sort. The identity used in the real case to establish the recurrence: $(xp_n^*, p_m^*) = (p_n^*, xp_m^*)$ does not carry over to the complex case where the inner product is Hermitian.

Theorem 10.1.3 tells us the location of the zeros of real orthogonal polynomials. L. Fejér found a remarkably simple proof of a theorem that covers a wide variety of cases, both real and complex.

LEMMA 10.2.1 (Fejér's Principle). *Let S designate a closed bounded convex set lying in the complex plane. z_1 is a point exterior to S. Then we may find a point z' such that*

$$|z - z'| < |z - z_1| \qquad \text{for all } z \in S. \tag{10.2.9}$$

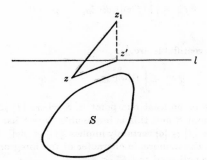

Figure 10.2.1.

Proof: We can find a line ℓ that separates z_1 and S, i.e., z_1 and S lie on opposite sides of ℓ. This geometrically evident fact can be established easily by the methods of the theory of convex bodies, but we shall not do so here (Cf. Prob. 23). Drop a perpendicular from z_1 to ℓ and call its foot z'. Then if z is any point on the "S side of ℓ," it is no trouble to show that $|z - z'| < |z - z_1|$. In particular, (10.2.9) follows.

DEFINITION 10.2.1. Let S be a set of points lying in the complex plane and F a family of functions of a complex variable defined on S. For each $f \in F$ let there be defined a real, nonnegative quantity designated by $|\!|\!|f|\!|\!|$ satisfying the following property: for distinct f and g, the condition

and

$$\left. \begin{array}{l} |f(z)| < |g(z)| \text{ whenever } g(z) \neq 0 \\[2mm] |f(z)| = |g(z)| \text{ whenever } g(z) = 0 \end{array} \right\} \tag{10.2.10}$$

implies

$$|\!|\!|f|\!|\!| < |\!|\!|g|\!|\!|. \tag{10.2.11}$$

The quantity $|\!|\!|f|\!|\!|$ will be known as a *Fejér* or *monotonic norm for F*. (It should be distinguished from the norms in a normed linear space.)

Here are some examples of monotonic norms. In each case F is the class of polynomials $z^n + a_1 z^{n-1} + \cdots + a_n$ with a_i complex.

Ex. 6. Let S be a closed bounded point set. If finite, it should contain at least n points. $\||f\|| = \max_{z \in S} |f(z)|$.

Ex. 7. Let S be a rectifiable curve or arc.

$$\||f\|| = \int_S |f(z)|^p \, ds. \qquad p > 0.$$

Ex. 8. Let S be a closed bounded region.

$$\||f\|| = \iint_S |f(z)|^p \, dx \, dy, \qquad p > 0.$$

Ex. 9. Let S be a rectifiable arc,

$$\||f\|| = \int_S e^{|f(z)|^p} \, ds, \qquad p > 0.$$

In Ex. 6, there must be at least one point in S where $|f| < |g|$, for otherwise $|f| = |g| = 0$ throughout S and this is impossible since S has at least n points. In the other examples, $|f| \leq |g|$ certainly implies $\||f\|| \leq \||g\||$. But since $|f| < |g|$ in at least one point, the monotonic character of the integrals tells us that the inequality may be strengthened to $\||f\|| < \||g\||$.

THEOREM 10.2.2 (Fejér's Convex Hull Theorem). *Let F designate the family of polynomials $z^n + a_1 z^{n-1} + \cdots + a_n$ with a_i complex. Let $\|| \, \||$ be a monotonic norm on F relative to a point set S. Let the problem*

$$\min_{a_i} \||z^n + a_1 z^{n-1} + \cdots + a_n\|| \qquad (10.2.12)$$

be solved by a polynomial $p(z)$. Then the zeros of $p(z)$ all lie in the closure of the convex hull of S. (Def. 7.3.3).

Proof: Let $p(z) = (z - z_1) \cdots (z - z_n)$. Assume that a typical root, z_1, lies exterior to the closure of the convex hull of S. This closure is also convex. By Fejér's principle, we can find a z' such that

$$|z - z'| < |z - z_1| \qquad \text{for all } z \in S.$$

Hence, if we set

$$q(z) = (z - z')(z - z_2) \cdots (z - z_n) \quad (\not\equiv p(z)), \qquad (10.2.13)$$

we have for z in S,

$$|q(z)| < |p(z)| \text{ whenever } p(z) \neq 0$$

and

$$|q(z)| = |p(z)| \text{ whenever } p(z) = 0.$$

$(10.2.14)$

Hence, $\||q\|| < \||p\||$, so that p could not possibly have been a minimal polynomial as asserted. Therefore z_1 must be in the closure of the convex hull of S.

COROLLARY 10.2.3.   *The zeros of complex orthogonal polynomials lie in the closure of the convex hull of the point sets over which the integration is performed. If the point set lies on the real axis and is contained in $[a, b]$, the orthogonal polynomials have real coefficients and the zeros lie in $[a, b]$.*

Proof: Cor. 8.5.7, Ex. 7, 8 of 10.2 with $p = 2$, and Theorem 10.2.2.

In contradistinction to the real case, no assertion can be made about the simplicity of the roots; indeed, z^n are orthogonal over circles $|z| = r$, but have an n-fold zero at $z = 0$.

The kernel polynomials of a real or complex system of orthogonal polynomials solve an important extremal problem. (Cf. Theorem 9.4.3.)

THEOREM 10.2.3.   *Let z_0 be an arbitrary point in the complex plane and $p(z)$ an arbitrary element of $\mathscr{P}_n$. The problem of finding*

$$\max_{p \in \mathscr{P}_n} |p(z_0)| \tag{10.2.15}$$

subject to

$$\|p\| = 1 \tag{10.2.16}$$

is solved by

$$q(z) = \frac{e^{i\theta} K_n(z, z_0)}{\sqrt{K_n(z_0, z_0)}} \tag{10.2.17}$$

where $K_n(z, z_0)$ is the kernel polynomial

$$K_n(z, z_0) = \sum_{k=0}^{n} p_k{}^*(z)\overline{p_k{}^*(z_0)} \tag{10.2.18}$$

and $\theta, 0 \le \theta < 2\pi$, is arbitrary. The maximum value of (10.2.15) is $\sqrt{K_n(z_0, z_0)}$.

Proof: Let $\{p_n{}^*\}$ be the orthonormal polynomials appropriate to the norm $\| \ \|$. Then an arbitrary $p \in \mathscr{P}_n$ can be written as $p(z) = \sum_{k=0}^{n} a_k p_k{}^*(z)$ with $\|p\|^2 = \sum_{k=0}^{n} |a_k|^2$, (Theorem 8.4.1). Hence

$$|p(z_0)|^2 = \left| \sum_{k=0}^{n} a_k p_k{}^*(z_0) \right|^2 \le \sum_{k=0}^{n} |a_k|^2 \sum_{k=0}^{n} |p_k{}^*(z_0)|^2.$$

Hence $|p(z_0)|^2 \le \sum_{k=0}^{n} |p_k{}^*(z_0)|^2 = K_n(z_0, z_0)$ whenever $\|p\| = 1$. On the other hand, $q(z) \in \mathscr{P}_n$, and

$$\|q\|^2 = \left(\frac{e^{i\theta} K_n(z, z_0)}{\sqrt{K_n(z_0, z_0)}}, \frac{e^{i\theta} K_n(z, z_0)}{\sqrt{K_n(z_0, z_0)}} \right)$$

$$= \frac{1}{K_n(z_0, z_0)} \sum_{j,k=0}^{n} (p_k{}^*(z)\overline{p_k{}^*(z_0)}, p_j{}^*(z)\overline{p_j{}^*(z_0)})$$

$$= \frac{1}{K_n(z_0, z_0)} \sum_{k=0}^{n} p_k{}^*(z_0)\overline{p_k{}^*(z_0)} = 1.$$

Furthermore,

$$|q(z_0)| = \sqrt{K_n(z_0, z_0)}.$$

Ex. 10. Let $\mathscr{P}_n$ be an inner product space of real or complex polynomials of degree $\leq n$. If z_0 is an arbitrary point in the complex plane, then $L(f) = f(z_0)$ is a bounded linear functional over $\mathscr{P}_n$ and $\|L\|^2 = K_n(z_0, z_0)$.

10.3   The Special Function Theory of the Jacobi Polynomials.

There is an extensive literature that contains many identities and interrelationships between the real orthogonal polynomials generated by simple weighting functions. We shall content ourselves with presenting the most important identities for the class of polynomials known as the Jacobi polynomials.

For simplicity, the fundamental interval is selected as $[-1, +1]$. The relevant weight function is

$$w(x) = (1 - x)^{\alpha}(1 + x)^{\beta}, \quad \alpha > -1, \quad \beta > -1. \tag{10.3.1}$$

If the exponents are between -1 and 0, then the weight has a singularity at the corresponding end point, but possesses a finite integral. Indeed,

$$\int_{-1}^{+1} (1 - x)^{\alpha}(1 + x)^{\beta} \, dx = 2^{\alpha+\beta+1} \frac{\Gamma(\alpha + 1)\Gamma(\beta + 1)}{\Gamma(\alpha + \beta + 2)}. \tag{10.3.2}$$

This is readily established by setting $t = \frac{1}{2}(1 + x)$ and using the standard integral for the beta function.

The orthonormal polynomials that result from orthonormalizing $1, x,$ $x^2, \ldots,$ with respect to the inner product $(f, g) = \displaystyle\int_{-1}^{1} f(x)g(x)w(x) \, dx$ by means of the Gram-Schmidt process are called the *Jacobi polynomials*, and will be designated by $p_n^{(\alpha,\beta)}(x)$.

The following special selections of α and β carry special names.

$\alpha = 0,\ \beta = 0$: Legendre polynomials
$\alpha = -\frac{1}{2},\ \beta = -\frac{1}{2}$: Tschebyscheff polynomials (of the first kind).
$\alpha = \frac{1}{2},\ \beta = \frac{1}{2}$: Tschebyscheff polynomials of the second kind.
$\alpha = \beta$: Ultraspherical polynomials.

We shall also employ orthogonal Jacobi polynomials, $P_n^{(\alpha,\beta)}(x)$, that have been standardized by requiring that

$$P_n^{(\alpha,\beta)}(1) = \frac{\Gamma(n + \alpha + 1)}{\Gamma(n + 1)\Gamma(\alpha + 1)}. \tag{10.3.3}$$

THEOREM 10.3.1 (Rodrigues' Formula).

$$P_n^{(\alpha,\beta)}(x) = \frac{(-1)^n}{2^n n!} (1 - x)^{-\alpha}(1 + x)^{-\beta} \frac{d^n}{dx^n} \{(1 - x)^{n+\alpha}(1 + x)^{n+\beta}\}. \tag{10.3.4}$$

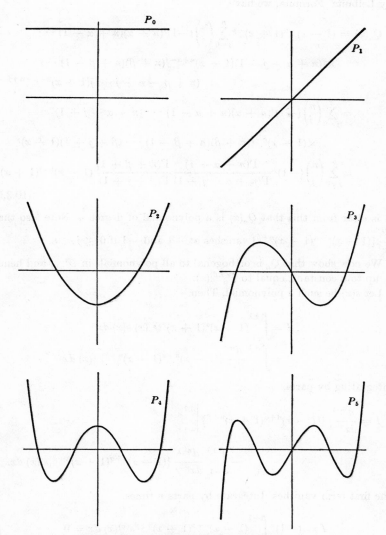

Figure 10.3.1 The Legendre Polynomials.

Proof: Consider first the expression

$$Q_n(x) = (1-x)^{-\alpha}(1+x)^{-\beta}\frac{d^n}{dx^n}\{(1-x)^{n+\alpha}(1+x)^{n+\beta}\}. \quad (10.3.5)$$

By Leibnitz' Formula, we have,

$$Q_n(x) = (1-x)^{-\alpha}(1+x)^{-\beta} \sum_{j=0}^{n} \binom{n}{j}(-1)^j(n+\alpha)(n+\alpha-1)\cdots$$

$$\times (n+\alpha-j+1)(1-x)^{n+\alpha-j}(n+\beta)(n+\beta-1)\cdots$$

$$\times (n+\beta-n+j+1)(1+x)^{n+\beta-n+j}$$

$$= \sum_{j=0}^{n} \binom{n}{j}(-1)^j(n+\alpha)(n+\alpha-1)\cdots(n+\alpha-j+1)$$

$$\times (1-x)^{n-j}(n+\beta)(n+\beta-1)\cdots(\beta+j+1)(1+x)^j$$

$$= \sum_{j=0}^{n} \binom{n}{j}(-1)^j \frac{\Gamma(n+\alpha+1)}{\Gamma(n+\alpha-j+1)} \frac{\Gamma(n+\beta+1)}{\Gamma(\beta+j+1)}(1-x)^{n-j}(1+x)^j.$$

$$(10.3.6)$$

It is clear from this that $Q_n(x)$ is a polynomial of degree n. Note also that $\dfrac{d^j}{dx^j}[(1-x)^{n+\alpha}(1+x)^{n+\beta}]$ vanishes at $+1$ and -1 if $0 \leq j < n$.

We now show that Q_n is orthogonal to all polynomials in $\mathscr{P}_{n-1}$ and hence is, up to a constant, equal to $P_n^{(\alpha,\beta)}(x)$.

Let $s(x)$ be such a polynomial. Then,

$$I = \int_{-1}^{+1}(1-x)^{\alpha}(1+x)^{\beta}Q_n(x)\,s(x)\,dx$$

$$= \int_{-1}^{+1} \frac{d^n}{dx^n}\{(1-x)^{n+\alpha}(1+x)^{n+\beta}\}\,s(x)\,dx.$$

Integrating by parts,

$$I = \frac{d^{n-1}}{dx^{n-1}}\{(1-x)^{n+\alpha}(1+x)^{n+\beta}\}\Big|_{-1}^{+1}$$

$$-\int_{-1}^{+1} \frac{d^{n-1}}{dx^{n-1}}\{(1-x)^{n+\alpha}(1+x)^{n+\beta}\}s'(x)\,dx.$$

The first term vanishes. Integrate by parts n times.

$$I = (-1)^n \int_{-1}^{+1}(1-x)^{n+\alpha}(1+x)^{n+\beta}s^{(n)}(x)\,dx = 0$$

inasmuch as $s(x) \in \mathscr{P}_{n-1}$ and $s^{(n)}(x) \equiv 0$. We can therefore write $P_n(x) = cQ_n(x)$. From (10.3.6), $Q_n(1) = (-1)^n \dfrac{\Gamma(n+\alpha+1)}{\Gamma(\alpha+1)}2^n$, and hence from (10.3.3), $c = \dfrac{(-1)^n}{n!\,2^n}$.

COROLLARY 10.3.2.

$$P_n^{(\alpha,\beta)}(x) = \sum_{j=0}^{n} \frac{\Gamma(n + \alpha + 1)}{j! \, \Gamma(n + \alpha - j + 1)} \frac{\Gamma(n + \beta + 1)}{(n-j)! \, \Gamma(\beta + j + 1)}$$

$$\times \left(\frac{x-1}{2}\right)^{n-j} \left(\frac{x+1}{2}\right)^{j}$$

$$= \sum_{j=0}^{n} \binom{n+\alpha}{j} \binom{n+\beta}{n-j} \left(\frac{x-1}{2}\right)^{n-j} \left(\frac{x+1}{2}\right)^{j}$$

where we extend the binomial symbol $\binom{y}{z}$ *to noninteger values in an obvious way.*

COROLLARY 10.3.3.

$$P_n^{(\alpha,\beta)}(-x) = (-1)^n P_n^{(\beta,\alpha)}(x).$$

$$P_n^{(\alpha,\beta)}(-1) = (-1)^n \frac{\Gamma(n + \beta + 1)}{\Gamma(n + 1)\Gamma(\beta + 1)}.$$

THEOREM 10.3.4. $P_n^{(\alpha,\beta)}(x) = K_n x^n + S_n x^{n-1} + \cdots$

where

$$K_n = \frac{1}{2^n} \binom{2n + \alpha + \beta}{n} \qquad (10.3.7)$$

and

$$S_n = \frac{\alpha - \beta}{2^n} \binom{2n + \alpha + \beta - 1}{n - 1}. \qquad (10.3.8)$$

Proof: We begin by establishing two identities for binomial coefficients. For $|x| < 1$ we have

$$(1 + x)^p = \sum_{n=0}^{\infty} \binom{p}{n} x^n, \quad (1 + x)^{\sigma} = \sum_{m=0}^{\infty} \binom{\sigma}{m} x^m$$

Hence, $(1 + x)^{p+\sigma} = \sum_{m,n=0}^{\infty} \binom{p}{n} \binom{\sigma}{m} x^{n+m}$

$$= \sum_{n=0}^{\infty} x^n \sum_{j=0}^{n} \binom{p}{j} \binom{\sigma}{n-j}.$$

Therefore

$$\sum_{j=0}^{n} \binom{p}{j} \binom{\sigma}{n-j} = \frac{1}{n!} \frac{d^n}{dx^n} (1 + x)^{p+\sigma} \Big|_{x=0} = \binom{p + \sigma}{n}. \qquad (10.3.9)$$

Furthermore, $p(1 + x)^{p-1} = \sum_{n=1}^{\infty} n \binom{p}{n} x^{n-1}$. Hence,

$$px(1 + x)^{p+\sigma-1} = \sum_{n=0}^{\infty} x^n \sum_{j=1}^{n} j \binom{p}{j} \binom{\sigma}{n-j}.$$

Therefore

$$\sum_{j=1}^{n} j \binom{p}{j} \binom{\sigma}{n-j} = \frac{1}{n!} \frac{d^n}{dx^n} px(1+x)^{p+\sigma-1} \bigg|_{x=0}$$

$$= \frac{p}{n!} \left(x \frac{d^n}{dx^n} (1+x)^{p+\sigma-1} + n \frac{d^{n-1}}{dx^{n-1}} (1+x)^{p+\sigma-1} \right) \bigg|_{x=0}$$

$$= p \binom{p+\sigma-1}{n-1}. \tag{10.3.10}$$

To prove the theorem, we use Corollary 10.3.2. Notice that

$$(x-1)^{n-j}(x+1)^j = [x^{n-j} - (n-j)x^{n-j-1} + \cdots][x^j + jx^{j-1} + \cdots]$$
$$= x^n + (2j-n)x^{n-1} + \cdots.$$

It is now clear that

$$K_n = \frac{1}{2^n} \sum_{j=0}^{n} \binom{n+\alpha}{j} \binom{n+\beta}{n-j} = \frac{1}{2^n} \binom{2n+\alpha+\beta}{n}. \tag{10.3.11}$$

The last equality follows from (10.3.9).

In the same way,

$$2^n S_n = \sum_{j=0}^{n} \binom{n+\alpha}{j} \binom{n+\beta}{n-j} (2j-n)$$

$$= 2 \sum_{j=0}^{n} j \binom{n+\alpha}{j} \binom{n+\beta}{n-j} - n \sum_{j=0}^{n} \binom{n+\alpha}{j} \binom{n+\beta}{n-j}$$

$$= 2(n+\alpha) \binom{2n+\alpha+\beta-1}{n-1} - n \binom{2n+\alpha+\beta}{n}$$

$$= (\alpha-\beta) \binom{2n+\alpha+\beta-1}{n-1}. \tag{10.3.12}$$

THEOREM 10.3.5.

$$\int_{-1}^{1} (1-x)^\alpha (1+x)^\beta (P_n^{(\alpha,\beta)}(x))^2 \, dx$$

$$= \frac{2^{\alpha+\beta+1}}{(2n+\alpha+\beta+1)} \frac{\Gamma(n+\alpha+1)\Gamma(n+\beta+1)}{\Gamma(n+1)\Gamma(n+\alpha+\beta+1)}.$$

Proof: Write $(1-x)^\alpha (1+x)^\beta = w(x)$ and $P_n^{(\alpha,\beta)} = P_n$.

$$I = \int_{-1}^{1} w(x)[P_n(x)]^2 \, dx = \int_{-1}^{1} w(x) P_n(x) P_n(x) \, dx.$$

Now, $P_n(x) = K_n x^n +$ polynomial of lower degree; hence,

$$I = K_n \int_{-1}^{1} w(x) P_n(x) x^n \, dx = \frac{(-1)^n K_n}{2^n n!} \int_{-1}^{+1} x^n \frac{d^n}{dx^n} \{(1-x)^{n+\alpha}(1+x)^{n+\beta}\} \, dx.$$

Integrate by parts n times and obtain

$$I = \frac{K_n}{2^n} \int_{-1}^{+1} (1-x)^{n+\alpha}(1+x)^{n+\beta}\, dx.$$

By (10.3.2) we have

$$I = \frac{K_n}{2^n} \cdot 2^{2n+\alpha+\beta+1} \frac{\Gamma(n+\alpha+1)\Gamma(n+\beta+1)}{\Gamma(2n+\alpha+\beta+2)},$$

and the theorem follows.

COROLLARY 10.3.6.   *The orthonormal Jacobi polynomials are given by*

$$p_n^{(\alpha,\beta)}(x) = \left\{ \frac{(2n+\alpha+\beta+1)}{2^{\alpha+\beta+1}} \frac{\Gamma(n+1)\Gamma(n+\alpha+\beta+1)}{\Gamma(n+\alpha+1)\Gamma(n+\beta+1)} \right\}^{\frac{1}{2}} P_n^{(\alpha,\beta)}(x).$$

$$(10.3.13)$$

COROLLARY 10.3.7.   *The kernel polynomial of the orthonormal system has the following expression:*

$$K_n(x,y) = \sum_{j=0}^{n} p_j^{(\alpha,\beta)}(x) p_j^{(\alpha,\beta)}(y)$$

$$= \frac{1}{\|P_n\|^2} \frac{K_n}{K_{n+1}} \frac{P_{n+1}^{(\alpha,\beta)}(x) P_n^{(\alpha,\beta)}(y) - P_n^{(\alpha,\beta)}(x) P_{n+1}^{(\alpha,\beta)}(y)}{x-y}$$

$$= \frac{(n+1)!\,\Gamma(n+\alpha+\beta+2)}{2^{\alpha+\beta}(2n+\alpha+\beta+2)\Gamma(n+\alpha+1)\Gamma(n+\beta+1)}$$

$$\cdot \frac{P_{n+1}^{(\alpha,\beta)}(x) P_n^{(\alpha,\beta)}(y) - P_n^{(\alpha,\beta)}(x) P_{n+1}^{(\alpha,\beta)}(y)}{x-y}.$$

Proof: This is an application of (10.1.7), (10.3.7), (10.3.13), and Theorem 10.3.5.

THEOREM 10.3.8.   *The Jacobi polynomials $P_n^{(\alpha,\beta)}(x)$ satisfy the following second order linear differential equation.*

$$(1-x^2)y'' + [(\beta-\alpha) - (\alpha+\beta+2)x]y' + n(n+\alpha+\beta+1)y = 0$$

$$(10.3.14)$$

Proof: Note that for any q,

$$\frac{d}{dx}[(1-x)^{\alpha+1}(1+x)^{\beta+1}q'] = (1-x)^{\alpha}(1+x)^{\beta}$$

$$\times [(1-x^2)q'' + \{(\beta-\alpha) - (\alpha+\beta+2)x\}q']. \quad (10.3.15)$$

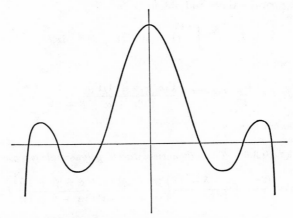

Figure 10.3.2 The Kernel Polynomial for $\alpha = \beta = 0$.

$$K_7(x, 0) = -\frac{35}{512}\{429x^6 - 693x^4 + 315x^2 - 35\}$$

Hence we may write the differential equation (10.3.14) in the alternate form

$$(1 - x)^{-\alpha}(1 + x)^{-\beta}\frac{d}{dx}[(1 - x)^{\alpha+1}(1 + x)^{\beta+1}y'] = -n(n + \alpha + \beta + 1)y.$$
$$(10.3.16)$$

Let $y = P_n^{(\alpha,\beta)}(x)$, $w(x) = (1 - x)^{\alpha}(1 + x)^{\beta}$ and $q(x) \in \mathscr{P}_{n-1}$. We shall show that the expression occurring on the left side of (10.3.16) is orthogonal to $q(x)$. Hence the left-hand side must coincide with y apart from a constant multiplier. Now,

$$I = \int_{-1}^{+1} w(x)(1 - x)^{-\alpha}(1 + x)^{-\beta}\frac{d}{dx}[(1 - x)^{\alpha+1}(1 + x)^{\beta+1}y']q(x)\,dx$$

$$= \int_{-1}^{+1} \frac{d}{dx}[(1 - x)^{\alpha+1}(1 + x)^{\beta+1}y']q(x)\,dx.$$

Integrate by parts,

$$I = -\int_{-1}^{+1}(1 - x)^{\alpha+1}(1 + x)^{\beta+1}y'q'\,dx. \qquad (10.3.17)$$

Integrate by parts once again,

$$I = \int_{-1}^{+1}\frac{d}{dx}[(1 - x)^{\alpha+1}(1 + x)^{\beta+1}q']y\,dx.$$

Now from (10.3.15) we observe that

$$\frac{d}{dx}[(1 - x)^{\alpha+1}(1 + x)^{\beta+1}q'] = (1 - x)^{\alpha}(1 + x)^{\beta}p(x) \quad \text{where} \quad p(x) \in \mathscr{P}_{n-1}.$$

Hence $\qquad I = \int_{-1}^{+1} (1 - x)^{\alpha}(1 + x)^{\beta} P_n^{(\alpha,\beta)}(x)p(x) \, dx = 0.$

Thus, certainly,

$(1 - x^2)y'' + \{(\beta - \alpha) - x(\alpha + \beta + 2)\}y' = cy$ for some constant c.

To determine c, set $P_n^{(\alpha,\beta)}(x) = y = K_n x^n + \cdots$. Then dealing only with the highest powers, $-n(n - 1)K_n x^n - n(\alpha + \beta + 2)K_n x^n = cK_n x^n$. Hence $c = -n(n + \alpha + \beta + 1)$ as required.

THEOREM 10.3.9. *The Jacobi polynomials $P_n^{(\alpha,\beta)}(x)$ satisfy the following three-term linear recurrence relation*

$$P_n^{(\alpha,\beta)}(x) = (A_n x + B_n)P_{n-1}^{(\alpha,\beta)}(x) - C_n P_{n-2}^{(\alpha,\beta)}(x) \qquad (10.3.18)$$

where

$$A_n = \frac{(2n + \alpha + \beta - 1)(2n + \alpha + \beta)}{2n(n + \alpha + \beta)} \qquad (10.3.19)$$

$$B_n = \frac{(\alpha^2 - \beta^2)(2n + \alpha + \beta - 1)}{2n(n + \alpha + \beta)(2n + \alpha + \beta - 2)} \qquad (10.3.20)$$

$$C_n = \frac{(n + \alpha - 1)(n + \beta - 1)(2n + \alpha + \beta)}{n(n + \alpha + \beta)(2n + \alpha + \beta - 2)}. \qquad (10.3.21)$$

Proof: We combine (10.1.6), (10.3.7), (10.3.8), and (10.3.13). If $p_n^{(\alpha,\beta)}(x) = \lambda_n P_n^{(\alpha,\beta)}(x)$, then

$$A_n = \frac{K_n}{K_{n-1}}, \qquad B_n = \frac{K_n}{K_{n-1}}\left(\frac{S_n}{K_n} - \frac{S_{n-1}}{K_{n-1}}\right), \qquad C_n = \left(\frac{\lambda_{n-2}}{\lambda_{n-1}}\right)^2 \frac{K_n K_{n-2}}{K_{n-1}^2}.$$

The expressions above result from this.

THEOREM 10.3.10. *In a neighborhood of $w = 0$, we have*

$$\sum_{n=0}^{\infty} P_n^{(\alpha,\beta)}(z)w^n = \frac{2^{\alpha+\beta}}{R(1 - w + R)^{\alpha}(1 + w + R)^{\beta}}, \qquad (10.3.22)$$

$$R = +\sqrt{1 - 2zw + w^2}.$$

The multivalued functions are taken positive for $w = 0$.

Proof: Cauchy's Theorem tells us that for analytic functions,

$$f^{(n)}(z) = \frac{n!}{2\pi i} \int_C \frac{f(t)}{(t - z)^{n+1}} \, dt \qquad (10.3.23)$$

where C is a closed curve containing z in its interior.

Consider the function $(1 - z)^{n+\alpha}(1 + z)^{n+\beta}$. In a simply connected region that does not contain ± 1 in its interior, it is possible to define a single valued analytic branch. Hence,

$$\frac{d^n}{dz^n}(1-z)^{n+\alpha}(1+z)^{n+\beta} = \frac{n!}{2\pi i}\int_C \frac{(1-t)^{n+\alpha}(1+t)^{n+\beta}}{(t-z)^{n+1}}\,dt$$

where C is a closed curve that contains z in its interior, but does not contain either $z=1$ or $z=-1$. From Rodrigues' formula (10.3.4),

$$P_n^{(\alpha,\beta)}(z) = \frac{(-1)^n}{2^n n!}\cdot\frac{n!}{2\pi i}\int_C \frac{(1-z)^{-\alpha}(1+z)^{-\beta}(1-t)^{n+\alpha}(1+t)^{n+\beta}}{(t-z)^{n+1}}\,dt$$

$$= \frac{1}{2\pi i}\int_C \left(\frac{1-t}{1-z}\right)^{\alpha}\left(\frac{1+t}{1+z}\right)^{\beta}\left(\frac{t^2-1}{2(t-z)}\right)^n\frac{dt}{t-z}.$$

The points $z=\pm 1$ are excluded from consideration here.

In this integral, make the change of variable

$$\frac{t^2-1}{2(t-z)} = \frac{1}{w}\,; \qquad t = \frac{1}{w}(1-\sqrt{1-2wz+w^2}). \tag{10.3.24}$$

Write

$$t = \frac{1}{w}(1-R) \quad\text{where}\quad R = \sqrt{1-2zw+w^2}. \tag{10.3.25}$$

Now $t-z = \frac{1}{2}w(z^2-1)+\cdots$ and so a neighborhood of $t=z$ is mapped conformally onto a neighborhood of $w=0$.

$$\frac{dt}{dw} = \frac{1}{w^2}\left(R-1-w\frac{dR}{dw}\right) \quad\text{and}\quad \frac{dR}{dw} = \frac{w-z}{R} \tag{10.3.26}$$

so that

$$\frac{dt}{dw} = \frac{t-z}{Rw}.$$

Moreover,

$$\frac{1+t}{1+z} = \frac{1+1/w-R/w}{1+z} = \frac{2}{1+w+R}$$

and

$$\frac{1-t}{1-z} = \frac{1-1/w+R/w}{1-z} = \frac{2}{1-w+R}.$$

Hence,

$$P_n^{(\alpha,\beta)}(z) = \frac{2^{\alpha+\beta}}{2\pi i}\int_{C_1}\frac{dw}{(1-w+R)^{\alpha}(1+w+R)^{\beta}w^{n+1}R}.$$

C_1 is the image of C under the conformal transformation and is a closed curve containing the origin $w=0$ in its interior.

Applying Cauchy's formula in the w-plane, we see that

$$P_n^{(\alpha,\beta)}(z) = \frac{1}{n!}\frac{d^n}{dw^n}\frac{2^{\alpha+\beta}}{R(1-w+R)^{\alpha}(1+w+R)^{\beta}}\bigg|_{w=0}.$$

Now, considered as a function of w, $R^{-1}(1-w+R)^{-\alpha}(1+w+R)^{-\beta}$ is analytic in a neighborhood of the origin, and we have (10.3.22). For $z = \pm 1$, the identity can be verified directly.

NOTES ON CHAPTER X

General references on orthogonal polynomials include Szegö [1], Tricomi [1], Sansone [1], Jackson [2], Alexits [1]. The bibliography referred to in the Foreword is the publication of Shohat, Hille, and Walsh [1].

10.1 For an analog of Theorem 10.1.4 for trigonometric Fourier series, see Tricomi [1] pp. 75–76. A converse of Theorem 10.1.1 was found by Favard and independently by Natanson. See Favard [1]. Also relevant is Dickinson, Pollak, and Wannier [1].

10.2 Ex. 3., Walsh [2] Chap. VI. Ex. 4., Nehari [1]. Ex. 5.: these and more extensive values were computed on the IBM 704 at the National Bureau of Standards. For Fejér's Principle see Fejér [1].

PROBLEMS

1. The powers x^m cannot be orthogonal on $[-1, 1]$ with respect to any weight function $w(x)$.

2. Let $w(-x) = w(x)$ and suppose that $p_n(x)$ are orthogonal with respect to

$$(f, g) = \int_{-1}^{+1} f(x)g(x)w(x)\, dx.$$ Then $p_n(-x) = (-1)^n p_n(x)$.

3. How do orthogonal polynomials change when the interval $[-1, 1]$ is shifted linearly to $[a, b]$ and the weight function changed correspondingly?

4. Let $p_0(z) = 1, p_n(z) = z^{n-1}(z - z_0)$ $n = 1, 2, \ldots$. Prove that for $m \neq n$,

$$\int_C \frac{p_m(z)\overline{p_n(z)}\, ds}{|z - z_0|^2} = 0 \text{ on all circles } C: |z| = R > |z_0|.$$

5. Let $|a_i| < 1$. Prove that the functions

$$r_1(z) = \left(\frac{1 - |a_1|^2}{2\pi}\right)^{\frac{1}{2}} \frac{1}{1 - \overline{a_1}z},$$

$$r_n(z) = \left(\frac{1 - |a_n|^2}{2\pi}\right)^{\frac{1}{2}} \frac{(z - a_1) \cdots (z - a_{n-1})}{(1 - \overline{a_1}z)(1 - \overline{a_2}z) \cdots (1 - \overline{a_n}z)} \qquad n = 2, 3, \ldots,$$

are orthonormal in the sense that $\displaystyle\int_{|z|=1} r_m(z)\, \overline{r_n(z)}\, ds = \delta_{m,n}$ (Takenaka, Walsh).

6. Let $p_n{}^*(x)$ be orthonormal with respect to the inner product $(f, g) = \int_a^b f(x)g(x)w(x)\, dx$. If $K_n(x, y)$ is the kernel polynomial, and if $x_0 \leq a$, the polynomials $K_n(x_0, x)$ are orthogonal with respect to the inner product $(f, g) = \int_a^b f(x)g(x)w(x)(x - x_0)\, dx$.

7. Let $p_n{}^*(x)$ be orthonormal with respect to the inner product $(f, g) = \int_{-1}^{+1} f(x)g(x)w(x)\, dx$, $w(x) > 0$. Prove that $\displaystyle\sum_{k=0}^{\infty} (p_k{}^*(x))^2 = \infty$ for $-1 \leq x \leq 1$.

Check this divergence directly for the Legendre polynomials at $x = 0$.

8. Using polynomials of degree 1 and $\|\|f\|\| = \max_{z \in B} |f(z)|$, prove that if C is the circle of smallest radius that contains a closed bounded set B, then the center of C lies in the convex hull of B.

9. In the previous problem, give an example to show that the center of C may fall on the boundary of the convex hull of B.

10. Check explicitly for $n = 1, 2, 3, 4$ that the polynomials orthonormal over the square have their zeros in the square. (Cf. 10.2, Ex. 5.)

11. Prove that
$$P_n(0) = 0, n \text{ odd}, \quad \text{and that} \quad P_n(0) = (-1)^{n/2} \frac{1 \cdot 3 \cdot 5 \cdots (n - 1)}{2 \cdot 4 \cdot 6 \cdots n}, n \text{ even}.$$
$P_n(x)$ are the Legendre polynomials with normalization (10.3.3).

12. The polynomial $p(x) = \sum_{k=0}^{n} \frac{2k + 1}{2} P_k(x)$ satisfies $\int_{-1}^{+1} x^j p(x) \, dx = 1$ for $j = 0, 1, \ldots, n$.

13. Show that the polynomial $p(x) = \sum_{j=0}^{M} (-1)^j \frac{4j + 1}{2^{2j+1}} \frac{(2j)!}{(j!)^2} P_{2j}(x)$ satisfies $\int_{-1}^{1} p(x) \, dx = 1$, $\int_{-1}^{1} p(x) x^j \, dx = 0$ $j = 1, 2, \ldots, 2M$.

14. Let the n zeros of $P_n^{(\alpha, \beta)}$ be $x_1, \ldots, x_n$. Then $x_1 + x_2 + \cdots + x_n = \dfrac{n(\beta - \alpha)}{2n + \alpha + \beta}$.

15. Verify (10.3.22) directly for $z = \pm 1$.

16. Show that in $\mathscr{P}_n[-1, 1]$ with $(f, g) = \int_{-1}^{+1} f(x)g(x) \, dx$, $L(p) = p'(x_0)$ is a bounded linear functional, and compute its norm. Generalize.

17. If
$$f(z) = \sum_{k=0}^{\infty} a_k T_k(x), |x| \leq 1 \quad \text{with} \quad \sum_{k=0}^{\infty} |a_k| < \infty,$$
then
$$\int_{-1}^{x} f(t) \, dt = \text{const} + \sum_{k=1}^{\infty} \left(\frac{a_{k-1} - a_{k+1}}{2k} \right) T_k(x), \quad |x| \leq 1.$$

18. The expansion of $f(x)$ in Tschebyscheff polynomials of the first kind is identical to the development of $f(\cos \theta)$ in a cosine series.

19. Arc $\cos x \sim \dfrac{\pi}{2} - \dfrac{4}{\pi} T_1(x) - \dfrac{16}{9\pi} T_2(x) - \cdots$.

20. $\dfrac{1}{1 + x^2} \sim \sqrt{2}[\frac{1}{2} - p^2 T_2(x) + p^4 T_4(x) - \cdots], \quad p = \sqrt{2} - 1$.

21. If curves C and D are related by the transformation $w = az + b$, what is the relationship between the orthonormal polynomials over C and those over D? What if the inner product is an area integral?

22. How do the symmetries of a curve influence the structure of its orthogonal polynomials (see Ex. 5, 10.2)?

23. Let S be a closed, bounded, convex set in the plane. Let z_1 be a point exterior to S. Show: 1. The problem of finding $\min_{z \in S} |z_1 - z|$ has a unique solution, z'. 2. At z' draw the line ℓ perpendicular to the segment $z_1 z'$. Then S lies on one side of ℓ. 3. The point z' satisfies the condition (10.2.9) (O. Shisha).

CHAPTER XI

The Theory of Closure and Completeness

11.1 The Fundamental Theorem of Closure and Completeness.
Theorem 8.9.1 related the concepts of closure and completeness for inner product spaces. In the present section, we shall do this for normed linear spaces.

DEFINITION 11.1.1. The sequence of elements $\{x_k\}$ is *complete* in a normed linear space X if $L(x_k) = 0$, $k = 1, 2, \ldots$, $L \in X^*$, implies $L = 0$. X^* is the normed conjugate space of X.

(In an inner product space, there are now two definitions of complete sequences, Definition 8.9.4 and Definition 11.1.1. If the space is complete, these definitions are equivalent.)

	Closed	Complete
Space	A subspace is closed if it contains all its limit points.	A space is complete if every Cauchy sequence has a limit in the space.
Sequence	A sequence is closed if every element of the space can be approximated arbitrarily closely by finite linear combinations of the elements of the sequence.	A sequence $\{x_n\}$ is complete if $L(x_n) = 0$, $n = 0, 1, \ldots$, $L \in X^*$, implies $L = 0$.

The fundamental theorem is that closure and completeness are equivalent concepts. This emerges as a consequence of the Hahn-Banach Extension Theorem, and it is to this that we now turn.

DEFINITION 11.1.2. Let X be a linear space and Y a linear subspace. Let L be a linear functional defined on Y. A linear functional L_1 is called an *extension* of L to X if $L_1(x)$ is defined for all $x \in X$ and if $L_1(x) = L(x)$ for $x \in Y$.

257

Ex. 1. X is the space of all functions defined on $[a, b]$. Let Y be the subspace $C[a, b]$. Let $a < x_1 < b$ and set $L(f) = \lim_{x \to x_1} f(x)$. Let $L_1(f) = f(x_1)$. Then L_1 is an extension of L from Y to X.

THEOREM 11.1.1. *Let X be a real normed linear space and Y a linear subspace. ($Y \neq X$). Let $p(x)$ be a real valued functional defined on the elements of X and possessing the following normlike properties*

$$
\begin{aligned}
p(x) &\geq 0; & x &\in X \\
p(x + y) &\leq p(x) + p(y); & x, y &\in X \\
p(\lambda x) &= \lambda p(x); & x &\in X, \quad \lambda \geq 0.
\end{aligned}
\tag{11.1.1}
$$

Let L be a real linear functional defined on Y that satisfies

$$
L(x) \leq p(x), \qquad x \in Y. \tag{11.1.2}
$$

Then L can be extended to be a linear functional L_1 defined on X and such that

$$
L_1(x) \leq p(x), \qquad x \in X. \tag{11.1.3}
$$

Proof: 1. Select an $x_0 \in X$ but $\notin Y$. Take $x, y \in Y$. Then

$$
L(y) - L(x) = L(y - x) \leq p(y - x).
$$

Now $p(y - x) \leq p(y + x_0) + p(-x - x_0)$ so that

$$
-p(-x - x_0) - L(x) \leq p(y + x_0) - L(y); \qquad x, y \in Y.
$$

Think of y as fixed in Y and x as varying in Y. Then the last inequality shows that $-p(-x - x_0) - L(x)$ is bounded above. Similarly, for varying y, $p(y + x_0) - L(y)$ is bounded below. If we set

$$
\begin{aligned}
r_1 &= \sup_{x \in Y} \left[-p(-x - x_0) - L(x) \right] \\
r_2 &= \inf_{y \in Y} \left[p(y + x_0) - L(y) \right]
\end{aligned}
$$

then we must have $-\infty < r_1 \leq r_2 < \infty$. Select a number r such that $r_1 \leq r \leq r_2$. Then,

$$
-p(-x - x_0) - L(x) \leq r \leq p(y + x_0) - L(y) \tag{11.1.4}
$$

for any $x, y \in Y$.

2. Consider the linear subspace Y_0 consisting of all elements y of the form $y = x + \lambda x_0$, $x \in Y$. Each element in Y_0 has a unique representation in this form. For suppose, $y \in Y_0$, and $y = x_1 + \lambda_1 x_0 = x_2 + \lambda_2 x_0$. Then $x_1 - x_2 = (\lambda_1 - \lambda_2)x_0$. If $\lambda_1 \neq \lambda_2$ then x_0 would be a linear combination of x_1, x_2 and hence in Y. This contradicts our selection of x_0. Therefore $\lambda_1 = \lambda_2$ and hence also $x_1 = x_2$.

Define L_1 on Y_0 by means of

$$
L_1(y) = L(x) + \lambda r.
$$

Now if $y \in Y$, $\lambda = 0$ and hence $L_1(y) = L(y)$, $y \in Y$. Since L is linear, it follows easily that L_1 is linear.

We wish to prove next that

$$L_1(y) \leq p(y) \qquad \text{for all} \quad y \in Y_0. \tag{11.1.5}$$

Decompose y into the form $y = x_1 + \lambda x_0$. We need only deal with the case when $\lambda \neq 0$. From (11.1.4) we have

$$-p\left(-\frac{x_1}{\lambda} - x_0\right) - L\left(\frac{x_1}{\lambda}\right) \leq r \leq p\left(\frac{x_1}{\lambda} + x_0\right) - L\left(\frac{x_1}{\lambda}\right).$$

Now if $\lambda > 0$, $p\left(\dfrac{x_1}{\lambda} + x_0\right) = \dfrac{1}{\lambda}\, p(x_1 + \lambda x_0)$ and the second inequality reduces to

$$r \leq \frac{1}{\lambda}\, p(x_1 + \lambda x_0) - \frac{1}{\lambda}\, L(x_1)$$

or

$$L_1(y) = L(x_1) + \lambda r \leq p(x_1 + \lambda x_0).$$

If $\lambda < 0$, the first inequality may be employed:

$$-p\left(-\frac{x_1}{\lambda} - x_0\right) = \frac{1}{\lambda}\, p(x_1 + \lambda x_0); \text{ hence, } p(x_1 + \lambda x_0) - L(x_1) \geq \lambda r \text{ and the}$$

conclusion is as before.

3. Consider, finally, all the linear functionals that extend L to some linear subspace containing Y and which satisfy the condition (11.1.5). A partial ordering $L' \leq L''$ is defined amongst these functionals by agreeing that $L' \leq L''$ means that L'' is an extension of L'. With this ordering, every totally ordered subset is seen to have an upper bound, i.e., the functional which is defined over the union of the domains of definition of the individual functionals and which takes on the values assigned by them. Zorn's Lemma (Theorem 1.13.1) tells us that there exists a maximal extension L_1. This linear functional is defined over the entire space X, for if not, it could have been further extended by the process described under 2.

If the space X is separable, the use of Zorn's Lemma (and hence the axiom of choice) can be avoided.

A functional $p(x)$ satisfying 11.1.1 is known as *a convex functional*.

THEOREM 11.1.2 (Hahn-Banach). *Let X be a real normed linear space and Y a subspace. Let L be defined on Y and have norm $\|L\|_Y$ there. Then there is a linear functional L_1 which extends L to X and such that $\|L_1\|_X = \|L\|_Y$.*

Proof: Set $p(x) = \|L\|_Y\, \|x\|$. The functional $p(x)$ is readily seen to fulfill the requirements (11.1.1). Therefore, by the previous theorem, we may extend L to L_1 so that

$$L_1(x) \leq \|L\|_Y\, \|x\|, \qquad x \in X.$$

Since also

$$-L_1(x) = L_1(-x) \leq \|L\|_Y \, \|-x\| = \|L\|_Y \, \|x\|,$$

it follows that

$$|L_1(x)| \leq \|L\|_Y \, \|x\|.$$

Hence,

$$\sup_{x \in X} \frac{|L_1(x)|}{\|x\|} \leq \|L\|_Y$$

so that

$$\|L_1\| \leq \|L\|_Y.$$

But $\|L\|_Y = \sup\limits_{x \in Y} \dfrac{|L(x)|}{\|x\|} = \sup\limits_{x \in Y} \dfrac{|L_1(x)|}{\|x\|} \leq \sup\limits_{x \in X} \dfrac{|L_1(x)|}{\|x\|} = \|L_1\|_X$ and therefore $\|L\|_Y \leq \|L_1\|_X$. Thus, finally $\|L_1\|_X = \|L\|_Y$.

This extension theorem also holds in complex normed linear spaces. To establish this, we make use of a simple device which associates a unique real normed linear space X_R to each complex normed linear space X. In this way, the burden of the proof is thrown back to the real situation.

DEFINITION 11.1.3. Let X be a complex normed linear space. The space X_R will consist of the same elements as X. Addition in X_R will be identical with addition in X. If a is real and $x \in X_R$ then ax will be the element $(a + i0)x = ax$ of X. $\|x\|$ in X_R will equal $\|x\|$ in X. If L is a bounded (and complex) linear functional defined on X, then by L_R we shall mean the real valued functional defined on X_R by means of

$$L_R(x) = \text{Real part of } L(x). \tag{11.1.6}$$

The x in the left-hand member is considered to lie in X_R while in the right it is considered to lie in X.

LEMMA 11.1.3. *If L is a bounded linear functional on X, then L_R is a bounded linear functional on X_R.*

Proof: Let $x, y \in X_R$ and a, b be real.

$$\begin{aligned}
L_R(ax + by) &= \text{Re } L(ax + by) \\
&= \text{Re } \{aL(x) + bL(y)\} = a \text{ Re } L(x) + b \text{ Re } L(y) \\
&= aL_R(x) + bL_R(y).
\end{aligned}$$

Therefore L_R is linear on X_R. Also

$$|L_R(x)| = |\text{Re } L(x)| \leq |L(x)| \leq \|L\| \, \|x\|_X = \|L\| \, \|x\|_{X_R}. \tag{11.1.7}$$

Therefore L_R is bounded on X_R, and $\|L_R\| \leq \|L\|$.

LEMMA 11.1.4.   *If L is a linear functional on X then*

$$L(x) = L_R(x) - iL_R(ix). \tag{11.1.8}$$

Conversely if Λ is a linear functional on X_R, then the equation

$$L(x) = \Lambda(x) - i\Lambda(ix) \tag{11.1.9}$$

defines a linear functional on X.

Proof: If $L(x) = \operatorname{Re} L(x) + i \operatorname{Im} L(x)$, $x \in X$, then

$$L(ix) = \operatorname{Re} L(ix) + i \operatorname{Im} L(ix)$$
$$= i \operatorname{Re} L(x) - \operatorname{Im} L(x).$$

Therefore $\operatorname{Im} L(x) = -\operatorname{Re} L(ix) = -L_R(ix)$, so (11.1.8) follows. Conversely, given $x, y \in X$, from (11.1.9) we see that

$$L(x + y) = \Lambda(x + y) - i\Lambda(ix + iy)$$
$$= \Lambda(x) + \Lambda(y) - i\Lambda(ix) - i\Lambda(iy)$$
$$= L(x) + L(y).$$

Moreover, if a is real,
$$L(ax) = \Lambda(ax) - i\Lambda(iax)$$
$$= a\Lambda(x) - ia\Lambda(ix) = aL(x).$$

Finally, $L(ix) = \Lambda(ix) - i\Lambda(-x)$
$$= i[\Lambda(x) - i\Lambda(ix)] = iL(x).$$

Thus, L is linear over X.

THEOREM 11.1.5 (Bohnenblust-Sobczyk-Suchomlinoff).   *Let X be a complex normed linear space and Y a subspace. Let L be a complex linear functional defined on Y and have norm $\|L\|_Y$ there. Then there is a linear functional L_1 that extends L to X and such that $\|L_1\|_X = \|L\|_Y$.*

Proof: Write $L(x) = L_R(x) + iL_I(x)$, $x \in Y$ where L_R and L_I are real valued. By Lemma 11.1.3, L_R is a bounded real valued linear functional defined on Y_R, the real normed linear space associated with Y. Extend L_R to X_R by Theorem 11.1.2, and obtain a real, bounded, linear functional $L_{1,R}$ for which $L_{1,R}(x) = L_R(x)$, $x \in Y_R$ and for which $\|L_{1,R}\| = \|L_R\|$. Define

$$L_1(x) = L_{1,R}(x) - iL_{1,R}(ix). \tag{11.1.10}$$

By Lemma 11.1.4, L_1 is a linear functional defined on the whole of X. It is an extension of L. For let $x \in Y$. Then, by (11.1.8) (taking the X as the present Y),

$$L(x) = L_R(x) - iL_R(ix) = L_{1,R}(x) - iL_{1,R}(ix) = L_1(x).$$

We must finally prove that $\|L_1\| = \|L\|$. Since L_1 is an extension of L, it is clear that $\|L\| \leq \|L_1\|$. On the other hand, suppose that $L_1(x) = re^{i\theta}$, $x \in X$. Then,

$$|L_1(x)| = r = \operatorname{Re} L_1(e^{-i\theta}x) = L_{1,R}(e^{-i\theta}x) \leq \|L_{1,R}\| \, \|e^{-i\theta}x\|$$

$$= \|L_{1,R}\| \, \|x\| = \|L_R\| \, \|x\| \leq \|L\| \, \|x\|.$$

The last inequality was observed after (11.1.7). Therefore, for all $x \in X$, $\dfrac{|L_1(x)|}{\|x\|} \leq \|L\|$, so that $\|L_1\| \leq \|L\|$. Thus, $\|L_1\| = \|L\|$.

Theorem 11.1.6. *Let X be a normed linear space and Y a linear subspace. Let $x_0 \in X$, but $x_0 \notin Y$ and suppose that $d = \inf_{y \in Y} \|y - x_0\| > 0$. Then we can find a bounded linear functional, L, on X such that*

$$\begin{aligned} L(x) &= 0 \qquad & x \in Y \\ L(x_0) &= 1 \\ \|L\| &= d^{-1}. \end{aligned} \qquad (11.1.11)$$

Proof: As in Theorem 11.1.1, let Y_0 be the linear subspace of elements of the form $x + \lambda x_0$, $x \in Y$. This decomposition is unique. Construct an L over Y_0 as follows:

$$L(y) = \lambda \quad \text{for} \quad y = x + \lambda x_0. \qquad (11.1.12)$$

In particular, $L(x) = 0$ whenever $x \in Y$, and $L(x_0) = L(0 + 1 \cdot x_0) = 1$. Now

$$\frac{|L(y)|}{\|y\|} = \frac{|\lambda|}{\|y\|} = \frac{|\lambda|}{\|x + \lambda x_0\|} = \frac{1}{\left\| \dfrac{x}{\lambda} + x_0 \right\|} = \frac{1}{\left\| x_0 - \left(-\dfrac{x}{\lambda} \right) \right\|}.$$

Since $-\dfrac{x}{\lambda} \in Y, \left\| x_0 - \left(-\dfrac{x}{\lambda} \right) \right\| \geq d$.

Hence

$$\|L\|_{Y_0} = \sup_{y \in Y_0} \frac{|L(y)|}{\|y\|} \leq \frac{1}{d}.$$

On the other hand, we can find a sequence of elements $\{x_n\} \in Y$ such that $\lim_{n \to \infty} \|x_n - x_0\| = d$. Now $x_n - x_0 \in Y_0$ so that

$$|L(x_n - x_0)| \leq \|L\|_{Y_0} \|x_n - x_0\|.$$

But $L(x_n) = 0$, $n = 1, 2, \ldots$, and $L(x_0) = 1$.

Hence, $\qquad\qquad\qquad 1 \leq \|L\|_{Y_0} \|x_n - x_0\|,$

so that $\qquad\qquad\qquad 1 \leq \|L\|_{Y_0} d.$

Therefore $\|L\|_{Y_0} \geq \dfrac{1}{d}$ and we must have

$$\|L\|_{Y_0} = \frac{1}{d}.$$

We now apply Theorem 11.1.5 to extend L from Y_0 to X with preservation of norm.

THEOREM 11.1.7 (Banach). *Let X be a normed linear space (real or complex). A sequence of elements $\{x_k\}$ is closed if and only if it is complete.*

Proof: Suppose $\{x_k\}$ is closed. Let $L \in X^*$ and suppose that $L(x_k) = 0$ $k = 1, 2, \ldots$. Given any $x \in X$, we may approximate x arbitrarily closely by finite combinations of x_k: $\|x - a_1 x_1 - a_2 x_2 - \cdots - a_n x_n\| \leq \varepsilon$ for some coefficients a_k. Then,

$$|L(x)| = |L(x - a_1 x_1 - \cdots - a_n x_n)|$$
$$\leq \|L\| \, \|x - a_1 x_1 - \cdots - a_n x_n\| \leq \|L\| \, \varepsilon.$$

Allow $\varepsilon \to 0$ and obtain $L(x) = 0$. Since x is arbitrary, $L = 0$.

Conversely, suppose that $L(x_k) = 0$, $k = 1, 2, \ldots$, implies $L = 0$. Let x_0 be an element of X and let Y be the linear subspace comprised of all finite linear combinations of $x_1, x_2, \ldots$. We wish to prove $d = \inf\limits_{y \in Y} \|x_0 - y\| = 0$. Suppose the contrary. Then by the previous theorem, we can find an L such that $L(y) = 0$, $y \in Y$ and $L(x_0) = 1$. In particular, $L(x_k) = 0$, $k = 1, 2, \ldots$. But by completeness, this implies $L = 0$ and contradicts $L(x_0) = 1$.

Ex. 2.   Let X be a complete inner product space. Any $L \in X^*$ has the representation $L(x) = (x, x_0)$ for some $x_0 \in X$. Hence, the definition of completeness of $\{x_k\}$ is that $(x_k, x) = 0$, $k = 1, 2, \ldots$, implies $x = 0$. In this case, the present theorem gives us the equivalence of A and E of Theorem 8.9.1.

Ex. 3.   Select $X = C[a, b]$, $\|f\| = \max\limits_{a \leq x \leq b} |f(x)|$. By Weierstrass' Theorem, the powers $1, x, x^2, \ldots$, are closed. For a given $g(x) \in C$ the linear functional $L(f) = \displaystyle\int_a^b f(x)g(x)\,dx$ is in X^*. Hence $\displaystyle\int_a^b x^n g(x)\,dx = 0$, $n = 0, 1, 2, \ldots$, implies $g(x) \equiv 0$.

Ex. 4.   Let $\{r_i\}$ be the set of all rational numbers lying in $a \leq x \leq b$. Let $S_i(x)$ be the step function defined by

$$S_i(x) = 1 \qquad a \leq x \leq r_i$$
$$S_i(x) = 0 \qquad r_i < x \leq b.$$

Then the system $\{S_i(x)\}$ is closed in $L^2[a, b]$.

Proof: Let $f(x) \in L^2[a, b]$. Suppose that $\int_a^b f(x)S_i(x) \, dx = 0, i = 1, 2, \ldots$.

Using the definition of S_i, $\int_a^{r_i} f(x) \, dx = 0, i = 1, 2, \ldots$, and hence the function

$F(t) = \int_a^t f(x) \, dx$ is zero at the rational points. But $F(t) \in C[a, b]$ and hence

$F(t) \equiv 0$. It follows that $f(x) \equiv 0$ almost everywhere. The system $\{S_i(x)\}$ is accordingly complete and hence, closed.

Ex. 5. Let D be a bounded multiply connected region and let $A(D)$ be the normed linear space of functions that are analytic in D and continuous in D plus its boundary. $\|f\| = \max\limits_{z \in \bar{D}} |f(z)|$. The sequence of functions $1, z, z^2, \ldots$, is not closed in $A(D)$. For if the sequence were closed, it would be complete. Now let z_0 be a point in one of the "holes" of D. Consider the linear functional $L(f) = \frac{1}{2\pi i} \int_C f(z) \, dz$, where C is a contour lying in D and containing z_0 in its interior. L is bounded over $A(D)$ and $L(z^n) = 0, n = 0, 1, \ldots$. Therefore $L = 0$ by completeness, but $\frac{1}{z - z_0}$ is in $A(D)$ and $L\left(\frac{1}{z - z_0}\right) = 1$. Hence we have a contradiction.

Though completeness and closure are equivalent, it is convenient to employ both terms so that attention may be called to the appropriate defining property.

The property of closure is transitive.

THEOREM 11.1.8 (Lauricella). *Let X be a normed linear space and let $\{x_n\}$ be a closed system. Then a second system $\{y_n\}$ is closed in X if and only if it is closed in $\{x_n\}$. By this we mean that each x_n can be approximated arbitrarily closely by linear combinations of the y_n.*

Proof: The necessity is trivial. To prove sufficiency, let $x \in X$ and prescribe $\varepsilon > 0$. Since $\{x_k\}$ is closed we may find constants $a_1, \ldots, a_N$ such that $\|x - a_1 x_1 - \cdots - a_N x_N\| \leq \varepsilon/2$. We may obviously assume that each $a_i \neq 0$, otherwise we simply ignore that coefficient. Since $\{y_k\}$ is closed in $\{x_k\}$, we may find constants $b_{i1}, b_{i2}, \ldots, b_{iN_i}, i = 1, 2, \ldots, N$ such that

$$\|x_i - b_{i1}y_1 - b_{i2}y_2 - \cdots - b_{iN_i}y_{N_i}\| \leq \frac{\varepsilon}{2N |a_i|} \quad i = 1, 2, \ldots, N.$$

If we set $w_i = a_i(b_{i1}y_1 + b_{i2}y_2 + \cdots + b_{iN_i}y_{N_i}), \quad i = 1, 2, \ldots, N,$

then, $\qquad \|a_i x_i - w_i\| \leq \dfrac{\varepsilon}{2N}, \qquad i = 1, 2, \ldots, N.$

Now, by the triangle inequality,

$$\|x - w_1 - w_2 - \cdots - w_n\|$$

$$= \|x - a_1 x_1 - \cdots - a_N x_N - w_1 - w_2 - \cdots - w_N + a_1 x_1 + \cdots + a_N x_N\|$$

$$\le \|x - a_1 x_1 - \cdots - a_N x_N\| + \|a_1 x_1 - w_1\| + \cdots + \|a_N x_N - w_N\|$$

$$\le \frac{\varepsilon}{2} + N \frac{\varepsilon}{2N} = \varepsilon.$$

In this way, we approximate x to within ε by $w_1 + \cdots + w_N$, which is a combination of the y's.

11.2   Completeness of the Powers and Trigonometric Systems for $L^2[a, b]$

THEOREM 11.2.1.   *The powers* $1, x, x^2, \ldots$, *are complete over* $L^2[a,b]$.

Proof: Let $f(x) \in L^2[a, b]$ and assume that

$$\int_a^b x^n f(x)\, dx = 0 \qquad n = 0, 1, 2, \ldots.$$

Set

$$F(x) = \int_a^x f(t)\, dt. \tag{11.2.1}$$

Then $F(x) \in C[a, b]$. In particular $F(a) = 0$, $F(b) = 0$. Integrating by parts,

$$0 = \int_a^b x^n f(x)\, dx = x^n F(x) \Big|_a^b - n \int_a^b x^{n-1} F(x)\, dx.$$

It follows that $\displaystyle\int_a^b x^{n-1} F(x)\, dx = 0 \quad n = 1, 2, \ldots.$

Now by Ex. 3, Section 11.1, the powers are complete for $C[a, b]$ with $\| f \| = \max_{a \le x \le b} |f(x)|$, and this implies that $F(x) = 0$. Therefore $f(x) = 0$ almost everywhere.

COROLLARY 11.2.2.   *The sequence of orthonormal polynomials on* $[a, b]$, $p_n^*(x) = k_n x^n + \cdots, k_n > 0$, *has all the properties* A–F *of Theorem* 8.9.1 *for* $f \in L^2[a, b]$.

A corresponding theorem for the trigononometric functions can be derived from the powers by a change of variable, but we prefer the following proof which makes use of an interesting analytic device.

LEMMA 11.2.3.   *Let* $f(x) \in C[-\pi, \pi]$ *and suppose that*

$$\int_{-\pi}^{\pi} f(x) \cos nx \, dx = 0 \qquad n = 0, 1, \ldots, \tag{11.2.2}$$

and

$$\int_{-\pi}^{\pi} f(x) \sin nx \, dx = 0 \qquad n = 1, 2, \ldots . \tag{11.2.3}$$

Then $f(x) \equiv 0$.

Proof: If $T_n(x)$ is an arbitrary trigonometric polynomial, it follows from (11.2.2) and (11.2.3) that $\int_{-\pi}^{\pi} f(x) T_n(x) \, dx = 0$. Assume that $f(x) \not\equiv 0$. Then there is a point x_0 interior to $[-\pi, \pi]$ at which $f(x_0) \neq 0$. For the sake of argument, assume that $f(x_0) = m > 0$. Then by continuity, we can find an interval $I: x_0 - \delta \leq x \leq x_0 + \delta$ contained in $(-\pi, \pi)$ throughout which $f(x) \geq m/2$. Construct the trigonometric function

$$t(x) = 1 - \cos \delta + \cos (x - x_0). \tag{11.2.4}$$

For $x_0 - \delta < x < x_0 + \delta$, $\cos (x - x_0) > \cos \delta$ and therefore $t(x) > 1$. For $x = x_0 \pm \delta$, $t(x) = 1$. Elsewhere in $[-\pi, \pi]$, $-1 \leq \cos (x - x_0) < \cos \delta$ so that

$$-\cos \delta \leq t(x) < 1 \quad \text{and therefore} \quad |t(x)| < 1.$$

Now consider the trigonometric polynomial of order n, $T_n(x) = [t(x)]^n$. It is clear that

$$T_n(x) > 1 \quad \text{for } I_1: \ x_0 - \delta < x < x_0 + \delta$$
$$T_n(x) = 1 \quad \text{for } \ x = x \pm \delta$$
$$|T_n(x)| < 1 \quad \text{for } I_2: \text{ the remaining portions of } [-\pi, \pi].$$

But

$$0 = \int_{-\pi}^{\pi} f(x) T_n(x) \, dx = \int_{I_1} f(x) T_n(x) \, dx + \int_{I_2} f(x) T_n(x) \, dx,$$

so that $\int_{I_1} = -\int_{I_2}$. Now $\left| \int_{I_2} f(x) T_n(x) \, dx \right| \leq \int_{I_2} |f(x)| \, dx$ and is therefore bounded as $n \to \infty$. Since $t(x) > 1$ on I_1, $t(x) \geq 1 + \varepsilon$ on, say,

$$I_3: x_0 - \frac{\delta}{2} \leq x \leq x_0 + \frac{\delta}{2}.$$

Therefore, $T_n(x) = [t(x)]^n \geq (1 + \varepsilon)^n$ on I_3 and

$$\int_{I_1} f(x) T_n(x) \, dx \geq \frac{m}{2} \int_{I_3} T_n(x) \, dx \geq \frac{m}{2} (1 + \varepsilon)^n.$$

This is a contradiction since $\int_{I_1} \to +\infty$ while $\int_{I_2}$ is bounded. The assumption that $f(x) \not\equiv 0$ cannot be maintained.

THEOREM 11.2.4.　*The system of functions* $\cos nx$, $n = 0, 1, \ldots$, $\sin nx$, $n = 1, 2, \ldots$, *is complete in* $L^2[-\pi, \pi]$.

Proof: Let $f(x) \in L^2[-\pi, \pi]$. We shall show that the conditions (11.2.2) and (11.2.3) imply $f(x) = 0$ almost everywhere. This will imply completeness for $L^2[-\pi, \pi]$ by Ex. 2 of Section 11.1. The function

$$F(x) = \int_{-\pi}^{x} f(t) \, dt \tag{11.2.5}$$

is in $C[-\pi, \pi]$ and $F(-\pi) = 0$, $F(\pi) = 0$. The last follows from (11.2.2) with $n = 0$. If $T(x)$ designates an arbitrary trigonometric polynomial, $\int_{-\pi}^{\pi} f(x)T(x) \, dx = 0$. But

$$\int_{-\pi}^{\pi} f(x)T(x) \, dx = F(x)T(x) \Big|_{-\pi}^{\pi} - \int_{-\pi}^{\pi} F(x)T'(x) \, dx.$$

Hence $\int_{-\pi}^{\pi} F(x)T'(x) \, dx = 0$ for all derivatives T' of trigonometric polynomials.

In particular, $\int_{-\pi}^{\pi} F(x) \dfrac{\sin nx}{\cos nx} \, dx = 0$　$n = 1, 2, \ldots$. Consider now

$$G(x) = F(x) - c, \quad c = \frac{1}{2\pi} \int_{-\pi}^{\pi} F(x) \, dx. \tag{11.2.6}$$

Then it is easily verified that

$$\int_{-\pi}^{\pi} G(x) \frac{\sin nx}{\cos mx} \, dx = 0 \quad \begin{array}{l} n = 1, 2, \ldots, \\ m = 0, 1, 2, \ldots. \end{array}$$

By Lemma 11.2.3, $G(x) \equiv 0$. Therefore $F(x) = c$. But $F(\pi) = 0$, so that $F(x) \equiv 0$. Accordingly $f(x) = 0$ almost everywhere.

COROLLARY 11.2.5.　*The sequence of* sines *and* cosines *satisfies all the conditions A–F of Theorem 8.9.1 for* $f \in L^2[-\pi, \pi]$. *In particular, we have the Parseval identity*

$$\frac{1}{\pi} \int_{-\pi}^{\pi} [f(x)]^2 \, dx = \tfrac{1}{2}a_0^2 + \sum_{n=1}^{\infty} (a_n^2 + b_n^2) \tag{11.2.7}$$

$$a_n = \frac{1}{\pi} \int_{-\pi}^{\pi} f(x) \cos nx \, dx, \quad b_n = \frac{1}{\pi} \int_{-\pi}^{\pi} f(x) \sin nx \, dx.$$

11.3　The Müntz Closure Theorem.　Suppose that one has been given a sequence of powers $\{x^{p_k}\}$. Under what circumstances can continuous functions or functions in L^2 be approximated by linear combinations of these powers? Müntz gave an extensive discussion of this problem and used a method that is a beautiful application of Theorem 8.7.4.

LEMMA 11.3.1 (Cauchy). *If*

$$D_n = \begin{vmatrix} \dfrac{1}{a_1 + b_1} & \dfrac{1}{a_1 + b_2} & \cdots & \dfrac{1}{a_1 + b_n} \\[2mm] \cdot & & & \cdot \\ \cdot & & & \cdot \\ \cdot & & & \cdot \\ \dfrac{1}{a_n + b_1} & \dfrac{1}{a_n + b_2} & \cdots & \dfrac{1}{a_n + b_n} \end{vmatrix}, \tag{11.3.1}$$

then

$$D_n = \frac{\displaystyle\prod_{i>j}^{n} (a_i - a_j)(b_i - b_j)}{\displaystyle\prod_{i,j=1}^{n} (a_i + b_j)}. \tag{11.3.2}$$

Proof: Regard the a_i's and the b_j's as $2n$ independent variables and think of D_n as expanded and put over a common denominator. This common denominator is $\displaystyle\prod_{i,j=1}^{n} (a_i + b_j)$. Each individual term is of degree -1 so that D_n is of degree $-n$. The common denominator has n^2 factors and hence is of degree n^2. It follows that the numerator must be a polynomial of degree $n^2 - n$ in the a's and b's.

Note that if $a_i = a_j$, the ith row and the jth row of D_n will be identical, and $D_n = 0$. A similar observation holds if $b_i = b_j$. It follows that the numerator must contain a factor of the form $\displaystyle\prod_{i>j}^{n} (a_i - a_j) \prod_{i>j}^{n} (b_i - b_j)$. Each product here contains $1 + 2 + \cdots + (n - 1) = \dfrac{n(n - 1)}{2}$ factors so that the complete product contains $n(n - 1)$ factors. The degree is therefore correct and we must have

$$D_n = c_n \frac{\displaystyle\prod_{i>j}^{n} (a_i - a_j)(b_i - b_j)}{\displaystyle\prod_{i,j=1}^{n} (a_i + b_j)} \tag{11.3.3}$$

where c_n is a constant independent of the a's and b's. We shall show that $c_n = 1$.

Note that $\quad a_n D_n = \begin{vmatrix} \dfrac{1}{a_1 + b_1} & \cdots & \dfrac{1}{a_1 + b_n} \\[2mm] \cdot & & \cdot \\ \cdot & & \cdot \\ \cdot & & \cdot \\ \dfrac{a_n}{a_n + b_1} & \cdots & \dfrac{a_n}{a_n + b_n} \end{vmatrix}.$

Therefore,

$$
\lim_{a_n \to \infty} a_n D_n =
\begin{vmatrix}
\dfrac{1}{a_1 + b_1} & \cdots & \dfrac{1}{a_1 + b_n} \\
\cdot & & \cdot \\
\cdot & & \cdot \\
\cdot & & \cdot \\
\dfrac{1}{a_{n-1} + b_1} & \cdots & \dfrac{1}{a_{n-1} + b_n} \\
1 & \cdots & 1
\end{vmatrix}.
\tag{11.3.4}
$$

Also

$$
\lim_{b_n \to \infty} \lim_{a_n \to \infty} a_n D_n =
\begin{vmatrix}
\dfrac{1}{a_1 + b_1} & \cdots & \dfrac{1}{a_1 + b_{n-1}} & 0 \\
\cdot & & \cdot & \cdot \\
\cdot & & \cdot & \cdot \\
\cdot & & \cdot & \cdot \\
\dfrac{1}{a_{n-1} + b_1} & \cdots & \dfrac{1}{a_{n-1} + b_{n-1}} & 0 \\
1 & \cdots & & 1
\end{vmatrix} = D_{n-1},
\tag{11.3.5}
$$

so that

$$
\lim_{b_n \to \infty} \lim_{a_n \to \infty} \frac{a_n D_n}{D_{n-1}} = 1.
\tag{11.3.6}
$$

But from (11.3.3),

$$
\frac{a_n D_n}{D_{n-1}} = \frac{c_n}{c_{n-1}} \frac{a_n \prod\limits_{j=1}^{n-1}(a_n - a_j) \prod\limits_{j=1}^{n-1}(b_n - b_j)}{\prod\limits_{j=1}^{n}(a_n + b_j) \prod\limits_{j=1}^{n-1}(b_n + a_j)}
\tag{11.3.7}
$$

it follows that

$$
\lim_{b_n \to \infty} \lim_{a_n \to \infty} \frac{a_n D_n}{D_{n-1}} = 1 = \frac{c_n}{c_{n-1}}.
\tag{11.3.8}
$$

Therefore, $c_n = c_{n-1}$. It is easily verified that $c_1 = 1$. Hence $c_n = 1$.

LEMMA 11.3.2. *Let* $p_i \neq p_j$. *Then, assuming* $p_i, q > -\frac{1}{2}$,

$$
\delta^2 = \min_{a_k} \int_0^1 |x^q - a_1 x^{p_1} - a_2 x^{p_2} - \cdots - a_n x^{p_n}|^2 \, dx
$$
$$
= \frac{1}{2q + 1} \prod_{i=1}^{n} \left\{ \frac{p_i - q}{p_i + q + 1} \right\}^2.
$$

Proof: From Theorem 8.7.4, $\delta^2 = \dfrac{g(x^q, x^{p_1}, x^{p_2}, \ldots, x^{p_n})}{g(x^{p_1}, x^{p_2}, \ldots, x^{p_n})}$. Now $(x^\alpha, x^\beta) = \displaystyle\int_0^1 x^\alpha x^\beta \, dx = \dfrac{1}{\alpha + \beta + 1}$. Therefore

$$g(x^{p_1}, \ldots, x^{p_n}) = \begin{vmatrix} \dfrac{1}{p_1 + p_1 + 1} & \cdots & \dfrac{1}{p_1 + p_n + 1} \\ \cdot & & \cdot \\ \cdot & & \cdot \\ \cdot & & \cdot \\ \dfrac{1}{p_n + p_1 + 1} & \cdots & \dfrac{1}{p_n + p_n + 1} \end{vmatrix}$$

$$= \prod_{i>j}^{n} (p_i - p_j)^2 \Big/ \prod_{i,j=1}^{n} (p_i + p_j + 1).$$

A similar expression is found for $g(x^q, x^{p_1}, \ldots, x^{p_n})$. We then have

$$\frac{g(x^q, x^{p_1}, \ldots, x^{p_n})}{g(x^{p_1}, \ldots, x^{p_n})}$$

$$= \frac{(q - p_1)^2 (q - p_2)^2 \cdots (q - p_n)^2}{(q + p_1 + 1)(q + p_2 + 1) \cdots (q + p_n + 1)\,(q + q + 1)(p_1 + 1 + q) \cdots (p_n + 1 + q)}$$

$$= \frac{1}{2q + 1} \prod_{i=1}^{n} \frac{(q - p_i)^2}{(q + p_i + 1)^2} \, .$$

THEOREM 11.3.3 (Müntz). *Let $\{x^p\}$ be a   given infinite set of distinct powers with $p > -\frac{1}{2}$. In order that this system be closed in $L^2(0, 1)$ it is necessary and sufficient that the exponents $\{p\}$ contains a sequence $\{p_i\}$ such that either*

$$\lim_{i \to \infty} p_i = -\tfrac{1}{2}, \quad \sum_{i=1}^{\infty} (p_i + \tfrac{1}{2}) = \infty, \qquad (11.3.9)$$

or

$$\lim_{i \to \infty} p_i = p, \quad -\tfrac{1}{2} < p < \infty, \qquad (11.3.10)$$

or

$$\lim_{i \to \infty} p_i = \infty, \quad p_i \neq 0, \quad \sum_{i=1}^{\infty} \frac{1}{p_i} = \infty. \qquad (11.3.11)$$

Proof: Note that the condition $p > -\frac{1}{2}$ insures that $x^p \in L^2(0, 1)$. The powers $1, x, x^2, \ldots$, are closed in $L^2(0, 1)$ by Theorem 11.2.1. Hence, $\{x^p\}$ will be closed in $L^2(0, 1)$ if and only if each power x^q, $q = 0, 1, \ldots$, is approximable. This follows from Theorem 11.1.8. Thus, for each q, we must be able to find a sequence $p_1, p_2, \ldots$, such that

$$\lim_{n \to \infty} \min_{a_i} \int_0^1 |x^q - a_1 x^{p_1} - a_2 x^{p_2} - \cdots - a_n x^{p_n}|^2 \, dx = 0.$$

Referring to Lemma 11.3.2, for each q we must be able to find a sequence $\{p_k\}$ such that

$$\lim_{n \to \infty} \prod_{i=1}^{n} \left(\frac{p_i - q}{p_i + q + 1} \right)^2 = 0. \tag{11.3.12}$$

Sufficiency. Suppose that $\{p\}$ has a finite limit point p with $p \neq -\frac{1}{2}$. Now

$$\lim_{i \to \infty} \frac{p_i - q}{p_i + q + 1} = \frac{p - q}{p + q + 1}. \tag{11.3.13}$$

Since $p > -\frac{1}{2}$ and $q \geq 0$, it is easily verified that $-1 < \dfrac{p - q}{p + q + 1} < 1$. Given an $\varepsilon > 0$, we have for $i \geq n_\varepsilon$, $\left| \dfrac{p_i - q}{p_i + q + 1} \right| < 1 - \varepsilon$. Hence, (11.3.12) holds.

Suppose we can select a sequence $\{p_i\}$ with (11.3.9) holding. Write

$$\prod_{i=1}^{n} \left(\frac{p_i - q}{p_i + q + 1} \right)^2 = \prod_{i=1}^{n} \left(1 - \frac{p_i + \frac{1}{2}}{q + \frac{1}{2}} \right)^2 \Big/ \prod_{i=1}^{n} \left(1 + \frac{p_i + \frac{1}{2}}{q + \frac{1}{2}} \right)^2. \tag{11.3.14}$$

Since $\lim\limits_{i \to \infty} p_i = -\frac{1}{2}$, we have, ultimately, $0 < \dfrac{p_i + \frac{1}{2}}{q + \frac{1}{2}} < 1$. Therefore,

$$0 < 1 - \frac{p_i + \frac{1}{2}}{q + \frac{1}{2}} < 1$$

and the numerator of the right hand of (11.3.14) remains bounded. Moreover, since $\sum\limits_{i=1}^{\infty} \dfrac{(p_i + \frac{1}{2})}{q + \frac{1}{2}} = \infty$, the denominator diverges to $+\infty$†. Condition (11.3.12) is therefore fulfilled.

Finally, suppose we can select a sequence with (11.3.11) holding. Write

$$\prod_{i=1}^{n} \left(\frac{p_i - q}{p_i + q + 1} \right)^2 = \prod_{i=1}^{n} \left(1 - \frac{q}{p_i} \right)^2 \Big/ \prod_{i=1}^{n} \left(1 + \frac{q + 1}{p_i} \right)^2. \tag{11.3.15}$$

In view of the hypothesis, the numerator converges to 0 while the denominator diverges to $+\infty$. Hence criterion (11.3.12) is fulfilled.

Necessity. Suppose that the set of exponents $\{p\}$ contains no sequence $\{p_i\}$ fulfilling either (11.3.9, 10, or 11). Then $\{p\}$ must be either

(A)   a sequence $\{p_i\}$ with $\lim\limits_{i \to \infty} p_i = -\frac{1}{2}$, $\sum\limits_{i=1}^{\infty} (p_i + \frac{1}{2}) < \infty$, or

(B)   a sequence $\{p_i\}$ with $\lim\limits_{i \to \infty} p_i = \infty$, $\sum\limits_{i=1}^{\infty} \dfrac{1}{p_i} < \infty$, or

(C)   a sequence $\{p_i\}$ which can be split into two sub-sequences $\{r_i\}$, $\{s_i\}$, one of type (A) and one of type (B).

† See, e.g., K. Knopp [1], p. 219.

In case (A), refer to (11.3.14). Select a $q \neq p_1, p_2, \ldots$ The numerator converges to zero if and only if one factor vanishes. This is impossible and hence the numerator has a positive limit. Similarly, the denominator converges to a positive limit. This means that (11.3.12) does not hold and x^q cannot be approximated arbitrarily closely by $x^{p_1}, x^{p_2}, \ldots$

In case (B), refer to (11.3.15). If $q \neq p_1, p_2, \ldots$, then the numerator and the denominator of its right hand side converge to a nonzero value. Again, arbitrarily close approximation of x^q is impossible.

In case (C), select $q \neq r_1, r_2, \ldots$; $s_1, s_2, \ldots$. Set $\prod_{i=1}^{\infty} \left(\dfrac{r_i - q}{r_i + q + 1} \right)^2 = a$, $\prod_{i=1}^{\infty} \left(\dfrac{s_i - q}{s_i + q + 1} \right)^2 = b$, where, as we know from the discussion, $0 < a < \infty$; $0 < b < \infty$. Then, $\prod_{i=1}^{\infty} \left(\dfrac{p_i - q}{p_i + q + 1} \right)^2 = ab$. Again, (11.3.12) does not hold.

THEOREM 11.3.4 (Müntz). *Let $\{p\}$ be a   sequence of distinct nonnegative numbers. In order that $\{x^p\}$ be closed in $C[0, 1]$ it is sufficient that*

$$
\begin{array}{l}
\text{One of the } p\text{'s is } 0 \text{ and } \{p\} \text{ contains a sequence } \{p_i\} \\
\text{for which } \lim_{i \to \infty} p_i = \infty \quad \text{and} \quad \sum_{\substack{i=1 \\ p_i \neq 0}}^{\infty} \frac{1}{p_i} = \infty,
\end{array}
\tag{11.3.16}
$$

or that

$$
\begin{array}{l}
\text{One of the } p\text{'s is } 0 \text{ and } \{p\} \text{ contains a sequence } \{p_i\} \\
\text{for which } \lim_{i \to \infty} p_i = p, \quad 0 < p < \infty.
\end{array}
\tag{11.3.17}
$$

Proof: Let $n > 0$, $p_i > 0$, then

$$
\begin{aligned}
\left| x^n - \sum_{i=1}^{N} a_i x^{p_i} \right| &= n \left| \int_0^x \left(t^{n-1} - \sum_{i=1}^{N} \frac{a_i p_i t^{p_i - 1}}{n} \right) dt \right| \\
&\leq n \int_0^1 \left| t^{n-1} - \sum_{i=1}^{N} \frac{a_i p_i t^{p_i - 1}}{n} \right| dt \\
&\leq n \left(\int_0^1 \left| t^{n-1} - \sum_{i=1}^{N} \frac{a_i p_i t^{p_i - 1}}{n} \right|^2 dt \right)^{\frac{1}{2}}.
\end{aligned}
\tag{11.3.18}
$$

If (11.3.16) holds, $p_i - 1 \to \infty$, and $\dfrac{1}{2} \sum_{1}^{\infty} \dfrac{1}{p_i} < \sum_{1}^{\infty} \dfrac{1}{p_i - 1} = \infty$. The set of functions $x^{p_1 - 1}, x^{p_2 - 1}, \ldots$, is therefore closed in $L^2[0, 1]$. By the inequality (11.3.18), for $n - 1 = 0, 1, 2, \ldots$, $\max_{0 \leq x \leq 1} \left| x^n - \sum_{i=1}^{N} a_i x^{p_i} \right|$ can be made arbitrarily small by appropriate selection of a_i. The set $x^{p_1}, x^{p_2}, \ldots$, is therefore closed in $x, x^2, \ldots$. By adjoining 1, the augmented set will be closed in $1, x, x^2, \ldots$ and hence in $C[0, 1]$.

Suppose next that (11.3.17) holds and that $p > \frac{1}{2}$. According to Theorem 11.3.3, $\{x^{p_i - 1}\}$ will be closed in $L^2[0, 1]$ and the remainder of the proof is as above. If, however, $p \leq \frac{1}{2}$, select a constant $c > 0$ such that $cp_i > \frac{1}{2}$, ultimately. Then, $\{x^{cp_i}\}$ is closed in $C[0, 1]$. Take an $f(x) \in C[0, 1]$ and let $g(x) = f(x^c)$. For appropriate constants a_i.

$$\max_{0 \leq x \leq 1} \left| f(x^c) - \sum_{i=1}^{n} a_i x^{cp_i} \right| \leq \varepsilon.$$

Setting $x' = x^c$,

$$\max_{0 \leq x' \leq 1} \left| f(x') - \sum_{i=1}^{n} a_i x'^{p_i} \right| \leq \varepsilon.$$

11.4 Closure Theorems for Classes of Analytic Functions

LEMMA 11.4.1. *Let C be a rectifiable arc (with end points a and b included) of length L. Let $f(z)$ be defined on C, be continuous, and have $w(\delta)$ as its modulus of continuity. Let $z_0 = a, z_1, \ldots, z_n = b$ be points of C taken in order along C. Suppose that $|z - z_i| \leq \delta$ for z in the arc $z_i z_{i+1}$, $i = 0, 1, \ldots, n - 1$. Then,*

$$\left| \int_C f(z)\, dz - \sum_{i=0}^{n-1} f(z_i)(z_{i+1} - z_i) \right| \leq w(\delta)L. \tag{11.4.1}$$

Proof: $\left| \int_C f(z)\, dz - \sum_{i=0}^{n-1} f(z_i)(z_{i+1} - z_i) \right| = \left| \sum_{i=0}^{n-1} \int_{z_i}^{z_{i+1}} (f(z) - f(z_i))\, dz \right|$

$$\leq \sum_{i=0}^{n-1} \int_{z_i}^{z_{i+1}} |f(z) - f(z_i)|\, ds.$$

Along C from z_i to z_{i+1}, we have $|z - z_i| \leq \delta$ so that $|f(z) - f(z_i)| \leq w(\delta)$.

Hence, $\sum_{i=0}^{n-1} \int_{z_i}^{z_{i+1}} |f(z) - f(z_i)|\, ds \leq \sum_{i=0}^{n-1} w(\delta) \int_{z_i}^{z_{i+1}} ds = w(\delta)L.$

DEFINITION 11.4.1. A *Jordan curve* in the plane is a homeomorphic image of a circle. That is, it is a point set whose points (x, y) can be represented parametrically $x = f(\theta)$, $y = g(\theta)$ where (a) f and g are continuous and periodic functions of period 2π and (b) $f(\theta_1) = f(\theta_2)$, $g(\theta_1) = g(\theta_2)$ implies that $\dfrac{1}{2\pi} (\theta_1 - \theta_2) = $ integer.

In the work that follows, some facts about Jordan curves will be used without proof.

THEOREM 11.4.2 (Runge). *Let C be a Jordan curve and let $f(z)$ be analytic in the interior of and on C. Given $\varepsilon > 0$, we may find a rational function $R(z)$ whose poles lie exterior to C and such that*

$$|f(z) - R(z)| \leq \varepsilon \tag{11.4.2}$$

for z inside and on C.

Proof: We may find a contour C' consisting of a finite number of analytic arcs which contains C in its interior and inside and on which $f(z)$ is still analytic. For z inside and on C, $f(z) = \dfrac{1}{2\pi i} \displaystyle\int_{C'} \dfrac{f(t)}{t-z}\, dt$. For arbitrary t_1 and t_2, $\dfrac{f(t_1)}{t_1 - z} - \dfrac{f(t_2)}{t_2 - z} = \dfrac{f(t_1) - f(t_2)}{t_1 - z} - \dfrac{f(t_2)(t_1 - t_2)}{(t_1 - z)(t_2 - z)}$, so that

$$\left| \frac{f(t_1)}{t_1 - z} - \frac{f(t_2)}{t_2 - z} \right| \leq \left[\frac{|f(t_1) - f(t_2)|}{|t_1 - z|} + \frac{|f(t_2)|\,|t_1 - t_2|}{|t_1 - z|\,|t_2 - z|} \right].$$

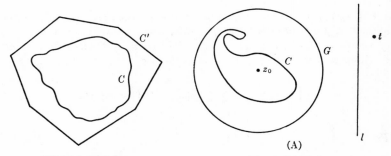

Figure 11.4.1.   Figure 11.4.2.

Now let t_1 and t_2 lie on C' and $|t_1 - t_2| \leq \delta$. Furthermore, set $w(\delta) =$ the modulus of continuity of f on C', $M = \max\limits_{z \in C'} |f(z)|$, and $\rho =$ minimum distance from C to C'. Then, for z in and on C and t on C',

$$\left| \frac{f(t_1)}{t_1 - z} - \frac{f(t_2)}{t_2 - z} \right| \leq \frac{w(\delta)}{\rho} + \frac{M\delta}{\rho^2}.$$

Therefore for t on C', $\dfrac{f(t)}{t - z}$ is uniformly continuous for z in and on C: It has a modulus of continuity $\Omega(\delta)$ independent of z.

Given an $\varepsilon > 0$. Let L' be the length of C'. Determine δ so small that $\Omega(\delta) \leq 2\pi\varepsilon/L'$. Determine sufficiently many points $t_0, t_1, \ldots, t_n = t_0$ on C' so that $|t - t_i| \leq \delta$, $t \in t_i t_{i+1}$, $i = 0, 1, \ldots, n-1$. By Lemma 11.4.1,

$$\left| \frac{1}{2\pi i} \int_{C'} \frac{f(t)\, dt}{t - z} - \frac{1}{2\pi i} \sum_{i=0}^{n-1} \frac{f(t_i)}{t_i - z} (t_{i+1} - t_i) \right| \leq \frac{\Omega(\delta)L'}{2\pi} \leq \varepsilon.$$ The rational func-

tion $R(z) = \dfrac{1}{2\pi i} \displaystyle\sum_{i=0}^{n-1} \dfrac{f(t_i)(t_{i+1} - t_i)}{t_i - z}$ fulfills the requirements of the theorem.

THEOREM 11.4.3 (Runge).   *Let C be a Jordan curve and let $A(C)$ designate the linear space of functions that are analytic inside and on C. Then the powers $1, z, z^2, \ldots$, are closed in $A(C)$.*

Proof: We shall show first that if t is a fixed point exterior to C, the particular function $\dfrac{1}{t - z}$ can be uniformly approximated inside and on C. There are two cases to consider. (A) The point t lies so far from C that a line ℓ can be drawn separating t and C. (B) No such line can be drawn. In case A, it is clear that we can draw a circle G that contains C in its interior and such that ℓ separates t from G.

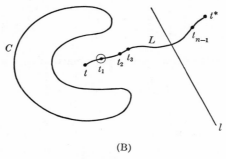

(B)

Figure 11.4.3.

Let z_0 be the center of G. Then we have,

$$\frac{1}{t - z} = \frac{1}{t - z_0} + \frac{z - z_0}{(t - z_0)^2} + \cdots. \qquad (11.4.3)$$

This series converges uniformly in z for $\left| \dfrac{z - z_0}{t - z_0} \right| \leq \rho < 1$. All points z in G and hence inside and on C satisfy this inequality. The convergence is therefore uniform inside and on C. If $\varepsilon > 0$ is prescribed, we need only take sufficiently many terms of (11.4.3) and arrive at a polynomial z which approximates $\dfrac{1}{t - z}$ uniformly inside and on C.

A simple modification of the expansion (11.4.3) allows us to conclude as much for the particular functions $\dfrac{c}{(t - z)^n}$, $n = 1, 2, \ldots$, and hence for any polynomial in $\dfrac{1}{t - z}$.

In case B, we proceed as follows: Since C is bounded, we can go out far enough and find a point t^* that can be separated from C by a line ℓ. Join t

and t^* by a curve L lying exterior to C. Let d be the minimum distance from L to C. Select a sequence of points on L, $t = t_0, t_1, \ldots, t_N = t^*$, such that $|t_{i+1} - t_i| < d$, $i = 0, \ldots, N - 1$. Now

$$\frac{1}{t - z} = \frac{1}{t_1 - z} - \frac{t - t_1}{(t_1 - z)^2} + \frac{(t - t_1)^2}{(t_1 - z)^3} - \cdots . \tag{11.4.4}$$

This series converges uniformly and absolutely for $\left| \dfrac{t - t_1}{t_1 - z} \right| \leq r < 1$; hence, in the exterior of every circle $|z - t_1| > \dfrac{1}{r} |t - t_1|$. Since $|t - t_1| < d$, the series converges uniformly inside and on C. If $\varepsilon > 0$ is prescribed, we can therefore find an integer N_1 such that

$$\left| \frac{1}{t - z} - P_{N_1}\!\left(\frac{1}{t_1 - z}\right) \right| \leq \frac{\varepsilon}{2N} , \quad z \text{ inside and on } C, \tag{11.4.5}$$

where P_{N_1} designates an appropriate polynomial of degree N_1:

$$P_{N_1}\!\left(\frac{1}{t_1 - z}\right) = \sum_{k=0}^{N_1} \frac{A_k}{(t_1 - z)^k} .$$

By a similar argument, we can approximate $\dfrac{A_k}{(t_1 - z)^k}$ by a polynomial in $\dfrac{1}{t_2 - z}$ uniformly inside and on C up to an error of $\dfrac{\varepsilon}{2N(N_1 + 1)}$. Combining these individual approximations, we arrive at a polynomial $P_{N_2}\!\left(\dfrac{1}{t_2 - z}\right)$ such that

$$\left| P_{N_1}\!\left(\frac{1}{t_1 - z}\right) - P_{N_2}\!\left(\frac{1}{t_2 - z}\right) \right| \leq \frac{\varepsilon}{2N} , \quad z \text{ inside and on } C. \tag{11.4.6}$$

We can therefore set up a chain of approximations,

$$\left| P_{N_k}\!\left(\frac{1}{t_k - z}\right) - P_{N_{k+1}}\!\left(\frac{1}{t_{k+1} - z}\right) \right| \leq \frac{\varepsilon}{2N} , \quad \begin{array}{l} z \text{ inside and on } C, \\ k = 1, 2, \ldots, N - 1. \end{array} \tag{11.4.7}$$

Once we have arrived at $t_n = t^*$, we use case A to change to an approximation in powers of z:

$$\left| P_{N_n}\!\left(\frac{1}{t_n - z}\right) - P(z) \right| \leq \frac{\varepsilon}{2} , \quad z \text{ inside and on } C.$$

The grand combination of these inequalities leads to

$$\left| \frac{1}{t - z} - P(z) \right| \leq \varepsilon, \quad z \text{ inside and on } C \tag{11.4.8}$$

for an appropriate polynomial $P(z)$.

The theorem is now completed by using Theorem 11.4.2. Let $f(z)$ be analytic inside and on C. Then we can find points $t_1, \ldots, t_M$ exterior to C and constants $a_1, \ldots, a_M$ such that

$$\left| f(z) - \sum_{k=1}^{M} \frac{a_k}{z - t_k} \right| \leq \frac{\varepsilon}{2}, \qquad z \text{ inside and on } C. \qquad (11.4.9)$$

But for each k, we can find a polynomial $P_{M_k}(z)$, of appropriate degree M_k, such that

$$\left| \frac{a_k}{z - t_k} - P_{M_k}(z) \right| \leq \frac{\varepsilon}{2M}, \qquad z \text{ inside and on } C, k = 1, 2, \ldots, M.$$

$$(11.4.10)$$

By combining these inequalities we obtain a polynomial $P(z)$ for which

$$|f(z) - P(z)| \leq \varepsilon, \qquad z \text{ inside and on } C.$$

Theorem 11.4.3 can be extended. It is sufficient to assume only that $f(z)$ is analytic inside C and continuous inside and on C. The proof of this extension depends upon a continuity theorem for mapping functions. We cannot go into this matter in detail. It must suffice to present the leading ideas.

Let B be a simply connected region whose boundary is a Jordan curve. A sequence of bounded simply connected regions B_n will be said to *converge to B from the outside* if

(A)   Each B_n contains $\bar{B}$ (the closure of B).

(B) B_n contains $\overline{B_{n+1}}$.

(C)   The set $B_1 \cap B_2 \cap \cdots$ contains no point exterior to B.

For each B, we can find such a convergent sequence. Let $z = 0$ be interior to B. Map B_n conformally onto the unit circle $|w| < 1$ by means of $\phi_n(z)$. B is mapped by $\phi(z)$. These mapping functions are fixed by requiring that $\phi_n(0) = \phi(0) = 0$; $\phi_n{}'(0) > 0$, $\phi(0) > 0$. Map B_n onto B by means of $w = m_n(z)$; $m_n(0) = 0$, $m_n{}'(0) > 0$.

THEOREM 11.4.4.   *With the above notation,*

$$\lim_{n \to \infty} \phi_n(z) = \phi(z)$$
$$\lim_{n \to \infty} m_n(z) = z \qquad (11.4.11)$$

the limits holding uniformly in $\bar{B}$.

For a proof of this theorem, the reader is referred to Walsh [1], p. 32.

DEFINITION 11.4.2.   Let C be a Jordan curve lying in the z-plane. $W(C)$ will designate the normed linear space of functions that are analytic in C and continuous inside and on C. The norm is defined by

$$\| f \| = \max_{z \in C} |f(z)|. \qquad (11.4.12)$$

THEOREM 11.4.5 (Walsh).   *The powers* $1, z, z^2, \ldots,$ *are closed in* $W(C)$.

Proof: Let C be the boundary of the finite region B. Let B_n be a sequence of simply connected regions that converge to B from the outside. Let $f(z)$ be analytic in B and continuous in $\bar{B}$. Then $f(m_n(z))$ will be analytic in $\bar{B}$. Given an ε, we can by Theorem 11.4.4 select an n such that

$$|f(m_n(z)) - f(z)| \leq \frac{\varepsilon}{2}, \qquad z \in B.$$

By Theorem 11.4.3, we can find a polynomial $P(z)$ such that

$$|f(m_n(z)) - P(z)| \leq \frac{\varepsilon}{2}, \qquad z \in \bar{B}.$$

Combining these two inequalities yields the required approximation.

Uniform approximation in the complex plane by polynomials has one feature that distinguishes it sharply from the real case. If a sequence of analytic functions converges uniformly in a region, the limit of the sequence is analytic. Thus, in regions, at least, only analytic functions can be approximated uniformly by polynomials. However, this does not rule out the possibility of more general functions being approximated uniformly on sets that lack interior points. Nor does it rule out the possibility of several distinct analytic functions (noncontinuable one to the other) from being simultaneously approximated over mutually exterior regions.

A half century of work on the problem of uniform approximation in the complex plane by such mathematicians as Runge, Walsh, Lavrentieff, Keldysch, and Mergelyan has led to the following definitive theorems.

DEFINITION 11.4.2.   A closed set S in the plane will be said to *separate* the plane if the complement of S is not connected.

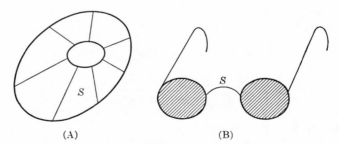

(A)   (B)

Figure 11.4.4.

THEOREM 11.4.6.   *Let S be a closed bounded set that does not separate the plane. Let $f(z)$ be continuous on S and be analytic at interior points of S. Then $f(z)$ may be uniformly approximated on S by polynomials.*

We formulate a converse as a separate theorem.

THEOREM 11.4.7.   *Let S be a closed bounded set. If every function $f(z)$ that is continuous on S and analytic at its interior points can be approximated uniformly on S by polynomials, then S does not separate the plane.*

Ex. 1.   Let $B_1, B_2, \ldots, B_n$ be n mutually exterior bounded regions, with contours $C_1, \ldots, C_n$ as boundaries. If f_k is analytic in B_k and continuous in $\overline{B}_k, k = 1, 2, \ldots, n$, then even though the functions f_k may have nothing to do with one another (i.e., are not analytic continuations of one another) we can find a sequence of polynomials that converges uniformly to f_k in $\overline{B}_k$ simultaneously for $k = 1, 2, \ldots, n$.

Ex. 2 (Walsh).   Let C be a Jordan arc. If $f(z)$ is continuous on C then it can be approximated uniformly by polynomials in z. When C is a segment of the real axis then this reduces to Weierstrass' theorem.

We terminate this section with a discussion of the closure of the powers $1, z, z^2, \ldots$, in the Hilbert space $L^2(B)$. Here again, the nature of the region B plays a crucial role.

A bounded region B of the complex plane whose boundary C is a Jordan curve has the following property: the complement of $B \cup C$ is a single simply connected region whose boundary is exactly C. The class of regions with this property is more extensive than the regions bounded by Jordan curves. For such regions B, the powers are closed in $L^2(B)$.

Ex. 3. B is the disc $|z| < 1$ with $0 \leq x < 1$ excluded. B does not have this property, for the complement of $B \cup C$ is $|z| > 1$ whose boundary $|z| = 1$ is only a part of C.

THEOREM 11.4.8 (Carleman-Farrell).   *Let B be a bounded simply connected region with boundary C. It is assumed that the complement of $B \cup C$ is a single region whose boundary is exactly C. Then, the set of functions $\{z^n\}$, $n = 0, 1, \ldots$ is closed in $L^2(B)$.*

Proof: As in Theorem 11.4.5, the proof depends upon the continuity of mapping functions.

Let B_n be a sequence of regions bounded by Jordan curves which converge to B from the outside. Let $w = m_n(z)$ map B_n conformally on B, $m_n(0) = 0$, $m_n'(0) > 0$. We assume that $z = 0$ is interior to B. (For the possibility of this type of convergence and for the uniform continuity of the mapping functions, (11.4.11), the reader is referred to Walsh [1], p. 35.)

Let $f(z) \in L^2(B)$. Consider the composite function

$$f_n(z) = f(m_n(z))m_n'(z). \tag{11.4.13}$$

$f_n(z)$ is analytic throughout the interior of B_n. We shall show first that

$$\lim_{n \to \infty} \iint_B |f(z) - f_n(z)|^2 \, dx \, dy = 0. \tag{11.4.14}$$

Let B' designate any closed Jordan region contained in the interior of B. Then,

$$\iint_B |f - f_n|^2 \, dx \, dy = \iint_{B'} |f - f_n|^2 \, dx \, dy + \iint_{B-B'} |f - f_n|^2 \, dx \, dy. \tag{11.4.15}$$

Now $m_n(z) \to z$, $m_n{}'(z) \to 1$ uniformly on B'. Hence, $f_n(z) \to f(z)$ uniformly on B' and this implies that $\iint_B |f - f_n|^2 \, dx \, dy \to 0$. Now, since $|f - f_n|^2 \le 2\,|f|^2 + 2\,|f_n|^2$,

$$\iint_{B-B'} |f - f_n|^2 \, dx \, dy \le 2 \iint_{B-B'} |f|^2 \, dx \, dy + 2 \iint_{B-B'} |f_n|^2 \, dx \, dy. \tag{11.4.16}$$

Given an $\varepsilon > 0$, since $f \in L^2(B)$, we may select a closed set $G \subset B$ such that $\iint_B |f|^2 \, dx \, dy - \iint_G |f|^2 \, dx \, dy \le \varepsilon/6$. If B' is now chosen so that $G \subset B'$ we have $2 \iint_{B-B'} |f|^2 \, dx \, dy \le \varepsilon/2$. Now,

$$\iint_{B-B'} |f_n|^2 \, dx \, dy = \iint_{B-B'} |f(m_n(z))|^2 \, |m_n{}'(z)|^2 \, dx \, dy = \iint_{S_n} |f(w)|^2 \, dA_w$$

where S_n is the image of $B - B'$ under $w = m_n(z)$ and where dA_w is the area element in the w variable. Now, by the continuity of the mapping function, for sufficiently large n, S_n lies in $B - B'$ and hence, as we have just seen,

$$2 \iint_{B-B'} |f_n|^2 \, dx \, dy \le \varepsilon/2. \tag{11.4.17}$$

Combining these inequalities leads to (11.4.14).

Since $f_n(z)$ is analytic throughout B_n, it is analytic in $\overline{B_{n+1}}$. Hence by Theorem 11.4.3, we can find a polynomial $p(z)$ such that

$$|f_n(z) - p(z)| \le \varepsilon, \qquad z \in \overline{B_{n+1}}, \tag{11.4.18}$$

and therefore

$$\iint_B |f_n(z) - p(z)|^2 \, dx \, dy \le \iint_{B_{n+1}} |f_n(z) - p(z)|^2 \, dx \, dy \tag{11.4.19}$$
$$\le \varepsilon^2 \cdot \text{area of } B_{n+1} \le \varepsilon^2 \cdot \text{area of } B_1.$$

By combining (11.4.14) and (11.4.19), we can find a polynomial $p(z)$ such that $\iint\limits_{B} |f(z) - p(z)|^2\, dx\, dy$ is arbitrarily small. The powers are therefore closed in $L^2(B)$.

COROLLARY 11.4.9.   *If B is a region as indicated, then orthonormal polynomials, $k_n z^n + \cdots$, $k_n > 0$, in $L^2(B)$ possess the properties A–F of Theorem 8.9.1.*

11.5 Closure Theorems for Normed Linear Spaces.   One closed system may be used to generate other closed systems.

THEOREM 11.5.1.   *Let X be a Banach space (i.e., a complete normed linear space). Let $\{x_k\}$ be a sequence that is closed in X and such that*

$$\limsup_{n \to \infty} \|x_n\|^{1/n} = \sigma, \quad 0 \le \sigma < \infty. \tag{11.5.1}$$

Let $\{z_n\}$ be a sequence of distinct complex numbers such that

$$0 < |z_n| \le \rho < \sigma^{-1}. \tag{11.5.2}$$

Then the sequence of elements

$$y_n = \sum_{k=1}^{\infty} z_n{}^k x_k, \quad n = 1, 2, \ldots, \tag{11.5.3}$$

is closed in X.

Proof: By equation (11.5.3) we mean, of course, that

$$\lim_{m \to \infty} \left\| y_n - \sum_{k=1}^{m} z_n{}^k x_k \right\| = 0, \quad n = 1, 2, \ldots.$$

To show that the series $\sum_{k=1}^{\infty} z^k x_k$ converges to an element y (we omit the subscripts), consider the sequence of elements

$$y^{(p)} = \sum_{k=1}^{p} z^k x_k. \tag{11.5.4}$$

Now,

$$\|y^{(p+q)} - y^{(p)}\| = \left\| \sum_{k=p+1}^{q} z^k x_k \right\| \le \sum_{k=p+1}^{q} \|x_k\|\, |z^k|. \tag{11.5.5}$$

In view of (11.5.1), the radius of convergence of the power series $\sum_{k=1}^{\infty} \|x_k\|\, z^k$ is σ^{-1}, so that for z in the range $|z| < \rho$, this series is convergent. For a given $\varepsilon > 0$, we can find N so large that for all $p \ge N$ and all $q \ge 1$,

$$\sum_{k=p+1}^{q} \|x_k\|\, |z^k| < \varepsilon.$$

The same holds for $\|y^{(p+q)} - y^{(p)}\|$. The sequence $\{y^{(p)}\}$ is therefore a Cauchy sequence. Since X is a complete space, there is a y such that

$$\lim_{p \to \infty} \|y - y^{(p)}\| = 0.$$

Let $L \in X^*$, then $|L(y) - L(y^{(p)})| \leq \|L\| \, \|y - y^{(p)}\|$. Hence,

$$L(y_n) = \sum_{k=0}^{\infty} L(x_k) z_n{}^k, \quad n = 1, 2, \dots.$$

Consider now the power series $f(z) = \sum_{k=0}^{\infty} L(x_k) z^k$. We have

$$|L(x_k)| \leq \|L\| \, \|x_k\|.$$

Therefore, by (11.5.1), $f(z)$ is analytic in $|z| < \sigma^{-1}$. Suppose that $L(y_n) = 0$, $n = 1, 2, \dots$; then $f(z_n) = 0$. Therefore the zeros of $f(z)$ have a limit point interior to $|z| < \sigma^{-1}$. By the uniqueness theorem for analytic functions, $f(z) \equiv 0$. This implies that $L(x_k) = 0$, $n = 1, 2, \dots$. Since $\{x_k\}$ is assumed complete, it follows that $L = 0$. Thus, the only solution of $L(y_k) = 0$, $k = 1, 2, \dots$, is $L = 0$. By Theorem 11.1.7, $\{y_k\}$ is closed in X.

COROLLARY 11.5.2 (Szász). *Let* $F(z) = \sum_{k=0}^{\infty} c_k z^k$, c_k *real, be a fixed power series and let* $r > 0$ *be its radius of convergence. Assume that* $\sum_{n=1}^{\infty} \dfrac{1}{k_n} = \infty$ *where* $k_1 < k_2 < \cdots$ *is the sequence of all those integers* ≥ 1 *for which* $c_k \neq 0$. *If now* $\{t_n\}$ *is a sequence of distinct real numbers satisfying*

$$0 < |t_n| \leq r_1 < r, \tag{11.5.6}$$

then the sequence of functions

$$f_n(x) = F(t_n x), \quad n = 1, 2, \dots, \tag{11.5.7}$$

is complete in $L^2(0, 1)$. *If* $c_0 \neq 0$, *it is also complete in* $C(0, 1)$.

Proof: Set $x_n = c_n x^n$. Theorems 11.3.3 and 11.3.4 imply that $\{c_n x^n\}$ is closed in $L^2[0, 1]$ and in $C[0, 1]$. Now, $\|c_n x^n\|_{C[0,1]} = |c_n|$ and $\|c_n x^n\|_{L^2[0,1]} = \dfrac{|c_n|}{\sqrt{2n + 1}}$. Hence, $\limsup\limits_{n \to \infty} \|c_n x^n\|^{1/n} = \limsup\limits_{n \to \infty} |c_n|^{1/n} = 1/r$. The corollary now follows.

Ex. 1. Let $F(z) = (1 + z)^s$ where s is real and $\neq 0, 1, 2, \dots$. Under these conditions, none of the Maclaurin coefficients of $F(z)$ vanishes, and $r = 1$. Let $\{t_n\}$ be a sequence of points satisfying $|t_n| \leq 1 - \varepsilon$, $n = 0, 1, \dots$. Then, $(1 + t_n x)^s$ is complete in $L^2[0, 1]$ and in $C[0, 1]$.

A sequence that is "sufficiently close" to a complete sequence is itself complete. We present two theorems to this effect.

Theorem 11.5.3 (Birkhoff). *Let H be a Hilbert space and $\{x_n{}^*\}$ be a complete orthonormal system. Let $\{y_n{}^*\}$ be a second orthonormal system. If*

$$\sum_{n=1}^{\infty} \|x_n{}^* - y_n{}^*\|^2 < \infty, \tag{11.5.8}$$

then $\{y_n{}^*\}$ is also complete.

Proof: (1) If for some $N \geq 0$,

$$\sum_{N+1}^{\infty} \|x_k{}^* - y_k{}^*\|^2 < 1, \tag{11.5.9}$$

then the system $x_1{}^*, x_2{}^*, \ldots, x_N{}^*, y_{N+1}^*, y_{N+2}^*, \ldots$, is complete in H. For, suppose there were an element $w \neq 0$ orthogonal to all these. Then

$$\|w\|^2 = \sum_{k=1}^{\infty} |(w, x_k{}^*)|^2 = \sum_{k=N+1}^{\infty} |(w, x_k{}^*)|^2 = \sum_{k=N+1}^{\infty} |(w, x_k{}^* - y_k{}^*)|^2$$

$$\leq \sum_{k=N+1}^{\infty} \|w\|^2 \|x_k{}^* - y_k{}^*\|^2 = \|w\|^2 \sum_{k=N+1}^{\infty} \|x_k{}^* - y_k{}^*\|^2 < \|w\|^2,$$

a contradiction.

(2) Suppose for some N, $\sum_{N+1}^{\infty} \|x_k{}^* - y_k{}^*\|^2 < 1$. Set

$$z_k = x_k{}^* - \sum_{j=N+1}^{\infty} (x_k{}^*, y_j{}^*)y_j{}^*, \quad k = 1, 2, \ldots, N.$$

Then the system, $z_1, z_2, \ldots, z_N, y_{N+1}^*, y_{N+2}^*, \ldots$, is complete in H. For let w be orthogonal to these elements. Then

$$0 = (w, z_k) = \left(w, x_k{}^* - \sum_{j=N+1}^{\infty} (x_k{}^*, y_j{}^*) y_j{}^*\right)$$

$$= (w, x_k{}^*) - \sum_{j=N+1}^{\infty} (x_k{}^*, y_j{}^*) (w, y_j{}^*) = (w, x_k{}^*).$$

Therefore $(w, x_k{}^*) = 0$, $k = 1, 2, \ldots, N$. By part (1), $w = 0$.

To prove the theorem, select N so large that $\sum_{N+1}^{\infty} \|x_k{}^* - y_k{}^*\|^2 < 1$. Let S designate the orthogonal complement of $\{y_{N+1}^*, y_{N+2}^*, \ldots\}$, i.e., the set of all elements orthogonal to these elements. For $j \geq N + 1$, and for $1 \leq k \leq N$,

$$(z_k, y_j{}^*) = (x_k{}^*, y_j{}^*) - \sum_{p=N+1}^{\infty} (x_k{}^*, y_p{}^*)(y_p{}^*, y_j{}^*) = (x_k{}^*, y_j{}^*) - (x_k{}^*, y_j{}^*) = 0.$$

Hence $z_1, z_2, \ldots, z_N \in S$. Since $z_1, \ldots, z_N, y_{N+1}^*, \ldots$, is complete, S cannot contain elements other than linear combinations of $z_1, \ldots, z_N$. Thus, S is a finite dimensional space of dimension $\leq N$. Note further that the elements $y_1{}^*, y_2{}^*, \ldots, y_N{}^*$ are in S. Since they are orthonormal, they are independent and hence span S. The elements $z_1, z_2, \ldots, z_N$ are therefore linear combinations of $y_1^*, y_2^*, \ldots, y_N^*$. If, therefore, $(w, y_k^*) = 0$, $k = 1, 2, \ldots$, it follows that $w \perp z_1, z_2, \ldots, z_N, y_{N+1}^*, \ldots$, and hence $w = 0$.

THEOREM 11.5.4 (Paley-Wiener-Boas). *Let X be a normed linear space and suppose that $\{x_n\}$ is closed in X. If $\{y_n\}$ is a sequence such that for some number λ, $0 \leq \lambda < 1$, and for all finite sequences of constants $a_1, a_2, \ldots, a_n$, we have*

$$\left\| \sum_{k=1}^{n} a_k(x_k - y_k) \right\| \leq \lambda \left\| \sum_{k=1}^{n} a_k x_k \right\|, \qquad (11.5.10)$$

then $\{y_n\}$ is also closed in X.

Proof: Let X_1 be the subspace of X spanned by $\{x_n\}$. That is, X_1 consists of all finite linear combinations $\sum_{k=1}^{n} a_k x_k$. Let $L \in X^*$ with $L(y_k) = 0$, $k = 1, 2, \ldots$. We shall estimate the norm of L on X_1.

$$\left| L\left(\sum_{k=1}^{n} a_k x_k \right) \right| = \left| L\left(\sum_{k=1}^{n} a_k(x_k - y_k) \right) \right| \leq \|L\| \left\| \sum_{k=1}^{n} a_k(x_k - y_k) \right\|$$

$$\leq \lambda \|L\| \left\| \sum_{k=1}^{n} a_k x_k \right\|.$$

Hence, over X_1, the norm of L does not exceed $\lambda\|L\|$.

By Theorem 11.1.5, we may extend L to the whole of X without increasing the norm. Call the extension F:

$$F(x_k) = L(x_k), \quad k = 1, 2, \ldots, \qquad (11.5.11)$$

$$\|F\| = \|L\|_{X_1}.$$

Now, since $\{x_k\}$ is complete, $F(x_k) - L(x_k) = 0$, $k = 1, 2, \ldots$, implies that $F = L$. Therefore $\|L\| = \|F\| = \|L\|_{X_1} \leq \lambda\|L\|$. Since $0 < \lambda < 1$, this implies that $L = 0$. Therefore $\{y_n\}$ is complete and closed in X.

COROLLARY 11.5.5. *Let H be a Hilbert space and let $\{x_k^*\}$ be a complete orthonormal sequence. Let $\{y_k\}$ be any sequence such that*

$$\left\| \sum_{k=1}^{n} a_k(x_k^* - y_k) \right\|^2 \leq \lambda \sum_{k=1}^{n} |a_k|^2. \quad 0 \leq \lambda < 1 \qquad (11.5.12)$$

for every sequence $\{a_k\}$ of complex members. Then $\{y_k\}$ is also complete. In particular, (11.5.12) holds if

$$\sum_{k=1}^{\infty} \|x_k^* - y_k\|^2 = \lambda < 1. \qquad (11.5.13)$$

Proof: $\sum_{k=1}^{n} |a_k|^2 = \left\| \sum_{k=1}^{n} a_k x_k^* \right\|^2$ so that (11.5.12) implies (11.5.10). Note that under the assumption (11.5.13),

$$\left\|\sum_{k=1}^{n} a_k(x_k{}^* - y_k)\right\| \le \sum_{k=1}^{n} |a_k|\, \|x_k{}^* - y_k\| \le \left(\sum_{k=1}^{n} |a_k|^2\right)^{\frac{1}{2}} \left(\sum_{k=1}^{n} \|x_k{}^* - y_k\|^2\right)^{\frac{1}{2}}$$
$$\le \lambda^{\frac{1}{2}}\left(\sum_{k=1}^{n} |a_k|^2\right)^{\frac{1}{2}}$$

so that (11.5.12) holds with $\lambda^{\frac{1}{2}}$ replacing λ.

COROLLARY 11.5.6 (Schäfke). *If* (11.5.10) *holds with* $0 < \lambda < \frac{1}{2}$, *then* $\{y_n\}$ *is closed if and only if* $\{x_n\}$ *is closed.*

Proof: We need only show that $\{y_n\}$ closed implies $\{x_n\}$ closed. From (11.5.10) and the triangle inequality, $\left\|\sum_{k=1}^{n} a_k x_k\right\| - \left\|\sum_{k=1}^{n} a_k y_k\right\| \le \lambda \left\|\sum_{k=1}^{n} a_k x_k\right\|$. Hence, $\left\|\sum_{k=1}^{n} a_k x_k\right\| \le \dfrac{1}{1-\lambda}\left\|\sum_{k=1}^{n} a_k y_k\right\|$. Combining this once again with (11.5.10), we have

$$\left\|\sum_{k=1}^{n} a_k(x_k - y_k)\right\| \le \frac{\lambda}{1-\lambda}\left\|\sum_{k=1}^{n} a_k y_k\right\|. \tag{11.5.14}$$

Now if $0 < \lambda < \frac{1}{2}$ then $0 < \dfrac{\lambda}{1-\lambda} < 1$, and (11.5.14) implies by Theorem 11.5.4 that $\{x_n\}$ is closed.

Ex. 2 (Duffin-Eachus). Let $L^2[-\pi, \pi]$ designate the space of complex valued functions of a real variable which are measurable on $[-\pi, \pi]$ and for which $(f, f) = \int_{-\pi}^{\pi} f(x)\overline{f(x)}\,dx < \infty$. Let $\{\lambda_n\}$, $n = 0, \pm 1, \pm 2, \ldots$, be a sequence of complex numbers such that

$$|\lambda_n - n| \le \sigma < \frac{\log 2}{\pi} = .22 \cdots, n = 0, \pm 1, \pm 2, \ldots. \tag{11.5.15}$$

Then the sequence of "nonharmonic" oscillations $\{e^{i\lambda_n x}\}$ is complete in $L^2[-\pi, \pi]$.

Proof: For simplicity of notation, think of the integers $n = 0, \pm 1, \pm 2, \ldots$, as indexed $I_1, I_2, \ldots$, and designate λ_{I_n} by λ_n. $|\lambda_n - I_n| \le \sigma$, $n = 1, 2, \ldots$. As one can verify using Theorem 11.2.4, $\left\{\dfrac{e^{iI_n x}}{\sqrt{2\pi}}\right\}$ is a complete orthonormal set for $L^2[-\pi, \pi]$. Furthermore, if $f \in L^2[-\pi, \pi]$, then

$$\|x^j f(x)\| = \left(\int_{-\pi}^{\pi} |x^j f(x)|^2\,dx\right)^{\frac{1}{2}} \le \pi^j \|f\|.$$

In Corollary 11.5.5, set

$$x_k{}^* = \frac{1}{\sqrt{2\pi}}\, e^{iI_k x}, \, y_k = \frac{e^{i\lambda_k x}}{\sqrt{2\pi}}.$$

Then

$$y_k - x_k{}^* = \frac{1}{\sqrt{2\pi}} e^{iI_k x} (e^{i(\lambda_k - I_k)x} - 1).$$

For arbitrary constants $a_1, \ldots, a_n$,

$$\left\| \sum_{k=1}^{n} a_k(y_k - x_k{}^*) \right\| = \left\| \sum_{k=1}^{n} a_k \frac{e^{iI_k x}}{\sqrt{2\pi}} \sum_{j=1}^{\infty} \frac{i^j(\lambda_k - I_k x^j)}{j!} \right\|$$

$$= \left\| \sum_{j=1}^{\infty} \frac{1}{j!} x^j \sum_{k=1}^{n} a_k i^j (\lambda_k - I_k)^j \frac{e^{iI_k x}}{\sqrt{2\pi}} \right\| \leq \sum_{j=1}^{\infty} \frac{1}{j!} \left\| x^j \sum_{k=1}^{n} a_k i^j (\lambda_k - I_k)^j \frac{e^{iI_k x}}{\sqrt{2\pi}} \right\|$$

$$\leq \sum_{j=1}^{\infty} \frac{\pi^j}{j!} \left\| \sum_{k=1}^{n} a_k i^j (\lambda_k - I_k)^j \frac{e^{iI_k x}}{\sqrt{2\pi}} \right\| = \sum_{j=1}^{\infty} \frac{\pi^j}{j!} \left(\sum_{k=1}^{n} |a_k|^2 |\lambda_k - I_k|^{2j} \right)^{\frac{1}{2}}$$

$$\leq \sum_{j=1}^{\infty} \frac{\sigma^j \pi^j}{j!} \left(\sum_{k=1}^{n} |a_k|^2 \right)^{\frac{1}{2}} = (e^{\sigma\pi} - 1) \left(\sum_{k=1}^{n} |a_k|^2 \right)^{\frac{1}{2}}.$$

Since $\sigma < \dfrac{\log 2}{\pi}$, $e^{\sigma\pi} - 1 < 1$, and condition (11.5.12) is fulfilled.

Ex. 3.   Let $\{h_n{}^*(z)\}$ be complete and orthonormal for $L^2(B)$. If $f_n(z) \in L^2(B)$ and if $\sum\limits_{n=1}^{\infty} \iint\limits_{B} |h_n{}^*(z) - f_n(z)|^2 \, dx \, dy < 1$, then $\{f_n\}$ is complete in $L^2(B)$. In particular, if B is the unit circle, $\sum\limits_{n=1}^{\infty} \iint\limits_{|z|<1} \left| \dfrac{\sqrt{n+1}}{\pi} z^n - f_n(z) \right|^2 dx \, dy < 1$ implies that $\{f_n\}$ is complete.

NOTES ON CHAPTER XI

11.1   Davis and Fan [1] has a generalization of Theorem 11.1.7.

11.2–11.3   Kaczmarz and Steinhaus [1], Natanson [1].

11.4   Walsh [2], Chapter I. Behnke and Sommer [1], pp. 244–249. For a proof of Theorems 11.4.6, 11.4.7 see Walsh [2], pp. 367–371.

11.5   For Theorem 11.5.1 Kaczmarz and Steinhaus [1], p. 145, Davis and Fan [1]. Theorem 11.5.3 is due to G. D. Birkhoff. The present formulation and proof is due to G. Birkhoff and G.-C. Rota [1]. For Theorem 11.5.4, R. P. Boas [1], Davis and Fan [1]. For additional references to this type of result see Buck [3], pp. 349–350.

PROBLEMS

1. If $\{x_n\}$ is complete in X and if $x_1 = a_{11}y_1$

$$x_2 = a_{21}y_1 + a_{22}y_2$$
$$\cdot$$
$$\cdot$$
$$\cdot$$

then $\{y_n\}$ is complete in X.

2. In ℓ^2, $x_1 = (a_{11}, 0, 0, 0, \ldots)$

$\qquad x_2 = (a_{21}, a_{22}, 0, 0, 0, \ldots)$

$\qquad x_3 = (a_{31}, a_{32}, a_{33}, 0, 0, 0, \ldots)$

$\qquad \cdot$

$\qquad \cdot$

$\qquad \cdot$

If $a_{ii} > 0$, prove that $\{x_i\}$ is closed.

3. If a sequence $\{x_n\}$ is complete for a space X, it is complete for every subspace.

4. Let $w(x) \in C[0, 1]$ and $\dfrac{1}{w(x)} \geq \varepsilon > 0$ there. The set $\{w(x)f_n(x)\}$ is closed in $L^2[0, 1]$ if and only if $\{f_n(x)\}$ is.

5. Discuss the second Riemann derivative (Prob. 15, Ch. 1) as an extension (in the sense of linear functionals) of the ordinary second derivative.

6. If $p(x)$ is a convex functional defined on a linear space X, the set of elements x defined by $p(x) \leq c$ is convex.

7. If C is a convex set in a linear space X, a boundary point of C may be defined as a point x for which we can find two line segments $x_1 x$ and $x x_2$ such that all the interior points of $x_1 x$ are in C while none of the interior points of $x x_2$ are in C. Prove that the boundary of $p(x) \leq c$ is given by $p(x) = c$.

8. Let X be a normed linear space and Y a linear subspace. Let $x_0 \in X$ but $\notin Y$. A necessary and sufficient condition that $\inf\limits_{y \in Y} \|y - x_0\| \leq d$ is that if $L \in X^*$ and $L(x_0) = 1$, $L(y) = 0$, $y \in Y$, then $\|L\| \geq \dfrac{1}{d}$.

9. Give a second proof of this theorem in a Hilbert space.

10. The interval $[a, b]$ is divided into n equal parts at $a = x_0 < x_1 < \cdots < x_n = b$. A function $f(x)$ is in $C[a, b]$, is linear between the x_i and x_{i+1}, and $f(x_i) = y_i = $ rational. The set of all such f's, $n = 1, 2, \ldots$, is denumerable and closed in $C[a, b]$.

11. Let $T_r(x) = \dfrac{x}{r}$, $0 \leq x \leq r$; $T_r(x) = 1$, $r \leq x \leq 1$. If r_n is the sequence of rationals lying in $0 < x \leq 1$, is $\{T_{r_n}(x)\}$ closed in $L^2[0, 1]$?

12. Two solutions in $L^2[0, 1]$ to the moment problem

$$\int_0^1 f(x)x^n = m_n \quad n = 0, 1, \ldots,$$

must be equal almost everywhere.

13. S consists of a finite or infinite set of disjoint closed intervals in $[0, 1]$. Prove that S is uniquely determined by its moments

$$m_n = \int_S x^n \, dx, \quad n = 0, 1, 2, \ldots.$$

14. Remove one term from the trigonometric set of Theorem 11.2.4. The resulting set is not complete in $L^2[-\pi, \pi]$.

15. Let $0 < a < \pi$. Set $f(x) = -1$ for $-\pi - a < x < -\pi + a$, $f(x) = 0$ for $-\pi + a < x < \pi - a$, $f(x) = 1$ for $\pi - a < x < \pi + a$. Show that

$$\int_{-\pi-a}^{\pi+a} f(x) \frac{\sin nx}{\cos nx} \, dx = 0.$$

Hence, show that $1, \sin x, \cos x, \ldots$, is not complete in $L^2[c, d]$ with $c < -\pi$, $d > \pi$.

16. Consider the space of complex valued functions of a real variable x that are measurable on $[-\pi, \pi]$ and for which $\int_{-\pi}^{\pi} |f(x)|^2 \, dx < \infty$. Prove that the set of functions $\{e^{inx}\}$ $n = 0, \pm 1, \pm 2, \ldots$, is complete in this space.

17. Let H_n be the $n \times n$ matrix whose i, jth element is $\dfrac{1}{i + j - 1}$. (The Hilbert matrix.) Show that $\det H_n = \dfrac{[1! \, 2! \cdots (n - 1)!]^4}{1! \, 2! \, 3! \cdots (2n - 1)!}$. Obtain an asymptotic expression for $\det H_n$ as $n \to \infty$ and compare with the exact value when $n = 4$.

18. Let a_i, b_i be distinct and set

$$A(x) = (x - a_1)(x - a_2) \cdots (x - a_n),$$
$$B(x) = (x - b_1)(x - b_2) \cdots (x - b_n),$$
$$A_i(x) = \frac{A(x)}{A'(a_i)(x - a_i)},$$
$$B_i(x) = \frac{B(x)}{B'(b_i)(x - b_i)}.$$

If C designates the matrix $\left(\dfrac{1}{a_i - b_j} \right)$, then $C^{-1} = (c_{ij})$ where

$$c_{ij} = (a_j - b_i)A_j(b_i)B_i(a_j).$$

19. If $H_n^{-1} = s_{ij}$, then $s_{ij} = \dfrac{(-1)^{i+j}}{i + j - 1} \dfrac{(n + i - 1)! \, (n + j - 1)!}{[(i - 1)! \, (j - 1)!]^2 (n - i)! \, (n - j)!}$. (Cf. Prob. 17.)

20. Is the set of functions $x, x^{\frac{1}{2}}, x^{\frac{1}{3}}, \ldots$, closed in $C[0, 1]$?

21. The set of functions $x^{10^{10}}, x^{10^{10}+1}, x^{10^{10}+2}, \ldots$, is closed in $L^2[0, 1]$.

22. Let p_n be the nth prime number. Then $x^{p_1}, x^{p_2}, \ldots$, is closed in $L^2[0, 1]$.

23. Does the Müntz theorem hold as stated for an arbitrary interval $[a, b]$?

24. Formulate a theorem as to when $\int_0^1 f(x)x^{\lambda_n} \, dx = 0$, $n = 0, 1, \ldots$, implies $f(x) \equiv 0$.

25. If $f \in L^2[0, 1]$ and $m_n = \int_0^1 f(x)x^n \, dx$, then $\{m_n\} \in \ell^2$. However, the converse doesn't necessarily hold.

26. The sequence $\dfrac{1}{x + 1}, \dfrac{1}{x + 2}, \dfrac{1}{x + 3}, \ldots$, is complete in $L^2[0, 1]$.

27. If $0 \le a < b$, the sequence $1, \log(x + 1), \log(x + 2), \ldots$, is complete in $L^2[a, b]$.

28. The sequence $f_n(z) = nz^n$ $n = 1, 2, \ldots$, converges to 0 uniformly on every closed subset of $|z| < 1$, but $\lim\limits_{n \to \infty} \iint\limits_{|z| < 1} |f_n(z)|^2 \neq 0$.

29. Let $A(C)$ be the space of functions that are analytic in C: $|z| < 1$ and continuous in $|z| \leq 1$ with $\|f\| = \max\limits_{|z| < 1} |f(z)|$. If one of the powers is omitted from the sequence $\{z^n\}$, $n = 0, 1, \ldots$, the resulting sequence is not closed in the space.

30. Let C designate a circle. The origin is assumed to be exterior to C. If $w(z)$ is in $A(C)$ and does not vanish in the closed circle, then the sequence $w(z)$, $zw(z)$, $z^2w(z)$, $\ldots$, is closed in $A(C)$. In particular, the sequence z^n, $z^{n+1}, \ldots$ is closed in $A(C)$ for any n. Generalize.

31. If B is bounded and multiply connected, then $1, z, z^2, \ldots$, are not closed in $L^2(B)$.

32. If $f(z)$ is continuous on $|z| = 1$, it can be approximated uniformly by linear combinations of $1, z, z^2, \ldots, \bar{z}, \bar{z}^2, \ldots$.

33. Let C be a smooth arc and let $f(z)$ be continuous on C. Show that

$$\int_C f(z)\bar{z}^n \, ds = 0, \quad n = 0, 1, \ldots,$$

implies $f(z) \equiv 0$. Generalize. What if C is a closed curve?

34. Let $\{x_k{}^*\}$ be a complete orthonormal system for a Hilbert space H. Show that $x_1{}^* - x_2{}^*, x_2{}^* - x_3{}^*, x_3{}^* - x_4{}^*, \ldots$, is also complete. The conclusion may be false if $\{x_k{}^*\}$ is merely complete.

35. In a Hilbert space, let $\{x_n{}^*\}$ be complete and orthonormal. Let $\{y_n\}$ be an arbitrary sequence for which $(x_n{}^*, y_n) \neq 0$ $n = 1, 2, \ldots$. If

$$\sum_{n=1}^{\infty} \left(1 - \left| \frac{(x_n{}^*, y_n)}{(y_n, y_n)} \right|^2 \right) < 1,$$

then $\{y_n\}$ is complete.

CHAPTER XII

Expansion Theorems for Orthogonal Functions

We have seen that under fairly general circumstances, Fourier expansions converge in the mean to the elements that gave rise to them. But for many purposes of mathematics, convergence in the mean is not sufficiently strong: we may want pointwise convergence, or even uniform convergence. In the present chapter we take three orthogonal systems and develop expansion theorems for them. They are (1) The sines and cosines, (2) The Legendre polynomials, (3) Complex orthogonal polynomials. In many ways, the behavior of (1) and (2) are very similar. The complex analytic case (3) is quite different and serves as a striking contrast to the first two.

12.1 The Historical Fourier Series. The system $\dfrac{1}{\sqrt{2\pi}}$, $\dfrac{1}{\sqrt{\pi}}\sin x$, $\dfrac{1}{\sqrt{\pi}}\cos x$, $\dfrac{1}{\sqrt{\pi}}\sin 2x$, $\dfrac{1}{\sqrt{\pi}}\cos 2x, \dots$, is orthonormal on $[-\pi, \pi]$ with respect to the inner product $(f, g) = \displaystyle\int_{-\pi}^{\pi} f(x)g(x)\,dx$. The Fourier series of $f(x)$ is therefore

$$f(x) \sim \frac{a_0}{2} + \sum_{k=1}^{\infty} a_k \cos kx + b_k \sin kx \qquad (12.1.1)$$

where

$$a_k = \frac{1}{\pi} \int_{-\pi}^{\pi} f(x) \cos kx\,dx, \quad k = 0, 1, \dots,$$

$$\qquad (12.1.2)$$

$$b_k = \frac{1}{\pi} \int_{-\pi}^{\pi} f(x) \sin kx\,dx \quad k = 1, 2, \dots.$$

Ex. 1.

$$f(x) = \begin{cases} 1 & 0 \le x \le \pi \\ -1 & -\pi \le x < 0 \end{cases}$$

$$f(x) \sim \frac{4}{\pi}\left(\sin x + \frac{\sin 3x}{3} + \frac{\sin 5x}{5} + \cdots\right).$$

Ex. 2.

$$f(x) = \begin{cases} \dfrac{\pi - x}{2} & 0 \leq x \leq \pi \\[2ex] \dfrac{-\pi - x}{2} & -\pi \leq x < 0 \end{cases}$$

$$f(x) \sim \sin x + \frac{\sin 2x}{2} + \frac{\sin 3x}{3} + \cdots.$$

Ex. 3.

$$f(x) = |x|, \qquad -\pi \leq x \leq \pi,$$

$$f(x) \sim \frac{\pi}{2} - \frac{4}{\pi}\left(\cos x + \frac{\cos 3x}{3^2} + \frac{\cos 5x}{5^2} + \cdots\right).$$

Ex. 4.

$$f(x) = x^2, \qquad -\pi \leq x \leq \pi,$$

$$f(x) \sim \frac{\pi^2}{3} - 4\left(\cos x - \frac{\cos 2x}{2^2} + \frac{\cos 3x}{3^2} - \cdots\right).$$

THEOREM 12.1.1. *Let $f(x)$ be continuous and periodic in $[-\pi, \pi]$, $(f(\pi) = f(-\pi))$, and suppose that the Fourier series of $f(x)$ converges uniformly there. Then it converges to $f(x)$.*

Proof: If $\dfrac{a_0}{2} + \sum\limits_{k=1}^{\infty} a_k \cos kx + b_k \sin bx$ converges uniformly, it will converge to a continuous function of period 2π. Call the sum $g(x)$:

$$g(x) = \frac{a_0}{2} + \sum_{k=1}^{\infty} a_k \cos kx + b_k \sin kx.$$

Since we may integrate a uniformly convergent series term by term, we have by orthogonality

$$(g, \cos kx) = a_k, \quad (g, \sin kx) = b_k.$$

But by definition of the Fourier coefficients,

$$(f, \cos kx) = a_k \quad (f, \sin kx) = b_k.$$

Hence $(f - g, \cos kx) = 0, k = 0, 1, 2, \ldots, (f - g, \sin kx) = 0, k = 1, 2, \ldots$. By the completeness of the sines and cosines (Theorem 11.2.3), $f - g \equiv 0$ and hence $f \equiv g$.

COROLLARY 12.1.2. *Let $f(x)$ be continuous and periodic in $[-\pi, \pi]$ and have Fourier coefficients a_k, b_k. Let $\sum\limits_{k=1}^{\infty} |a_k| + |b_k| < \infty$. Then the Fourier series of f converges absolutely and uniformly to $f(x)$, $-\pi \leq x \leq \pi$.*

Proof: Since $|a_k \cos kx + b_k \sin kx| \leq |a_k| + |b_k|$, it follows from the Weierstrass "M test" that the Fourier series converges uniformly and absolutely. By Theorem 12.1.1, it converges to $f(x)$.

Ex. 5. In Examples 3 and 4 just considered, the sum of the Fourier coefficients converges absolutely. Hence we may replace the "$\sim$" by "$=$".

The smoothness of the function drastically affects the size of the Fourier coefficients: the smoother the function the more rapid is the decrease of the coefficients. The study of the convergence of Fourier series is largely the study of the interplay between assumptions of smoothness and conclusions about convergence.

THEOREM 12.1.3. *Let $f(x)$ be periodic in $[-\pi, \pi]$, $(f(\pi) = f(-\pi), f'(\pi) = f'(-\pi))$ and be of class C^2 there. Then, the Fourier series of $f(x)$ converges uniformly and absolutely to $f(x)$.*

Proof: $a_n = \dfrac{1}{\pi} \displaystyle\int_{-\pi}^{\pi} f(x) \cos nx \, dx$, $n = 1, 2, \ldots$. Integrate by parts.

$$a_n = f(x) \frac{\sin nx}{\pi n} \Big|_{-\pi}^{\pi} - \frac{1}{\pi n} \int_{-\pi}^{\pi} f'(x) \sin nx \, dx = -\frac{1}{\pi n} \int_{-\pi}^{\pi} f'(x) \sin nx \, dx$$

$$= \frac{f'(x) \cos nx}{\pi n^2} \Big|_{-\pi}^{\pi} - \frac{1}{\pi n^2} \int_{-\pi}^{\pi} f''(x) \cos nx \, dx = -\frac{1}{\pi n^2} \int_{-\pi}^{\pi} f''(x) \cos nx \, dx.$$

We use the fact that $f(\pi) = f(-\pi)$, $f'(\pi) = f'(-\pi)$. Since f'' is continuous on $[-\pi, \pi]$, we have $|f''(x)| \leq M$, $-\pi \leq x \leq \pi$, so that

$$|a_n| \leq \frac{2M}{n^2} .$$

A similar inequality can be derived for b_n. The conclusion now follows by applying Corollary 12.1.2.

In the above proof, the convergence of $\sum |a_n| + |b_n|$ is guaranteed by the convergence of $\displaystyle\sum_1^{\infty} \frac{1}{n^2}$. The strength of the power n^2 is unnecessary for convergence and can be reduced to $n^{1+\varepsilon}$, $\varepsilon > 0$. The appropriate smoothness is that the derivative of the function satisfy a Lipschitz condition.

THEOREM 12.1.4. *Let $f(x) \in C^n[-\pi, \pi]$, for some $n \geq 1$, and have period 2π. $(f(-\pi) = f(\pi), f'(-\pi) = f'(\pi), \ldots, f^{(n)}(-\pi) = f^{(n)}(\pi).)$ Suppose that $f^{(n)}(x)$ satisfies a Lipschitz condition of order α: $0 < \alpha \leq 1$. Then the Fourier coefficients of f satisfy $|a_k|, |b_k| \leq \dfrac{\text{const}}{k^{n+\alpha}}$, $k = 1, 2, \ldots$, and the Fourier series converges uniformly to $f(x)$ in $[-\pi, \pi]$.*

Proof: We prove the case $n = 1$. The cases $n > 1$ follow from this case by integration by parts as in Theorem 12.1.3.

Consider $a_n = \dfrac{1}{\pi} \displaystyle\int_{-\pi}^{\pi} f(x) \cos nx \, dx = - \dfrac{1}{\pi n} \displaystyle\int_{-\pi}^{\pi} f'(x) \sin nx \, dx$. In this last integral, set $x = x' + \dfrac{\pi}{n}$. Then, for $n \geq 1$,

$$a_n = - \frac{1}{\pi n} \int_{-\pi - \pi/n}^{\pi - \pi/n} f'(x' + \pi/n) \sin\left(x' + \frac{\pi}{n}\right) dx'$$

$$= \frac{1}{\pi n} \int_{-\pi - \pi/n}^{\pi - \pi/n} f'(x' + \pi/n) \sin nx' \, dx'$$

$$= \frac{1}{\pi n} \int_{-\pi}^{\pi} f'(x + \pi/n) \sin nx \, dx,$$

where we have extended the definition of $f(x)$ by periodicity. Thus,

$$a_n = \frac{1}{2\pi n} \int_{-\pi}^{\pi} [f'(x + \pi/n) - f'(x)] \sin nx \, dx.$$

Now $|f'(x + \pi/n) - f'(x)| \leq C \left(\dfrac{\pi}{n}\right)^{\alpha}$, all x. Hence $|a_n| \leq \dfrac{1}{2\pi n} C \left(\dfrac{\pi}{n}\right)^{\alpha} \cdot 2\pi = \dfrac{\text{const}}{n^{1+\alpha}}$. A similar inequality can be established for b_n. Since

$$\sum_{n=1}^{\infty} |a_n| + |b_n| < \infty,$$

the series converges uniformly.

THEOREM 12.1.5. *Let $f(x)$ be continuous and periodic in $[-\pi, \pi]$, and have a derivative $f'(x)$ that is piecewise continuous in $[-\pi, \pi]$. Then the Fourier series of $f(x)$ converges uniformly and absolutely in $[-\pi, \pi]$ to $f(x)$.*

Proof: By the above work, for $n \geq 1$, $a_n = - \dfrac{1}{\pi n} \displaystyle\int_{-\pi}^{\pi} f'(x) \sin nx \, dx$. Write $a_n' = \dfrac{1}{\pi} \displaystyle\int_{-\pi}^{\pi} f'(x) \sin nx \, dx$. Since f' is piecewise continuous, it is in $L^2[-\pi, \pi]$. By Corollary 11.2.5, $\displaystyle\sum_{n=1}^{\infty} (a_n')^2 < \infty$. Now

$$\sum_{n=1}^{\infty} |a_n| = \sum_{n=1}^{\infty} \frac{1}{n} |a_n'| \leq \left(\sum_{n=1}^{\infty} \frac{1}{n^2}\right)^{\frac{1}{2}} \left(\sum_{n=1}^{\infty} (a_n')^2\right)^{\frac{1}{2}} < \infty.$$

Hence $\displaystyle\sum_{n=1}^{\infty} |a_n| < \infty$. A similar argument shows that $\displaystyle\sum_{n=1}^{\infty} |b_n| < \infty$. The theorem now follows by Corollary 12.1.2.

We note in passing that it is sufficient to assume that $f'(x) \in L^2[-\pi, \pi]$. One can show that the integration by parts is still valid, and the remainder of the proof holds as before.

Ex. 6.   If $f(x)$ is a continuous, periodic, piecewise linear function, its Fourier series converges uniformly and absolutely to $f(x)$.

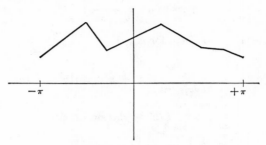

$-\pi$ $+\pi$

Figure 12.1.1.

More penetrating analyses of the convergence of Fourier series are based upon the study of its partial sums, rather than its coefficients. It is assumed that f is defined on $-\pi \le x < \pi$ and then extended over the whole axis periodically.

LEMMA 12.1.6.   *The following is a trigonometric identity:*

$$\tfrac{1}{2} + \cos x + \cos 2x + \cdots + \cos nx = \frac{\sin (n + \tfrac{1}{2})x}{2 \sin \dfrac{x}{2}}. \qquad (12.1.3)$$

Proof:

$$\sin (k + \tfrac{1}{2})x - \sin (k - \tfrac{1}{2})x = 2 \sin \frac{x}{2} \cos kx.$$

Hence

$$2 \sin \frac{x}{2} \sum_{k=0}^{n} \cos kx = \sum_{k=0}^{n} \{\sin (k + \tfrac{1}{2})x - \sin (k - \tfrac{1}{2})x\}$$
$$= \sin (n + \tfrac{1}{2})x - \sin (-\tfrac{1}{2})x$$

or,

$$\tfrac{1}{2} + \sum_{k=1}^{n} \cos kx = \frac{\sin (n + \tfrac{1}{2})x}{2 \sin \dfrac{x}{2}}.$$

COROLLARY 12.1.7.

$$K_n(x, t) = \frac{1}{2\pi} + \frac{1}{\pi} \sum_{k=1}^{n} \cos nx \cos nt + \sin nx \sin nt$$

$$= \frac{1}{\pi} \frac{\sin (n + \tfrac{1}{2})(x - t)}{2 \sin \tfrac{1}{2}(x - t)}$$

COROLLARY 12.1.8.

$$\frac{1}{\pi} \int_{0}^{\pi} \frac{\sin (n + \tfrac{1}{2})t \, dt}{\sin \tfrac{1}{2}t} = 1.$$

The function $K_n(x, t)$ is called the *Dirichlet kernel* and stands in the same relation to the orthogonal system of sines and cosines as does the Kernel Polynomial to systems of orthogonal polynomials. The Dirichlet kernel will reproduce, under integration, finite trigonometric sums of the form

$$\sum_{k=0}^{n} A_k \cos kx + B_k \sin kx.$$

DEFINITION 12.1.1.   Let $f(x) \sim \dfrac{a_0}{2} + \displaystyle\sum_{k=0}^{\infty} a_k \cos kx + b_k \sin kx$. Then

$$S_n(f; x) = S_n(x) = \frac{a_0}{2} + \sum_{k=1}^{n} a_k \cos kx + b_k \sin kx.$$

LEMMA 12.1.9.

$$S_n(x) - f(x) = \frac{1}{2\pi} \int_0^{\pi} \frac{f(x + t) + f(x - t) - 2f(x)}{\sin \dfrac{t}{2}} \sin (n + \tfrac{1}{2})t \, dt \tag{12.1.4}$$

Proof:

$$S_n(x) = \int_{-\pi}^{\pi} K_n(x, t) f(t) \, dt = \frac{1}{2\pi} \int_{-\pi}^{\pi} \frac{\sin (n + \tfrac{1}{2})(x - t)}{\sin \tfrac{1}{2}(x - t)} f(t) \, dt.$$

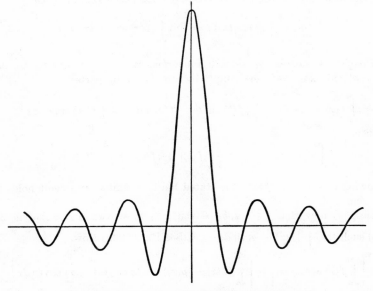

Figure 12.1.2   The Dirichlet Kernel $K_6(x, 0)$.

Set $t = t' + x$.

$$S_n(x) = \frac{1}{2\pi} \int_{-\pi-x}^{\pi-x} f(t' + x) \frac{\sin (n + \frac{1}{2})t'}{\sin \frac{1}{2}t'} \, dt'$$

$$= \frac{1}{2\pi} \int_{-\pi}^{\pi} f(x + t) \frac{\sin (n + \frac{1}{2})t}{\sin \frac{1}{2}t} \, dt,$$

where the f has now been extended by periodicity. Thus,

$$S_n(x) = \frac{1}{2\pi} \int_0^{\pi} f(x + t) \frac{\sin (n + \frac{1}{2})t}{\sin \frac{1}{2}t} \, dt + \frac{1}{2\pi} \int_{-\pi}^0 f(x + t) \frac{\sin (n + \frac{1}{2})t}{\sin \frac{1}{2}t} \, dt.$$

In the last integral, set $t' = -t$, and obtain,

$$S_n(x) = \frac{1}{2\pi} \int_0^{\pi} [f(x + t) + f(x - t)] \frac{\sin (n + \frac{1}{2})t}{\sin \frac{1}{2}t} \, dt. \qquad (12.1.5)$$

Now, from Corollary 12.1.8,

$$f(x) = \frac{1}{2\pi} \int_0^{\pi} 2f(x) \frac{\sin (n + \frac{1}{2})t}{\sin \dfrac{t}{2}} \, dt. \qquad (12.1.6)$$

The lemma follows by subtracting (12.1.6) from (12.1.5).

THEOREM 12.1.10 (Riemann-Lebesgue).   *Let $f(x) \in L[a, b]$. Then*

$$\lim_{t \to \infty} \int_a^b f(x) \cos tx \, dx = \lim_{t \to \infty} \int_a^b f(x) \sin tx \, dx = 0 \qquad (12.1.7)$$

Proof: For sufficiently smooth functions, say those belonging to class $C^1[a, b]$, this is proved very simply by integration by parts:

$$\int_a^b f(x) \cos tx \, dx = \frac{1}{t} \left[f(b) \sin tb - f(a) \sin ta - \int_a^b f'(x) \sin tx \, dx \right].$$

Hence,

$$\left| \int_a^b f(x) \cos tx \, dx \right| \le \frac{1}{t} \left[|f(b)| + |f(a)| + \int_a^b |f'(x)| \, dx \right].$$

Allowing $t \to \infty$, we obtain the stated limit. A similar argument holds for the sine.

Suppose now that $f \in L[a, b]$. Given $\varepsilon > 0$, we can find a polynomial $p(x)$ such that $\int_a^b |f(x) - g(x)| \, dx \le \varepsilon$. (See Ex. 4, 8.9.) Now,

$$\left| \int_a^b f(x) \cos tx \, dx \right| = \left| \int_a^b (f(x) - p(x)) \cos tx \, dx + \int_a^b p(x) \cos tx \, dx \right|$$

$$\le \int_a^b |f(x) - p(x)| \, dx + \left| \int_a^b p(x) \cos tx \, dx \right|.$$

As argued above, $\lim\limits_{t\to\infty} \int_a^b p(x) \cos tx\, dx = 0$, and hence,

$$\limsup_{t\to\infty} \left| \int_a^b f(x) \cos tx\, dx \right| \le \varepsilon.$$

Since ε is arbitrary, $\lim\limits_{t\to\infty} \int_a^b f(x) \cos tx\, dx = 0$.

LEMMA 12.1.11. *Let $f \in L[-\pi, \pi]$. For any δ such that $0 < \delta < \pi$,*

$$S_n(x) - f(x) = \frac{1}{2\pi} \int_0^\delta \frac{f(x+t) + f(x-t) - 2f(x)}{\sin\frac{t}{2}} \sin(n + \tfrac{1}{2})t\, dt$$

$$+ \ \varepsilon_n \ where \ \varepsilon_n \to \infty \ as \ n \to \infty. \qquad (12.1.8)$$

Proof: Write the integral (12.1.4) in the form $\int_0^\delta + \int_\delta^\pi$. Notice that over $[\delta, \pi]$, the integrand is an integrable function and hence by the Riemann-Lebesgue Theorem, $\int_\delta^\pi \to 0$ as $n \to \infty$.

LEMMA 12.1.12. *Let $f(x) \in L[-\pi, \pi]$. We have $\lim\limits_{n\to\infty} S_n(x) - f(x) = 0$ if and only if*

$$\lim_{n\to\infty} \int_0^\delta [f(x+t) + f(x-t) - 2f(x)] \frac{\sin(n+\tfrac{1}{2})t}{t}\, dt = 0. \qquad (12.1.9)$$

Proof: Consider x as fixed and write

$$\varphi(t) = f(x+t) + f(x-t) - 2f(x). \qquad (12.1.10)$$

By Lemma 12.1.11, $S_n(x) - f(x) \to 0$ if and only if

$$\lim_{n\to\infty} \int_0^\delta \varphi(t) \frac{\sin(n+\tfrac{1}{2})t}{\sin\frac{t}{2}}\, dt = 0.$$

Now,

$$2\int_0^\delta \varphi(t) \frac{\sin(n+\tfrac{1}{2})t}{t}\, dt$$

$$= \int_0^\delta \varphi(t) \frac{\sin(n+\tfrac{1}{2})t}{\sin\frac{t}{2}}\, dt + \int_0^\delta \varphi(t) \left(\frac{2}{t} - \frac{1}{\sin\frac{t}{2}}\right) \sin(n+\tfrac{1}{2})t\, dt.$$

Inasmuch as $\dfrac{2}{t} - \dfrac{1}{\sin\dfrac{t}{2}}$ has no singularity at the origin, it is integrable over $[0, \delta]$. Hence, by the Riemann-Lebesgue Theorem, this last integral $\to 0$ as $n \to \infty$, and the lemma follows.

THEOREM 12.1.13 (Riemann's Principle of Localization).
Let $f(x) \in L[-\pi, \pi]$ and have a Fourier Series

$$f(x) \sim \frac{a_0}{2} + \sum_{n=1}^{\infty} a_n \cos nx + b_n \sin nx.$$

The convergence of this series to $f(x)$ at a fixed point x depends only upon the behavior of $f(x)$ in an arbitrarily small neighborhood of x.

Proof: By Lemma 12.1.12, convergence to $f(x)$ depends upon

$$\lim_{n \to \infty} \int_0^\delta [f(x+t) + f(x-t) - 2f(x)] \frac{\sin (n + \frac{1}{2})t}{t} \, dt.$$

Now this integral utilizes the values of $f(x)$ only in the interval $(x - \delta, x + \delta)$.

THEOREM 12.1.14 (Dini's Criterion).   Let x be fixed and suppose that

$$\int_0^\delta \left| \frac{\varphi(t)}{t} \right| dt < \infty. \tag{12.1.11}$$

Then the Fourier series of $f(x)$ converges at x to $f(x)$.

Proof: Under the hypothesis, $\int_0^\delta \frac{\varphi(t)}{t} \sin (n + \frac{1}{2})t \, dt \to 0$ by the Riemann-Lebesgue Theorem. The theorem follows from Lemma 12.1.12.

COROLLARY 12.1.15.   Let $f(x)$ be differentiable at x. Then the Fourier series of $f(x)$ converges at x to $f(x)$.
Proof:

$$\lim_{t \to 0} \frac{f(x+t) - f(x)}{t} = \lim_{t \to 0} \frac{f(x-t) - f(x)}{t} = f'(x).$$

In a neighborhood of x, these two fractions, and consequently their sum, $\frac{\varphi(t)}{t}$, are bounded functions of t. Thus, $\int_0^\delta \left| \frac{\varphi(t)}{t} \right| dt < \infty$.
Actually, still less is required for convergence.

COROLLARY 12.1.16.   Let f satisfy a Lipschitz condition of order $\alpha > 0$ at x. Then the Fourier series of $f(x)$ converges at x to $f(x)$.

Proof: $|f(x+t) - f(x)| \leq Ct^\alpha$; hence $\left| \frac{f(x+t) - f(x)}{t} \right| \leq Ct^{\alpha-1}$. Thus $\left| \frac{\varphi(t)}{t} \right| < 2Ct^{\alpha-1}$, and $\int_0^\delta \left| \frac{\varphi(t)}{t} \right| dt < \infty$.

If f has a simple jump discontinuity at a point, the Fourier Series converges to a value half way between the ends of the jump. To make this precise, we shall suppose that at a point x, the two limits

$$\lim_{t \to 0^+} f(x+t) = f(x^+), \quad \lim_{t \to 0^-} f(x+t) = f(x^-) \qquad (12.1.12)$$

exist. Suppose moreover, that at x, f has both a right-hand derivative

$$\lim_{t \to 0^+} \frac{f(x+t) - f(x^+)}{t} = f_+'(x) \qquad (12.1.13)$$

and a left-hand derivative

$$\lim_{t \to 0^-} \frac{f(x+t) - f(x^-)}{t} = f_-'(x). \qquad (12.1.14)$$

THEOREM 12.1.17.   *Let $f \in L[-\pi, \pi]$ and at the point x satisfy the above conditions. Then the Fourier Series for $f(x)$ converges to the value*

$$\tfrac{1}{2}(f(x^+) + f(x^-)).$$

Proof: As in Lemma 12.1.9, we have

$$S_n(x) - \tfrac{1}{2}[f(x^+) + f(x^-)] = \frac{1}{2\pi} \int_0^\pi \frac{\varphi(t) \sin(n + \tfrac{1}{2})t}{\sin \dfrac{t}{2}} \, dt$$

where we now write

$$\varphi(t) = f(x+t) + f(x-t) - 2[\tfrac{1}{2}(f(x^+) + f(x^-))]. \qquad (12.1.15)$$

A parallel argument shows that Dini's criterion (12.1.11) is valid with this $\varphi(t)$. However, $\left| \dfrac{\varphi(t)}{t} \right| \leq \left| \dfrac{f(x+t) - f(x^+)}{t} \right| + \left| \dfrac{f(x-t) - f(x^-)}{t} \right|$ and in view of (12.1.13) and (12.1.14), we can find a δ sufficiently small so that (12.1.11) holds.

COROLLARY 12.1.18.   *Let $f \in L[-\pi, \pi]$ and be piecewise smooth (each piece is in C^1) in $I = [a, b]$, $-\pi \leq a < b \leq \pi$. Then the Fourier series of f converges at all points of I. Its sum is $f(x)$ at points of continuity and*

$$\tfrac{1}{2}(f(x^+) + f(x^-))$$

at points of discontinuity.

12.2 Fejér's Theory of Fourier Series. The theory of divergent series makes much of a particular mode of summation introduced by E. Cesàro. If an infinite series $\sum_{n=0}^{\infty} a_n$ diverges, its partial sums

$$s_n = a_0 + \cdots + a_n,$$

of course, do not possess a limit. But it is quite possible that the averages of the partial sums $\dfrac{1}{n+1}(s_0 + s_1 + \cdots + s_n)$ have a limit s. In such a case,

the series $\sum\limits_{n=0}^{\infty} a_n$ is said to be $(C, 1)$ *summable* to the value s and we write

$$s = \sum_{n=0}^{\infty} a_n, \quad (C, 1). \tag{12.2.1}$$

Ex. 1. $a_n = (-1)^n$. The series $1 - 1 + 1 - 1 + 1 - \cdots$ is divergent. The partial sums are $1, 0, 1, 0, 1, 0, 1, 0 \cdots$. Their averages are $1, \frac{1}{2}, \frac{2}{3}, \frac{1}{2}, \frac{3}{5}, \frac{1}{2}, \ldots$. The sequence of averages converges to $\frac{1}{2}$. Thus $\frac{1}{2} = 1 - 1 + 1 - 1 + \ldots$, $(C, 1)$.

Ex. 2. $a_n = (-1)^n n$. The series $-1 + 2 - 3 + 4 - 5 + \cdots$ is divergent. The partial sums are $-1, 1, -2, 2, -3, 3, -4, 4 \cdots$. Their averages are -1, $0, -\frac{2}{3}, 0, -\frac{3}{5}, 0, -\frac{4}{7}, 0, \ldots$. This sequence does not converge and hence the series is not summable $(C, 1)$.

As we have already proved (cf. 4.4), if $\sum\limits_{n=0}^{\infty} a_n$ is convergent, then it is $(C, 1)$ summable to the same value.

A family of summability methods of increasing strength is provided by Cesàro summability of rth order $(r > -1)$.

DEFINITION 12.2.1.   Given a series $\sum\limits_{n=0}^{\infty} a_n$. Associate with it the formal power series $f(x) = \sum\limits_{n=0}^{\infty} a_n x^n$ and define constants $s_n^{(r)}$ by means of the formal equation

$$\frac{f(x)}{(1-x)^{r+1}} = \sum_{n=0}^{\infty} s_n^{(r)} x^n . \tag{12.2.2}$$

If

$$\lim_{n \to \infty} \frac{s_n^{(r)}}{\binom{n+r}{n}} = s, \tag{12.2.3}$$

then we shall write

$$s = \sum_{n=0}^{\infty} a_n, (C, r). \tag{12.2.4}$$

In 1904 L. Fejér made the remarkable discovery that the Fourier series of a continuous function $f(x)$ is uniformly $(C, 1)$ summable to $f(x)$. This stands in strong contrast to known fact (duBois-Reymond, 1876, cf. Theorem 14.4.15) that there are continuous functions whose Fourier series are divergent at a point. Corresponding to the $(C, 1)$ sums there is a kernel analogous to the Dirichlet kernel. The latter is oscillating, but Fejér's kernel is positive.

LEMMA 12.2.1. *The following is a trigonometrical identity:*

$$\sin \frac{x}{2} + \sin \frac{3}{2}x + \cdots + \sin (n - \tfrac{1}{2})x = \frac{\sin^2 \dfrac{nx}{2}}{\sin \dfrac{x}{2}}. \qquad (12.2.5)$$

Proof: $\sin (k - \tfrac{1}{2})x \sin \dfrac{x}{2} = \tfrac{1}{2}[\cos (k - 1)x - \cos kx]$. Hence,

$$\sum_{k=1}^{n} \sin (k - \tfrac{1}{2})x \sin \frac{x}{2} = \tfrac{1}{2} \sum_{k=1}^{n} (\cos (k - 1)x - \cos kx)$$

$$= \sin \frac{x}{2} \sum_{k=1}^{n} \sin (k - \tfrac{1}{2})x = \tfrac{1}{2}[1 - \cos nx] = \sin^2 \frac{nx}{2}.$$

Therefore

$$\sum_{k=1}^{n} \sin (k - \tfrac{1}{2})x = \frac{\sin^2 \dfrac{nx}{2}}{\sin \dfrac{x}{2}}.$$

LEMMA 12.2.2. *Let* $S_n(x) = \dfrac{a_0}{2} + \sum_{k=1}^{n} a_k \cos kx + b_k \sin kx$ *be the partial sum of the Fourier series of* $f(x)$. *Let* $\sigma_n(x) = \dfrac{1}{n} (S_0(x) + S_1(x) + \cdots + S_{n-1}(x))$. *Then,*

$$\sigma_n(x) = \frac{1}{2\pi n} \int_0^{\pi} [f(x + t) + f(x - t)] \frac{\sin^2 \dfrac{n}{2} t}{\left(\sin \dfrac{t}{2}\right)^2} \, dt. \qquad (12.2.6)$$

Proof: From (12.1.5), $S_k(x) = \dfrac{1}{2\pi} \int_0^{\pi} [f(x + t) + f(x - t)] \dfrac{\sin (k + \tfrac{1}{2})t}{\sin \dfrac{t}{2}} \, dt.$

Hence,

$$\sigma_n(x) = \frac{1}{2\pi n} \int_0^{\pi} \frac{[f(x + t) + f(x - t)]}{\sin \dfrac{t}{2}} \sum_{k=0}^{n-1} \sin (k + \tfrac{1}{2})t \, dt$$

$$= \frac{1}{2\pi n} \int_0^{\pi} [f(x + t) + f(x - t)] \frac{\sin^2 \dfrac{nt}{2}}{\left(\sin \dfrac{t}{2}\right)^2} \, dt.$$

COROLLARY 12.2.3. $\dfrac{1}{\pi n} \displaystyle\int_0^{\pi} \dfrac{\sin^2 \dfrac{n}{2} t}{\sin^2 \dfrac{t}{2}} \, dt = 1.$

Proof: Take $f(x) \equiv 1$. Then the above integral is $\sigma_n(x)$ for the Fourier series $1 + 0 + 0 + \cdots$. It is readily verified that $\sigma_n(x) \equiv 1$.

COROLLARY 12.2.4. *The Fejér sums $\sigma_n(x)$ are bounded by the bounds of $f(x)$ itself; that is to say, if $m \leq f(x) \leq M$ then $m \leq \sigma_n(x) \leq M$.*

Proof: If $f(x) \leq M$ then $f(x + t) + f(x - t) \leq 2M$ so that

$$\sigma_n(x) \leq \frac{2M}{2\pi n} \int_0^\pi \frac{\sin^2 \dfrac{nt}{2}}{\sin^2 \dfrac{t}{2}}\, dt = M.$$

Similarly for the lower bound.

LEMMA 12.2.5. *For a fixed x set $\varphi(t) = f(x + t) + f(x - t) - 2f(x)$ and*

$$K_n(t) = \frac{1}{2\pi n} \frac{\sin^2 \dfrac{n}{2} t}{\sin^2 \dfrac{t}{2}}. \tag{12.2.7}$$

Then we have

$$\sigma_n(x) - f(x) = \int_0^\pi K_n(t)\varphi(t)\, dt \tag{12.2.8}$$

Proof: By Corollary 12.2.3

$$f(x) = \frac{1}{2\pi n} \int_0^\pi \frac{\sin^2 \dfrac{n}{2} t}{\sin^2 \dfrac{t}{2}} f(x)\, dt. \tag{12.2.9}$$

By subtracting (12.2.9) from (12.2.6) we obtain (12.2.8).

The function $K_n(t) = \dfrac{\sin^2 \dfrac{n}{2} t}{2\pi n \sin^2 \dfrac{t}{2}}$ is known as the *Fejér kernel*. It satisfies not only

$$K_n(t) \geq 0 \tag{12.2.10}$$

but also

$$\int_0^\pi K_n(t)\, dt = \text{constant}, \quad n = 1, 2, \ldots, \quad \text{(Cor. 11.2.3)}, \tag{12.2.11}$$

and

$$\lim_{n \to \infty} M_n(\delta) = 0 \tag{12.2.12}$$

where

$$M_n(\delta) = \max_{\delta \leq t \leq \pi} K_n(t).$$

This last follows from $\dfrac{\sin^2 \dfrac{nt}{2}}{2\pi n \sin^2 \dfrac{t}{2}} \leq \dfrac{1}{2\pi n \left(\sin \dfrac{\delta}{2}\right)^2}$ in $\delta \leq t \leq \pi$.

Continuous functions that satisfy the three conditions (12.2.10, 11, 12) are known as *general Fejér kernels*.

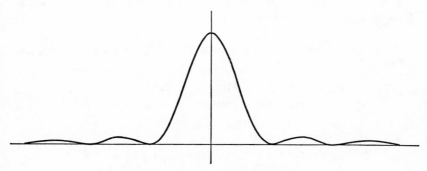

Figure 12.2.1 Fejér's Kernel $K_6(t)$.

LEMMA 12.2.6. $\displaystyle \limsup_{n \to \infty} |\sigma_n(x) - f(x)| \leq \limsup_{n \to \infty} \int_0^\delta K_n(t) |\varphi(t)| \, dt.$

Proof: From (12.2.8),

$$|\sigma_n(x) - f(x)| \leq \int_0^\pi K_n(t) |\varphi(t)| \, dt$$

$$\leq \int_0^\delta K_n(t) |\varphi(t)| \, dt + \int_\delta^\pi K_n(t) |\varphi(t)| \, dt$$

$$\leq \int_0^\delta K_n(t) |\varphi(t)| \, dt + M_n(\delta) \int_\delta^\pi |\varphi(t)| \, dt.$$

Now, in view of (12.2.12), we have

$$\limsup_{n \to \infty} |\sigma_n(x) - f(x)| \leq \limsup_{n \to \infty} \int_0^\delta K_n(t) |\varphi(t)| \, dt.$$

THEOREM 12.2.7. *Let $f(x) \in L[-\pi, \pi]$ and be continuous at the point x. Then*

$$\lim_{n \to \infty} \sigma_n(x) = f(x). \tag{12.2.13}$$

Proof: $\varphi(t) = f(x + t) + f(x - t) - 2f(x)$. If f is continuous at x, given an $\varepsilon > 0$, we can find a $\delta > 0$ such that $|\varphi(t)| < \varepsilon$ for all $0 \leq t \leq \delta$. Now, with this $\delta, \displaystyle \int_0^\delta K_n(t) |\varphi(t)| \, dt \leq \varepsilon \int_0^\delta K_n(t) \, dt \leq \varepsilon \int_0^\pi K_n(t) \, dt = \dfrac{\varepsilon}{2}$. By Lemma

12.2.6, $\lim\limits_{n \to \infty} \sup |\sigma_n(x) - f(x)| \leq \dfrac{\varepsilon}{2}$. But ε is arbitrary and hence

$$\lim_{n \to \infty} \sigma_n(x) = f(x).$$

This conclusion can be strengthened under the assumption that $f(x)$ is continuous in some closed interval $I : [a, b]$.

THEOREM 12.2.8. *Let* $f(x) \in L[-\pi, \pi]$ *and* $\in C[a, b]$ *where* $-\pi \leq a < b \leq \pi$. *Then* $\lim\limits_{n \to \infty} \sigma_n(x) = f(x)$ *uniformly in* $I: [a, b]$.

Proof: $f(x)$ is uniformly continuous in I. Hence, given an ε, we can find a δ such that

$$|\varphi(t)| = |f(x + t) + f(x - t) - 2f(x)|$$
$$\leq |f(x + t) - f(x)| + |f(x - t) - f(x)| \leq \varepsilon$$

for $0 \leq t \leq \delta$ and for all x in I. As before,

$$|\sigma_n(x) - f(x)| \leq \int_0^\delta K_n(t) |\varphi(t)| \, dt + M_n(\delta) \int_\delta^\pi |\varphi(t)| \, dt.$$

Now,

$$\int_\delta^\pi |\varphi(t)| \, dt \leq \int_0^\pi |\varphi(t)| \, dt \leq \int_0^\pi |f(x + t)| \, dt + \int_0^\pi |f(x - t)| \, dt + 2 \int_0^\pi |f(x)| \, dt$$
$$= \int_{-\pi}^\pi |f(x + t)| \, dt + 2\pi |f(x)| = \int_{-\pi}^\pi |f(t)| \, dt + 2\pi |f(x)|.$$

For $x \in I$, f is continuous and hence $|f| \leq M$. Thus,

$$\int_\delta^\pi |\varphi(t)| \, dt \leq \int_{-\pi}^\pi |f(t)| \, dt + 2\pi M = \text{const.}, \qquad x \in I.$$

Finally, for all $x \in I$,

$$|\sigma_n(x) - f(x)| \leq \varepsilon \int_0^\delta K_n(t) \, dt + M_n(\delta) \cdot \text{const.} \leq \frac{\varepsilon}{2} + M_n(\delta) \cdot \text{const.}$$

This bound is independent of x, and hence the convergence is uniform.

COROLLARY 12.2.9. *Let* $f(x)$ *be continuous and periodic in* $[-\pi, \pi]$. *Then it may be approximated uniformly by trigonometric polynomials.*

Proof: The trigonometric polynomials $\sigma_n(x)$ will serve as approximants.

THEOREM 12.2.10. *Let* $f \in L[-\pi, \pi]$ *and suppose that at a point* x, $f(x)$ *has left-hand and right-hand limits,* $f(x^+)$ *and* $f(x^-)$; *then*

$$\lim_{n \to \infty} \sigma_n(x) = \tfrac{1}{2}[f(x^+) + f(x^-)]. \tag{12.2.14}$$

Proof: As before, we can write

$$\sigma_n(x) - \tfrac{1}{2}[f(x^+) + f(x^-)] = \int_0^\pi K_n(t)\varphi(t)\,dt$$

where we now write $\varphi(t) = f(x + t) + f(x - t) - 2[\tfrac{1}{2}(f(x^+) + f(x^-))]$. The proof now proceeds as in Theorem 12.2.7.

12.3 Fourier Series of Periodic Analytic Functions.

We turn from periodic functions of great generality to periodic functions that are also analytic. Naturally, convergence must be uniform (Theorem 12.1.3), but the Fourier series emerges as a transformed version of the Laurent expansion, and convergence takes place over an entire strip of the complex plane.

It will be more convenient to deal with the complex form of the Fourier series and to assume that we are dealing with a complex function that has the complex period p:

$$f(z) = f(z + p). \tag{12.3.1}$$

In the real Fourier series

$$f(x) \sim \frac{a_0}{2} + \sum_{n=1}^{\infty} a_n \cos nx + b_n \sin nx \tag{12.3.2}$$

place $\cos nx = \tfrac{1}{2}(e^{inx} + e^{-inx})$, $\sin nx = \dfrac{i}{2}(-e^{inx} + e^{-inx})$, and obtain

$$f(x) \sim \frac{a_0}{2} + \sum_{n=1}^{\infty} \left(\frac{a_n - ib_n}{2}\right) e^{inx} + \left(\frac{a_n + ib_n}{2}\right) e^{-inx}. \tag{12.3.3}$$

This may be rewritten formally as

$$f(x) \sim \sum_{-\infty}^{+\infty} c_n e^{inx} \tag{12.3.4}$$

where

$$c_0 = \frac{a_0}{2}, \quad c_n = \frac{a_n - ib_n}{2}, \quad c_{-n} = \frac{a_n + ib_n}{2} = \bar{c}_n. \tag{12.3.5}$$

In general, a complex Fourier series with a complex period p is one that can be written in the form

$$f(z) \sim \sum_{-\infty}^{\infty} a_n e^{2\pi i n z/p} \tag{12.3.6}$$

where

$$a_n = \frac{1}{p} \int_{z_0}^{z_0 + p} f(z) e^{-2\pi i n z/p}\, dz.$$

If z_0 and p are real and if $a_n = \bar{a}_{-n}$, then (12.3.6) reduces to a real Fourier series.

Let us suppose that we have a function $f(z)$ defined on a line $L: z_0 + \sigma p$, $-\infty < \sigma < \infty$, lying in the complex plane. $f(z)$ will be assumed to have period p and be analytic on L. We can continue the function $f(z)$ from the line into the complex plane, and in view of periodicity, it is clear that we can find an infinite strip of constant width parallel to L and containing L in which $f(z)$ may be assumed to be both periodic and analytic. There is a largest such strip. In certain instances it may degenerate to a half-plane or the entire plane. The maximum strip of analyticity of $f(z)$ will be called S and can be described in the following way

$$S: \quad -\infty \le t_1 < \mathrm{Im}\left(\frac{z}{p}\right) < t_2 \le \infty. \qquad (12.3.7)$$

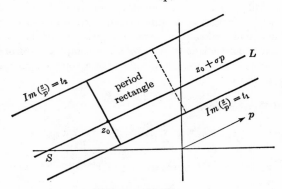

Figure 12.3.1.

THEOREM 12.3.1. *Let $f(z)$ be analytic in S and have period p. Then*

$$f(z) = \sum_{k=-\infty}^{+\infty} a_k e^{2\pi i n z/p} \qquad (12.3.8)$$

where

$$a_k = \frac{1}{p} \int_{z_0}^{z_0+p} f(z) e^{-2\pi i n z/p} \, dz. \qquad (12.3.9)$$

The series (12.3.8) *converges uniformly and absolutely in every substrip* $S': t_1 < t_1' \le \mathrm{Im}\,(z/p) \le t_2' < t_2$.

Proof: Make the change of variable

$$w = e^{(2\pi i z/p)}, \quad z = \frac{p}{2\pi i} \log w. \qquad (12.3.10)$$

This function maps the strip S into the annulus

$$A: \quad e^{-2\pi t_2} < |w| < e^{-2\pi t_1} \qquad (12.3.11)$$

in the w-plane.

As a matter of fact, each period rectangle is mapped into the annulus so that the infinity of points of S, $z \pm mp$, m integer, have the same image in A. In view of the analyticity and periodicity of $f(z)$, the function $g(w) =$

$f\left(\dfrac{p}{2\pi i}\log w\right)$ will be single valued and analytic in the annulus. It therefore has a Laurent expansion

$$g(w) = \sum_{n=-\infty}^{+\infty} b_n w^n \qquad (12.3.12)$$

with

$$b_n = \frac{1}{2\pi i} \int_C \frac{g(w)}{w^{n+1}}\, dw \qquad n = 0, \pm 1, \pm 2, \ldots \qquad (12.3.13)$$

C is any circle $|w| = \rho$ contained in A. In particular, select $\rho = e^{-2\pi \,\mathrm{Im}(z_0/p)}$. The series (12.3.12) converges uniformly and absolutely in any sub-annulus. Passing back to the w variable,

$$f(z) = g(e^{2\pi i z/p}) = \sum_{n=-\infty}^{\infty} b_n e^{2\pi i n z/p}. \qquad (12.3.14)$$

The integrals for b_n become

$$b_n = \frac{1}{2\pi i} \int_{z_0}^{z_0+p} f(z) e^{-(n+1)(2\pi i z/p)} \cdot \frac{2\pi i}{p} e^{2\pi i z/p}\, dz \qquad (12.3.15)$$

This is identical to (12.3.9).

THEOREM 12.3.2. *Let*

$$f(z) = \sum_{n=-\infty}^{\infty} a_n e^{2\pi i n z/p} \qquad (12.3.16)$$

where

$$\lim_{n\to+\infty} \sup \frac{1}{2\pi n} \log |a_n| = A \qquad (12.3.17)$$

$$\lim_{n\to+\infty} \inf \frac{1}{2\pi n} \log |a_{-n}| = B \qquad (12.3.18)$$

and

$$-\infty \le A < B \le \infty.$$

Then $f(z)$ is analytic and periodic in the strip $A < \mathrm{Im}\left(\dfrac{z}{p}\right) < B$, but is analytic in no larger strip $A < \mathrm{Im}\left(\dfrac{z}{p}\right) < B_1$ or $A_1 < \mathrm{Im}\left(\dfrac{z}{p}\right) < B$; $A_1 < A$, $B_1 > B$.

Proof: Consider the series $\displaystyle\sum_{n=0}^{\infty} a_n w^n + \sum_{n=1}^{\infty} a_{-n} w^{-n} \equiv g_1(w) + g_2(w)$. The radius of convergence r_1 of g_1 is given by (Theorem 1.9.4)

$$\frac{1}{r_1} = \lim_{n\to\infty} \sup |a_n|^{1/n} = e^{2\pi A}.$$

The radius of convergence r_2 of g_2 is

$$r_2 = \lim_{n\to\infty} \sup |a_{-n}|^{1/n} = e^{-2\pi B}.$$

Since $A < B$, $g_1 + g_2$ is regular in the annulus $e^{-2\pi B} < |w| < e^{-2\pi A}$ and cannot be continued analytically into a larger annulus over either the inner or outer circle. Applying (12.3.10), $f(z)$ is analytic in the strip

$$A < \text{Im } (z/p) < B,$$

but in no larger strip.

Ex. 1.   Let $0 < r < 1$.

$$\frac{1}{1 - rw} = \sum_{k=0}^{\infty} r^k w^k \quad \text{for} \quad |rw| < 1.$$

Hence

$$\frac{1}{1 - re^{iz}} = \sum_{k=0}^{\infty} r^k e^{ikz} \quad \text{for} \quad |re^{iz}| < 1, \quad \text{i.e., for} \quad \text{Im } z > \log r.$$

Similarly

$$\frac{1}{1 - re^{-iz}} = \sum_{k=0}^{\infty} r^k e^{-ikz} \quad \text{for} \quad \text{Im } z < -\log r.$$

Adding,

$$\frac{1 - r^2}{1 + r^2 - 2r \cos z} = \frac{1}{1 - re^{iz}} + \frac{1}{1 - re^{-iz}} - 1 = 1 + 2 \sum_{k=1}^{\infty} r^k \cos kz$$

convergent uniformly and absolutely for $\log r < \text{Im } z < -\log r$. This is the *Poisson kernel* of Potential Theory.

Ex. 2. $\text{Cot } z = \dfrac{\cos z}{\sin z} = i \cdot \dfrac{e^{2iz} + 1}{e^{2iz} - 1}$. The only singularities are on the real axis, and hence we anticipate two Fourier expansions, one valid in $\text{Im } z > 0$ and the other in $\text{Im } z < 0$. For instance,

$$\tfrac{1}{2}(i \cot z - 1) = \frac{e^{2iz}}{1 - e^{2iz}} = \sum_{k=1}^{\infty} e^{2ikz} \text{ valid for } |e^{2iz}| < 1, \text{ i.e., for } \text{Im } z > 0.$$

12.4 Convergence of the Legendre Series for Analytic Functions.

In this section, P_n designates the Legendre polynomials standardized by $P_n(1) = 1$.

LEMMA 12.4.1.   *Let $w = \tfrac{1}{2}(z + z^{-1})$; then,*

$$P_n(w) = \sum_{j=0}^{n} a_j a_{n-j} z^{n-2j}; \quad a_j = \frac{(2j)!}{(j!)^2 2^{2j}}. \tag{12.4.1}$$

Proof: From the generating identity (10.3.22) we have

$$\frac{1}{\sqrt{1 - 2wt + t^2}} = \sum_{n=0}^{\infty} P_n(w) t^n.$$

Now $(1 - tz)^{-\frac{1}{2}}(1 - t/z)^{-\frac{1}{2}} = (1 - 2wt + t^2)^{-\frac{1}{2}}$. Moreover,

$$(1 - tz)^{-\frac{1}{2}} = \sum_{n=0}^{\infty} a_n t^n z^n, \quad (1 - t/z)^{-\frac{1}{2}} = \sum_{n=0}^{\infty} a_n \frac{t^n}{z^n}. \tag{12.4.2}$$

Hence,

$$(1 - 2wt + t^2)^{-\frac{1}{2}} = \sum_{j,k=0}^{\infty} a_j a_k t^{j+k} z^{k-j} = \sum_{n=0}^{\infty} t^n \sum_{j=0}^{n} a_j a_{n-j} z^{n-2j}.$$

Comparing coefficients, we obtain (12.4.1).

COROLLARY 12.4.2. $|P_n(x)| \leq 1,$ $-1 \leq x \leq 1.$

Proof: Set $z = e^{i\theta}$, then $w = \cos\theta$. Hence, $P_n(\cos\theta) = \sum_{j=0}^{n} a_j a_{n-j} e^{i(n-2j)\theta}$. Therefore, $|P_n(\cos\theta)| \leq \sum_{j=0}^{n} a_j a_{n-j} = P_n(\cos 0) = P_n(1) = 1.$

LEMMA 12.4.3.   *The quantities a_j defined by (12.4.1) are positive and decreasing. The quantities $\dfrac{a_{n-j}}{a_n} - 1$ are positive and increase with j. Moreover,*

$$\lim_{n \to \infty} \sqrt{\pi n} a_n = 1. \tag{12.4.3}$$

$$\lim_{n \to \infty} \frac{a_{n-j}}{a_n} = 1, \quad \text{for } j \text{ fixed,} \tag{12.4.4}$$

$$a_j a_{n-j} \leq a_n, \qquad j = 0, 1, 2, \dots, n. \tag{12.4.5}$$

Proof: (12.4.3) and (12.4.4) are proved by Stirling's formula. The remaining statements are clear from inspecting the formula

$$a_0 = 1, \quad a_n = \frac{1 \cdot 3 \cdots 2n - 1}{2 \cdot 4 \cdots 2n}.$$

LEMMA 12.4.4.   *Let*

$$R_n(z) = \sum_{j=0}^{n} \left(\frac{a_{n-j}}{a_n} - 1\right) a_j z^{-2j}. \tag{12.4.6}$$

Then

$$\lim_{n \to \infty} R_n(z) = 0 \tag{12.4.7}$$

uniformly in the exterior of every circle $|z| \geq \rho > 1$.

Proof: Given an $\varepsilon > 0$. Select $N = N(\varepsilon)$ so large that $\sum_{j=N+1}^{\infty} \rho^{-2j} < \varepsilon$. Now,

$$R_n(z) = \sum_{j=0}^{N} \left(\frac{a_{n-j}}{a_n} - 1\right) a_j z^{-2j} + \sum_{N+1}^{n} \left(\frac{a_{n-j}}{a_n} - 1\right) a_j z^{-2j}.$$

We have

$$\left|\sum_{j=0}^{N} \left(\frac{a_{n-j}}{a_n} - 1\right) a_j z^{-2j}\right| \leq \sum_{j=0}^{N} \left(\frac{a_{n-j}}{a_n} - 1\right) a_j |z^{-2j}|$$

$$\leq \left(\frac{a_{n-N}}{a_n} - 1\right) \sum_{j=0}^{N} a_j |z^{-2j}| \leq \left(\frac{a_{n-N}}{a_n} - 1\right) \sum_{j=0}^{\infty} a_j |z^{-2j}|$$

$$= \left(\frac{a_{n-N}}{a_n} - 1\right) \frac{|z|}{\sqrt{|z|^2 - 1}} \leq \left(\frac{a_{n-N}}{a_n} - 1\right) \frac{\rho}{\sqrt{\rho^2 - 1}}.$$

The second inequality here is due to the fact that $\dfrac{a_{n-j}}{a_n} - 1$ increases with j. The last follows since $r/\sqrt{r^2 - 1}$ is decreasing for $r > 1$. Again, since

$$0 \le \left(\frac{a_{n-j}}{a_n} - 1\right) a_j \le \frac{a_{n-j} a_j}{a_n} \le 1,$$

$$\left| \sum_{N+1}^{n} \left(\frac{a_{n-j}}{a_n} - 1\right) a_j z^{-2j} \right| \le \sum_{N+1}^{n} |z|^{-2j} \le \sum_{N+1}^{\infty} |z|^{-2j}.$$

Thus, for $|z| \ge \rho > 1$,

$$|R_n(z)| \le \left(\frac{a_{n-N}}{a_n} - 1\right) \frac{\rho}{\sqrt{\rho^2 - 1}} + \varepsilon.$$

Now since $\lim\limits_{n \to \infty} \dfrac{a_{n-N}}{a_n} = 1$, we can find an $N_1 = N_1(N, \varepsilon)$ such that $\dfrac{a_{n-N}}{a_n} - 1 \le \varepsilon$ for all $n \ge N_1$. Thus, for all $n \ge N_1$, and for all $|z| \ge \rho > 1$, $|R_n(z)| \le \varepsilon(1 + \rho(\rho^2 - 1)^{-\frac{1}{2}})$. (12.4.7) follows from this inequality.

THEOREM 12.4.5 (Laplace-Heine).   *Let* $w = \frac{1}{2}(z + z^{-1})$. *(Cf. 1.13.) Then,*

$$\lim_{n \to \infty} \frac{\sqrt{\pi n}(1 - z^{-2})^{\frac{1}{2}} P_n(w)}{z^n} = 1, \tag{12.4.8}$$

uniformly in $|z| \ge \rho > 1$, *and*

$$\lim_{n \to \infty} \frac{\sqrt{2\pi n}(w^2 - 1)^{\frac{1}{4}} P_n(w)}{(w + \sqrt{w^2 - 1})^{n+\frac{1}{2}}} = 1 \tag{12.4.9}$$

uniformly in the exterior of any region that contains $-1 \le w \le 1$. *Furthermore,*

$$\lim_{n \to \infty} |P_n(z)|^{1/n} = \rho, \quad z \in \mathscr{E}_\rho, \tag{12.4.10}$$

and uniformly on all $\mathscr{E}_\rho$, $\rho \ge \rho' > 1$.

Proof: $\dfrac{P_n(w)}{a_n z^n} = \sum\limits_{j=0}^{n} \dfrac{a_j a_{n-j}}{a_n} z^{-2j}$. Hence, $\dfrac{P_n(w)}{a_n z^n} - \sum\limits_{j=0}^{n} a_j z^{-2j} = R_n(z)$. By (12.4.2) we have $\sum\limits_{j=0}^{n} a_j z^{-2j} \to (1 - z^{-2})^{-\frac{1}{2}}$ and $R_n(z) \to 0$, both uniformly in any $|z| \ge \rho > 1$. Therefore

$$\frac{P_n(w)}{a_n z^n} \to (1 - z^{-2})^{-\frac{1}{2}}$$

uniformly. Since $\sqrt{n\pi} a_n \to 1$, (12.4.8) follows. Now with appropriate interpretations of the square root, a simple calculation shows that $(1 - z^{-2}) = \dfrac{2\sqrt{w^2 - 1}}{w + \sqrt{w^2 - 1}}$ and (12.4.10) follows from (12.4.8).

Finally,

$$\frac{\sqrt{\pi n}\,|(1 - z^{-2})^{\frac{1}{2}}|\,|P_n(w)|}{|z|^n} = 1 + \varepsilon_n(z) \tag{12.4.11}$$

where $\varepsilon_n(z) \to 0$ uniformly in $|z| \geq \rho > 1$. Thus, also $(1 + \varepsilon_n(z))^{1/n} \to 1$ uniformly there. Extracting the nth root of both sides of (12.4.11) we obtain

$$\lim_{n \to \infty} |P_n(w)|^{1/n} = |z| \tag{12.4.12}$$

uniformly in $|z| \geq \rho > 1$, and this is equivalent to (12.4.10).

LEMMA 12.4.6. *If*

$$Q_n(z) = \tfrac{1}{2} \int_{-1}^{+1} \frac{P_n(t)}{z - t}\,dt, \quad n = 0, 1, 2, \ldots, \tag{12.4.13}$$

and $w = \tfrac{1}{2}(z + z^{-1})$, *then*

$$Q_n(w) = \sum_{k=n+1}^{\infty} \frac{\sigma_{nk}}{z^k} \tag{12.4.14}$$

where

$$|\sigma_{nk}| \leq \pi \qquad n = 0, 1, 2, \ldots; \quad k = n + 1, n + 2, \ldots. \tag{12.4.15}$$

The function $Q_n(z)$ is known as the *Legendre function of the second kind*. It is also a solution of the differential equation (10.3.14) with $\alpha = \beta = 0$, but is linearly independent of $P_n(z)$. The integral (12.4.13) defines $Q_n(z)$ as a single valued function for all z in the plane with $-1 \leq z \leq 1$ deleted. Actually $Q_n(z)$ may be continued analytically to the whole finite plane save at -1, $+1$ as a multivalued function, but we shall not need this continuation.

Proof: From (12.4.13) we have $Q_n(w) = z\displaystyle\int_{-1}^{+1} \frac{P_n(t)}{z^2 - 2zt + 1}\,dt$, and setting

$$t = \cos\theta, \quad Q_n(w) = \tfrac{1}{2}\int_0^{\pi} \frac{P_n(\cos\theta)\sin\theta\,d\theta}{1 - \dfrac{2\cos\theta}{z} + \dfrac{1}{z^2}}.$$

Now

$$\frac{\sin\theta}{z}\left\{\frac{1}{1 - \dfrac{2\cos\theta}{z} + \dfrac{1}{z^2}}\right\} = \frac{1}{2i}\left\{\frac{\dfrac{e^{i\theta}}{z}}{1 - \dfrac{e^{i\theta}}{z}} - \frac{\dfrac{e^{-i\theta}}{z}}{1 - \dfrac{e^{-i\theta}}{z}}\right\}$$

$$= \frac{1}{2i}\sum_{m=1}^{\infty}\left(\frac{e^{i\theta}}{z}\right)^m - \left(\frac{e^{-i\theta}}{z}\right)^m = \sum_{m=1}^{\infty} \frac{\sin m\theta}{z^m}.$$

The last series converges uniformly and absolutely for $0 \leq \theta \leq \pi$ and for all

$$|z| \geq \rho > 1. \text{ Hence, } Q_n(w) = \int_0^\pi P_n(\cos \theta) \sum_{m=1}^\infty \frac{\sin m\theta}{z^m} = \sum_{m=1}^\infty \frac{\sigma_{nm}}{z^m}$$

where

$$\sigma_{nm} = \int_0^\pi P_n(\cos \theta) \sin m\theta \, d\theta. \tag{12.4.16}$$

Since by Corollary 12.4.2, $|P_n(\cos \theta)| \leq 1$, (12.4.15) follows. Again setting $\cos \theta = t$, we have

$$\sigma_{nm} = \int_{-1}^{+1} P_n(t) \frac{\sin (m \cos^{-1} t)}{\sqrt{1 - t^2}} \, dt. \tag{12.4.17}$$

But $\dfrac{\sin (m \cos^{-1} t)}{\sqrt{1 - t^2}} = U_{m-1}(t)$ is a polynomial of degree $m - 1$. Hence, by orthogonality, for $m - 1 < n$, i.e., for $m < n + 1$, $\sigma_{nm} = 0$. This establishes (12.4.14).

THEOREM 12.4.7 (K. Neumann). *Let $f(z)$ be analytic in the interior of $\mathscr{E}_\rho$, $\rho > 1$, but not in the interior of any $\mathscr{E}_{\rho'}$ with $\rho' > \rho$. Then,*

$$f(z) = \sum_{n=0}^\infty a_n P_n(z) \tag{12.4.18}$$

with

$$a_n = \frac{2n + 1}{2} \int_{-1}^{+1} f(x) P_n(x) \, dx \tag{12.4.19}$$

or

$$a_n = \frac{2n + 1}{2^{n+1} n!} \int_{-1}^{+1} f^{(n)}(x)(1 - x^2)^n \, dx. \tag{12.4.20}$$

The series converges absolutely and uniformly on any closed set in the interior of $\mathscr{E}_\rho$. The series diverges exterior to $\mathscr{E}_\rho$. Moreover,

$$\limsup_{n \to \infty} |a_n|^{1/n} = \frac{1}{\rho}. \tag{12.4.21}$$

Conversely, let $\{a_n\}$ be a sequence of constants satisfying (12.4.21) with $\rho > 1$. Then the series (12.4.18) converges uniformly and absolutely on any closed set in the interior of $\mathscr{E}_\rho$ to a function $f(z)$ for which (12.4.18–20) holds. The series diverges outside $\mathscr{E}_\rho$.

Proof: Let $\{a_n\}$ be a sequence of constants for which $\limsup\limits_{n \to \infty} |a_n|^{1/n} = \dfrac{1}{\rho}$, $1 < \rho \leq \infty$. Then by Lemma 4.4.2 and (12.4.10), the series $\sum\limits_{n=0}^\infty a_n P_n(z)$

converges in the interior of $\mathscr{E}_\rho$ to an analytic function $f(z)$. The convergence is uniform inside every $\mathscr{E}_{\rho'}$, $\rho' < \rho$. The series diverges outside $\mathscr{E}_\rho$. We have,

$$\frac{2k+1}{2} \int_{-1}^{+1} f(x) P_k(x)\, dx = \frac{2k+1}{2} \sum_{n=0}^{\infty} a_n \int_{-1}^{+1} P_n(x) P_k(x)\, dx = a_k.$$

This follows from uniform convergence, the orthogonality of the P_n's, and (10.3.13). The alternate expression (12.4.20) is formed from (12.4.19) by using Rodrigues' formula and integrating by parts.

Suppose, conversely, that $f(z)$ is analytic in the interior of $\mathscr{E}_\rho$, $\rho > 1$. We will estimate the size of the constants

$$a_n = \frac{2n+1}{2} \int_{-1}^{+1} f(t) P_n(t)\, dt. \tag{12.4.22}$$

Select a ρ' with $1 < \rho' < \rho$. Then $f(z)$ is analytic in and on $\mathscr{E}_{\rho'}$. Hence we may write for $-1 \le t \le 1$,

$$f(t) = \frac{1}{2\pi i} \int_{\mathscr{E}_{\rho'}} \frac{f(z)}{z-t}\, dz. \tag{12.4.23}$$

Combining the two equations,

$$a_n = \frac{2n+1}{4\pi i} \int_{-1}^{1} P_n(t) \int_{\mathscr{E}_{\rho'}} \frac{f(z)}{z-t}\, dz\, dt$$

$$= \frac{2n+1}{2\pi i} \int_{\mathscr{E}_{\rho'}} f(z) \int_{-1}^{+1} \frac{P_n(t)}{z-t}\, dt\, dz = \frac{2n+1}{2\pi i} \int_{\mathscr{E}_{\rho'}} f(z) Q_n(z)\, dz. \tag{12.4.24}$$

Therefore

$$|a_n| \le \frac{2n+1}{2\pi} L(\mathscr{E}_{\rho'}) \max_{z \in \mathscr{E}_{\rho'}} |f(z)| \max_{z \in \mathscr{E}_{\rho'}} |Q_n(z)| \tag{12.4.25}$$

where $L(\mathscr{E}_{\rho'})$ designates the length of $\mathscr{E}_{\rho'}$. Now from Lemma 12.4.6, $Q_n(z) = \sum_{k=n+1}^{\infty} \frac{\sigma_{nk}}{v^k}$ where $z = \frac{1}{2}(v + v^{-1})$. As z describes the ellipse $\mathscr{E}_{\rho'}$, v describes the circle $|v| = \rho'$. Hence, from inequality (12.4.15),

$$\max_{z \in \mathscr{E}_{\rho'}} |Q_n(z)| \le \sum_{k=n+1}^{\infty} \frac{\pi}{(\rho')^k} = \frac{\pi(\rho')^{-n}}{\rho' - 1}.$$

Combining this with (12.4.25), $|a_n| \le c(2n+1)(\rho')^{-n}$ where c is a constant that depends upon ρ' but not on n. Therefore $\limsup_{n \to \infty} |a_n|^{1/n} \le \frac{1}{\rho'}$. Since $\rho' < \rho$ is arbitrary, it follows that

$$\limsup_{n \to \infty} |a_n|^{1/n} \le \frac{1}{\rho}. \tag{12.4.26}$$

A slight modification of the first part of the proof now tells us that $\sum\limits_{n=0}^{\infty} a_n P_n(z)$ converges absolutely and uniformly in the interior of $\mathscr{E}_\rho$. If now,

$$\limsup_{n\to\infty} |a_n|^{1/n} > \frac{1}{\rho}, \qquad (12.4.27)$$

then $\sum\limits_{n=0}^{\infty} a_n P_n(z)$ would converge uniformly in the interior of a larger ellipse $\mathscr{E}_{\rho''}$, $\rho'' > \rho$, and hence provide an analytic continuation of $f(z)$ there. By hypothesis this is impossible, since $\mathscr{E}_\rho$ is the largest ellipse in which $f(z)$ is analytic. Hence we must have

$$\limsup_{n\to\infty} |a_n|^{1/n} = \frac{1}{\rho}. \qquad (12.4.28)$$

12.5   Complex Orthogonal Expansions.   Let B be a region of the complex plane for which $L^2(B)$ is a Hilbert space. (Cf. Theorem 9.2.10.)

LEMMA 12.5.1. *Let* $\{h_n{}^*(z)\}$ *be a sequence of functions of class* $L^2(B)$ *that are orthonormal with respect to the inner product* $(f, g) = \iint\limits_{B} f\bar{g}\,dx\,dy$. *Then,*

$$\sum_{n=1}^{\infty} |h_n{}^*(z)|^2 \le \frac{1}{\pi r^2(z)}, \qquad (12.5.1)$$

where $z \in B$, *and* $r(z)$ *is the distance from* z *to the boundary of* B.

Proof: For fixed z, $L(f) = f(z)$ is a bounded linear functional over $L^2(B)$ (cf. Ex. 4, 9.3). Consider the problem of minimizing $\|f\|^2$ from among those combinations $f = a_1 h_1{}^* + \cdots + a_n h_n{}^*$ for which $f(z) = 1$. By Theorem 9.4.3, the minimum value is

$$\left(\sum_{i=1}^{n} |h_i{}^*(z)|^2 \right)^{-1}.$$

By (9.2.28), $\pi r^2(z) \cdot 1 \le \|f\|^2 = \left(\sum\limits_{i=1}^{n} |h_i{}^*(z)|^2 \right)^{-1}$, where f is the minimizing function. Then, $\sum\limits_{i=1}^{n} |h_i{}^*(z)|^2 \le \frac{1}{\pi r^2(z)}$. This inequality is independent of n and hence (12.5.1) follows.

THEOREM 12.5.2.   *Let* $\{h_n{}^*(z)\}$ *be a complete orthonormal system for* $L^2(B)$. *If the sequence of constants* $\{a_n\}$ *satisfies*

$$\sum_{n=1}^{\infty} |a_n|^2 < \infty, \qquad (12.5.2)$$

the series

$$f(z) = \sum_{n=1}^{\infty} a_n h_n{}^*(z) \qquad (12.5.3)$$

converges absolutely and uniformly in every closed bounded set contained in B

to a function $f(z)$ of class $L^2(B)$ and for which

$$a_n = (f, h_n^*), \qquad n = 1, 2, \ldots. \tag{12.5.4}$$

Conversely, if $f(z) \in L^2(B)$ and coefficients a_n are defined by (12.5.4), then the series (12.5.3) converges to $f(z)$ absolutely and uniformly in every closed bounded set contained in B.

Proof: Assume (12.5.2). By the Schwarz inequality and (12.5.1),

$$\sum_{k=m}^{n} |a_k h_k^*(z)| \le \left[\sum_{k=m}^{n} |a_k|^2 \right]^{\frac{1}{2}} \left[\sum_{k=m}^{n} |h_k^*(z)|^2 \right]^{\frac{1}{2}} \le \left[\sum_{k=m}^{\infty} |a_k|^2 \right]^{\frac{1}{2}} \cdot \frac{1}{\sqrt{\pi r(z)}} \cdot$$

Let B' designate a closed bounded set contained in B. Then there is a minimum distance r from B' to the boundary of B. Therefore,

$$\sum_{k=m}^{n} |a_k h_k^*(z)| \le \frac{1}{\sqrt{\pi r}} \left[\sum_{k=m}^{\infty} |a_k|^2 \right]^{\frac{1}{2}}$$

throughout B'. In view of (12.5.1), for m sufficiently large the left-hand side can be made arbitrarily small uniformly for all $n > m$ and for all z in B'. Therefore (12.5.3) converges uniformly and absolutely to a function $f(z)$ which must be analytic in B'. Since B' is an arbitrary subregion of B, $f(z)$ must be analytic in B.

We must show that $f(z) \in L^2(B)$. Set $f_n(z) = \sum_{k=1}^{n} a_k h_k^*(z)$. Then

$$\|f_m(z) - f_n(z)\|^2 = \sum_{m}^{n} |a_k|^2.$$

In view of (12.5.1), $\{f_m(z)\}$ is a Cauchy sequence and hence by the completeness of $L^2(B)$ (Lemma 9.2.9) it converges in norm to a function $g(z) \in L^2(B)$. But precisely as in the proof of Lemma 9.2.9, convergence in norm implies uniform convergence on closed bounded subregions of B. Hence $f_n(z) \to g(z)$ on B'. Therefore $f(z) \equiv g(z)$ and is in $L^2(B)$.

Consider $(f, h_k^*) - (f_n, h_k^*) = (f - f_n, h_k^*)$ for $n \ge k$. We have,

$$|(f, h_k^*) - (f_n, h_k^*)|^2 \le \|f - f_n\| \, \|h_k^*\| = \|f - f_n\|.$$

Since $\lim_{n \to \infty} \|f - f_n\| = 0$, $(f_n, h_k^*) \to (f, h_k^*)$. But for $n \ge k$ we have by orthogonality, $(f_n, h_k^*) = a_k$. Therefore, $a_k = (f, h_k^*)$.

Conversely, let $f(z) \in L^2(B)$. Define constants $a_k = (f, h_k^*)$. Since $\{h_k^*\}$ is closed, it follows from Theorem 8.9.1C that $\|f\|^2 = \sum_{k=1}^{\infty} |a_k|^2 < \infty$. Now, apply the first part of the present theorem. The series $\sum_{k=1}^{\infty} a_k h_k^*(z)$ converges uniformly and absolutely to a function, call it $g(z)$, which is in $L^2(B)$ and for which $(g, h_k^*) = a_k = (f, h_k^*)$. Thus $(g - f, h_k^*) = 0$ $k = 1, 2, \ldots$. By Theorem 8.9.1, this implies that $g - f \equiv 0$.

COROLLARY 12.5.3. *Let B be a bounded simply connected region whose boundary is C. Suppose that the complement of $B + C$ is a single region whose boundary is exactly C. Then, any $f \in L^2(B)$ may be expanded in a series of orthogonal polynomials that is uniformly and absolutely convergent in closed subsets of B.*

Proof: Under these assumptions, the powers $1, z, z^2, \ldots$, are complete in $L^2(B)$, and hence there is a complete orthonormal system of polynomials. (Cf. Theorem 11.4.8.)

Note the implication of Theorem 12.5.2. If $f(z)$ is single valued and analytic in the closed region B, it may not necessarily possess a power series expansion that converges to f in B. But it can be expanded in a series of orthogonal functions that converge uniformly in arbitrary bounded subregions.

THEOREM 12.5.4. *Let L be a linear functional defined on $L^2(B)$ and which is applicable term by term to series of analytic functions that converge uniformly in closed bounded subsets of B. Then L is bounded over $L^2(B)$.*

Proof: Let $\{h_n{}^*(z)\}$ be a complete orthonormal sequence for $L^2(B)$. For any $f \in L^2(B)$, $f(z) = \sum\limits_{n=1}^{\infty} (f, h_n{}^*)h_n{}^*(z)$, converging uniformly in every closed bounded subset of B. Hence for all $f \in L^2(B)$, $L(f) = \sum\limits_{n=1}^{\infty} (f, h_n{}^*)L(h_n{}^*)$. The proof is completed by application of Theorem 9.3.8.

COROLLARY 12.5.5. *The following linear functionals are bounded over $L^2(B)$.*

(1) $L(f) = f^{(n)}(z_0)$, $n = 0, 1, \ldots$, *where z_0 is a point of B.*

(2) $L(f) = \displaystyle\int_C w(z) f(z)\, dz$ *where C is a rectifiable arc lying in B and $w(z)$ is continuous.*

(3) $L(f) = \displaystyle\iint_G w(z) f(z)\, dx\, dy$ *where G is a closed region contained in B and $w(z)$ is continuous there.*

Proof: Each of these linear functionals can be applied term by term to uniformly convergent series of analytic functions in B.

EX. 1. Any finite linear combination of linear functionals mentioned above is bounded over $L^2(B)$.

12.6 Reproducing Kernel Functions.

Let S designate a point set lying in the space of one or more real or complex variables. We shall designate points of S by P, Q, . . . , etc. X will denote a complete inner product space of functions defined on S.

DEFINITION 12.6.1.   A function of two variables P and Q in S, $K(P, Q)$, is called *a reproducing kernel function for the space X* if
   (a) For each fixed $Q \in S$, $K(P, Q)$, considered as a function of P is in X.
   (b) For every function $f(P) \in X$ and for every point $Q \in S$, *the reproducing property*

$$f(Q) = (f(P), K(P, Q))_P \tag{12.6.1}$$

holds. The subscript outside the last parenthesis (which will be frequently omitted) indicates that Q is held constant and the inner product is formed on the variable P.

THEOREM 12.6.1 (Aronszajn).   *A necessary and sufficient condition that X have a reproducing kernel function is that for each fixed $Q \in S$, the linear functional*

$$L(f) = f(Q) \tag{12.6.2}$$

be bounded:

$$|L(f)| \le c_Q \|f\| \tag{12.6.3}$$

for some constant c_Q and for all $f \in X$.

Proof: Suppose first that $K(P, Q)$ is a reproducing kernel. Then,

$$f(Q) = (f(P), K(P, Q))_P.$$

By the Schwarz inequality,

$$|f(Q)|^2 \le (f(P), f(P))(K(P, Q), K(P, Q))_P \tag{12.6.4}$$
$$= \|f\|^2 K(Q, Q).$$

The last identity follows by the reproducing property of K. Hence (12.6.3) holds with $c_Q = \sqrt{K(Q, Q)}$.

Suppose, conversely, that (12.6.3) holds for each fixed Q. By Theorem 9.3.3, we can find a function $g(P) = g_Q(P) \in X$ such that

$$f(Q) = L(f) = (f(P), g_Q(P)). \tag{12.6.5}$$

Now set $K(P, Q) = g_Q(P)$ and it is clear that K is a reproducing kernel.

THEOREM 12.6.2.   *If X possesses a reproducing kernel, the kernel is unique.*

Proof: Suppose we have two reproducing kernels $K(P, Q)$, $J(P, Q)$. Let Q be fixed; then

$$\|K(P, Q) - J(P, Q)\|^2 = (K(P, Q) - J(P, Q), K(P, Q) - J(P, Q))_P$$
$$= (K(P, Q) - J(P, Q), K(P, Q))_P - (K(P, Q)$$
$$- J(P, Q), J(P, Q))_P.$$

Since K and J are both reproducing kernels, each of the two inner products is equal to $K(Q, Q) - J(Q, Q)$. Hence $\|K(P, Q) - J(P, Q)\|^2 = 0$ and hence $K(P, Q) = J(P, Q)$ for all P. Since Q is arbitrary, it holds for all P and Q.

THEOREM 12.6.3. *If X possesses a reproducing kernel $K(P, Q)$ then*

$$K(R, Q) = \overline{K(Q, R)}. \tag{12.6.6}$$

Proof: Let P, Q, and R be three points in S. Then, by the reproducing property,

$$(K(P, Q), K(P, R))_P = K(R, Q). \tag{12.6.7}$$

Similarly,

$$(K(P, R), K(P, Q))_P = K(Q, R). \tag{12.6.8}$$

But by the Hermitian symmetry of the inner product, the left-hand sides must be complex conjugate quantities.

If X has a reproducing kernel, then convergence in norm implies pointwise convergence. More precisely,

THEOREM 12.6.4. *Let X have a reproducing kernel and let $\lim_{n \to \infty} \|f - f_n\| = 0$. Then, for each $P \in S$,*

$$\lim_{n \to \infty} f_n(P) = f(P). \tag{12.6.9}$$

The convergence holds uniformly in every subset S' of S over which $K(Q, Q)$ is bounded.

Proof:
$$\begin{aligned} |f(Q) - f_n(Q)|^2 &= |(f(P) - f_n(P), K(P, Q))_P|^2 \\ &\leq \|f - f_n\|^2 (K(P, Q), K(P, Q))_P \\ &= \|f - f_n\|^2 K(Q, Q) \leq \|f - f_n\|^2 M \end{aligned}$$

where M is a bound for $K(Q, Q)$ in S'. The theorem follows by allowing $n \to \infty$.

COROLLARY 12.6.5. *If X has a reproducing kernel, then the Fourier expansion of a function in a complete orthonormal system converges pointwise to the function and uniformly over subsets of S for which $K(Q, Q)$ is bounded.*

Proof: Theorem 8.9.1(b).

If X has a reproducing kernel, the representer of a bounded linear functional has a very simple expression.

THEOREM 12.6.6. *Let X have a reproducing kernel $K(P, Q)$, and let L be a bounded linear functional defined on X. Then the function*

$$h(Q) = \overline{L_P K(P, Q)} \tag{12.6.10}$$

is in X and for all $f \in X$,

$$L(f) = (f(Q), h(Q)). \tag{12.6.11}$$

(The subscript in L_P indicates that Q is held fixed and L is applied to $K(P, Q)$ as a function of P.)

Proof: By Theorem 9.3.3, there is an $h(P) \in X$ such that $L(f) = (f, h)$ for all $f \in X$. Hence,

$$\overline{L_P K(P, Q)} = \overline{(K(P, Q), h(P))_P} = (h(P), K(P, Q))_P = h(Q).$$

COROLLARY 12.6.7.

$$\|L\|^2 = L_Q \overline{L_P K(P, Q)}. \qquad (12.6.12)$$

Proof: From Corollary 9.3.4 we have,

$$\|L\|^2 = (h(Q), h(Q)) = L_Q(\overline{h(Q)}) = L_Q \overline{L_P K(P, Q)}.$$

Ex. 1.   Let $\mathscr{P}_n$ be an inner product space of real or complex polynomials of degree $\leq n$ as in 10.1 or 10.2. The linear functional $L(f) = f(z_0)$ is bounded. (Ex. 10, 10.2.) The kernel polynomial is a reproducing kernel for $\mathscr{P}_n$. (Theorem 10.1.5.)

Ex. 2.   Let B be a region of the complex plane. If $z \in B$, then $L(f) = f(z)$ is a bounded linear functional over $L^2(B)$ (Ex. 4, 9.3). Hence by Lemma 9.2.9 and Theorem 12.6.1, $L^2(B)$ has a reproducing kernel. If $L^2(B)$ is a Hilbert space and if $\{h_n{}^*(z)\}$ is a complete orthonormal sequence, the reproducing kernel has the representation

$$K(z, w) = \sum_{n=1}^{\infty} h_n{}^*(z)\overline{h_n{}^*(w)}. \qquad (12.6.13)$$

$K(z, w)$ is known as the *Bergman kernel* for the region B. This representation rests upon the following theorems.

THEOREM 12.6.8.   *Let $\{h_n{}^*(z)\}$ be a complete orthonormal system in $L^2(B)$. For fixed $w \in B$, the series (12.6.13) converges uniformly and absolutely in every closed bounded subregion B' contained in B. The sum is analytic in $z \in B$ and of class $L^2(B)$. Furthermore, $(K(z, w), h_k{}^*(z))_z = \overline{h_k{}^*(w)}$.*

Proof: By Lemma 12.5.1, $\sum_{n=1}^{\infty} |h_n{}^*(w)|^2 < \infty$. The theorem now follows by applying Theorem 12.5.2.

THEOREM 12.6.9.   *Let $\{h_n{}^*(z)\}$ be a complete orthonormal system for $L^2(B)$. Then (12.6.13) is a (and hence the) reproducing kernel for $L^2(B)$; that is, for all $f(z) \in L^2(B)$ we have*

$$f(w) = (f(z), K(z, w))_z = \iint_B f(z) \, \overline{K(z, w)} \, dx \, dy. \qquad (12.6.14)$$

Proof: Since $f \in L^2(B)$, $f(z) = \sum\limits_{k=1}^{\infty} (f, h_k{}^*) h_k{}^*(z)$. Now from Theorem 12.6.8, $(K(z, w), h_k{}^*(z))_z = \overline{h_k{}^*(w)}$. From Theorem 8.9.1(C′),

$$(f(z), K(z, w))_z = \sum_{k=1}^{\infty} (f, h_k{}^*)(h_k{}^*(z), K(z, w))_z = \sum_{k=1}^{\infty} (f, h_k{}^*) h_k{}^*(w) = f(w).$$

COROLLARY 12.6.10. *Let* $\{h_n{}^*(z)\}$ *be a complete orthonormal sequence for* $L^2(B)$. *Suppose that* L *is a bounded linear functional defined on* $L^2(B)$. *Then*

$$h(w) = \sum_{n=1}^{\infty} h_n{}^*(w) \overline{L(h_n{}^*(z))} \tag{12.6.15}$$

is the representer of L, *and*

$$\|L\|^2 = \sum_{n=1}^{\infty} |L(h_n{}^*(z))|^2 = L_w \overline{L_z K(z, w)}. \tag{12.6.16}$$

Proof: Use Theorems 9.3.5 and 12.6.9.

Ex. 3. Let B designate the circle $|z| < 1$. The functions

$$\sqrt{\frac{n+1}{\pi}} z^n \quad (n = 0, 1, 2, \ldots)$$

are complete and orthonormal in $L^2(B)$. Hence the reproducing kernel for $L^2(B)$ is

$$K(z, w) = \sum_{n=0}^{\infty} \frac{n+1}{\pi} z^n \bar{w}^n = \frac{1}{\pi(1 - z\bar{w})^2}. \tag{12.6.17}$$

Ex. 4. Consider an ellipse $\mathscr{E}_\rho$ (Definition 1.13.1). Designate its inside by $\hat{\mathscr{E}}_\rho$. As we have seen in Ex. 4, 10.2, the polynomials

$$p_n{}^*(z) = 2\sqrt{\frac{n+1}{\pi}} (\rho^{2n+2} - \rho^{-2n-2})^{-\frac{1}{2}} U_n(z),$$

$$U_n(z) = (1 - z^2)^{-\frac{1}{2}} \sin [(n+1) \text{ arc cos } z]$$

are orthonormal for $L^2(\hat{\mathscr{E}}_\rho)$. They are also complete (Theorem 11.4.8). Hence the reproducing kernel for $L^2(\hat{\mathscr{E}}_\rho)$ is

$$K(z, w) = \frac{4}{\pi} \sum_{n=0}^{\infty} (n+1) \frac{U_n(z) \overline{U_n(w)}}{\rho^{2n+2} - \rho^{-2n-2}}. \tag{12.6.18}$$

Ex. 5. Let z_0 be interior to B: $|z| < 1$. If $z_0 \in B$, the linear functional

$$L(f) = f^{(j)}(z_0) \qquad j, \text{ integer} \geq 0, \tag{12.6.19}$$

is bounded over $L^2(B)$. (Ex. 4, 9.3.) Its representer $g(w)$ is given by

$$\overline{g(w)} = \frac{d^{(j)}}{dz^{(j)}} \frac{1}{\pi(1 - z\bar{w})^2} \bigg|_{z = z_0} = \frac{(-1)^j (j+1)! \, \bar{w}^j}{\pi(1 - z_0 \bar{w})^{j+2}}.$$

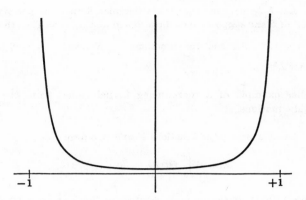

Figure 12.6.1 Kernel Function for Unit Circle, $K(x, x) = \dfrac{1}{\pi(1 - x^2)^2}$.

In particular, the conjugate representer for $L(f) = f(z_0)$ is $\dfrac{1}{\pi(1 - z_0\bar{w})^2}$. Its norm is $\|L\|^2 = \dfrac{1}{\pi(1 - |z_0|^2)^2}$.

It should be observed that the bilinear expression (12.6.13) is not necessarily convergent for an arbitrary Hilbert space of functions. Thus, over $L^2[-\pi, \pi]$ we are led to the formal series

$$\frac{1}{2\pi} + \frac{1}{\pi} \sum_{k=1}^{\infty} \cos kx \cos kt + \sin kx \sin kt.$$

According to Corollary 12.1.7, convergence would be equivalent to the existence and finiteness of $\lim\limits_{n \to \infty} K_n(x, t) = \lim\limits_{n \to \infty} \dfrac{1}{\pi} \dfrac{\sin (n + \frac{1}{2})(x - t)}{2 \sin \frac{1}{2}(x - t)}$. But for any selection of x and t, this limit does not exist.

In a space with a reproducing kernel, complete sets of functions can be generated conveniently from complete sets of functionals.

THEOREM 12.6.11. *Let X be a complete inner product space of functions that has a reproducing kernel $K(P, Q)$. If $\{L_k\}$ is a sequence of bounded linear functionals on X such that $f \in X$, $L_k(f) = 0$, $k = 1, 2, \ldots$, implies $f \equiv 0$, then the functions*

$$h_n(Q) = \overline{L_{n,P}K(P, Q)} \qquad n = 1, 2, \ldots, \tag{12.6.20}$$

form a complete set for X.

Proof: By Theorem 12.6.6, $L_n(f) = (f(Q), h_n(Q))$. Hence $\{h_n(Q)\}$ is a complete sequence of functions, by Definition 8.9.4.

Ex. 6. Let B be a region that has a Bergman kernel function $K(z, w)$. Let z' be a point of B and suppose that z_i lie in B and $\lim\limits_{i \to \infty} z_i = z'$. Then the functions $\overline{K(z_i, w)}$ $i = 1, 2, \ldots$, and the functions $\left.\dfrac{d^n}{dz^n} K(z, w)\right|_{z = z'}$ $n = 0, 1, \ldots$, are complete for $L^2(B)$.

A further example of a reproducing kernel comes from the topic of differential equations.

Ex. 7. Let X consist of all functions $F(x)$ of the form

$$F(x) = \int_0^x f(t)\, dt, \qquad 0 \le x \le 1 \tag{12.6.21}$$

where $f(t) \in L^2[0, 1]$. If $G(x) = \int_0^x g(t)\, dt$, then an inner product (F, G) will be defined by

$$(F, G) = \int_0^1 f(x)g(x)\, dx = \int_0^1 F'(x)G'(x)\, dx. \tag{12.6.22}$$

Now

$$F(x) = \int_0^x F'(t)\, dt = \int_0^1 F'(t)K'(t, x)\, dt, \tag{12.6.23}$$

where

$$\begin{aligned} K'(t, x) &= 1 \qquad 0 \le t \le x \\ K'(t, x) &= 0 \qquad x < t \le 1. \end{aligned} \tag{12.6.24}$$

Hence, if we write

$$\begin{aligned} K(t, x) &= t \qquad 0 \le t \le x \\ K(t, x) &= x \qquad x \le t \le 1, \end{aligned} \tag{12.6.25}$$

we have

$$F(x) = (F(t), K(t, x))_t. \tag{12.6.26}$$

$K(t, x)$ is the reproducing kernel for X.

The reader may recognize $K'(t, x)$ as the *Green's function* for the differential system $y' = 0$, $y(0) = 0$. This is an instance of the following more inclusive situation. Let X denote the space of functions defined on $[a, b]$ that can be written as an n-fold iterated integral

$$F(x) = \int_a^x dx \overset{(n)}{\cdots} \int_a^x f(x)\, dx \tag{12.6.27}$$

with $f(x) \in L^2[a, b]$. Introduce the bilinear differential expression

$$(F, G) = \int_a^b \sum_{j=0}^n a_j(x)F^{(j)}(x)G^{(j)}(x)\, dx \tag{12.6.28}$$

$$a_0(x) > 0,\, a_n(x) > 0;\quad a_i(x) \ge 0, \qquad i = 1, 2, \ldots, n - 1.$$

For $F = G$, this leads to the norm (or energy integral)

$$\|F\|^2 = (F, F) = \int_a^b \sum_{j=0}^n a_j(x)[F^{(j)}(x)]^2\, dx. \tag{12.6.29}$$

The energy integral has associated with it the Euler-Lagrange differential equation

$$\sum_{j=0}^{n} (-1)^j (a_j(x) F^{(j)}(x))^{(j)} = 0. \qquad (12.6.30)$$

Consider also the $2n$ boundary conditions

$$\sum_{j=m}^{n} (-1)^j (a_j(x) F^{(j)}(x))^{(j-m)} = 0 \qquad (12.6.31)$$

at $x = a$ and $x = b$, $m = 1, 2, \ldots, n$.

Then, if $K(x, t)$ is the Green's function for the differential system (12.6.30, 31), it is also the reproducing kernel for the function space X. (Compare also the Peano kernel of 3.7.)

Ex. 8. In Ex. 7, $K(x, x) = x$, and from (12.6.4),

$$|F(x)|^2 \le x \,\|F\|^2 = x \int_0^1 |F'(t)|^2 \, dt.$$

This exhibits the boundedness of the linear functional $L(F) = F(x)$ in this space.

Ex. 9. Let $X = L^2[-\pi, \pi]$. This is not, properly speaking a function space, but a space of equivalence classes in which two functions are equivalent if and only if they differ on a set of at most zero measure. The functional $L(f) = f(x)$, relevant to the definition of a reproducing kernel, is not defined.

However, let us not ignore the most famous reproducing kernel of them all: the *Dirac delta function* $K(x, y) = \delta(x - y)$. The reproducing integral expression

$$f(x) = \int_{-\pi}^{\pi} f(y) \, \delta(x - y) \, dy \qquad (12.6.32)$$

and the (divergent) orthogonal expansion

$$\delta(x - y) = \frac{1}{2\pi} + \frac{1}{\pi} \sum_{k=1}^{\infty} \cos kx \cos ky + \sin kx \sin ky = \frac{1}{2\pi} + \frac{1}{\pi} \sum_{k=1}^{\infty} \cos k(x - y) \qquad (12.6.33)$$

are basic to the theory of the δ function. Though many suggestive things can be done formally, a proper relation between the δ function and reproducing kernels must be sought within a different framework of ideas.

The intimate relationship between reproducing kernels and completeness may be strengthened.

Theorem 12.6.12. *Let $L^2(B)$ be a Hilbert space and let $K(z, w)$ be its reproducing kernel. Let $\{h_n{}^*(z)\}$ be an orthonormal system. This system is complete if and only if*

$$K(z, z) = \sum_{k=1}^{\infty} |h_k{}^*(z)|^2, \qquad z \in B. \qquad (12.6.34)$$

Proof: If $\{h_n{}^*(z)\}$ is complete, (12.6.34) follows directly from Theorem 12.6.8 and (12.6.13). Suppose the system is incomplete. Then by Theorem 9.3.12 we may find nonzero functions $g_n{}^*(z)$ such that the combined system $\{h_n{}^*\}$, $\{g_n{}^*\}$ is complete and orthonormal. By (12.6.13),

$$K(z, z) = \sum_{k=1}^{\infty} |h_k{}^*(z)|^2 + \sum_{k=1}^{\infty} |g_k{}^*(z)|^2.$$

Hence $K(z, z) > \sum_{k=1}^{\infty} |h_k{}^*(z)|^2$ at some point of B.

A similar criterion holds for $L^2[a, b]$.

THEOREM 12.6.13 (Vitali). *The orthonormal sequence $\{f_n{}^*(x)\}$ is complete for $L^2[a, b]$ if and only if*

$$\sum_{n=1}^{\infty} \left(\int_a^t f_n{}^*(x) \, dx \right)^2 = t - a, \qquad a \le t \le b. \tag{12.6.35}$$

Proof: For $S(x) \in L^2[a, b]$, and assuming the sequence complete, we have by Theorem 8.9.1(C), $\int_a^b S^2(x) \, dx = \sum_{n=1}^{\infty} \left(\int_a^b S(x) f_n{}^*(x) \, dx \right)^2$. Fix t in $[a, b]$ and select $S(x) = 1$, $a \le x \le t$, $S(x) = 0$, $t < x \le b$. This yields (12.6.35) immediately.

Conversely, suppose that (12.6.35) holds. As we have seen, this tells us that Parseval's Identity holds for step functions $S(x)$ of the above type. As in the proof of Theorem 8.9.1 (C $\rightarrow$ B), the sequence $\{f_n{}^*(x)\}$ is therefore closed in the set of these S's. By Ex. 4, 11.1, the step functions S_i are closed in $L^2[a, b]$. By the transitivity of closure, (Theorem 11.1.8), $\{f_n{}^*\}$ is closed in $L^2[a, b]$.

The relationship between (12.6.34) and (12.6.35) is this: whereas the former is the kernel identity, the latter is an integrated form of this identity. Starting with the symbolic equation,

$$\delta(x - y) = \sum_{n=1}^{\infty} f_n{}^*(x) f_n{}^*(y)$$

as suggested by (12.6.33), a single integration yields

$$1 = \int_a^u \delta(x - y) \, dy = \sum_{n=1}^{\infty} f_n{}^*(x) \int_a^u f_n{}^*(y) \, dy.$$

A second integration yields

$$u - a = \int_a^u dx = \sum_{n=1}^{\infty} \int_a^u f_n{}^*(x) \, dx \int_a^u f_n{}^*(y) \, dy = \sum_{n=1}^{\infty} \left(\int_a^u f_n{}^*(x) \, dx \right)^2.$$

We conclude this chapter by giving a useful alternative expression for the kernel function for $L^2(B)$ in a case where B is fairly simple.

THEOREM 12.6.14. *Let $w = m(z)$ map a region B $1-1$ conformally onto the unit circle C. Then, the reproducing kernel for $L^2(B)$ is*

$$K(z, t) = \frac{m'(z)\overline{m'(t)}}{\pi(1 - m(z)\overline{m(t)})^2}, \quad z, t \in B. \tag{12.6.36}$$

Proof: Let the map inverse to m be $z = M(w)$. Then, $m(M(w)) \equiv w$ and $m'(M(w))M'(w) \equiv 1$. For any $f, g \in L^2(B)$, we have

$$(f, g) = \iint_B f(z)\overline{g(z)}\, dx\, dy = \iint_C f(M(w))\overline{g(M(w))}\, |M'(w)|^2\, du\, dv,$$
$$(w = u + iv).$$

This follows from (9.2.35) and the method of Lemma 9.2.4. The system of functions $u_n(z) = \sqrt{\dfrac{n+1}{\pi}}\,(m(z))^n m'(z)$, $n = 0, 1, \ldots$, is complete and orthonormal for $L^2(B)$. For

$$\sqrt{\frac{\pi}{n+1}}\,(f, u_n) = \iint_B f(z)\overline{(m(z))^n m'(z)}\, dx\, dy$$

$$= \iint_C f(M(w))\overline{w^n m'(M(w))}\, |M'(w)|^2\, du\, dv$$

$$= \iint_C f(M(w))M'(w)\overline{w^n}\, du\, dv.$$

Hence $(f, u_n) = 0$ $n = 0, 1, \ldots$, implies, by the completeness of $1, w, w^2$, $\ldots$, over $L^2(C)$, that $f(M(w))M'(w) \equiv 0$. Since $M' \neq 0, f \equiv 0$. Therefore $\{u_n\}$ is complete for $L^2(B)$. Similarly for orthonormality.

By Theorem 12.6.9,

$$K(z, t) = \frac{1}{\pi} \sum_{n=0}^{\infty} (n+1)(m(z))^n m'(z)\overline{(m(t))^n m'(t)}.$$

In view of the identity $(1 - x)^{-2} = \sum_{n=0}^{\infty} (n+1)x^n$, we obtain (12.6.36).

COROLLARY 12.6.15. *Let $w = m(z)$ map B onto the unit circle. The mapping is normalized by requiring that $m(t) = 0, m'(t) > 0$ where t is a fixed point in B. Then,*

$$m(z) = \text{const} \cdot \int_t^z K(z, t)\, dz. \tag{12.6.37}$$

Proof: Since $m(t) = 0$, $K(z, t) = \dfrac{1}{\pi} m'(z)m'(t)$. Hence

$$m(z) = \frac{\pi}{m'(t)} \int_t^z K(z, t)\, dz.$$

If B has a sufficiently smooth boundary, (see Theorem 11.4.8), there is a complete orthonormal set of polynomials in $L^2(B)$, and $K(z, t)$ can, in principle, be computed directly by (12.6.13). Equation (12.6.37) then gives an elementary representation for the mapping function.

Ex. 10.   Let C designate the unit circle $|z| < 1$ and C_S be C with the radius $0 \leq x < 1$ deleted. Note that C_S does not fulfill the condition of Theorem 11.4.8. The powers $1, z, z^2, \ldots$, cannot be complete in $L^2(C_S)$. For assume they were. The removal of $0 \leq x \leq 1$ does not affect the double integrals. Hence,

$$\sqrt{\frac{n+1}{\pi}} z^n \quad n = 0, 1, \ldots, \text{ is a complete orthonormal system for both } L^2(C)$$

and $L^2(C_S)$. This implies that $K_C(z, t) = K_{C_S}(z, t)$. By (12.6.37), this is absurd.

NOTES ON CHAPTER XII

12.1–12.2   Zygmund [1] is an exhaustive treatise on this topic.

12.3   See, e.g., Behnke and Sommer [1] pp. 249–256.

12.4   Jackson [2] pp. 63–68. Szegö [1], Chapter 9; Hobson [1], Chapter 2; Rainville [1], Chapter 10.

12.5   Bergman [2]. Nehari [1].

12.6   Bergman [2]. Aronszajn [2]. Bergman and Schiffer [1], Part B, Chapter II. Example 5: Golomb and Weinberger [1]. For further criteria as in Theorem 12.6.12, see Tricomi [1] pp. 26–30.

PROBLEMS

1. How do the symmetries (1) $f(-x) = f(x)$, (2) $f(-x) = -f(x)$, (3) $f(x + \pi) = f(x)$, (4) $f(x + \pi) = -f(x)$ influence the structure of the Fourier coefficients of $f(x)$?

2. What is the Fourier series of a trigonometric polynomial?

3. What is the Fourier series for $\cos^n x$?

4. $\text{Sgn} \sin kt = \frac{4}{\pi} \sum_{j=0}^{\infty} \frac{\sin(2j+1)kt}{2j+1}$ $\quad k = 1, 2, \ldots$. Discuss the convergence.

5. $|\sin x| = \frac{8}{\pi} \sum_{n=1}^{\infty} \frac{\sin^2 nx}{4n^2 - 1}$.

6. Let $0 < h < \pi$ and $f(x) = 1$ for $|x| \leq h, f(x) = 0$ for $h < |x| \leq \pi$.

$$f(x) \sim \frac{2h}{\pi} \left[\frac{1}{2} + \sum_{k=1}^{\infty} \frac{\sin kh}{kh} \cos kx \right].$$

Convergence?

7. Discuss the convergence of $\sum_{n=1}^{\infty} r^n \sin nx = \frac{r \sin x}{1 - 2r \cos x + r^2}$.

8. Prove that for $|r| < 1$, $\dfrac{1 - r \cos z}{1 - 2r \cos z + r^2} = 1 + r \cos z + r^2 \cos 2z + \cdots$.

The expansion is convergent in an infinite strip parallel to the x axis.

9. Prove that arc tan $\dfrac{r \sin z}{1 - r \cos z} = r \sin z + \tfrac{1}{2} r^2 \sin 2z + \cdots$, and discuss the convergence.

10. Set $f(x) = e^{zx}$ in Parseval's Theorem on $[0, 2\pi]$ and derive the identity

$$\pi \frac{e^{2\pi z} + 1}{e^{2\pi z} - 1} = \frac{1}{z} + 2z \sum_{n=1}^{\infty} \frac{1}{z^2 + n^2}.$$

11. Prove that $\dfrac{1}{p} \displaystyle\int_{z_0}^{z_0+p} |f(z)|^2 \, dz = \sum_{n=-\infty}^{\infty} |a_n|^2 \, e^{-4\pi n \, \mathrm{Im}\,(z_0/p)}$.

12. Use Theorem 12.2.7 to prove that if $f(x)$ is continuous at x and if the Fourier series of $f(x)$ converges at x, it converges to $f(x)$.

13. Is there a $(C, 1)$ sum of the formal kernel

$$\frac{1}{2\pi} + \frac{1}{\pi} \sum_{k=1}^{\infty} \cos nx \cos nt + \sin nx \sin nt \ ?$$

14.
$$|x| \sim \sum_{k=0}^{\infty} a_n P_n(x)$$

where $a_n = 0$, n odd, $a_0 = \tfrac{1}{2}$, $a_2 = \tfrac{5}{8}$

$$a_n = \frac{2n + 1}{(n - 1)(n + 2)} \, P_n(0) \quad n = 3, 4, \ldots.$$

15. If $f(x) = 1$, $0 \le x \le 1$, $f(x) = -1$, $-1 \le x < 0$, then

$$f(x) \sim \angle \sum_{k=0}^{\infty} \frac{4k + 3}{2k + 1} \binom{-\tfrac{1}{2}}{k + 1} P_{2k+1}(x).$$

16. Let $0 < a < 1$. Show by Rodrigues' Formula and integration by parts that

$$\frac{1}{(1 - x)^a} \sim \sum_{n=0}^{\infty} a_n P_n(x)$$

where

$$a_0 = \frac{1}{2^a(1 - a)}, \, a_1 = \frac{3a}{2^a(1 - a)(2 - a)}, \, a_n = \frac{(2n + 1)a(a + 1) \cdots (a + n - 1)}{2^a(1 - a)(2 - a) \cdots (n + 1 - a)}.$$

Discuss convergence.

17. $\dfrac{1 - h^2}{(1 - 2hz + h^2)^{\frac{3}{2}}} = \displaystyle\sum_{n=0}^{\infty} (2n + 1)h^n P_n(z)$. Discuss convergence.

18. Use Theorem 12.4.7 to conclude that if $f \in C^{\infty}[-1, 1]$ and if the derivatives are "not too large," then f can be continued analytically into the complex plane.

19. The Bergman kernel for $|z| < r$ is $\dfrac{r^2}{\pi(r^2 - z\bar{w})^2}$.

20. Verify by Vitali's Theorem that $\cos nx$, $n = 0, 1, 2, \ldots$, is complete for $L^2[0, \pi]$.

21. Apply Vitali's Theorem to $P_n(x)$ and obtain an identity for the Legendre polynomials.

CHAPTER XIII

Degree of Approximation

13.1 The Measure of Best Approximation. Let X be a normed linear space and $\{x_i\}$ a sequence of independent elements. For an arbitrary $y \in X$, there is a best approximation to y from among the linear combinations of $x_1, x_2, \ldots, x_n$, and we have already defined the measure of best approximation by means of

$$E_n(y) = \min_{a_i} \left\| y - \sum_{i=1}^{n} a_i x_i \right\|. \tag{13.1.1}$$

(See (7.4.11).) We know that $E_1 \geq E_2 \geq \cdots \geq 0$. The limit is 0 if and only if y can be approximated arbitrarily closely by combinations of x_1, $x_2, \ldots$. Hence we have (cf. Theorem 7.6.1)

THEOREM 13.1.1. $\lim_{n \to \infty} E_n(y) = 0$ *for all* $y \in X$ *if and only if* $\{x_i\}$ *is closed in* X.

In the present chapter, the rapidity with which $E_n(y)$ approaches zero will be studied.

Ex. 1. The quantities $E_n(y)$ are computed easily when X is an inner product space. Assuming, as we may, that we have orthonormalized the $\{x_i\}$ to yield $\{x_i{}^*\}$, then by Corollary 8.5.3,

$$E_n{}^2(y) = \|y\|^2 - \sum_{k=1}^{n} |(y, x_k{}^*)|^2.$$

If X is a Hilbert space and $\{x_i{}^*\}$ is complete and orthonormal, then by Theorem 8.9.1 (C),

$$E_n{}^2(y) = \sum_{k=n+1}^{\infty} |(y, x_k{}^*)|^2.$$

Note that in the Hilbert space situation, $E_n(y)$ may approach zero with arbitrary rapidity. For let $\varepsilon_1, \varepsilon_2$ be an arbitrary nonincreasing sequence of nonnegative numbers with $\lim_{n \to \infty} \varepsilon_n = 0$. Set

$$a_1 = \sqrt{\varepsilon_1{}^2 - \varepsilon_2{}^2},\, a_2 = \sqrt{\varepsilon_2{}^2 - \varepsilon_3{}^2},\, \ldots.$$

Then,

$$\sum_{i=1}^{\infty} a_i{}^2 = (\varepsilon_1{}^2 - \varepsilon_2{}^2) + (\varepsilon_2{}^2 - \varepsilon_3{}^2) + \cdots = \varepsilon_1{}^2 < \infty.$$

Hence, there is an element y with Fourier coefficients $\{a_i\}$. Now,

$$E_n^2(y) = \sum_{k=n+1}^{\infty} (\varepsilon_k^2 - \varepsilon_{k+1}^2) = \varepsilon_{n+1}^2.$$

The arbitrary approach to zero holds in a Banach space as well. The proof is somewhat more involved and preliminary work is necessary.

THEOREM 13.1.2. *The quantities $E_n(y)$ have the following properties*

$$E_n\left(y + \sum_{i=1}^{n} b_i x_i\right) = E_n(y). \tag{13.1.2}$$

$$E_n(y + z) \le E_n(y) + E_n(z) \tag{13.1.3}$$

$$E_n(\sigma y) = |\sigma|\, E_n(y). \tag{13.1.4}$$

$$E_n(y - z) \ge |E_n(y) - E_n(z)| \tag{13.1.5}$$

$E_n(y + \sigma z)$ *is a continuous function of σ. If σ is real, it is a convex function. If $E_n(z) > 0$, $\lim\limits_{|\sigma| \to \infty} E_n(y + \sigma z) = \infty$.* $\tag{13.1.6}$

Proof:

(2). Let $E_n(y) = \left\| y - \sum\limits_{i=1}^{n} a_i x_i \right\|$.

$$E_n\left(y + \sum_{i=1}^{n} b_i x_i\right) = \min_{c_i} \left\| y + \sum_{i=1}^{n} b_i x_i - \sum_{i=1}^{n} c_i x_i \right\|$$

$$= \min_{c_i} \left\| y - \sum_{i=1}^{n} (c_i - b_i) x_i \right\| = E_n(y).$$

(3). If $E_n(y) = \left\| y - \sum\limits_{i=1}^{n} b_i x_i \right\|$ and $E_n(z) = \left\| z - \sum\limits_{i=1}^{n} c_i x_i \right\|$, then by the triangle inequality for norms, $\left\| (y + z) - \sum\limits_{i=1}^{n} (b_i + c_i) x_i \right\| \le E_n(y) + E_n(z)$. But $E_n(y + z) = \min\limits_{a_i} \left\| (y + z) - \sum\limits_{i=1}^{n} a_i x_i \right\| \le \left\| (y + z) - \sum\limits_{i=1}^{n} (b_i + c_i) x_i \right\|$.

(4). Let $E_n(y) = \left\| y - \sum\limits_{i=1}^{n} a_i x_i \right\|$. Then,

$$|\sigma|\, E_n(y) = \left\| \sigma y - \sum_{i=1}^{n} \sigma a_i x_i \right\| \ge E_n(\sigma y).$$

It follows that if $\sigma \ne 0$, $E_n(y) = E_n\left(\dfrac{\sigma y}{\sigma}\right) \le \dfrac{1}{|\sigma|} E_n(\sigma y)$. Hence $E_n(\sigma y) \ge |\sigma|\, E_n(y)$. Combining these yields (13.1.4). If $\sigma = 0$ the statement is trivial.

(5). $E_n(y) = E_n(z + y - z) \le E_n(z) + E_n(y - z)$. Hence, $E_n(y - z) \ge E_n(y) - E_n(z)$. Similarly, $E_n(y - z) = E_n(z - y) \ge E_n(z) - E_n(y)$.

(6). By (13.1.5),

$$|E_n(y + \sigma_1 z) - E_n(y + \sigma_2 z)| \leq E_n((\sigma_1 - \sigma_2)z) = |\sigma_1 - \sigma_2| E_n(z).$$

Continuity follows from this. Let $0 \leq t \leq 1$. Then

$$E_n(y + (t\sigma_1 + (1 - t)\sigma_2)z) = E_n(t(y + \sigma_1 z) + (1 - t)(y + \sigma_2 z))$$
$$\leq tE_n(y + \sigma_1 z) + (1 - t)E_n(y + \sigma_2 z),$$

and this implies convexity. Finally,

$$E_n(y + \sigma z) \geq E_n(\sigma z) - E_n(y) = |\sigma| E_n(z) - E_n(y).$$

If $\qquad\qquad E_n(z) \neq 0, \lim_{|\sigma| \to \infty} E_n(y + \sigma z) = \infty.$

LEMMA 13.1.3.   *Let X be a normed linear space and $x_1, x_2, \ldots,$ a sequence of independent elements. For arbitrary $x \in X$ write*

$$E_j(x) = \min_{a_i} \left\| x - \sum_{i=1}^{j} a_i x_i \right\|. \qquad (13.1.7)$$

If e is a number that satisfies $e \geq E_{j+1}(x)$, we can find a σ such that

$$E_j(x + \sigma x_{j+1}) = e. \qquad (13.1.8)$$

Proof: Suppose that $E_{j+1}(x) = \left\| x - \sum_{i=1}^{j+1} a_i x_i \right\|.$ Then,

$$E_j(x - a_{j+1}x_{j+1}) = \min_{b_i} \left\| x - a_{j+1}x_{j+1} - \sum_{i=1}^{j} b_i x_i \right\| \leq \left\| x - \sum_{i=1}^{j+1} a_i x_i \right\| = E_{j+1}(x)$$

$$= E_{j+1}(x - a_{j+1}x_{j+1}) \leq (\text{by } (7.4.12)) \, E_j(x - a_{j+1}x_{j+1}).$$

Hence $E_j(x - a_{j+1}x_{j+1}) = E_{j+1}(x)$. Now $E_j(x_{j+1}) = \min_{a_i} \left\| x_{j+1} - \sum_{i=1}^{j} a_i x_i \right\| > 0,$ otherwise the x's would be dependent. Hence, by (13.1.6) we can find a σ' such that $E_j(x + \sigma' x_{j+1}) > e$. Thus, $E_j(x - a_{j+1}x_{j+1}) \leq e < E_j(x + \sigma' x_{j+1}).$ By the continuity property (13.1.6), the proper σ can be found.

THEOREM 13.1.4.   *Let X be a normed linear space and $x_1, x_2, \ldots, x_n, x_{n+1}$ a sequence of independent elements. If $e_1 \geq e_2 \geq \cdots \geq e_n \geq 0$ are prescribed, we can find a linear combination*

$$y = b_1 x_1 + b_2 x_2 + \cdots + b_n x_n + b_{n+1} x_{n+1}$$

such that

$$E_k(y) = e_k \qquad k = 1, 2, \ldots, n \qquad (13.1.9)$$

and

$$\|y\| = e_1. \qquad (13.1.10)$$

Proof: Since $E_{n+1}(0) = 0$, there is, by Lemma 13.1.3, a constant b_{n+1} such that

$$E_n(0 + b_{n+1}x_{n+1}) = e_n.$$

Since $e_{n-1} \geq e_n$, there is, by the Lemma, a b_n such that

$$E_{n-1}(b_n x_n + b_{n+1}x_{n+1}) = e_{n-1}.$$

By (13.1.2), $E_n(b_n x_n + b_{n+1}x_{n+1}) = E_n(b_{n+1}x_{n+1}) = e_n$. Since $e_{n-2} \geq e_{n-1}$, there is a b_{n-1} such that

$$E_{n-2}(b_{n-1}x_{n-1} + b_n x_n + b_{n+1}x_{n+1}) = e_{n-2}.$$

Then, $E_{n-1}(b_{n-1}x_{n-1} + b_n x_n + b_{n+1}x_{n+1}) = E_{n-1}(b_n x_n + b_{n+1}x_{n+1}) = e_{n-1}$, and

$$E_n(b_{n-1}x_{n-1} + b_n x_n + b_{n+1}x_{n+1}) = E_n(b_{n+1}x_{n+1}) = e_n.$$

Continuing in this way, we can find an element $z = b_2 x_2 + \cdots + b_{n+1}x_{n+1}$ such that

$$E_n(z) = e_n, E_{n-1}(z) = e_{n-1}, \ldots, E_1(z) = e_1.$$

Now $e_1 = E_1(z) = \min_b \|b_2 x_2 + \cdots + b_{n+1}x_{n+1} - bx_1\|$. Select a b that yields the minimum value. Call it $-b_1$. Then if $y = b_1 x_1 + b_2 x_2 + \cdots + b_{n+1}x_{n+1}$, conditions (13.1.9–10) hold.

LEMMA 13.1.5. *Let K be a bounded subset of a normed linear space X. Suppose that X has finite dimension N. If $\{y_k\}$ is an arbitrary sequence of elements of K, we may find a subsequence $y_{k_1}, y_{k_2}, \ldots,$ that converges to an element $y \in X$: $\lim_{n \to \infty} \|y - y_{k_n}\| = 0$. If K is closed, $y \in K$, and K is said to be sequentially compact.*

Proof: Let $x_1, x_2, \ldots, x_N$ be a set of independent elements of X. Then the elements x of K satisfy $\|x\| = \|a_1 x_1 + \cdots + a_N x_N\| \leq M$ for some M. Employing the notation of the proof of Theorem 7.4.1, we have from (7.4.6), $M \geq \|x\| \geq mr$, $r = (|a_1|^2 + \cdots + |a_N|^2)^{\frac{1}{2}}$, for every $x \in K$. The coefficients of elements of K satisfy $(|a_1|^2 + \cdots + |a_N|^2)^{\frac{1}{2}} \leq M/m$, and in particular, $|a_j| \leq M/m$. If $y_k = \sum_{i=1}^{N} a_i^{(k)} x_i$, then $|a_j^{(k)}| \leq M/m, j = 1, \ldots, N$ and the points P_k: $(a_1^{(k)}, a_2^{(k)}, \ldots, a_N^{(k)})$ lie in a bounded portion of R_N or C_N. By the Bolzano-Weierstrass Theorem, we can find a convergent subsequence $\{P_{k_n}\}$ of $\{P_k\}$ with $\lim_{n \to \infty} P_{k_n} = P$. Let P: $(b_1, \ldots, b_N)$. We have $\lim_{n \to \infty} a_j^{(k_n)} = b_j$ for $j = 1, \ldots, N$. Set $y = b_1 x_1 + \cdots + b_N x_N$. Then,

$$\|y - y_{k_n}\| \leq \sum_{j=1}^{N} |b_j - a_j^{(k_n)}| \|x_j\| \quad \text{and so} \quad \lim_{n \to \infty} \|y - y_{k_n}\| = 0.$$

If K is closed, it contains all its limit points and hence $y \in K$.

THEOREM 13.1.6 (Bernstein). *Let X be a complete normed linear space. Let $\{x_i\}$ be a sequence of independent elements. Given an arbitrary sequence $e_1 \geq e_2 \geq \cdots > 0$, $\lim\limits_{n \to \infty} e_n = 0$, we can find a $y \in X$ such that*

$$E_n(y) = e_n, \qquad n = 1, 2, \ldots. \tag{13.1.11}$$

Proof: For each n, use Theorem 13.1.4 to determine a linear combination $y_n = \sum\limits_{j=1}^{n+1} b_j^{(n)} x_j$ such that $E_k(y_n) = e_k$, $k = 1, 2, \ldots, n$ and $\|y_n\| = e_1$. We will show that a suitable subsequence of $\{y_n\}$ converges to the required y.

For each k, $k = 1, 2, \ldots$, let $z_{n,k} = \sum\limits_{j=1}^{k} b_j^{(n,k)} x_j$ be a linear combination of $x_1, \ldots, x_k$ yielding a best approximation to y_n. Then, $\|y_n - z_{n,k}\| \leq \|y_n\|$. Hence, $\|z_{n,k}\| \leq \|y_n\| + \|z_{n,k} - y_n\| \leq \|y_n\| + \|y_n\| = 2e_1$. Let k be fixed. The set of elements $z_{n,k}$ is bounded. By Lemma 13.1.5, we can find a subsequence $z_{n_p,k}$ converging to an element

$$w_k = \sum_{j=1}^{k} c_{j,k} x_j \quad \text{i.e.,} \quad \lim_{p \to \infty} \|w_k - z_{n_p,k}\| = 0.$$

For sufficiently large p (depending upon k), $\|w_k - z_{n_p,k}\| < e_k$. Consequently,

$$\|w_k - y_{n_p}\| \leq \|w_k - z_{n_p,k}\| + \|z_{n_p,k} - y_{n_p}\| < e_k + e_k = 2e_k.$$

Again, for sufficiently large p and q,

$$\|y_{n_p} - y_{n_q}\| \leq \|w_k - y_{n_p}\| + \|w_k - y_{n_q}\| \leq 2e_k + 2e_k = 4e_k.$$

Now $\lim\limits_{k \to \infty} e_k = 0$; hence $\{y_{n_p}\}$ is a Cauchy sequence. Since X is complete, $\{y_{n_p}\}$ converges to an element $y \in X$. For sufficiently large p,

$$e_k = E_k(y_{n_p}) = \|y_{n_p} - z_{n_p,k}\|.$$

By the continuity of the norm, $e_k = \|y - w_k\| = \left\| y - \sum\limits_{j=1}^{k} c_{j,k} x_j \right\|$. Thus, for each k, $E_k(y) \leq e_k$. Now if there were a linear combination $\sum\limits_{j=1}^{k} d_j x_j$ for which $\left\| y - \sum\limits_{j=1}^{k} d_j x_j \right\| < e_k$, then for sufficiently large p, $\left\| y_{n_p} - \sum\limits_{j=1}^{k} d_j x_j \right\| < e_k$. However for $n_p > k$, $E_k(y_{n_p}) = e_k$ and this is a contradiction. Therefore, $E_k(y) = e_k$, $k = 1, 2, \ldots$.

COROLLARY 13.1.7. *Let $E_n(f) = \min\limits_{a_i} \max\limits_{a \leq x \leq b} \left| f(x) - \sum\limits_{i=0}^{n} a_i x^i \right|$. Then given $e_0 \geq e_1 \geq \cdots > 0$, $\lim\limits_{n \to \infty} e_n = 0$, we can find an $f \in C[a, b]$ with $E_n(f) = e_n$ $n = 0, 1, \ldots$.*

This theorem tells us that in order to obtain nontrivial asymptotic theorems for $E_n(x)$, we must operate in a normed linear space that is not complete. As an example, when $f \in C[a, b]$ and polynomial approximation is considered, $E_n(f)$ can go to zero with arbitrary rapidity. This is not the case for $f \in C^1[a, b]$. As subspaces of smoother and smoother functions are considered, $E_n(f)$ goes to zero more and more rapidly. This is a phenomenon that pervades the theory of the asymptotics of approximation.

13.2   Degree of Approximation with Some Square Norms

THEOREM 13.2.1.   *Let B be the unit circle, and set*

$$E_n{}^2(f) = \min_{a_i} \iint\limits_B |f(z) - a_0 - a_1 z - \cdots - a_n z^n|^2 \, dx \, dy, \qquad (13.2.1)$$

$$f \in L^2(B).$$

If f is analytic in $|z| < \rho$, $\rho \geq 1$, but not in $|z| < \rho'$ with $\rho' > \rho$, then

$$\lim_{n \to \infty} \sup \, [E_n(f)]^{1/n} = 1/\rho. \qquad (13.2.2)$$

Proof: Let $f(z)$ have the Taylor expansion $f(z) = \sum\limits_{k=0}^{\infty} a_k z^k$ where

$$\lim_{n \to \infty} \sup \, |a_k|^{1/k} = 1/\rho.$$

Since $\left\{\sqrt{\dfrac{n+1}{\pi}} \, z^n\right\}$ are orthonormal in $L^2(B)$,

$$f(z) = \sum_{k=0}^{\infty} \left(\sqrt{\frac{\pi}{k+1}} \, a_k\right) \sqrt{\frac{k+1}{\pi}} \, z^k$$

is the Fourier expansion of f. By Ex. 1, 13.1,

$$|E_n(f)|^2 = \sum_{k=n+1}^{\infty} \frac{\pi}{k+1} \, |a_k|^2.$$

Given $\varepsilon > 0$, we have for $k \geq N(\varepsilon)$, $|a_k|^{1/k} \leq 1/\rho + \varepsilon$, so that $|a_k|^2 \leq (1/\rho + \varepsilon)^{2k}$. Therefore, for $n \geq N(\varepsilon)$,

$$|E_n(f)|^2 \leq \sum_{k=n+1}^{\infty} \frac{\pi}{k+1} (1/\rho + \varepsilon)^{2k} \leq \pi \sum_{k=n+1}^{\infty} (1/\rho + \varepsilon)^{2k} = \frac{\pi(1/\rho + \varepsilon)^{2n+2}}{1 - (1/\rho + \varepsilon)^2}.$$

It follows from this that

$$\lim_{n \to \infty} \sup \, [E_n(f)]^{1/n} \leq 1/\rho + \varepsilon.$$

Since ε is arbitrary, $\lim\limits_{n \to \infty} \sup \, [E_n(f)]^{1/n} \leq 1/\rho$.

On the other hand, to an arbitrarily given $\varepsilon > 0$, we have $|a_k|^{1/k} \geq \dfrac{1}{\rho} - \varepsilon$ for an increasing sequence of integers $k = k_1, k_2, \ldots$. Hence,

$$\frac{\pi}{k_j + 1} \, |a_{k_j}|^2 \geq \frac{\pi}{k_j + 1} (1/\rho - \varepsilon)^{2k_j}.$$

But,

$$[E_{k_j-1}(f)]^2 \geq \frac{\pi}{k_j + 1} \, |a_{k_j}|^2 \geq \frac{\pi}{k_j + 1} (1/\rho - \varepsilon)^{2k_j}.$$

Hence,

$$[E_{k_j-1}(f)]^{1/(k_j-1)} \geq \left(\frac{\pi}{k_j+1}\right)^{1/[2(k_j-1)]} (1/\rho - \varepsilon)^{k_j/(k_j-1)}.$$

Now,

$$\lim_{j\to\infty} \frac{k_j}{k_j-1} = 1 \text{ and } \lim_{j\to\infty} \left(\frac{\pi}{k_j+1}\right)^{1/[2(k_j-1)]} = 1.$$

Therefore

$$\lim_{n\to\infty} \sup [E_n(f)]^{1/n} \geq (1/\rho - \varepsilon).$$

Since ε is arbitrary,

$$\lim_{n\to\infty} \sup [E_n(f)]^{1/n} \geq 1/\rho.$$

Consequently (13.2.1) holds.

A similar theorem can be proved for least square approximation of analytic functions on $[-1, 1]$ by polynomials. The same technique of proof is used and is based on Corollary 10.3.6 and Theorem 12.4.7.

THEOREM 13.2.2. *Let*

$$E_n{}^2(f) = \min_{a_i} \int_{-1}^{+1} |f(x) - a_0 - a_1 x - \cdots - a_n x^n|^2 \, dx. \qquad (13.2.3)$$

If f is analytic in $\mathscr{E}_\rho$, but not in any $\mathscr{E}_{\rho'}$ with $\rho' > \rho$, then

$$\lim_{n\to\infty} \sup [E_n(f)]^{1/n} = 1/\rho. \qquad (13.2.4)$$

Note that the larger $\mathscr{E}_\rho$ is assumed to be, the more rapid is the approximation.

EX. 1. If f is an entire function and $E_n(f)$ is defined either by (13.2.1) or (13.2.3), then $\lim_{n\to\infty} [E_n(f)]^{1/n} = 0$.

13.3 Degree of Approximation with Uniform Norm. We begin with an easily established theorem that shows the relationship between smoothness and degree of approximation.

THEOREM 13.3.1. *Let $f(x) \in C^p[-\pi, \pi]$ for some $p \geq 1$ and have period 2π.*

$$(f(-\pi) = f(\pi), f'(-\pi) = f'(\pi), \ldots, f^{(p)}(-\pi) = f^{(p)}(\pi).)$$

Suppose that $f^{(p)}(x)$ satisfies a Lipschitz condition of order α, $0 < \alpha \leq 1$. If

$$E_n(f) = \min_{c_k, d_k} \max_{-\pi \leq x \leq \pi} \left| f(x) - \sum_{k=0}^{n} c_k \cos kx + d_k \sin kx \right|, \text{ then}$$

$$E_n(f) \leq \frac{\text{const}}{n^{p-1+\alpha}}. \qquad (13.3.1)$$

Proof: Let $S_n(x)$ be the nth partial sum of the Fourier expansion of f and a_k, b_k its Fourier coefficients. Then,

$$E_n(f) \leq \max_{-\pi \leq x \leq \pi} |f(x) - S_n(x)| = \max_{-\pi \leq x \leq \pi} \left| \sum_{k=1}^{\infty} \cdots \right| \leq \sum_{k=n+1}^{\infty} \cdots$$

By Theorem 12.1.4, $|a_k|, |b_k| \leq \dfrac{c}{k^{p+\alpha}}$ $k = 1, 2, \ldots$.

Hence

$$E_n(f) \leq 2c \sum_{k=n+1}^{\infty} \frac{1}{k^{p+\alpha}} < 2c \int_n^{\infty} \frac{dx}{x^{p+\alpha}} < \frac{\text{const}}{n^{p-1+\alpha}}.$$

The estimate in (13.3.1) was obtained by using the partial sums of the Fourier series of f as a comparison function. There is no reason to suppose that these are the most efficient trigonometric polynomials of order n to use, and, indeed, D. Jackson has found that other polynomials lead to a better estimate.

LEMMA 13.3.2.   *Let $f(x)$ be periodic on $[-\pi, \pi]$ and let a_k, b_k be its Fourier coefficients. Then, for arbitrary constants $\rho_{n,k}$, the linear combination*

$$\sigma_n(f; x) = \frac{a_0}{2} + \sum_{k=1}^{n} \rho_{n,k}(a_k \cos kx + b_k \sin kx), \tag{13.3.2}$$

can be expressed as

$$\sigma_n(f; x) = \frac{1}{\pi} \int_{-\pi}^{\pi} f(x + t)K_n(t)\, dt \tag{13.3.3}$$

where the kernel $K_n(t)$ is given by

$$K_n(t) = \tfrac{1}{2} + \sum_{k=1}^{n} \rho_{n,k} \cos kt. \tag{13.3.4}$$

Proof: See the proof of Lemma 12.1.9.

Ex. 1.   If $\rho_{n,k} \equiv 1$, then $\sigma_n(f; x) = S_n(f; x)$, and

$$K_n(t) = \tfrac{1}{2} + \sum_{k=1}^{n} \cos kt = \frac{\sin (n + \tfrac{1}{2})t}{2 \sin \dfrac{t}{2}} = \text{the Dirichlet kernel.}$$

Ex. 2.   If $\rho_{n,k} = \dfrac{n - k}{n}$, then $\sigma_n(f; x)$ are the Fejér sums (Lemma 12.2.2) and $K_n(t)$ is the Fejér kernel.

LEMMA 13.3.3.   *If $K_n(t) \geq 0$ for $-\pi \leq t \leq \pi$, then*

$$I = \frac{1}{\pi} \int_{-\pi}^{\pi} |t|\, K_n(t)\, dt \leq \frac{\pi}{\sqrt{2}} \sqrt{1 - \rho_{n1}}. \tag{13.3.5}$$

Proof: Since $\dfrac{2t}{\pi} \le \sin t,\ 0 \le t \le \dfrac{\pi}{2}$, (draw a graph),

$$I = \frac{2}{\pi}\int_{-\pi}^{\pi} \left|\frac{t}{2}\right| K_n(t)\,dt \le \int_{-\pi}^{\pi} \sin\left|\frac{t}{2}\right| K_n(t)\,dt$$

$$\le \left(\int_{-\pi}^{\pi} \sin^2\frac{t}{2}\,K_n(t)\,dt\right)^{\frac{1}{2}}\left(\int_{-\pi}^{\pi} K_n(t)\,dt\right)^{\frac{1}{2}}.$$

Now,

$$\int_{-\pi}^{\pi} K_n(t)\,dt = \pi,$$

and

$$\int_{-\pi}^{\pi} \sin^2\frac{t}{2}\,K_n(t)\,dt = \frac{1}{2}\int_{-\pi}^{\pi}(1-\cos t)\left(\frac{1}{2}+\sum_{k=1}^{n}\rho_{nk}\cos kt\right)dt$$

$$= \frac{1}{2}\int_{-\pi}^{\pi}\left(\frac{1}{2}-\rho_{n1}\cos^2 t\right)dt = \frac{\pi}{2}(1-\rho_{n1}).$$

Combining this information, we obtain (13.3.5).

LEMMA 13.3.4 (Korovkin). *If* $K_n(t) \ge 0$ *and if* $f \in C[-\pi,\pi]$ *and is periodic, then*

$$|\sigma_n(f;x) - f(x)| \le w\left(\frac{1}{m}\right)\left(1 + m\,\frac{\pi}{\sqrt{2}}\,\sqrt{1-\rho_{n1}}\right), \qquad (13.3.6)$$

for any integer $m > 0$. $w(\delta)$ *is the modulus of continuity of* $f(x)$.

Proof: $\sigma_n(f;x) - f(x) = \dfrac{1}{\pi}\displaystyle\int_{-\pi}^{\pi}\{f(x+t)-f(x)\}K_n(t)\,dt.$

$$|\sigma_n(f;x)-f(x)| \le \frac{1}{\pi}\int_{-\pi}^{\pi}|f(x+t)-f(x)|\,K_n(t)\,dt \le \frac{1}{\pi}\int_{-\pi}^{\pi}w(|t|)K_n(t)\,dt.$$

Notice that if $\lambda > 0$,

$$w(\lambda\delta) \le (\lambda+1)w(\delta). \qquad (13.3.7)$$

For set $\lambda = n + \theta$, where n is an appropriate integer and $0 \le \theta < 1$. Then

$$w(\lambda\delta) = w(n\delta + \theta\delta) \le w(n\delta) + w(\theta\delta) \le nw(\delta) + w(\delta)$$
$$= (n+1)w(\delta) \le (\lambda+1)w(\delta).$$

These inequalities follow from (1.4.8)–(1.4.10). Let $\delta = \dfrac{1}{m}$ and $\lambda = m\,|t|$; then $w(|t|) \le (m\,|t| + 1)w\left(\dfrac{1}{m}\right)$. Hence,

$$|\sigma_n(f;x) - f(x)| \le \frac{1}{\pi}\,w\left(\frac{1}{m}\right)\int_{-\pi}^{\pi}(m\,|t| + 1)K_n(t)\,dt.$$

Applying (13.3.5), and $\dfrac{1}{\pi}\displaystyle\int_{-\pi}^{\pi} \tilde{K}_n(t)\, dt = 1$, we obtain (13.3.6).

LEMMA 13.3.5. *For* $n = 1, 2, \ldots$, *we can find kernels* $\tilde{K}_n(t)$ *such that*

$$\tilde{K}_n(t) = \tfrac{1}{2} + \sum_{k=1}^{n} \rho_{nk} \cos kt, \tag{13.3.8}$$

$$\tilde{K}_n(t) \geq 0 \qquad -\pi \leq t \leq \pi, \tag{13.3.9}$$

$$\rho_{n1} = \cos \frac{\pi}{n+2}. \tag{13.3.10}$$

Proof: For every fixed $n \geq 1$, set

$$\tilde{K}_n(t) = A_n\, |a_0 + a_1 z + \cdots + a_n z^n|^2 \tag{13.3.11}$$

where

$$a_k = \sin \frac{(k+1)\pi}{n+2} \qquad k = 0, 1, \ldots, n, \tag{13.3.12}$$

$$A_n = [2(a_0{}^2 + a_1{}^2 + \cdots + a_n{}^2)]^{-1}, \tag{13.3.13}$$

and $z = e^{it}$.

Property (13.3.9) is obvious.

$$\tilde{K}_n(t) = A_n \overline{\sum_{j=0}^{n} a_j z^j} \sum_{j=0}^{n} a_j z^j = A_n \sum_{j,k=0}^{n} a_j a_k\, e^{i(j-k)t}$$

$$= A_n\left(\sum_{j=0}^{n} a_j a_j + 2\left(\sum_{j=0}^{n-1} a_j a_{j+1}\right) \cos t + 2\left(\sum_{j=0}^{n-2} a_j a_{j+2}\right) \cos 2t + \cdots + 2(a_0 a_n) \cos nt\right).$$

Hence (13.3.8) follows by inspection. Finally, observe that

$$\cos \frac{\pi}{n+2} \sin \frac{k\pi}{n+2} = \frac{1}{2}\left(\sin \frac{k+1}{n+2}\,\pi + \sin \frac{k-1}{n+2}\,\pi\right),$$

or,

$$a_{k-1} \cos \frac{\pi}{n+2} = \tfrac{1}{2}(a_{k-2} + a_k).$$

Hence,

$$\cos \frac{\pi}{n+2}\, a_{k-1}^2 = \tfrac{1}{2}(a_{k-2} a_{k-1} + a_{k-1} a_k).$$

Therefore, $\cos \dfrac{\pi}{n+2} \displaystyle\sum_{k=0}^{n+1} a_{k-1}^2 = \tfrac{1}{2}\displaystyle\sum_{k=0}^{n+1} (a_{k-2} a_{k-1} + a_{k-1} a_k)$. Since

$$a_{-1} = \sin 0 = a_{n+1} = \sin \frac{n+2}{n+2}\,\pi = 0,$$

$\cos \dfrac{\pi}{n+2}\, (a_0{}^2 + a_1{}^2 + \cdots + a_n{}^2) = a_0 a_1 + a_1 a_2 + \cdots + a_{n-1} a_n$. Now the coefficient of $\cos t$ in $\tilde{K}_n(t)$ is $2A_n(a_0 a_1 + \cdots + a_{n-1} a_n)$, and therefore (13.3.13) implies (13.3.10).

THEOREM 13.3.6 (Jackson).   *Let $f(x) \in C[-\pi, \pi]$ and be periodic. If*

$$E_n(f) = \min_{a_k, b_k} \ \max_{-\pi \leq x \leq \pi} |f(x) - \sum_{k=0}^{n} a_k \cos kx + b_k \sin kx|, \quad (13.3.14)$$

then

$$E_n(f) \leq \left(1 + \frac{\pi^2}{2}\right) w\left(\frac{1}{n}\right). \quad (13.3.15)$$

$w(\delta)$ is the modulus of continuity of f.

Proof: Use the kernels $\tilde{K}_n(t)$ to form $\sigma_n(f; x)$. Then,

$$E_n(f) \leq \max_{-\pi \leq x \leq \pi} |f(x) - \sigma_n(f; x)| \leq w\left(\frac{1}{n}\right)\left(1 + \frac{n\pi}{\sqrt{2}}\sqrt{1 - \cos\frac{\pi}{n+2}}\right).$$

$$(13.3.16)$$

Now, $\sqrt{1 - \cos\dfrac{\pi}{n+2}} = \sqrt{2}\sin\dfrac{\pi}{2(n+2)}$. Since $\sin x \leq x$ for $0 \leq x \leq \dfrac{\pi}{2}$,

$$1 + \frac{n\pi}{\sqrt{2}}\sqrt{1 - \cos\frac{\pi}{n+2}} \leq 1 + \frac{n\pi^2}{2n+4} \leq 1 + \frac{\pi^2}{2}.$$

The study of the degree of approximation to continuous functions by polynomials can be referred to Theorem 13.3.6 by a change of variable.

THEOREM 13.3.7 (Jackson).   *Let $f(x) \in C[-1, 1]$ and set*

$$E_n(f) = \min_{a_i} \ \max_{-1 \leq x \leq 1} |f(x) - \sum_{i=0}^{n} a_i x^i|. \quad (13.3.17)$$

Then,

$$E_n(f) \leq \left(1 + \frac{\pi^2}{2}\right) w\left(\frac{1}{n}\right) \quad (13.3.18)$$

where $w(\delta)$ is the modulus of continuity of f.

Proof: Set $g(x) = f(\cos x)$. Then g is continuous, periodic, and even in $[-\pi, \pi]$. Using $\tilde{K}_n(t)$, form $\sigma_n(f; x)$. By (13.3.4),

$$\sigma_n(g; x) = \frac{a_0}{2} + \sum_{k=1}^{n} \rho_{n,k} a_k \cos kx,$$

where a_k are the Fourier coefficients of $g(x)$. From the proof of Theorem 13.3.6,

$$\max_{-\pi \leq x \leq \pi} |g(x) - \sigma_n(g; x)| \leq \left(1 + \frac{\pi^2}{2}\right) w\left(\frac{1}{n}; g\right).$$

Now,

$$w(\delta; g) = \max_{|x_1 - x_2| \leq \delta} |g(x_1) - g(x_2)| = \max_{|x_1 - x_2| \leq \delta} |f(\cos x_1) - f(\cos x_2)|.$$

Since

$$\left|\frac{d}{dx}\cos x\right| \leq 1, \ |\cos x_1 - \cos x_2| \leq |x_1 - x_2|.$$

Hence,

$$\max_{|x_1-x_2|\leq\delta}|f(\cos x_1) - f(\cos x_2)| \leq \max_{|x_1-x_2|\leq\delta}|f(x_1) - f(x_2)| = w(\delta;f).$$

Therefore,

$$\max_{-\pi\leq x\leq\pi}\left|f(\cos x) - \frac{a_0}{2} - \sum_{k=1}^{n}\rho_{nk}\cos kx\right| \leq \left(1 + \frac{\pi^2}{2}\right)w\left(\frac{1}{n};f\right).$$

Setting $y = \cos x$,

$$\max_{-1\leq y\leq 1}\left|f(y) - \frac{a_0}{2} - \sum_{k=1}^{n}\rho_{nk}\cos(k \text{ arc cos } y)\right| \leq \left(1 + \frac{\pi^2}{2}\right)w\left(\frac{1}{n};f\right).$$

Since $\cos(k \text{ arc cos } y) = T_k(y)$,

$$p_n(y) = \frac{a_0}{2} + \sum_{k=1}^{n}\rho_{nk}\cos(k \text{ arc cos } y) \in \mathscr{P}_n.$$

Hence,

$$E_n(f) \leq \max_{-1\leq y\leq 1}|f(y) - p_n(y)| \leq \left(1 + \frac{\pi^2}{2}\right)w\left(\frac{1}{n}\right).$$

NOTES ON CHAPTER XIII

13.1   Tieman [1] pp. 50–55.

13.2–13.3   Much is known about the degree of approximation. For analytic functions, consult Walsh [2], [3]. For real functions, Jackson [1], Bernstein [1], Natanson [1], Tieman [1], Korovkin [2], Alexits [1], Zygmund [1], vol. I.

PROBLEMS

1. Let the radius of convergence of $f(z) = \sum\limits_{n=0}^{\infty} a_n z^n$ be R. Set $S_n(z) = \sum\limits_{k=0}^{n} a_k z^k$. Show that

$$\limsup_{n\to\infty}\left[\max_{|z|\leq r}|f(z) - S_n(z)|\right]^{1/n} = \frac{r}{R}, \quad \text{for} \quad r < R$$

and

$$\limsup_{n\to\infty}\left[\max_{|z|\leq r}|S_n(z)|\right]^{1/n} = \frac{r}{R}, \quad \text{for} \quad r \geq R.$$

2. Let ρ be a fixed number >1, and let X be the space of functions that are analytic in $|z| < \rho$. For $f\in X$, set $\|f\|^2 = \iint\limits_{|z|\leq 1}|f(z)|^2\,dx\,dy$. Show that X is not complete.

3. Discuss the degree of approximation of a periodic analytic function by its Fourier series.

4. Discuss the rapidity of convergence in $L^2(\mathscr{E}_\rho)$ of the best polynomial approximants to functions that are analytic in $\mathscr{E}_{\rho'}$ $\rho' > \rho$.

5. The kernel

$$K_n(t) = \frac{3}{2n(2n^2 + 1)}\left(\frac{\sin\dfrac{nt}{2}}{\sin\dfrac{t}{2}}\right)^4$$

is known as Jackson's kernel. Show that

$$\left(\frac{\sin\dfrac{nt}{2}}{\sin\dfrac{t}{2}}\right)^2 = n + 2[(n-1)\cos t + (n-2)\cos 2t + \cdots + \cos(n-1)t]$$

and hence that

$$K_n(t) = \tfrac{1}{2} + \sum_{k=1}^{2n-2} \rho_{2n-2,k}\cos kt \quad\text{where}\quad \rho_{2n-2,1} = 1 - \frac{3}{2n^2}.$$

6. Show that Jackson's kernel is of Fejér's type on $[-\pi, \pi]$. (See 12.2.)

CHAPTER XIV

Approximation of Linear Functionals

14.1 Rules and Their Determination. Numerical analysis has given rise to many approximate formulas for interpolation, extrapolation, differentiation, and integration. These formulas, frequently called "rules," are not only of practical importance, but very frequently have an unusual interest in their own right. Integrals, derivatives at a point, etc., are linear functionals defined over appropriate linear spaces of functions, and the problem of rule formation is equivalent to the approximation of these functionals by linear combinations of prescribed linear functionals.

Ex. 1. $\displaystyle\int_a^{a+2h} f(x)\, dx \approx \frac{h}{3}[f(a) + 4f(a + h) + f(a + 2h)]$ (Simpson's Rule).

Ex. 2. $f(a) \approx \frac{1}{2}[f(a - h) + f(a + h)]$ (Linear Interpolation).

Ex. 3. $f(a + h) \approx f(a) + \dfrac{h}{2}[f'(a) + f'(a + h)]$ (Trapezoidal Rule for Differential Equations).

In general, the prescribed linear functionals are values of a function at a point. Occasionally, as in Ex. 3, derivatives are employed. Integrals over subintervals have not been used to any extent as approximating linear functionals, but they, too, are possible, and might conceivably become important.

Two roads may be taken to develop rules: (a) The method of interpolation and (b) The method of approximation. Let $L_1, L_2, \ldots, L_n$ be n prescribed "elementary" linear functionals. It is desired to approximate a given linear functional L by linear combinations of the L_i:

$$L \approx a_1 L_1 + a_2 L_2 + \cdots + a_n L_n. \tag{14.1.1}$$

The remainder or error

$$E = L - (a_1 L_1 + \cdots + a_n L_n) \tag{14.1.2}$$

is itself a linear functional. (Note that in this chapter, the symbols E, E_n will be used with a different meaning than in Chapters 7, 13.) The method of interpolation selects constants a_i in such a way that E vanishes on n

341

prescribed elements $x_1, \ldots, x_n$:

$$E(x_i) = 0 \qquad i = 1, 2, \ldots, n. \tag{14.1.3}$$

We may say that the constants a_i are determined so that E is orthogonal to $x_1, \ldots, x_n$. The theory of this method has been discussed extensively in 2.2.

Ex. 4.   The rules of Ex. 1, Ex. 2, Ex. 3 are exact for $\mathscr{P}_3$, $\mathscr{P}_1$, and $\mathscr{P}_2$, respectively.

Method (a) has had a long history and by far the bulk of the approximate formulas of numerical analysis are of this type. The method of approximation (b) is a recent one and has certain advantages and disadvantages over (a). Here we assume that we are dealing with a normed linear space X of functions and that the functionals in question L, L_i, belong to the normed conjugate space X^*. We wish then to determine constants a_i such that

$$\|L - (a_1 L_1 + a_2 L_2 + \cdots + a_n L_n)\| = \text{minimum}. \tag{14.1.4}$$

In the case where X is a Hilbert space so that X and X^* are essentially the same (Theorem 9.3.9) the problem of approximation of functionals can be handled by orthogonalization methods. We therefore note:

THEOREM 14.1.1.   *The problem* (14.1.4) *possesses a solution.*

Proof: Apply Theorem 7.4.1 to X^*.

THEOREM 14.1.2.   *Let H be a Hilbert space and $L, L_1, \ldots, L_n$ be bounded linear functionals, the last n assumed independent. Let $x, x_1, \ldots, x_n$ be their representers. If* $\left\| x - \sum_{i=1}^{n} a_i x_i \right\| = minimum$, *then* $\left\| L - \sum_{i=1}^{n} \bar{a}_i L_i \right\| = minimum$.

14.2   The Gauss-Jacobi Theory of Approximate Integration.   One of the finest examples of rules of interpolatory type is that of Gauss-Jacobi. Here the approximating linear functionals are not prescribed in advance, but the problem is to determine them so that the rule will be exact for polynomials of maximal degree. Let $w(x) > 0$ be a weight function defined on $[a, b]$. If n distinct points $x_1, \ldots, x_n$ are specified in advance, then we know that we can find coefficients $w_1, \ldots, w_n$ such that the rule

$$L(f) = \int_a^b w(x) f(x)\, dx \approx \sum_{k=1}^{n} w_k f(x_k) \tag{14.2.1}$$

will be exact for $\mathscr{P}_{n-1}$. If we treat *both* the x_k and the w_k as $2n$ unknowns, then, perhaps, we can arrange matters in such a way that the rule will be exact for $\mathscr{P}_{2n-1}$, (i.e., linear combinations of the $2n$ powers $1, x, x^2, \ldots, x^{2n-1}$). For such a rule to have practical importance, the points x_i must lie in the interval $[a, b]$, although for analytic functions, this condition could be waived.

This is indeed possible as was found by Gauss and generalized by Jacobi. The solution exhibits a surprising relation to the orthogonal polynomials generated by the weight $w(x)$.

THEOREM 14.2.1 (Gauss-Jacobi).   *Let $w(x) > 0$ be a weight function defined on $[a, b]$ with corresponding orthogonal polynomials $p_n(x)$, $n \geq 1$. Let the zeros of $p_n(x)$ be $a < x_1 < x_2 < \cdots < x_n < b$. Then, we can find positive constants $w_1, w_2, \ldots, w_n$ such that*

$$\int_a^b w(x)p(x)\,dx = \sum_{k=1}^n w_k p(x_k) \tag{14.2.2}$$

whenever $p(x) \in \mathscr{P}_{2n-1}$.

Proof: Given a $p(x)$ as above. Let $q(x) \in \mathscr{P}_{n-1}$ take on the values of $p(x)$ at $x_1, \ldots, x_n$. We can write it in the Lagrange form

$$q(x) = \sum_{k=1}^n p(x_k)\ell_k(x),\, \ell_k(x) = \frac{w(x)}{(x - x_k)w'(x_k)}\,,\, w(x) = \prod_{k=1}^n (x - x_k).$$

The polynomial $p(x) - q(x)$ therefore has zeros at $x_1, \ldots, x_n$ and consequently $p(x) - q(x) = p_n(x)r_{n-1}(x)$ for some polynomial $r_{n-1}(x) \in \mathscr{P}_{n-1}$. Therefore, in view of the orthogonality of $p_n(x)$ to polynomials of lower degree,

$$\int_a^b w(x)p(x)\,dx = \int_a^b w(x)[q(x) + p_n(x)r_{n-1}(x)]\,dx = \int_a^b w(x)q(x)\,dx$$

$$= \int_a^b w(x) \sum_{k=1}^n p(x_k)\ell_k(x)\,dx = \sum_{k=1}^n \left\{ \int_a^b w(x)\ell_k(x)\,dx \right\} p(x_k).$$

If we set

$$w_k = \int_a^b w(x)\ell_k(x)\,dx \tag{14.2.3}$$

then the identity (14.2.2) holds. Now $\ell_k(x) \in \mathscr{P}_{n-1}$ and vanishes at $x_1, \ldots,$ $x_{k-1}, x_{k+1}, \ldots, x_n$. Therefore $(\ell_k(x))^2 \in \mathscr{P}_{2n-2}$ and vanishes at these same points. Moreover $\ell_k(x_k) = 1$. Hence,

$$w_k = \sum_{j=1}^n w_j(\ell_k(x_j))^2 = \int_a^b w(x)[\ell_k(x)]^2\,dx > 0.$$

The abscissas x_i are the zeros of certain polynomials, and are generally irrational numbers. If computing is done by hand, it is a nuisance to deal with many digits, and so in years gone by the Gauss rule was not popular. High speed digital computers, on the other hand, do not distinguish between "simple" numbers such as .500000000 and more "complicated" ones such as .577350269. The Gauss rules, which are excellent for large classes of functions arising in practise, have therefore been rehabilitated in the eyes of computers and are much employed.

The following remainder theorem holds for the Gauss-Jacobi rule.

THEOREM 14.2.2 (A. Markoff). *Let* $w(x)$ *and* $x_1, \ldots, x_n$ *be as in the previous theorem. Let* $f(x) \in C^{2n}[a, b]$. *Then,*

$$E_n(f) = \int_a^b w(x) f(x)\, dx - \sum_{k=1}^n w_k f(x_k) = \frac{f^{(2n)}(\eta)}{(2n)!\, k_n^2}, \qquad (14.2.4)$$

where k_n *is the leading coefficient of the orthonormal polynomial* $p_n^*(x)$ *associated with* $w(x)$, *and where* $a < \eta < b$.

Proof: The device here is to employ Hermite interpolation with each abscissa repeated once. Let $h_{2n-1}(x)$ be that polynomial of $\mathscr{P}_{2n-1}$ for which $h_{2n-1}(x_k) = f(x_k)$, $h'_{2n-1}(x_k) = f'(x_k)$, $k = 1, 2, \ldots, n$. Then according to Theorem 3.5.1, we have

$$f(x) = h_{2n-1}(x) + \frac{f^{(2n)}(\xi(x))}{(2n)!}(x - x_1)^2(x - x_2)^2 \cdots (x - x_n)^2 \quad (14.2.5)$$

for $a \leq x \leq b$ and $a < \xi(x) < b$. Notice that by Theorem 1.6.6 applied to $f(x) - h_{2n-1}(x)$, $\dfrac{f(x) - h_{2n-1}(x)}{(x - x_1)^2 \cdots (x - x_n)^2} \in C[a, b]$ and hence $f^{(2n)}(\xi(x))$ is also continuous. Multiply (14.2.5) by $w(x)$:

$$w(x) f(x) = w(x) h_{2n-1}(x) + \frac{f^{(2n)}(\xi(x))}{(2n)!}\, w(x) \frac{[p_n^*(x)]^2}{k_n^2} \qquad (14.2.6)$$

where $p_n^*(x)$ is the orthonormal polynomial of degree n associated with $w(x)$. Integrate (14.2.6), and employ the mean value theorem for integrals:

$$\int_a^b w(x) f(x)\, dx = \int_a^b w(x) h_{2n-1}(x)\, dx + \frac{1}{(2n)!\, k_n^2} \int_a^b f^{(2n)}(\xi(x)) w(x) [p_n^*(x)]^2\, dx$$

$$= \int_a^b w(x) h_{2n-1}(x)\, dx + \frac{f^{(2n)}(\eta)}{(2n)!\, k_n^2} \int_a^b w(x) [p_n^*(x)]^2\, dx$$

$$= \sum_{k=1}^n w_k f(x_k) + \frac{f^{(2n)}(\eta)}{(2n)!\, k_n^2}.$$

COROLLARY 14.2.3. *For the Jacobi weight* $w(x) = (1 - x)^\alpha(1 + x)^\beta$, $\alpha > -1, \beta > -1$, *and for* $f(x) \in C^{2n}[-1, 1]$, *then*

$$E_n(f) = \frac{2^{2n+\alpha+\beta+1}\Gamma(n + \alpha + 1)\Gamma(n + \beta + 1)\Gamma(n + \alpha + \beta + 1) n!\, f^{(2n)}(\eta)}{\Gamma(2n + \alpha + \beta + 1)\Gamma(2n + \alpha + \beta + 2)(2n)!}. \qquad (14.2.7)$$

For $\alpha = \beta = 0$, this reduces to

$$E_n(f) = \frac{2^{2n+1}(n!)^4}{(2n + 1)[(2n)!]^3} f^{(2n)}(\eta). \qquad (14.2.8)$$

14.3 Norms of Functionals as Error Estimates. Let X be a normed linear space and L a bounded linear functional defined on X. Then, as in (9.3.2),

$$|L(x)| \leq \|L\|\, \|x\|, \ x \in X. \tag{14.3.1}$$

$\|L\|$ is independent of x. This fundamental inequality may be made the basis of error estimates for the linear rules of numerical analysis. Let L be a given functional and let a rule R be an approximation to L. Then construct the error

$$E = L - R. \tag{14.3.2}$$

The smaller the norm $\|E\| = \|L - R\|$, the better the approximation of R to L.

Ex. 1. Let X be the linear space of continuous functions on $[a, b]$ that have a bounded, piecewise continuous derivative. Let $\|f\| = \sup\limits_{a \leq x \leq b} |f'(x)| + |f(a)|$. Then X is a normed linear space. If $K(x)$ is a bounded, piecewise continuous function on $[a, b]$, then $L(f) = \int_a^b f'(x)K(x)\,dx$ is a bounded linear functional on X. Since $|L(f)| \leq \sup\limits_{a \leq x \leq b} |f'(x)| \int_a^b |K(x)|\,dx \leq \|f\| \int_a^b |K(x)|\,dx$, it follows that $\|L\| \leq \int_a^b |K(x)|\,dx$. On the other hand, the function sgn $K(x)$ is bounded and piecewise continuous in $[a, b]$. Hence the function $g(x) = \int_a^x$ sgn $K(t)\,dt$ is in X, and $g(a) = 0$. Also $L(g) = \int_a^b (\text{sgn } K(x))K(x)\,dx = \int_a^b |K(x)|\,dx$. If $K(x) \not\equiv 0$, then $\|g\| = \sup\limits_{a \leq x \leq b} |\text{sgn } K(x)| = 1$. Hence, $|L(g)| = \|g\| \int_a^b |K(x)|\,dx$. Therefore $\|L\| \geq \int_a^b |K(x)|\,dx$, and hence $\|L\| = \int_a^b |K(x)|\,dx$.

As a specific instance, the error in the trapezoidal rule (cf. Ex. 4, 3.7) is given by

$$E(f) = \int_a^b f(x)\,dx - \frac{b-a}{n}\left[\frac{f(a)}{2} + f(a + \sigma) + f(a + 2\sigma) + \cdots \right.$$
$$\left. + f(a + (n-1)\sigma) + \frac{f(b)}{2}\right], \sigma = \frac{b-a}{n}.$$

Therefore (cf. (3.7.21)),

$$E(f) = -\sum_{k=0}^{n-1} \int_{a+k\sigma}^{a+(k+1)\sigma} (t - (a + (k + \tfrac{1}{2})\sigma))f'(t)\,dt.$$

Hence, by the above work,

$$\|E\| = \sum_{k=0}^{n-1} \int_{a+k\sigma}^{a+(k+1)\sigma} |t - (a + (k + \tfrac{1}{2})\sigma)|\,dt = \frac{n\sigma^2}{4} = \frac{(b-a)^2}{4n}.$$

Therefore, $|E(f)| \leq \dfrac{\|f\|\,(b-a)^2}{4n}$.

Ex. 2.   In the space $L^2(B)$ where B is the unit circle, consider the interpolation formula $f(0) \approx \frac{1}{2}(f(h) + f(-h))$, where $0 < h < 1$. Set

$$E(f) = f(0) - \frac{1}{2}(f(h) + f(-h)).$$

By Corollary 12.5.5, E is bounded over $L^2(B)$. From Corollary 12.6.7,

$$\|E\|^2 = E_w \overline{E_z K(z, w)},$$

and from (12.6.17),

$$K(z, w) = \frac{1}{\pi(1 - z\bar{w})^2}.$$

Hence,

$$E_z K(z, w) = \frac{1}{\pi} - \frac{1}{2\pi}\left[\frac{1}{(1 - h\bar{w})^2} + \frac{1}{(1 + h\bar{w})^2}\right] = \frac{1}{\pi}\left[\frac{-3h^2\bar{w}^2 + h^4\bar{w}^4}{(1 - h^2\bar{w}^2)^2}\right],$$

and

$$\|E\|^2 = \frac{1}{\pi}\left[\frac{3h^4 - h^8}{(1 - h^4)^2}\right].$$

Therefore,

$$\|E\| = \frac{h^2}{\sqrt{\pi}}\frac{\sqrt{3 - h^4}}{(1 - h^4)}.$$

Hence,

$$|E(f)| \leq \frac{h^2}{\sqrt{\pi}}\frac{\sqrt{3 - h^4}}{(1 - h^4)}\left[\iint\limits_{|z| < 1} |f(z)|^2 \, dx \, dy\right]^{\frac{1}{2}}.$$

Note that as $h \to 0$, $\|E\| \sim \sqrt{\frac{3}{\pi}} h^2$. As $h \to 1^-$, $\|E\| \to \infty$. This reflects the fact that $f(0) - \frac{1}{2}(f(1) + f(-1))$ is not bounded over $L^2(B)$.

Ex. 3.   Let $\mathscr{E}_\rho$ be the ellipse of Def. 1.13.1. Let E be a bounded linear functional over $L^2(\hat{\mathscr{E}}_\rho)$ and suppose that $E(1) = E(z) = \cdots = E(z^n) = 0$. From (12.6.18),

$$\|E\|^2 = \frac{4}{\pi} \sum_{k=n+1}^{\infty} (k + 1) \frac{|E(U_k)|^2}{\rho^{2k+2} - \rho^{-2k-2}}.$$

As concrete examples, the following values may be cited. Select the ellipse $\mathscr{E}_\rho$ with $a = 2, b = \sqrt{3}, \rho = 3.732$. Approximate integration over $[-1, 1]$ is considered.

Rule	n	Norm of Error
Trapezoidal	1	5.08×10^{-2}
Simpson	3	3.72×10^{-3}
Gauss 7 point	13	8.75×10^{-9}
Gauss 10 point	19	3.87×10^{-12}
Gauss 16 point	31	6.70×10^{-19}

14.4   Weak* Convergence.

Numerous modes of convergence have been defined and studied for normed linear spaces. One mode, known as weak* convergence, is particularly relevant to the problem of approximation of linear functionals. In order to see how it fits into the scheme of things, it is convenient to make several preliminary definitions, some of which have been met before.

DEFINITION 14.4.1.   (Strong or ordinary convergence in X.) A sequence of elements $\{x_n\}$ is said to converge strongly to x if $\lim\limits_{n\to\infty} \|x - x_n\| = 0$ (cf. 7.2).

Ex. 1.   If H is a Hilbert space and $\{x_k{}^*\}$ is a complete orthonormal sequence, then for arbitrary $x \in H$, the Fourier segments $\sum\limits_{k=1}^{n} (x, x_k{}^*)x_k{}^*$ converge strongly to x (Theorem 8.9.1, B).

DEFINITION 14.4.2.   (Weak convergence in X.) A sequence of elements $\{x_n\}$ is said to converge weakly to x if $\lim\limits_{n\to\infty} |L(x) - L(x_n)| = 0$ for all $L \in X^*$ (X^* is the normed conjugate space of X).

Ex. 2.   Let H be a Hilbert space. Then, any orthonormal sequence $\{x_k{}^*\}$ converges weakly to 0. For by Theorem 9.3.3, $L(x_k{}^*) = (x_k{}^*, x)$ for an appropriate $x \in H$. By Theorem 8.9.1, C, $\lim\limits_{k\to\infty} |L(x_k{}^*) - L(0)| = 0$. On the other hand, $\{x_k{}^*\}$ does not converge strongly to 0, for $\|x_k{}^* - 0\| = 1$.

In the normed linear space X^*, there are also several modes of convergence.

DEFINITION 14.4.3.   Strong convergence in X^*. A sequence of elements $\{L_n\}$ of X^* is said to converge strongly to L if $\lim\limits_{n\to\infty} \|L - L_n\| = 0$. This parallels Definition 14.4.1.

Ex. 3. $X = C[a, b]$ with $\|f\| = \max\limits_{a \le x \le b} |f(x)|$. Let $L_n(f) = \sum\limits_{k=1}^{n} a_{kn}f(x_{kn})$ where $a \le x_{kn} \le b$ and $\lim\limits_{n\to\infty} \sum\limits_{k=1}^{n} |a_{kn}| = 0$. Then, in view of Ex. 5, 9.3, $L_n \to 0$ strongly.

Less useful is

DEFINITION 14.4.4.   Weak convergence in X^*. A sequence of elements $\{L_n\}$ of X^* is said to converge weakly in X^* to L if $\lim\limits_{n\to\infty} |L^*(L_n) - L^*(L)| = 0$ for every element $L^* \in (X^*)^*$.

An intermediate mode is

DEFINITION 14.4.5.   Weak* convergence: A sequence of elements $\{L_n\}$ of X^* has a weak* limit L if

$$\lim_{n\to\infty} |L_n(x) - L(x)| = 0 \qquad (14.4.1)$$

for all $x \in X$.

Weak* convergence is precisely what is required for the convergence of approximate rules, for we would like the approximate answers obtained, $L_n(x)$, to approach the true answer $L(x)$, for any element x.

Ex. 4. $X = C[a, b]$, $\|f\| = \max\limits_{a \le x \le b} |f(x)|$. Let

$$x_{00}$$
$$x_{10}, \; x_{11}$$
$$x_{20}, \; x_{21}, \; x_{22}$$
$$\cdot$$
$$\cdot$$
$$\cdot$$

be a triangular sequence of points of $[a, b]$ such that

$$x_{n0} = a, \; x_{nn} = b, \; x_{nk} < x_{n,k+1} \quad \text{and} \quad \lim_{n \to \infty} \max_k (x_{n,k+1} - x_{n,k}) = 0.$$

Set $L_n(f) = \sum\limits_{k=1}^{n} f(x_{nk})(x_{n,k} - x_{n,k-1})$. If $L(f) = \int_a^b f(x)\,dx$, then by the properties of the Riemann integral, $L_n(f) \to L(f)$ for every $f \in X$. L is the weak* limit of $\{L_n\}$.

Ex. 5. $X = C[-1, 1]$, $\|f\| = \max\limits_{-1 \le x \le 1} |f(x)|$. Let the functions $K_n(x)$ satisfy Fejér's conditions (cf. 12.2.10–.12): $K_n(x)$ integrable and ≥ 0, $\int_{-1}^{+1} K_n(x)\,dx = 1$, and if $M_n(\delta) = \max\limits_{\delta \le |x| \le 1} K_n(x)$, then $\lim\limits_{n \to \infty} M_n(\delta) = 0$ for all $0 < \delta < 1$. Set $L_n(f) = \int_{-1}^{+1} f(x) K_n(x)\,dx$, $L(f) = f(0)$. Then, L_n converges weakly to L. For,

$$L(f) - L_n(f) = \int_{-1}^{+1} (f(0) - f(x)) K_n(x)\,dx = \int_{-\delta}^{+\delta} (f(0) - f(x)) K_n(x)\,dx$$

$$+ \int_{\delta}^{1} (f(0) - f(x)) K_n(x)\,dx + \int_{-1}^{-\delta} (f(0) - f(x)) K_n(x)\,dx.$$

Hence

$$|L(f) - L_n(f)| \le |f(0) - f(\xi)| \int_{-\delta}^{+\delta} K_n(x)\,dx$$

$$+ M_n(\delta)\left[\int_{\delta}^{1} |f(0) - f(x)|\,dx + \int_{-1}^{-\delta} |f(0) - f(x)|\,dx\right]$$

$$\le |f(0) - f(\xi)| \int_{-1}^{+1} K_n(x)\,dx + M_n(\delta) \int_{-1}^{+1} |f(0) - f(x)|\,dx, \quad \text{where } -\delta \le \xi \le \delta.$$

Allow $n \to \infty$; then $\limsup\limits_{n \to \infty} |L(f) - L_n(f)| \le |f(0) - f(\xi)|$.

Since δ is arbitrary and f is continuous, we conclude that $\lim\limits_{n \to \infty} L_n(f) = L(f)$.

Employing the Dirac δ function: $f(0) = \int_{-1}^{+1} f(x)\delta(x)\,dx$, we may speak of the Fejér kernels $K_n(x)$ as "converging" to $\delta(x)$ in the above sense.

A sequence of functionals can have only one weak* limit. For suppose that L and M are both weak* limits of $\{L_n\}$. Then, for all $x \in X$, $L_n(x) \to L(x)$ and $L_n(x) \to M(x)$. Hence $L(x) = M(x)$ for all $x \in X$, and this means that $L = M$.

Strong convergence in X^* implies weak* convergence. For, assume that $\lim\limits_{n \to \infty} \|L - L_n\| = 0$. Then, for any $x \in X$, $|L(x) - L_n(x)| = |(L - L_n)(x)| \le \|L - L_n\| \, \|x\| \to 0$.

On the other hand, weak* convergence does not necessarily imply strong convergence. As in Ex. 2, this is most easily seen in a Hilbert space. If $\{x_n{}^*\}$ is an orthonormal system, set $L_n(x) = (x, x_n{}^*)$. Then, $L_n(x) \to 0$ for all $x \in H$ so that 0 is the weak* limit of $\{L_n\}$. On the other hand, $\|0 - L_n\| = \|x_n{}^*\| = 1$.

However, in a Hilbert space, if L is the weak* limit of $\{L_n\}$ and if $\|L_n\| \to \|L\|$, then $\|L - L_n\| \to 0$. For let $L_n(y) = (y, x_n)$ and $L(y) = (y, x)$. Then $(y, x_n) \to (y, x)$ for all $y \in H$. In particular, $(x, x_n) \to (x, x)$. Moreover $(x_n, x_n) \to (x, x)$. Now,

$$\|L - L_n\|^2 = \|x - x_n\|^2$$
$$= (x, x) - (x, x_n) - (x_n, x) + (x_n, x_n) \to (x, x) - (x, x) - (x, x)$$
$$+ (x, x) = 0.$$

Lemma 14.4.1. *Let X be a complete metric space (cf. 7.2). Suppose that $U_n = U(x_n, \varepsilon_n)$ is a sequence of closed balls such that $U_1 \supseteq U_2 \supseteq \cdots$, and $\lim\limits_{n \to \infty} \varepsilon_n = 0$. Then, there is an $x \in X$ with $x \in U_n$, $n = 1, 2, \ldots$.*

Proof: Given ε, we can find N such that $\varepsilon_n < \varepsilon$ for all $n \ge N$. If $m \ge n$, then $U_m \subseteq U_n$ and hence $x_m \in U_n$. Therefore $d(x_m, x_n) < \varepsilon$ for all m, $n \ge N$, and $\{x_m\}$ is a Cauchy sequence. Now X is a complete space and so there is an $x \in X$ with $\lim\limits_{n \to \infty} d(x, x_n) = 0$. Let U_j be one of the balls. The elements $x_j, x_{j+1}, \ldots,$ all belong to U_j. Since U_j is closed, the limit $x \in U_j$ and hence belongs to all the balls.

Theorem 14.4.2. *Let X be a complete normed linear space. Suppose that $\{L_n\}$ is a sequence of bounded linear functionals such that for all $x \in X$, the sequence of constants $\{L_n(x)\}$ is convergent. Then, we can find an M such that*

$$\|L_n\| \le M \qquad n = 1, 2, \ldots \qquad (14.4.2)$$

Proof: Suppose that (14.4.2) were not true. Then, $\limsup\limits_{n \to \infty} \|L_n\| = \infty$. For a given $x_0 \in X$, consider the closed ball $U(x_0, \varepsilon)$ consisting of those elements x for which $\|x - x_0\| \leq \varepsilon$. It is now claimed that

$$|L_n(x)| \leq K \qquad n = 1, 2, \ldots ; x \in U(x_0, \varepsilon) \qquad (14.4.3)$$

is impossible. For take any $y \in X$, $y \neq 0$. Then

$$z = \frac{\varepsilon y}{\|y\|} + x_0 \qquad (14.4.4)$$

is clearly in $U(x_0, \varepsilon)$. Now, $L_n(z) = \dfrac{\varepsilon}{\|y\|} L_n(y) + L_n(x_0)$. Hence,

$$\frac{\varepsilon}{\|y\|} |L_n(y)| - |L_n(x_0)| \leq \left| \frac{\varepsilon}{\|y\|} L_n(y) + L_n(x_0) \right| = |L_n(z)| \leq K.$$

This implies that $|L_n(y)| \leq (K + |L_n(x_0)|) \dfrac{\|y\|}{\varepsilon}$. Since $\{L_n(x_0)\}$ converges, the sequence $|L_n(x_0)|$ is bounded. For some constant K_1, therefore, $|L_n(y)| \leq K_1 \|y\|$ for all $y \in X$. This tells us that $\|L_n\| \leq K_1$, contrary to our assumption.

Take a $U(x_0, \varepsilon)$. As has just been established, we can find an index n_1 and an element $x_1 \in U(x_0, \varepsilon)$ for which

$$|L_{n_1}(x_1)| > 1. \qquad (14.4.5)$$

Since L_{n_1} is continuous, we can assume that x_1 is in the interior of $U(x_0, \varepsilon)$ and can therefore find a second ball centered at x_1, $U(x_1, \varepsilon_1)$, with $U(x_1, \varepsilon_1) \subset U(x_0, \varepsilon)$ such that $|L_{n_1}(x)| > 1$ for all $x \in U(x_1, \varepsilon_1)$. But again, the quantities $|L_n(x)|$ cannot be bounded for $x \in U(x_1, \varepsilon_1)$. Hence, we can find an $n_2 > n_1$ for which $|L_{n_2}(x_2)| > 2$. We proceed in this way. A sequence of balls $U(x_n, \varepsilon_n)$, each contained in the previous one, can be found for which $|L_{n_j}(x)| > j$, $x \in U(x_j, \varepsilon_j)$. Moreover we may choose $\varepsilon_n \to 0$. Since X is complete, we can by Lemma 14.4.1 find an element x' lying in each of the nested sequence of balls. Then $|L_{n_j}(x')| > j, j = 1, 2, \ldots$. This contradicts the hypothesis that $\{L_n(x)\}$ is convergent for all x. The assumption that

$$\lim_{n \to \infty} \sup \|L_n\| = \infty$$

must have been fallacious.

COROLLARY 14.4.3.   *Under the hypothesis of the last theorem, we can find a bounded linear functional L for which*

$$\lim_{n \to \infty} L_n(x) = L(x), x \in X. \qquad (14.4.6)$$

Proof: For a given $x \in X$, define $L(x)$ by means of

$$L(x) = \lim_{n \to \infty} L_n(x). \qquad (14.4.7)$$

Now

$$L(ax + by) = \lim_{n \to \infty} L_n(ax + by) = \lim_{n \to \infty} \{aL_n(x) + bL_n(y)\}$$

$$= a \lim_{n \to \infty} L_n(x) + b \lim_{n \to \infty} L_n(y) = aL(x) + bL(y).$$

The functional L is therefore linear.

By the theorem, there is an M such that $\|L_n\| \leq M$, $n = 1, 2, \ldots$. Take an $x \in X$. Then, $|L_n(x)| \leq \|L_n\| \, \|x\| \leq M \, \|x\|$. From (14.4.7), $|L(x)| \leq M \, \|x\|$. Accordingly, L is bounded and $\|L\| \leq M$.

A similar theorem holds when the L_n are linear *operators* which send the elements of one complete normed linear space X into a second such space Y.

We now come to the fundamental theorem of weak* convergence. Many men have made contributions toward the final statement. Among them are Osgood, Vitali, Lebesgue, Pólya, and Banach.

THEOREM 14.4.4.   *Let X be a complete normed linear space, and let $L, L_1,$ $L_2, \ldots,$ be bounded linear functionals defined on X. In order that L be the weak* limit of the sequence $\{L_k\}$, it is necessary and sufficient that*

$$\|L_k\| \leq M \qquad k = 1, 2, \ldots, \tag{14.4.8}$$

for some constant M, and that

$$\lim_{k \to \infty} L_k(x_j) = L(x_j) \qquad j = 1, 2, \ldots, \tag{14.4.9}$$

for some sequence of elements $\{x_j\}$ that is closed in X.

In this connection, the constants $\|L_k\|$ are known as the *Lebesgue Constants*.

Proof: Sufficiency. We shall show that (14.4.8–9) implies that

$$\lim_{n \to \infty} L_n(x) = L(x)$$

for all $x \in X$. Take an $x \in X$. Given an ε, we can, in view of the closure of $\{x_j\}$, find constants $a_1, \ldots, a_k$ such that $\|x - y_k\| \leq \varepsilon$ where

$$y_k = a_1 x_1 + \cdots + a_k x_k.$$

Now $L(x) - L_n(x) = L(x) - L(y_k) + L(y_k) - L_n(y_k) + L_n(y_k) - L_n(x)$ so that

$$|L(x) - L_n(x)| \leq |L(x) - L(y_k)| + |L(y_k) - L_n(y_k)| + |L_n(y_k) - L_n(x)|$$

$$\leq \|L\| \, \|x - y_k\| + |L(y_k) - L_n(y_k)| + \|L_n\| \, \|x - y_k\|.$$

Keep k fixed and allow $n \to \infty$. By (14.4.9), we have $\lim_{n \to \infty} L_n(y_k) = L(y_k)$. Using this and (14.4.8), we have $\limsup_{n \to \infty} |L(x) - L_n(x)| \leq \|L\| \, \varepsilon + M\varepsilon$. Since ε is arbitrary, this implies that $\lim_{n \to \infty} L_n(x) = L(x)$.

Necessity. (14.4.9) is trivial. (14.4.8) follows from Theorem 14.4.2.

The fundamental theorem of weak* convergence can be applied usefully in two ways. Since conditions (14.4.8–9) are sufficient, they can be used to demonstrate the convergence of a sequence of functionals. Since they are also necessary, they can also be used to demonstrate the impossibility of convergence. Examples of both types of application follow. Some of the results have been obtained previously.

Ex. 6.   Convergence of Bernstein Polynomials.

Let $X = C[0, 1]$, $\|f\| = \max\limits_{0 \le x \le 1} |f(x)|$. Let x be a fixed point in $[0, 1]$ and set

$$L_n(f) = \sum_{k=0}^{n} f\left(\frac{k}{n}\right) \binom{n}{k} x^k (1 - x)^{n-k} \equiv B_n(f; x) \qquad (14.4.10)$$

and

$$L(f) = f(x). \qquad (14.4.11)$$

Now, from 9.3, Ex. 5, $\|L_n\| = \sum_{k=0}^{n} \binom{n}{k} x^k (1 - x)^{n-k} = 1$. Furthermore (6.2, Ex. 5) $L_n(e^{\lambda x}) = (1 - x + x e^{\lambda/n})^n$ so that $\lim\limits_{n \to \infty} L_n(e^{\lambda x}) = e^{\lambda x} = L(e^{\lambda x})$. The system of exponentials $1, e^x, e^{2x}, \ldots$, is closed in $C[0, 1]$. This follows from Weierstrass' Theorem (on the interval $[1, e]$) by a change of variable. Applying Theorem 14.4.4, we learn that $B_n(f; x) \to f(x)$.

Ex. 7.   Fejér Summability of Fourier Series.

Let X be the set of functions that are continuous and periodic on $[-\pi, \pi]$, $\|f\| = \max\limits_{-\pi \le x \le \pi} |f(x)|$. Let x be a fixed point in the interval and set $L(f) = f(x)$, $L_n(f) = \sigma_n(f; x) = $ the Fejér sums for f (cf. Lemma 12.2.2). For an integer p, $\sin px$ has the Fourier expansion $0 + 0 + \cdots + \sin px + 0 + 0 + \cdots$, and hence, for n sufficiently large, its Fejér sums (the averages of the partial sums of the series) must be $\dfrac{(n - p) \sin px}{n} \to \sin px$. A similar remark holds for $\cos px$.

Thus, $L_n\begin{pmatrix} \sin px \\ \cos px \end{pmatrix} \to L\begin{pmatrix} \sin px \\ \cos px \end{pmatrix}$. Now, $L_n(f) = \dfrac{1}{2\pi n} \displaystyle\int_{-\pi}^{\pi} f(x + t) \dfrac{\sin^2 nt/2}{(\sin t/2)^2} \, dt$. Since the kernel is positive, $\|L_n\| = \dfrac{1}{2\pi n} \displaystyle\int_{-\pi}^{\pi} \dfrac{\sin^2 nt/2}{(\sin t/2)^2} \, dt = 1$. An application of the Theorem 14.4.4 yields $\sigma_n(f; x) \to f(x)$.

Ex. 8.   Convergence of Quadrature Processes.

Let $X = C[a, b]$, $\|f\| = \max\limits_{a \le x \le b} |f(x)|$. Suppose that we have been given a triangular system of abscissas and weights

x_{11}			a_{11}		
x_{21}	x_{22}		a_{21}	a_{22}	
x_{31}	x_{32}	x_{33}	a_{31}	a_{32}	a_{33}

$$(14.4.12)$$

. .

. .

. .

and we construct a sequence of quadrature rules from them:

$$L_n(f) = \sum_{k=1}^{n} a_{nk} f(x_{nk}) \qquad n = 1, 2, \ldots \qquad (14.4.13)$$

Under what circumstances can we assert that

$$\lim_{n \to \infty} L_n(f) = \int_a^b f(x) \, dx, f \in C[a, b]? \qquad (14.4.14)$$

Inasmuch as $\|L_n\| = \sum_{k=1}^{n} |a_{nk}|$, an application of Theorem 14.4.4 yields:

THEOREM 14.4.5 (Pólya).   *Let there be given a sequence of quadrature rules* $\{L_n\}$. *We have*

$$\lim_{n \to \infty} L_n(f) = \int_a^b f(x) \, dx \quad \text{for all } f \in C[a, b] \qquad (14.4.14)$$

if and only if

$$\lim_{n \to \infty} L_n(x^k) = \int_a^b x^k \, dx \qquad k = 0, 1, \ldots, \qquad (14.4.15)$$

and

$$\sum_{k=1}^{n} |a_{nk}| \leq M \qquad n = 1, 2, \ldots \qquad (14.4.16)$$

for some constant M.

COROLLARY 14.4.6 (Stekloff).   *Let* $a_{nk} > 0$. *Then* (14.4.14) *holds if and only if* (14.4.15) *holds.*

Proof: $\sum_{k=1}^{n} |a_{nk}| = \sum_{k=1}^{n} a_{nk} = L_n(1)$. Now if (14.4.15) holds then $L_n(1)$ is bounded. Hence (14.4.16) holds and (14.4.14) follows. Conversely if (14.4.14) holds, (14.4.15) holds trivially.

COROLLARY 14.4.7 (Stieltjes).   *Let* $w(x)$ *be a weight function on* $[a, b]$. *Let* x_{ni} *be the abscissas and* w_{ni} *be the weights in the Gauss-Jacobi quadrature formula. Then,*

$$\int_a^b w(x) f(x) \, dx = \lim_{n \to \infty} \sum_{i=1}^{n} w_{ni} f(x_{ni}) \qquad (14.4.17)$$

for all $f(x) \in C[a, b]$.

Proof: Here we take $L(f) = \int_a^b w(x) f(x) \, dx$. If $L_n(f) = \sum_{i=1}^{n} w_{ni} f(x_{ni})$ then $L_n(x^k) = \int_a^b x^k \, dx$ for $0 \leq k \leq 2n - 1$. Moreover the weights w_{ni} are all positive. Hence $\sum_{i=1}^{n} |w_{ni}| = b - a$.

Actually, more may be proved: $L_n(f) \to \int_a^b w(x) f(x) \, dx$ for any bounded $f(x)$ for which the Riemann integral $\int_a^b w(x) f(x) \, dx$ exists.

Ex. 9.   Equidistributed Sequences.

A particular instance of the scheme (14.4.12) is

$$
\begin{array}{lll}
x_1 & h & \\[6pt]
x_1\,x_2 & \dfrac{h}{2} & \dfrac{h}{2} \\[10pt]
x_1\,x_2\,x_3 & \dfrac{h}{3} \quad \dfrac{h}{3} \quad \dfrac{h}{3} & \\[8pt]
\vdots & \vdots & \\[8pt]
& h = b - a,
\end{array}
\tag{14.4.18}
$$

leading to the equation

$$
\lim_{n \to \infty} \frac{1}{n} \sum_{k=1}^{n} f(x_k) = \frac{1}{b-a} \int_a^b f(x)\, dx, \quad f \in C[a,b]. \tag{14.4.19}
$$

The theory of this equation will now be developed.

LEMMA 14.4.8.   *Let F and G be two families of bounded, Riemann integrable functions on $[a,b]$. If (14.4.19) holds for all $f \in F$, and if for each $g \in G$ and $\varepsilon > 0$, we can find $f_1, f_2 \in F$ such that*

$$
f_1 \le g \le f_2 \qquad a \le x \le b
$$
$$
\int_a^b (f_2 - f_1)\, dx \le \varepsilon, \tag{14.4.20}
$$

then (14.4.19) holds for all $g \in G$.

Proof:

$$
\frac{1}{n}[f_1(x_1) + f_1(x_2) + \cdots + f_1(x_n)] \le \frac{1}{n}[g(x_1) + \cdots + g(x_n)]
$$
$$
\le \frac{1}{n}[f_2(x_1) + \cdots + f_2(x_n)].
$$

Hence,

$$
\int_a^b f_1(x)\, dx \le \lim_{n \to \infty} \inf \frac{1}{n}[g(x_1) + \cdots + g(x_n)]
$$
$$
\le \lim_{n \to \infty} \sup \frac{1}{n}[g(x_1) + \cdots + g(x_n)] \le \int_a^b f_2(x)\, dx.
$$

In view of (14.4.20), the lim inf and the lim sup can be made arbitrarily close, and hence $\lim_{n \to \infty} \frac{1}{n}[g(x_1) + \cdots + g(x_n)]$ exists. Call it I. Then,

$$
\int_a^b f_1(x)\, dx \le I \le \int_a^b f_2(x)\, dx.
$$

Then,

$$0 \le \left| I - \int_a^b g(x)\, dx \right| \le \int_a^b f_2(x)\, dx - \int_a^b f_1(x)\, dx \le \varepsilon.$$

Since ε is arbitrary, $I = \int_a^b g(x)\, dx.$

COROLLARY 14.4.9.   *Let $R[a, b]$ designate the space of bounded, Riemann integrable functions on $[a, b]$. The limit (14.4.19) holds for all $f \in C[a, b]$ if and only if it holds for all $f \in R[a, b]$.*

Proof: Since $C[a, b] \subset R[a, b]$, it remains to prove the "only if" part. Let $F = C[a, b]$ and $G = R[a, b]$. Take a $g \in R[a, b]$. We can obviously assume that g is nonnegative. (For otherwise add a sufficiently large constant.) Observe first that we can find piecewise constant functions f_1 and f_2 such that (14.4.20) holds. For, by an elementary property of the Riemann integral, we can find a sub-division $a = \xi_0 < \xi_1 < \cdots < \xi_{n-1} < \xi_n = b$ such that the upper sum $U = \sum_{k=1}^{n} M_k(\xi_k - \xi_{k-1})$ and the lower sum $L = \sum_{k=1}^{n} m_k(\xi_k - \xi_{k-1})$ differ by less than ε: $U - L < \varepsilon$. Here we have written $M_k = \sup_{\xi_{k-1} \le x \le \xi_k} f(x)$, $m_k = \inf_{\xi_{k-1} \le x \le \xi_k} f(x)$. Now set $f_2(x) = M_k$ on $\xi_{k-1} \le x < \xi_k$, $k = 1, 2, \ldots, n-1$ and $f_2(x) = M_n$ on $\xi_{n-1} \le x \le \xi_n$. Use a similar definition for f_1.

We wish to show next that we can find two functions f_1 and $f_2 \in F = C[a, b]$ such that the approximation (14.4.20) holds. This type of approximation has a transitivity property, so that by what we have just proved, it suffices to take g as positive and piecewise constant. Such functions g are linear combinations with positive coefficients of functions of the type $h(x) = 1$ for $c \le x \le d$, $h(x) = 0$ elsewhere in $[a, b]$ $(a \le c < d \le b)$. For sufficiently small δ, define a continuous function h_δ by means of $h_\delta(x) = 1 + \delta$ for $c \le x \le d$, $h_\delta(x) = \delta$ for $a \le x \le c - \delta$ and $d + \delta \le x \le b$, $h_\delta(x) = \text{linear for } c - \delta \le x \le c, d \le x \le d + \delta$. Then it is easily verified that $h_\delta(x) \ge h(x) + \delta$ and $0 < \int_a^b (h_\delta(x) - h(x))\, dx$ can be made arbitrarily small for δ sufficiently small. A similar process can be carried out on the under side of h.

DEFINITION 14.4.6.   A sequence of points $\{x_k\}$ lying in $[a, b]$ is called *equidistributed* in $[a, b]$ if (14.4.19) holds for all $f \in C[a, b]$.

The word "equidistributed" arises from the following property of such sequences. Let $[\sigma_1, \sigma_2]$ be any interval contained in $[a, b]$ and let $N_n[\sigma_1, \sigma_2]$ designate the number of points among the $x_1, x_2, \ldots, x_n$ that lie in

$$\sigma_1 \le x \le \sigma_2.$$

Then,

$$\lim_{n \to \infty} \frac{1}{n} N_n[\sigma_1, \sigma_2] = \frac{\sigma_2 - \sigma_1}{b - a}. \tag{14.4.21}$$

In other words, each interval contains, asymptotically, a fraction of the terms of the sequence in direct proportion to its length. This is a simple consequence of Corollary 14.4.9 by selecting $f = 1$ on $\sigma_1 \le x \le \sigma_2$ and $f = 0$ elsewhere.

COROLLARY 14.4.10. *Let $C_p[a, b]$ designate the space of continuous and periodic (i.e., $f(a) = f(b)$) functions on $[a, b]$. The limit (14.4.19) holds for all $f \in C_p[a, b]$ if and only if it holds for all $f \in C[a, b]$.*

Proof: Since $C_p[a, b] \subseteq C[a, b]$, it remains to prove the "only if" part. Let $F = C_p[a, b]$ and $G = C[a, b]$. Take $g \in C[a, b]$. We may evidently suppose it to be positive. Let $M = \max_{a \le x \le b} g(x)$. Select K so that $\dfrac{K}{g(a)} > 1$, $\dfrac{K}{g(b)} > 1$. Pick an η so that $0 < \eta < \min\left[b - a,\ \varepsilon M^{-1} K^{-1} \left(\dfrac{1}{g(a)} + \dfrac{1}{g(b)} \right)^{-1} \right]$. Define a continuous function $\Phi(x)$ by means of $\Phi(a) = \dfrac{K}{g(a)}$, $\Phi(b) = \dfrac{K}{g(b)}$, $\Phi(x) = 1$ for $a + \eta \le x \le b - \eta$, and $\Phi(x) = $ linear elsewhere. Then $\Phi(x) \ge 1$. Set $f_2(x) = \Phi(x) g(x)$. Then $f_2(x) \ge g(x)$ and $f_2(x)$ is continuous. Moreover, $f_2(a) = \Phi(a) g(a) = K = \Phi(b) g(b) = f_2(b)$, so that $f_2 \in C_p[a, b]$. Finally,

$$\int_a^b (f_2(x) - g(x))\, dx \le \int_a^{a+\eta} \frac{KM}{g(a)}\, dx + \int_{b-\eta}^b \frac{KM}{g(b)}\, dx = \eta K M \left(\frac{1}{g(a)} + \frac{1}{g(b)} \right) < \varepsilon.$$

A similar process can be carried out on the under side of g to yield an f_1.

THEOREM 14.4.11. *The sequence $\{x_k\}$ is equidistributed in $[a, b]$ if and only if (14.4.19) holds for a sequence of functions $\{f_k(x)\}$ that is closed in $C[a, b]$ or $C_p[a, b]$ with $\|f\| = \max_{a \le x \le b} |f(x)|$.*

Proof: Set $L(f) = \displaystyle\int_a^b f(x)\, dx$, $L_n(f) = \dfrac{1}{n} \sum_{k=1}^n f(x_k)$. Then, $\|L_n\| = 1$. Apply Theorem 14.4.4 and Corollary 14.4.10.

As yet, we have not exhibited an equidistributed sequence. The simplest one is provided by

THEOREM 14.4.12 (Bohl-Sierpínski-Weyl). *Let θ be an irrational number, and set $x_n = $ the fractional part of $n\theta$, i.e., $x_n = n\theta - [n\theta]$, where $[n\theta]$ is the largest integer $< n\theta$. Then $\{x_n\}$ is equidistributed in $[0, 1]$.*

Proof: The functions $e^{2\pi ikx}$, $k = 0, \pm 1, \pm 2, \ldots$, are closed in $C_p[0, 1]$ with $\|f\| = \max_{0 \le x \le 1} |f(x)|$. Hence by Theorem 14.4.11, it suffices to show that $\lim_{n \to \infty} \frac{1}{n} [e^{2\pi ikx_1} + \cdots + e^{2\pi ikx_n}] = \int_0^1 e^{2\pi ikx} \, dx$, $k = 0, \pm 1, \pm 2, \ldots$. For $k = 0$, the limit holds trivially. For $k \ne 0, \int_0^1 e^{2\pi ikx} \, dx = 0$. Since $e^{2\pi ikx_j} = e^{2\pi ik(j\theta - [j\theta])} = e^{2\pi ikj\theta}$,

$$\frac{1}{n} [e^{2\pi ikx_1} + \cdots + e^{2\pi ikx_n}] = \frac{1}{n} [e^{2\pi ik} + (e^{2\pi ik\theta})^2 + \cdots + (e^{2\pi ik\theta})^n]$$

$$= \frac{1}{n} (e^{2\pi ik\theta}) \frac{e^{2\pi ikn\theta} - 1}{e^{2\pi ik\theta} - 1}.$$

Since θ is irrational, $e^{2\pi ik\theta} \ne 1$. For $k = \pm 1, \pm 2, \ldots$, the exponential expression is bounded as $n \to \infty$ and the limit 0 is obtained.

Ex. 10. A nonexistence theorem for quadratures.
We have seen that it is indeed possible to have

$$\sum_{k=1}^n a_{nk} f(x_k) \to \int_a^b f(x) \, dx \quad \text{for} \quad f \in C[a, b].$$

Notice that the left-hand member involves a double array of weights. Would it be possible to replace it with a single array and have

$$\sum_{k=1}^\infty a_k f(x_k) = \int_a^b f(x) \, dx \tag{14.4.22}$$

for all $f \in C[a, b]$? Such a formula would obviously be much simpler. In (14.4.22) we assume that $a_k \ne 0$, $a \le x_k \le b$, the x_k are distinct, and both the a_k and the x_k are independent of f. An application of Theorem 14.4.4 will show this is impossible:

THEOREM 14.4.13. *It is impossible to have* $\int_a^b f(x) \, dx = \sum_{k=1}^\infty a_k f(x_k)$ *for all* $f \in C[a, b]$ *under the above conditions on* a_k *and* x_k.

Proof: Note first that if (14.4.22) holds, then the x_k must be dense in $[a, b]$. For, suppose there were an interval I contained in $[a, b]$ which is free of abscissas x_k. Let I' be an interval interior to I. Then it is clear that we may find a continuous function $f(x)$ which is zero exterior to I and is positive interior to I'. Then, $f(x_k) = 0$, $k = 1, 2, \ldots$, since all the x_k lie outside I. Thus, $0 < \int_a^b f(x) \, dx = \sum_{k=1}^\infty a_k f(x_k) = 0$.

Work in $X = C[a, b]$, $\|f\| = \max_{a \le x \le b} |f(x)|$. Set $L(f) = \int_a^b f(x) \, dx$ and $L_n(f) = \sum_{k=1}^n a_k f(x_k)$. Then, $\|L\| = b - a$, $\|L_n\| = \sum_{k=1}^n |a_k|$. By Theorem 14.4.4, if $L_n(f) \to L(f)$ for all $f \in X$, we can find an M such that $\sum\limits_{k=1}^{n} |a_k| < M$ for all n. That is, $\sum\limits_{k=1}^{\infty} a_k$ must converge absolutely.

Select a k for which $a < x_k < b$. Designate by I_ε the interval

$$x_k - \varepsilon \leq x \leq x_k + \varepsilon.$$

We can choose ε sufficiently small so that I_ε will be contained in $a < x < b$. The abscissas x_i lying in I_ε will form a subset of all abscissas, and of the former, by choosing ε small enough, we can also guarantee that x_k will be the one which possesses the minimum subscript. In order of increasing subscripts the abscissas lying in I_ε will be designated by $x_k = x_{n_1(\varepsilon)}, x_{n_2(\varepsilon)}, \cdots,$ where, moreover, $\lim\limits_{\varepsilon \to 0} n_2(\varepsilon) = \infty$.

For each ε construct the following continuous triangular shaped function:

$$f_\varepsilon(x) = 0 \qquad \text{for } x \text{ exterior to } I_\varepsilon$$
$$f_\varepsilon(x) = \text{linear} \qquad x_k - \varepsilon \leq x \leq x_k + \varepsilon$$
$$f_\varepsilon(x_k) = 1$$
$$f_\varepsilon(x) = \text{linear} \qquad x_k \leq x \leq x_k + \varepsilon.$$

Then $|f_\varepsilon(x)| \leq 1$ and $\int_a^b f_\varepsilon(x)\,dx = \varepsilon$. Also,

$$\varepsilon = \sum_{j=0}^{\infty} a_j f_\varepsilon(x_j) = a_k f_\varepsilon(x_k) + \sum_{j=2}^{\infty} a_{n_j(\varepsilon)} f_\varepsilon(x_{n_j(\varepsilon)}).$$

Therefore, $|\varepsilon - a_k| \leq \sum\limits_{j=2}^{\infty} |a_{n_j(\varepsilon)}|$. Now let $\varepsilon \to 0$. The left side approaches $a_k \neq 0$. Since $n_2(\varepsilon) \to \infty$, the right side approaches zero. This is a contradiction and (14.4.22) is impossible.

Ex. 11. Divergence of Fourier Series of a Continuous Function. This is developed in the final two theorems.

THEOREM 14.4.14 (Fejér). *Let*

$$\rho_n = \frac{1}{2\pi} \int_{-\pi}^{\pi} \left| \frac{\sin (n + \frac{1}{2})t}{\sin \dfrac{t}{2}} \right| dt. \tag{14.4.23}$$

Then,

$$\lim_{n \to \infty} \frac{\rho_n}{\log n} = \frac{4}{\pi^2}. \tag{14.4.24}$$

Proof: We have

$$\rho_n = \frac{1}{\pi} \int_{-\pi/2}^{\pi/2} \left| \frac{\sin (2n + 1)t}{\sin t} \right| dt = \frac{2}{\pi} \int_{0}^{\pi/2} \frac{|\sin (2n + 1)t|}{\sin t}\,dt.$$

Hence,

$$\rho_n = \frac{2}{n} \int_0^{\pi/2} \frac{|\sin{(2n+1)t}|}{t} \, dt + \frac{2}{\pi} \int_0^{\pi/2} \left(\frac{1}{\sin t} - \frac{1}{t} \right) |\sin{(2n+1)t}| \, dt$$

$$= \text{I}_n + \text{II}_n$$

Now the function $\dfrac{1}{\sin t} - \dfrac{1}{t}$ is nonnegative on $\left[0, \dfrac{\pi}{2} \right]$ and $0 \le |\sin{(2n+1)t}| \le 1$.

Therefore, $0 \le \text{II}_n \le \dfrac{2}{\pi} \displaystyle\int_0^{\pi/2} \left(\dfrac{1}{\sin t} - \dfrac{1}{t} \right) \, dt$, and the integrals II_n are bounded. Again,

$$\text{I}_n = \frac{2}{\pi} \int_0^{(n+1)\pi/(2n+1)} \frac{|\sin{(2n+1)t}|}{t} \, dt - \frac{2}{\pi} \int_{\pi/2}^{(n+1)\pi/(2n+1)} \frac{|\sin{(2n+1)t}|}{t} \, dt$$

$$= \text{III}_n - \text{IV}_n.$$

But,

$$0 \le \text{IV}_n = \frac{2}{\pi} \int_{\pi/2}^{(n+1)\pi/(2n+1)} \frac{|\sin{(2n+1)t}|}{t} \, dt \le \frac{2}{\pi} \int_{\pi/2}^{(n+1)\pi/(2n+1)} \frac{dt}{t}.$$

Since $\displaystyle\lim_{n \to \infty} \frac{(n+1)\pi}{2n+1} = \frac{\pi}{2}$, $\displaystyle\lim_{n \to \infty} \text{IV}_n = 0$. Changing variables,

$$\text{III}_n = \frac{2}{\pi} \int_0^{(n+1)\pi/(2n+1)} \frac{|\sin{(2n+1)t}|}{t} \, dt = \frac{2}{\pi} \int_0^{(n+1)\pi} \frac{|\sin t|}{t} \, dt$$

$$= \frac{2}{\pi} \int_0^{\pi} \frac{|\sin t|}{t} \, dt + \frac{2}{\pi} \int_{\pi}^{(n+1)\pi} \frac{|\sin t|}{t} \, dt$$

$$= \text{V} + \text{VI}_n.$$

Now,

$$\text{VI}_n = \frac{2}{\pi} \int_0^{\pi} \left(\frac{1}{t+\pi} + \frac{1}{t+2\pi} + \cdots + \frac{1}{t+n\pi} \right) \sin t \, dt$$

$$= \frac{2}{\pi} \int_0^{\pi} S_n(t) \sin t \, dt,$$

where $S_n(t) = \dfrac{1}{t+\pi} + \dfrac{1}{t+2\pi} + \cdots + \dfrac{1}{t+n\pi}$. On $0 \le t \le \pi$ we have,

$$\frac{1}{2\pi} + \frac{1}{3\pi} + \cdots + \frac{1}{(n+1)\pi} \le S_n(t) \le \frac{1}{\pi} + \frac{1}{2\pi} + \cdots + \frac{1}{n\pi}.$$ We now employ the well known property of the harmonic series:

$$1 + \frac{1}{2} + \frac{1}{3} + \cdots + \frac{1}{n} - \log n = \gamma_n, \lim_{n \to \infty} \gamma_n = \gamma = .577 \cdots.$$

Thus,

$$\frac{1}{\pi} (\log (n + 1) + \gamma_{n+1} - 1) \leq S_n(t) \leq \frac{1}{\pi} (\log n + \gamma_n),$$

and

$$\frac{2}{\pi^2} (\log (n + 1) + \gamma_{n+1} - 1) \int_0^\pi \sin t \, dt \leq VI_n \leq \frac{2}{\pi^2} (\log n + \gamma_n) \int_0^\pi \sin t \, dt.$$

Therefore, $\lim\limits_{n \to \infty} \dfrac{VI_n}{\log n} = \dfrac{4}{\pi^2}$. Now $\rho_n = V + VI_n - IV_n + II_n$ and the theorem follows from the individual limiting behaviors.

THEOREM 14.4.15 (du Bois-Reymond).   *There exists a continuous function whose Fourier series diverges at $x = 0$.*

Proof: From (12.1.5), the partial sum of the Fourier series evaluated at $x = 0$ is $L_n(f) = \dfrac{1}{2\pi} \displaystyle\int_{-\pi}^\pi \dfrac{\sin (n + \frac{1}{2})t}{\sin \dfrac{t}{2}} f(t) \, dt$. Work in the space $X: C[-\pi, \pi]$,

$f(-\pi) = f(\pi)$, $\|f\| = \max\limits_{-\pi \leq x \leq \pi} |f(x)|$. Since, $\|L_n\| = \rho_n \to \infty$, the theorem follows from Theorem 14.4.4.

NOTES ON CHAPTER XIV

14.1   For a glance at the integration rules available, see Stroud [1].

14.2   Markoff [1], Chapters V-VII, Szegö [1], pp. 47–48, Hobson [1] pp. 76–83.

14.3   Sard [1], Nikolsky [1], Davis [2], [3], Davis and Rabinowitz [1], Golomb and Weinberger [1], Krilov [1] Chapter 8.

14.4   Ljusternik and Sobolew [1] pp. 131–136. Feldheim [1], Krilov [1], Chapter 12. For equidistributed sequences, see Pólya and Szegö [1] vol. I, pp. 67–77. Theorem 14.4.13, Davis [1]. On the other hand, if duplications in the abscissas are allowed, (14.4.22) is possible. See John [1].

PROBLEMS

1. Can constants a_1, a_2, x_1, x_2 be determined so that the differentiation rule of "Gauss type" $f'(0) = a_1 f(x_1) + a_2 f(x_2)$ is valid for $f = 1, x, x^2, x^3$?

2. Integrate the Bernstein polynomial $B_n(f; x)$ over $[0, 1]$ and interpret what is obtained.

3. Let $\{p_n{}^*(x)\}$ be orthonormal polynomials with respect to the inner product $(f, g) = \displaystyle\int_{-1}^{+1} w(x) f(x) g(x) \, dx$. Let $x_1, \ldots, x_n$ and $w_1, \ldots, w_n$ be the abscissas

and weights of the corresponding Gauss-Jacobi integration rule. The polynomials $p_0^*, \ldots, p_{n-1}^*$ are also orthonormal with respect to the inner product

$$(f, g) = \sum_{k=1}^{n} w_k f(x_k) g(x_k).$$

4. The error in the Gauss rule of order n over $[a, b]$ is

$$\frac{(b - a)^{2n+1} 2^{2n+1} (n!)^4 f^{(2n)}(\xi)}{(2n + 1)[(2n)!]^3}.$$

5. In problems 5, 6, 7, and 8, $E_n(f)$ designates the error in the Gauss rule of order n over $[-1, 1]$. Determine $E_n(x^{2n})$.

6. If f is analytic, then $E_n(f) = \dfrac{1}{\pi i} \displaystyle\int_C \frac{f(z) Q_n(z)}{P_n(z)} \, dz.$ (Cf. (12.4.13).)

7. Prove that $\lim\limits_{n \to \infty} E_n(f) = 0$ if f is a bounded Riemann integrable function on $[-1, 1]$.

8. Study the rapidity of convergence to 0 of $E_n(f)$ where f is analytic on $[-1, 1]$.

9. Let $Q_n(f) = \displaystyle\sum_{k=1}^{n} w_{nk} f(x_{nk})$, $n = 1, \ldots$, be a sequence of quadrature rules for $\displaystyle\int_a^b f(x) \, dx$. Suppose that Q_n is exact for $\mathscr{P}_n$. If for some ε, the abscissas satisfy $a + \varepsilon \leq x_{nk} \leq b - \varepsilon$, then for sufficiently large n, $w_{n1}, \ldots, w_{nn}$ must have a change in sign.

10. Derive Landau's Theorem (Lemma 9.3.7) as a consequence of Theorem 14.4.4.

11. Study the behavior as $n \to \infty$ of the sequence

$$\sigma_n = |\sin 1 \sin 2 \sin 3 \cdots \sin n|.$$

12. Study the behavior as $n \to \infty$ of the sequence

$$\sigma_n = [\theta] + [2\theta] + \cdots + [n\theta]$$

for θ rational and irrational.

13. A triangular sequence $x_{n1}, x_{n2}, \ldots, x_{nn}$ $(n = 1, 2, \ldots)$ is called equidistributed on $[a, b]$ if $\lim\limits_{n \to \infty} \dfrac{1}{n} \displaystyle\sum_{k=1}^{n} f(x_{nk}) = \int_a^b f(x) \, dx$ for all $f \in C[a, b]$. Give simple examples of equidistributed triangular sequences.

14. Find necessary and sufficient conditions that a triangular sequence be equidistributed.

Appendix

Short Guide to the Orthogonal Polynomials

I

Name: Legendre   *Symbol:* $P_n(x)$ *Interval;* $[-1, 1]$

Weight: 1   *Standardization:* $P_n(1) = 1$

Norm: $\displaystyle\int_{-1}^{+1} (P_n(x))^2 \, dx = \frac{2}{2n + 1}$

Explicit Expression: $\displaystyle P_n(x) = \frac{1}{2^n} \sum_{m=0}^{[n/2]} (-1)^m \binom{n}{m} \binom{2n - 2m}{n} x^{n-2m}$

Recurrence Relation: $(n + 1)P_{n+1}(x) = (2n + 1)xP_n(x) - nP_{n-1}(x)$

Differential Equation: $(1 - x^2)y'' - 2xy' + n(n + 1)y = 0$
$$y = P_n(x)$$

Rodrigues' Formula: $\displaystyle P_n(x) = \frac{(-1)^n}{2^n n!} \frac{d^n}{dx^n} \{(1 - x^2)^n\}$

Generating Function: $\displaystyle R^{-1} = \sum_{n=0}^{\infty} P_n(x)z^n; \quad -1 < x < 1, |z| < 1,$
$$R = \sqrt{1 - 2xz + z^2}.$$

Inequality: $|P_n(x)| \leq 1, \quad -1 \leq x \leq 1.$

II

Name: Tschebysheff, First Kind   *Symbol:* $T_n(x)$ *Interval:* $[-1, 1]$

Weight: $(1 - x^2)^{-\frac{1}{2}}$ *Standardization:* $T_n(1) = 1$

Norm: $\displaystyle\int_{-1}^{+1} (1 - x^2)^{-\frac{1}{2}} (T_n(x))^2 \, dx = \begin{cases} \pi/2, & n \neq 0 \\ \pi, & n = 0 \end{cases}$

Explicit Expression: $\displaystyle \frac{n}{2} \sum_{m=0}^{[n/2]} (-1)^m \frac{(n - m - 1)!}{m! \, (n - 2m)!} (2x)^{n-2m} = \cos(n \arccos x)$
$$= T_n(x)$$

Recurrence Relation: $T_{n+1}(x) = 2xT_n(x) - T_{n-1}(x)$

Differential Equation: $(1 - x^2)y'' - xy' + n^2 y = 0$
$$y = T_n(x)$$

Rodrigues' Formula: $\displaystyle \frac{(-1)^n (1 - x^2)^{\frac{1}{2}} \sqrt{\pi}}{2^{n+1} \Gamma(n + \frac{1}{2})} \frac{d^n}{dx^n} \{(1 - x^2)^{n - \frac{1}{2}}\} = T_n(x)$

Generating Function: $\displaystyle \frac{1 - xz}{1 - 2xz + z^2} = \sum_{n=0}^{\infty} T_n(x)z^n, \quad -1 < x < 1, |z| < 1.$

Inequality: $|T_n(x)| \leq 1, \quad -1 \leq x \leq 1.$

III

Name: Tschebysheff, Second Kind   *Symbol:* $U_n(x)$ *Interval:* $[-1, 1]$

Weight: $(1 - x^2)^{\frac{1}{2}}$ *Standardization:* $U_n(1) = n + 1$

Norm: $\displaystyle\int_{-1}^{+1} (1 - x^2)^{\frac{1}{2}} [U_n(x)]^2 \, dx = \frac{\pi}{2}$

Explicit Expression: $\displaystyle U_n(x) = \sum_{m=0}^{[n/2]} (-1)^m \frac{(m - n)!}{m! \, (n - 2m)!} (2x)^{n-2m}$

$$U_n(\cos \theta) = \frac{\sin (n + 1)\theta}{\sin \theta}$$

Recurrence Relation: $U_{n+1}(x) = 2xU_n(x) - U_{n-1}(x)$

Differential Equation: $(1 - x^2)y'' - 3xy' + n(n + 2)y = 0$

$$y = U_n(x)$$

Rodrigues' Formula: $\displaystyle U_n(x) = \frac{(-1)^n(n + 1)\sqrt{\pi}}{(1 - x^2)^{\frac{1}{2}} 2^{n+1}\Gamma(n + \frac{3}{2})} \frac{d^n}{dx^n} \{(1 - x^2)^{n+\frac{1}{2}}\}$

Generating Function: $\displaystyle \frac{1}{1 - 2xz + z^2} = \sum_{n=0}^{\infty} U_n(x)z^n, \; -1 < x < 1, |z| < 1.$

Inequality: $|U_n(x)| \leq n + 1, \; -1 \leq x \leq 1.$

IV

Name: Jacobi   *Symbol:* $P_n^{(\alpha,\beta)}(x)$ *Interval:* $[-1, 1]$

Weight: $(1 - x)^\alpha(1 + x)^\beta; \; \alpha, \beta > -1$

Standardization: $\displaystyle P_n^{(\alpha,\beta)}(x) = \binom{n + \alpha}{n}$

Norm: $\displaystyle\int_{-1}^{+1} (1 - x)^\alpha(1 + x)^\beta [P_n^{(\alpha,\beta)}(x)]^2 \, dx$

$$= \frac{2^{\alpha+\beta+1}\Gamma(n + \alpha + 1)\Gamma(n + \beta + 1)}{(2n + \alpha + \beta + 1)n! \, \Gamma(n + \alpha + \beta + 1)}$$

Explicit Expression:

$$P_n^{(\alpha,\beta)}(x) = \frac{1}{2^n} \sum_{m=0}^{n} \binom{n + \alpha}{m} \binom{n + \beta}{n - m} (x - 1)^{n-m}(x + 1)^m$$

Recurrence Relation: $2(n + 1)(n + \alpha + \beta + 1)(2n + \alpha + \beta)P_{n+1}^{(\alpha,\beta)}(x)$

$$= (2n + \alpha + \beta + 1)[(\alpha^2 - \beta^2) + (2n + \alpha + \beta + 2)$$

$$\times (2n + \alpha + \beta)x]P_n^{(\alpha,\beta)}(x)$$

$$- 2(n + \alpha)(n + \beta)(2n + \alpha + \beta + 2)P_{n-1}^{(\alpha,\beta)}(x)$$

Differential Equation:

$$(1 - x^2)y'' + (\beta - \alpha - (\alpha + \beta + 2)x)y' + n(n + \alpha + \beta + 1)y = 0$$

$$y = P_n^{(\alpha,\beta)}(x)$$

Rodrigues' Formula:

$$P_n^{(\alpha,\beta)}(x) = \frac{(-1)^n}{2^n n!\,(1-x)^\alpha(1+x)^\beta} \frac{d^n}{dx^n}\{(1-x)^{n+\alpha}(1+x)^{n+\beta}\}$$

Generating Function:

$$R^{-1}(1 - z + R)^{-\alpha}(1 + z + R)^{-\beta} = \sum_{n=0}^{\infty} 2^{-\alpha-\beta} P_n^{(\alpha,\beta)}(x)z^n,$$

$$R = \sqrt{1 - 2xz + z^2},\ |z| < 1$$

Inequality: $\displaystyle\max_{-1 \le x \le 1} |P_n^{(\alpha,\beta)}(x)| = \begin{cases} \dbinom{n+q}{n} \sim n^q \text{ if } q = \max(\alpha, \beta) \ge -\tfrac{1}{2} \\[2mm] |P_n^{(\alpha,\beta)}(x')| \sim n^{-\frac{1}{2}} \text{ if } q < -\tfrac{1}{2}. \\ x' \text{ is one of the two maximum points nearest} \\[2mm] \dfrac{\beta - \alpha}{\alpha + \beta + 1} \end{cases}$

V

Name: Generalized Laguerre   *Symbol:* $L_n^{(\alpha)}(x)$ *Interval:* $[0, \infty]$

Weight: $x^\alpha e^{-x},\ \alpha > -1$ *Standardization:* $L_n^{(\alpha)}(x) = \dfrac{(-1)^n}{n!} x^n + \cdots$

Norm: $\displaystyle\int_0^\infty x^\alpha e^{-x}(L_n^{(\alpha)}(x))^2\,dx = \frac{\Gamma(n + \alpha + 1)}{n!}$

Explicit Expression: $L_n^{(\alpha)}(x) = \displaystyle\sum_{m=0}^{n} (-1)^m \binom{n+\alpha}{n-m} \frac{1}{m!} x^m$

Recurrence Relation:

$$(n + 1)L_{n+1}^{(\alpha)}(x) = [(2n + \alpha + 1) - x]L_n^{(\alpha)}(x) - (n + \alpha)L_{n-1}^{(\alpha)}(x)$$

Differential Equation: $xy'' + (\alpha + 1 - x)y' + ny = 0$

$$y = L_n^{(\alpha)}(x)$$

Rodrigues' Formula: $L_n^{(\alpha)}(x) = \dfrac{1}{n!\,x^\alpha e^{-x}} \dfrac{d^n}{dx^n}\{x^{n+\alpha}e^{-x}\}$

Generating Function: $(1 - z)^{-\alpha-1} \exp\left(\dfrac{xz}{z - 1}\right) = \displaystyle\sum_{n=0}^{\infty} L_n^{(\alpha)}(x)z^n$

Inequality: $|L_n^{(\alpha)}(x)| \le \dfrac{\Gamma(n + \alpha + 1)}{n!\,\Gamma(\alpha + 1)} e^{x/2};\quad \begin{array}{l} x \ge 0 \\ \alpha > 0 \end{array}$

$$|L_n^{(\alpha)}(x)| \le \left[2 - \frac{\Gamma(\alpha + n + 1)}{n!\,\Gamma(\alpha + 1)}\right] e^{x/2};\quad \begin{array}{l} x \ge 0 \\ -1 < \alpha < 0 \end{array}$$

VI

Name: Hermite   *Symbol:* $H_n(x)$ *Interval:* $[-\infty, \infty]$

Weight: e^{-x^2} *Standardization:* $H_n(x) = 2^n x^n + \cdots$

Norm: $\displaystyle\int_{-\infty}^{+\infty} e^{-x^2} [H_n(x)]^2 \, dx = \sqrt{\pi} \, 2^n n!$

Explicit Expression: $\displaystyle H_n(x) = n! \sum_{m=0}^{[n/2]} (-1)^m \frac{(2x)^{n-2m}}{m! \, (n-2m)!}$

Recurrence Relation: $H_{n+1}(x) = 2x H_n(x) - 2n H_{n-1}(x)$

Differential Equation: $y'' - 2xy' + 2ny = 0$
$$y = H_n(x)$$

Rodrigues' Formula: $\displaystyle H_n(x) = (-1)^n e^{x^2} \frac{d^n}{dx^n} (e^{-x^2})$

Generating Function: $\displaystyle e^{2xz - z^2} = \sum_{n=0}^{\infty} \frac{H_n(x) z^n}{n!}$

Inequality: $\displaystyle |H_{2m}(x)| \leq e^{x^2/2} \, 2^{2m} \, m! \left[2 - \frac{1}{2^{2m}} \binom{2m}{m} \right], \quad x \geq 0$

$$|H_{2m+1}(x)| \leq x e^{x^2/2} \frac{(2m+2)!}{(m+1)!}, \quad x \geq 0$$

Table of the Tschebyscheff Polynomials

$T_0(x) = 1$

$T_1(x) = x$

$T_2(x) = 2x^2 - 1$

$T_3(x) = 4x^3 - 3x$

$T_4(x) = 8x^4 - 8x^2 + 1$

$T_5(x) = 16x^5 - 20x^3 + 5x$

$T_6(x) = 32x^6 - 48x^4 + 18x^2 - 1$

$T_7(x) = 64x^7 - 112x^5 + 56x^3 - 7x$

$T_8(x) = 128x^8 - 256x^6 + 160x^4 - 32x^2 + 1$

$T_9(x) = 256x^9 - 576x^7 + 432x^5 - 120x^3 + 9x$

$T_{10}(x) = 512x^{10} - 1280x^8 + 1120x^6 - 400x^4 + 50x^2 - 1$

$T_{11}(x) = 1024x^{11} - 2816x^9 + 2816x^7 - 1232x^5 + 220x^3 - 11x$

$T_{12}(x) = 2048x^{12} - 6144x^{10} + 6912x^8 - 3584x^6 + 840x^4 - 72x^2 + 1$

Table of Powers as Combinations of the Tschebyscheff Polynomials

$$1 = T_0$$

$$x = T_1$$

$$x^2 = \frac{1}{2}(T_0 + T_2)$$

$$x^3 = \frac{1}{4}(3T_1 + T_3)$$

$$x^4 = \frac{1}{8}(3T_0 + 4T_2 + T_4)$$

$$x^5 = \frac{1}{16}(10T_1 + 5T_3 + T_5)$$

$$x^6 = \frac{1}{32}(10T_0 + 15T_2 + 6T_4 + T_6)$$

$$x^7 = \frac{1}{64}(35T_1 + 21T_3 + 7T_5 + T_7)$$

$$x^8 = \frac{1}{128}(35T_0 + 56T_2 + 28T_4 + 8T_6 + T_8)$$

$$x^9 = \frac{1}{256}(126T_1 + 84T_3 + 36T_5 + 9T_7 + T_9)$$

$$x^{10} = \frac{1}{512}(126T_0 + 210T_2 + 120T_4 + 45T_6 + 10T_8 + T_{10})$$

$$x^{11} = \frac{1}{1024}(462T_1 + 330T_2 + 165T_5 + 55T_7 + 11T_9 + T_{11})$$

$$x^{12} = \frac{1}{2048}(462T_0 + 792T_2 + 495T_4 + 220T_6 + 66T_8 + 12T_{10} + T_{12})$$

Table of the Legendre Polynomials

$P_0 = 1$

$P_1 = x$

$P_2 = \dfrac{1}{2}(3x^2 - 1)$

$P_3 = \dfrac{1}{2}(5x^3 - 3x)$

$P_4 = \dfrac{1}{8}(35x^4 - 30x^2 + 3)$

$P_5 = \dfrac{1}{8}(63x^5 - 70x^3 + 15x)$

$P_6 = \dfrac{1}{16}(231x^6 - 315x^4 + 105x^2 - 5)$

$P_7 = \dfrac{1}{16}(429x^7 - 693x^5 + 315x^3 - 35x)$

$P_8 = \dfrac{1}{128}(6435x^8 - 12012x^6 + 6930x^4 - 1260x^2 + 35)$

Table of Powers as Linear Combinations of the Legendre Polynomials

$x^0 = P_0$

$x^1 = P_1$

$x^2 = \dfrac{1}{3}\,(2P_2 + P_0)$

$x^3 = \dfrac{1}{5}\,(2P_3 + 3P_1)$

$x^4 = \dfrac{1}{35}\,(8P_4 + 20P_2 + 7P_0)$

$x^5 = \dfrac{1}{63}\,(8P_5 + 28P_3 + 27P_1)$

$x^6 = \dfrac{1}{231}\,(16P_6 + 72P_4 + 110P_2 + 33P_0)$

$x^7 = \dfrac{1}{429}\,(16P_7 + 88P_5 + 182P_3 + 143P_1)$

$x^8 = \dfrac{1}{6435}\,(128P_8 + 832P_6 + 2160P_4 + 2600P_2 + 715P_0)$

Bibliography

Achieser, N. I.
1. Theory of Approximation, New York, 1956.

Alexits, G.
1. Konvergenzprobleme der Orthogonalreihen, Budapest, 1960.

Aronszajn, N.
1. Introduction to the Theory of Hilbert Spaces, Stillwater, Oklahoma, 1950.
2. Theory of Reproducing Kernels, Trans. Amer. Math. Soc., vol. 68 (1950), pp. 337–404.

Banach, S.
1. Théorie des Opérations Linéaires, New York, 1932.

Behnke, H. and Sommer, F.
1. Theorie der Analytischen Funktionen einer Komplexen Veränderlichen, Berlin, 1955.

Bergman, S.
1. Sur les Fonctions Orthogonales, New York, 1941.
2. The Kernel Function and Conformal Mapping, New York, 1950.

Bergman, S. and Schiffer, M.
1. Kernel Functions and Elliptic Differential Equations in Mathematical Physics, New York, 1953.

Bernstein, S.
1. Leçons sur les Propriétés Extremales, Paris, 1926.

Birkhoff, G. and Rota, G. C.
1. On the Completeness of Sturm-Liouville Expansions, Amer. Math. Mon., vol. 67 (1960), pp. 835–841.

Boas, R. P., Jr.
1. Expansions of Analytic Functions, Trans. Amer. Math. Soc., vol. 48 (1940), pp. 467–487.
2. Representation of Functions by Lidstone Series, Duke Math. J., vol. 10 (1943), pp. 239–245.
3. Entire Functions, New York, 1954.
4. A Primer of Real Functions, Rahway, New Jersey, 1960.

Boas, R. P., Jr. and Buck, R. C.
1. Polynomial Expansions of Analytic Functions, Berlin, 1958.

Buck, R. C.
1. Interpolation Series, Trans. Amer. Math. Soc., vol. 64 (1948), pp. 283–298.
2. Linear Spaces and Approximation Theory, pp. 11–23 of "On Numerical Approximation," R. E. Langer, ed., Madison, 1959.
3. Survey of Recent Russian Literature on Approximation, in "On Numerical Approximation," R. E. Langer, ed., Madison, 1959.
4. Bounded Continuous Functions on a Locally Compact Space, Michigan Math. Journal, vol. 5 (1958) pp. 95–104.
5. Zero Sets for Continuous Functions, Proc. Amer. Math. Soc., vol. 11, (1960), pp. 630–633.
6. Studies in Modern Analysis, Vol. 1, The Mathematical Association of America, 1962.

Clenshaw, C. W.
1. Chebyshev Series for Mathematical Functions, Math. Tables Series No. 5, National Physical Laboratories, Her Majesty's Stationery Office, London, 1962.

Cooke, R. G.
1. Infinite Matrices and Sequence Spaces, London, 1950.

Curry, H. B.
1. Abstract Differential Operators and Interpolation Formulas, Portugaliae Mathematica, vol. 10 (1951), pp. 135–162.

Davis, P. J.
1. On Simple Quadratures, Pro. Amer. Math. Soc., vol. 4 (1953), pp. 127–136.
2. Errors of Numerical Approximation for Analytic Functions, Journal of Rational Mechanics and Analysis, vol. 2 (1953), pp. 303–313.
3. On a Problem in the Theory of Mechanical Quadratures, Pac. J. Math., vol. 5 (1955), pp. 669–674.

Davis, P. J. and Fan, K.
1. Complete Sequences and Approximations in Normed Linear Spaces, Duke Math. J., vol. 24 (1957), pp. 183–192.

Davis, P. J. and Rabinowitz, P.
1. On the Estimation of Quadrature Errors for Analytic Functions, Math. Tables and Other Aids to Computation, vol. VIII (1954), pp. 193–203.
2. Advances in Orthonormalizing Computation, Advances in Computers, vol. II, F.L. Alt, ed., New York, 1961.

Day, M. M.
1. Normed Linear Spaces, Berlin, 1958.

Dickinson, D. J., Pollak, H. O., and Wannier, G. H.
1. On a Class of Polynomials Orthogonal Over a Denumerable Set, Pac. J. Math., vol. 6 (1956), pp. 234–247.

Dieudonné, J.
1. Foundations of Modern Analysis, New York, 1960.

Dunford, N. and Schwartz, J. T.
1. Linear Operators, New York, 1958.

Eggleston, H. G.
1. Convexity, Cambridge, 1958.

Favard, J.
1. Sur les Polynomes de Tschebicheff, C. R. Acad. Sci. Paris, vol. 200 (1935), pp. 2052–2053.

Fejér, L.
1. Über die Lage der Nullstellen von Polynomen, die aus Minimumforderungen gewisser Art entspringen, Math. Ann., vol. 85 (1922), pp. 41–48.

Fekete, M.
1. Über den Transfiniten Durchmesser ebener Punktmengen, 3te Mitteilung Math. Zeitschrift, vol. 37 (1933), pp. 635–646.

Feldheim, M. E.
1. Théorie de la Convergence des Procédes d'interpolation et de Quadrature Mécanique, Paris, 1939.

Franklin, P.
1. Functions of a Complex Variable with Assigned Derivatives at an Infinite Number of Points . . . , Acta Mathematica, vol. 47 (1926), pp. 371–385.

Gantmacher, F. R.
1. The Theory of Matrices, New York, 1959.

Gelfond, A. O.
1. Calculus of Finite Differences, Moscow-Leningrad, 1952. (In Russian)

Golomb, M. and Weinberger, H. F.
1. Optimal Approximation and Error Bounds, pp. 117–190 of "On Numerical Approximation," R. E. Langer, ed., Madison, 1959.

Gontscharoff, V. L.
1. Theory of Interpolation and Approximation of Functions, Moscow, 1954. (In Russian)

Halmos, P. R.
1. Introduction to Hilbert Space, New York, 1951.
2. Naive Set Theory, New York, 1960.

Hardy, G. H., Littlewood, J. E., and Pólya, G.
1. Inequalities, Cambridge, 1934.

Hirschfeld, R. A.
1. On Best Approximations in Normed Vector Spaces, I, Nieuw Archief voor Wiskunde, vol. 3 (1958), pp. 41–51.
2. On Best Approximations in Normed Vector Spaces, II, Nieuw Archief voor Wiskunde, vol. 3 (1958), pp. 99–107.

Hobson, E. W.
1. The Theory of Spherical and Ellipsoidal Harmonics, Cambridge, 1931.

Jackson, D.
1. The Theory of Approximation, New York, 1930.
2. Fourier Series and Orthogonal Polynomials, Menasha, Wisconsin, 1941.

John, F.
1. A representation of Stieltjes integrals by conditionally convergent series, Amer. J. Math., vol. 59 (1937), pp. 379–384.

Kaczmarz, S. and Steinhaus, H.
1. Theorie der Orthogonalreihen, New York, 1951.

Knopp, K.
1. Theory and Application of Infinite Series, New York, 1948.

Kolmogorov, A. N. and Fomin, S. V.
1. Elements of the Theory of Functions and Functional Analysis, Rochester, New York, 1957.

378 BIBLIOGRAPHY

Korovkin, P.
  1. Sur une généralization de la série de Taylor, C. R. (Doklady) de l'Ac. des Sciences de l'URSS, vol. 14 (1937).
  2. Linear Operators and Approximation Theory, Delhi, 1960.

Kowalewski, G.
  1. Interpolation und Genäherte Quadratur, Berlin, 1932.
  2. Einführung in die Determinantentheorie, New York, 1948.

Krilov, V. I.
  1. Approximate computation of Integrals, Moscow, 1959. (In Russian)

Kuntzmann, J.
  1. Méthodes Numériques, Paris, 1959.

Landau, E.
  1. Differential and Integral Calculus, New York, 1951.

Langer, R. E., ed.
  1. On Numerical Approximation, Madison, 1959.

Ljusternik, L. A. and Sobolew, W. I.
  1. Elemente der Funktionalanalysis, Berlin, 1955.

Lorentz, G. G.
  1. Bernstein Polynomials, Toronto, 1953.

Maehly, H. and Witzgall, C.
  1. Tschebyscheff—Approximationen in Kleinen Intervallen I, Numerische Mathematik, vol. 2 (1960), pp. 142–150.
  2. Tschebyscheff—Approximationen in Kleinen Intervallen II, Numerische Mathematik, vol. 2 (1960), pp. 293–307.

Markoff, A. A.
  1. Differenzenrechnung, Leipzig, 1896.

Mathematics Tables Project
  1. Tables of Lagrangian Coefficients, New York, 1944.

McShane, E. J. and Botts, T.
  1. Real Analysis, Princeton, 1959.

Milne, W. E.
  1. The Remainder in Linear Methods of Approximation, N.B.S. Jour. of Research, vol. 43 (1949), pp. 501–511.

Motzkin, T. S.
  1. Approximation by Curves of a Unisolvent Family, Bull. Amer. Math. Soc., vol. 55 (1949), pp. 789–793.

Motzkin, T. S. and Walsh, J. L.
  1. Least pth power polynomials on a real finite point set, Trans. Amer. Math. Soc., vol. 78 (1955), pp. 67–81.

Murnaghan, F. D. and Wrench, J. W., Jr.
1. The Approximation of Differentiable Functions by Polynomials, David Taylor Model Basin: Report 1175, 1958, pp. 1–52.
2. The Determination of the Chebyshev Approximating Polynomial for a Differentiable Function, Math. Tables and Other Aids to Comp., vol. XIII (1959), pp. 185–193.

Natanson, I. P.
1. Konstruktive Funktionentheorie, Berlin, 1955.

National Bureau of Standards
1. Tables of Chebyshev Polynomials, Washington, 1952.

Nehari, Zeev.
1. Conformal Mapping, New York, 1952.

Nikolsky, S. M.
1. On the problem of estimating the remainder in approximate quadrature formulas, Uspechi Mat. Nauk, vol. 5 (1950), pp. 165–177.

Nörlund, N. E.
1. Vorlesungen über Differenzenrechnung, Berlin, 1924.

Olmsted, J. M. H.
1. Completeness and Parseval's Equation, Amer. Math. Mon., vol. LXV (1958), pp. 343–345.

Paley, R. E. A. C. and Wiener, N.
1. Fourier Transforms in the Complex Domain, New York, 1934.

Peano, G.
1. Resto Nelle Formule di Quadratura Espresso con un integrale Definito, Atti Della Reale Accademia Dei Lincei, Rendiconti (5), vol. 22 (1913), pp. 562–569.
2. Residuo In Formulas de Quadratura, Mathesis (4), vol. 34 (1914), pp. 5–10.

Pólya, G.
1. Eine einfache, mit funktionentheoretischen Aufgaben verknüpfte, hinreichende Bedingung fur die Auflösbarkeit eines Systems unendlich vieler linearer Gleichungen. Commentarii Math. Helv., vol. 11 (1938–39), pp. 234–252.
2. Sur l'existence de fonctions entières satisfaisant á certaines conditions linéaires, Trans. Amer. Math. Soc., vol. 50 (1941), pp. 129–139.

Pólya, G. and Szegö, G.
1. Aufgaben und Lehrsätze aus der Analysis, New York, 1945.

Rainville, E. D.
1. Special Functions, New York, 1960.

Remez, E. Y.
1. Computational Methods for Tschebyscheff Approximation, Kiev, 1957. (In Russian)

Rice, J. R.
1. The Characterization of Best Nonlinear Tchebycheff Approximations, Trans. Amer. Math. Soc., vol. 96 (1960), pp. 322–340.
2. Chebyshev Approximation by $ab^x + c$, J. Soc. Indust. Appl. Math., vol. 8 (1960), pp. 691–702.
3. Tchebycheff Approximations by Functions Unisolvent of Variable Degree, Trans. Amer. Math. Soc., vol. 99 (1961), pp. 298–302.

Riesz, F. and Sz.-Nagy, B.
1. Functional Analysis, New York, 1955.

Ritt, J. F.
1. On the Derivatives of a Function at a Point, Annals of Math., vol. 18 (1916), p. 18.

Rivlin, T. J.
1. A Note on Smooth Interpolation, Siam Review, vol. 2 (1960), pp. 27–30.

Rivlin, T. J. and Shapiro, H. S.
1. A Unified Approach to Certain Problems of Approximation and Minimization, Journal of Soc. for Industrial and Applied Math., vol. 9, (1961), pp. 670–699.

Robin, L.
1. Fonctions Sphériques de Legendre et Fonctions Sphéroïdales, vol. I-III, Paris, 1957–1959.

Rosenbloom, P. C. and Warschawski, S. E.
1. Approximation by Polynomials, in "Lectures on Functions of a Complex Variable," W. Kaplan, ed., Ann Arbor, 1955.

Rudin, W.
1. Principles of Mathematical Analysis, New York, 1953.

Salzer, H. E.
1. Some New Divided Difference Algorithms for Two Variables, in "On Numerical Approximation," R. E. Langer, ed., Madison, 1959.

Sansone, G.
1. Orthogonal Functions, New York, 1959.

Sard, A.
1. Integral Representations of Remainders, Duke Math. J., vol. 15 (1948), pp. 333–345.
2. Best Approximate Integration formulas; best approximation formulas, Amer. J. Math., vol. 71 (1949), pp. 80–91.

Schoenberg, I. J.
1. On Variation Diminishing Approximation Methods, pp. 249–274 of "On Numerical Approximation," R. E. Langer, ed., Madison, 1959.

Schreier, O. and Sperner, E.
1. Modern Algebra and Matrix Theory. New York, 1951.

Sheffer, I. M.
1. Concerning some methods of best Approximation and the theorem of Birkhoff, Amer. J. Math., vol. 57 (1935), pp. 587–614.

Shisha, O., Sternin, C., and Fekete, M.
1. On the Accuracy of Approximation to Given Functions By Certain Inter-polatory Polynomials of Given Degree, Riveon Lematimatika, vol. 8 (1954), pp. 59–64.

Shohat, J. A.
1. Théorie Générale des Polynomes Orthogonaux de Tchebichef, Mémorial des Sciences Mathématiques, Fasc. 66, Paris, 1934.

Shohat, J. A., Hille, E., and Walsh, J. L.
1. A Bibliography on Orthogonal Polynomials, Bull. Natl. Research Council (U.S.), vol. 103 (1940).

Sommerville, D. M. Y.
1. An Introduction to the Geometry of n-dimensions, New York, 1958.

Steffensen, J. F.
1. Interpolation, New York, 1950.

Stiefel, E. L.
1. Numerical Methods of Tchebycheff Approximation, pp. 217–232 of "On Numerical Approximation," R. E. Langer, ed., Madison, 1959.

Stroud, A. H.
1. A Bibliography on Approximate Integration, Math. of Computation, vol. 15 (1961), pp. 52–80.

Synge, J. L.
1. The Hypercircle in Mathematical Physics, Cambridge, 1957.

Szegö, G.
1. Orthogonal Polynomials, New York, 1959.

Taylor, A. E.
1. Banach Spaces of Functions Analytic in the Unit Circle, I, Studia Mathematica, T. XI (1949), pp. 145–170.
2. Banach Spaces of Functions Analytic in the Unit Circle, II, Studia Mathematica, T. XII (1951), pp. 25–50.
3. Introduction to Functional Analysis, New York, 1958.

Thacher, H. C., Jr.
1. Numerical Properties of Functions of More than one Independent Variable, Annals of the New York Academy of Sciences, vol. 86 (1960), pp. 677–874.

Tieman, A. F.
1. Theory of Approximation of Functions of a Real Variable, Moscow, 1960. (In Russian)

Titchmarsh, E. C.
1. The Theory of Functions, London, 1939.

Todd, J., ed.,
1. A Survey of Numerical Analysis, New York, 1962.

Tornheim, L.
1. On n-parameter families of Functions and Associated Convex Functions, Trans. Amer. Math. Soc., vol. 69 (1950), pp. 457–467.

Tricomi, F. G.
1. Vorlesungen über Orthogonalreihen, Berlin, 1955.

de la Vallée Poussin, G.
1. Leçons sur l'approximation des Fonctions d'une variable réelle, Paris, 1919.

Walsh, J. L.
1. Interpolation and Functions Analytic Interior to the Unit Circle, Trans. Amer. Math. Soc., vol. 34 (1932), pp. 523–556.
2. Interpolation and Approximation by Rational Functions in the Complex Domain, Providence, Rhode Island, 1956.
3. Approximation by Bounded Analytic Functions, Mémorial des Sciences Math., vol. 144, Paris, 1960.

Walsh, J. L. and Davis, P. J.
1. Interpolation and Orthonormal Systems, J. d'Analyse Mathématique, vol. 2 (1952), pp. 1–28.

Walsh, J. L. and Motzkin, T. S.
1. Best Approximations within a Linear Family on an Interval, Proc. Natl. Acad. Sci., vol. 46 (1960), pp. 1225–1233.

Ward, L. E., Jr.
1. Linear Programming and Approximation Problems, Amer. Math. Mon., vol. 68 (1961), pp. 46–52.

Watson, G. N.
1. Theory of Bessel Functions, 2nd ed., Cambridge, 1958.

Whittaker, J. M.
1. Interpolatory Function Theory, Cambridge, 1935.

Widder, D. V.
1. A Generalization of Taylor's Series, Trans. Amer. Math. Soc., vol. 30 (1927), pp. 126–154.
2. On the Expansion of Analytic Functions of the Complex Variable in Generalized Taylor's Series, Trans. Amer. Math. Soc., vol. 31 (1928), pp. 43–52.
3. Completely convex functions and Lidstone series, Trans. Amer. Math. Soc., vol. 51 (1942), pp. 387–398.

Young, J. W.
1. General Theory of Approximation by functions involving a given number of arbitrary parameters, Trans. Amer. Math. Soc., vol. 8 (1907), pp. 331–344.

Zaanen, A. C.
1. Linear Analysis, New York, 1953.

Zygmund, A.
1. Trigonometric Series, Cambridge, 1959.

Index